SOTHEBY'S
INSTITUTE

The Encyclopaedic Dictionary in the Eighteenth Century
Architecture, Arts and Crafts

Volume Two

Ephraim Chambers
Cyclopaedia

The Encyclopaedic Dictionary in the Eighteenth Century

Architecture, Arts and Crafts

Volume Two

Ephraim Chambers Cyclopaedia

Terence M. Russell

Ashgate

Published by

Ashgate Publishing Limited
Gower House
Croft Road
Aldershot
Hants GU11 3HR
England

Ashgate Publishing Limited
Old Post Road
Brookfield
Vermont 05036-9704
USA

British Library Cataloguing in Publication Data
Russell, Terence M.
The encyclopaedic dictionary in the eighteenth century: architecture, arts and crafts
Vol. 2: Ephraim Chambers Cyclopaedia
1. Architecture – Terminology – History and criticism
2. Encyclopedias and dictionaries – History and criticism
3. English language – Lexicography – History – 18th century
I. Title II. Chambers, Ephraim, c. 1680-1740. Cyclopaedia
720.1'4

Library of Congress Cataloging-in-Publication Data
Russell, Terence M.
The encyclopaedic dictionary in the eighteenth century: architecture, arts and crafts / Terence M. Russell.
Includes bibliographical references and indexes.
Contents: v. 1. John Harris and the Lexicon Technicum: incorporating works of Sir Francis Bacon and Sir Henry Wotton – v. 2. Ephraim Chambers Cyclopaedia – v. 3. The builders dictionary – v. 4. Samuel Johnson: a dictionary of the English language – v. 5. A society of gentlemen: the Encyclopaedia Britannica.
1. Architecture. 2. Building. I. Title
NA2540.R88 1997
720'.941'09033–DC21 96-54482
CIP

Printed on acid-free paper

ISBN 1 85928 063 3

Printed in Great Britain at the University Press, Cambridge

To

The Central Technical School Sheffield
1954-1959
In recognition of the formative influence it exerted upon me.

TMR

THEOLOGIA
CYCLOPÆDIA

Frontispiece

The engraving portrays an idealized concourse of scholarly learning and inquiry. A young man is receiving instruction from an elderly sage amidst a gathering of about 150 individuals who are occupied in the pursuit of numerous arts, sciences, trades and crafts – see page seven. Isaac Newton, who died in the year that *Cyclopaedia* was published, has already been elevated to this hall of fame – his sculpture can be found adorning the balustrade above the Corinthian order together with celebrations of Pythagoras, Epicurus and Plato.

This highly accomplished illustration manifests a considerable understanding of eighteenth-century natural philosophy and was subsequently used in early editions of *Encyclopaedia Britannica.*

"Merit and good works is the end of man's motion."

Sir Francis Bacon

Contents

Illustrations

The following plates are incorporated at the end of the text:

Acknowledgements

This work has its origins in my companion publication *Architecture in the Encyclopédie of Diderot and d'Alembert: The Letterpress Articles and Selected Engravings*. On completion of this book I discussed with the publishers, Ashgate Publishing Company, the possibility of further exploring eighteenth-century writings on architectural subjects as manifest in other encyclopaedic-dictionaries of the period. Initially, the proposed project was to be a survey of the *Lexicon Technicum* of John Harris and the *Cyclopaedia* of Ephraim Chambers but the concept was eventually enlarged to incorporate *The Builder's Dictionary*, *A Dictionary of the English Language* of Samuel Johnson and *Encyclopaedia Britannica*. The result is that within the compass of a single work, albeit in multiple-volume format, we are now able to offer the reader selected writings on many facets of architecture and allied arts and crafts. I wish to express my gratitude to Mr Nigel Farrow, Chairman of the Ashgate Publishing Company, for endorsing my enthusiasm and for giving his consent for this large enterprise to proceed with the support of Ashgate Publishing. Ms Rachel Lynch, Editor at Scolar Press within the Ashgate Publishing Group, has supervised the progress of the work from its inception with diligence and unfailing courtesy and has frequently helped me to solve various procedural problems.

My orientation to many of the themes discussed in these volumes has its origins in my secondary education at a technical school and for this reason I have given expression to my debt of gratitude by inscribing each part of the work with an appropriate dedication. The scholarly impulse to the subjects here presented was nurtured by my professional training in architecture which served also to kindle my interest in the science of reference documentation and architectural bibliographical compilation. This disposition has exerted a considerable influence on the gestation and creation of my new study and for the benefit of the reader with an interest in works of reference, I have taken the liberty of listing details of my other published works in the field of architectural and historical bibliography – see *References*.

I have compiled my text from the first editions, or authoritative facsimile reproductions, of the original works which form the subject of the various volumes in the series. The researcher is well served in Scotland by a number of very fine archival collections, and I am indebted particularly to the following for their permission to cite edited passages from the original works: The National Library of Scotland, The University of Aberdeen, The University of Edinburgh and The University of St Andrews. In the case of each institution, the staff of 'Special

Collections' have been most helpful in providing me with literary treasures from their fire-proof vaults.

I am grateful to Mr Peter Hood, Convenor of the Technical Panel in Scotland of the Society for the Protection of Ancient Buildings, for allowing me to consult and to cite edited passages from his splendid copy of *The Builder's Dictionary*.

I am indebted to the trustees of the Earl of Moray Endowment Fund at the University of Edinburgh for financial assistance towards the creation of the archive of writings from which my text has been compiled. The conversion of the mass of accumulated papers into a coherent text has been a difficult undertaking that has involved the manipulation of hundreds of paper slips bearing the definitions of terms and edited abstracts of the writings selected for publication. This has imposed the need for considerable vigilance and accuracy in endeavouring to preserve the orthography, style and idiom of the original texts. In this context, I wish to record my appreciation of the efforts of Rosalind M. Russell for her contribution to the early stages of the work in the form of establishing several guiding principles and for typing an extensive trial section of the manuscript. For continuing this work and for typing the entire text and preparing it for publication with unswerving scholarly assiduity, I am immensely grateful to Thomas G. Russell.

A number of my colleagues have read sections of the manuscript and have contributed welcome suggestions, for all of which I herewith express my appreciation.

TMR

Preface

The Encyclopaedic Dictionary in the Eighteenth Century: Architecture, Arts and Crafts is the thematic title of the series of publications of which this book forms a part. The individual titles of the volumes in the series are as follows:

John Harris *Lexicon Technicum*:
Incorporating works of Sir Francis Bacon and Sir Henry Wotton
Ephraim Chambers *Cyclopaedia*
The Builder's Dictionary
Samuel Johnson and *A Dictionary of the English Language*
Encyclopaedia Britannica

The principal theme of the series is the manner in which architecture and allied arts and crafts are portrayed in the major British encyclopaedic-dictionaries of the eighteenth century. Before describing the aims of this work in greater detail it will assist the reader to say a few words first about our previous researches which have led directly to the writing of this series of encyclopaedic-dictionary studies.

The present work follows on from and is complimentary to our study *Architecture in the Encyclopédie of Diderot and d'Alembert: The Letterpress Articles and Selected Engravings.* Scolar Press, Aldershot 1993. Our new work is also a development of the author's earlier bibliographical researches, namely: *BIBLIOGRAPHICAL INDEX: A Subject Guide to Printed Books in the Fields of Architecture, Building, Constructional Engineering, Environment, Technology, Planning, Landscape Architecture, Design and allied Arts and Crafts.* University Microfilms International (UMI) Ann Arbor, Michigan USA 1982; and, *THE BUILT ENVIRONMENT: A Subject Index 1800-1960.* Gregg Publishing, London 1988.

For the reader interested in these works, additional details are provided under the heading *References*.

From the preceding information it will be seen that our new study *The Encyclopaedic Dictionary in the Eighteenth Century: Architecture, Arts and Crafts* is the consequence of a long period of gestation, in point of fact a timescale extending over more than twenty years. The primary intention of our two studies *Bibliographical Index* and *The Built Environment* has been to create a bibliographical structure for the considerable literature of what in recent years has come to be called the *built environment* or the man-made context within which we live our lives. In these two works we estimate to have collated about 75,000 bibliographical sources, bearing upon the systematic description of books with regard to their authorship, printing, publication and editions. Following on from these bibliographical researches, the opportunity arose to extend our programme of work, by undertaking a more focused investigation of historically significant encyclopaedic dictionaries.

Having provided the reader with this contextual information, we can now examine in greater detail the aims of the present work.

The principal theme of our series is the portrayal of architecture and allied arts and crafts in the major British encyclopaedic-dictionaries of the eighteenth century. The eighteenth-century impulse to compile encyclopaedic works of reference, the purpose of which was to explain the achievements of mankind to a wide readership, is well exemplified in the intellectual aspirations of the polymath and systematic thinker Wilhelm Gottfried Leibniz who conceived a project 'to extract the quintessence of the best books, to add to them the unwritten observations of the most tried in every profession, and in this manner to build systems of knowledge, based upon experience and demonstration for the further progress of mankind'. The spirit of these intentions is also captured in the views of the philosopher and statesman Sir Francis Bacon who, a generation earlier, had affirmed the value of what he considered could be gained to both philosophy and art through an increase of mutual understanding. 'Philosophy', Bacon argued, 'would be improv'd by having the Secrets of all Trades Lye open not only because such Experimental Philosophy is coucht amongst them; but also that the trades themselves might by a Philosopher be Improv'd.' Bacon's wider intellectual theories, particularly concerning the ordering of knowledge and the division of the sciences into different branches, greatly impressed the French *philosphes* Denis Diderot and Jean le Rond d'Alembert: indeed, they respectfully refer to Bacon as 'the immortal Chancellor of England'. They embodied his views in their celebrated *Preliminary Discourse* to the *Encyclopédie* in which they argued the Age-of-Enlightenment case for the advancement of learning. We consider these philosophical, intellectual and historical aspects which surround our study in greater detail in the context of the introductory chapters to the separate volumes in the series.

In our survey of the selected texts we look at the aims of each work, we consider its structure, we survey its system of arrangement and we provide a comprehensive inventory of all the writings (letterpress articles) on architectural subjects including all the allied arts and crafts. Our compilations encompass architectural terms, building construction, building materials, cabinet work, carpentry, the classical orders, decorative art, design principles including aesthetics, domes, drainage, fortification, furnishings, gardening, joinery, landscape terms, military art, mouldings, ornament, painting, perspective, sanitation, sculpture, surveying, terminology of the building trades and crafts, and the tools and equipment used in building construction.

Each article is presented under its headword and original spellings and archaisms have been retained within the orthography of the eighteenth century. Articles of all kinds have been incorporated including short, definition-like texts alongside more extended treatises. In this context particular mention may be made of our inclusion of Sir Francis Bacon's Civil Essays, *Of Building* and *Of Gardens* and Sir Henry Wotton's treatise *The Elements of Architecture* with a commentary in the form of an adaptation of Isaac Walton's celebrated *The Life of Sir Henry Wotton*.

The volumes in the series are thematically linked but each has been conceived so as to be independent of the rest with the minimum of cross referencing.

A noteworthy feature of each study is the incorporation of the author's original Preface. The inclusion of these texts preserves for the reader the guiding principles by means of which the editor-compilers created their great works of reference bearing upon such considerations as their aims and intentions; philosophy and outlook; the range and scope of their work; working method; and the many contextual considerations determined by time and place.

We believe that the writings we have extracted from *Lexicon Technicum*, *Cyclopaedia*, *The Builder's Dictionary*, *A Dictionary of the English Language* and *Encyclopaedia Britannica* provide the reader with a wealth of information about eighteenth-century building-trade practices and procedures, a conspectus of contemporary knowledge of the period, a survey of terminology and nomenclature, and a systematically presented series of theoretical and practical writings about numerous facets of architecture, building and the attendant decorative arts and crafts. The following inventory will give the reader an idea of the scope of our investigations. The number of letterpress articles extracted from their original sources is as indicated in brackets: *Harris*, Vol. I (361), Vol. II (212); *Chambers*, Vol. I (371), Vol. II (453); *Builder's Dictionary*, Vol. I (214), Vol. II (254); *Johnson*, Vol. I (349), Vol. II (395); and *Encyclopaedia Britannica*, Vol. I (167), Vol. II (230), Vol. III (216) – a total of 3,222 articles. In addition, an examination of these texts will reveal to the reader that many of our articles are in the form of extended writings which consist of numerous subordinate and secondary articles on subjects related to the themes of the principal texts. Consequently, such entries as, for example, *architecture*, *building*, *column*, *fortification*, *mosaic*, *order* and *scenography* greatly extend, through their diversity, the overall total of actual articles.

It is our hope that the volumes in our series *The Encyclopaedic Dictionary in the Eighteenth Century*: *Architecture*, *Arts and Crafts* will be of benefit to architects, artists and craftsmen, designers, restorers, preservationists, conservationists, librarians, historians, antiquarians and students. Furthermore, by incorporating within the scope of our work a systematic survey of Samuel Johnson's celebrated *A Dictionary of the English Language*, we trust our study will also be of value to the reader interested in the development of the English language bearing upon Johnson's lexicographic methodology and his definitions and descriptions of architectural and technical terms. More generally, we hope also that the books in our series, the subject matter of which ranges across architecture, history, biography and philosophy, will be regarded as having made a useful contribution to eighteenth-century, inter-disciplinary studies within the framework of the scholarship of the European Enlightenment.

It remains to take leave of the reader with a brief reference to our working method. Our first step was to individually consult each of the several hundred thousand articles embracing all manner of subjects contained within *Lexicon Technicum*, *Cyclopaedia*, The *Builder's Dictionary*, *A Dictionary of the English Language* and *Encyclopaedia Britannica* – a considerable undertaking necessitating a line-by-line scrutiny of each page of the original works from which our studies have been derived. From this vast horde of writings, inventories were compiled consisting of the headwords of all the texts relating to architecture and allied arts and crafts contained within each encyclopaedic-dictionary: these inventories are preserved in the body of our completed text in order to serve the reader as an index of contents. Working systematically through these lists, an abstract was then made on to paper slips of each text selected for inclusion in the final work. This activity, which required a number of years to bring to completion, necessitated the constant application of rigorous editorial judgement bearing upon the retention of essential information and the exclusion of extraneous material. In making our final compilations, we have erred on the side of caution in order to give the reader the maximum possible information within the scope of the limits set by the publisher. We have endeavoured throughout to be true and accurate to the original texts and to faithfully preserve the original eighteenth-century idiom If we have been negligent in places concerning such matters, we ask the reader to bear in mind the eloquent words of Theodore Besterman, doyen of editorial compilers:

> May the noble minded scholars instead of cherishing ill feeling kindly correct whatever errors have been committed . . . in the way of wrong interpretations and misstatements.

Terence M. Russell
Edinburgh, 1996.

Introduction

The second part of our survey is concerned with the portrayal of architecture and allied arts and crafts in the *Cyclopaedia* of Ephraim Chambers.

Chambers planned his *Cyclopaedia* to be 'a course of antient and modern learning . . . compiled from the best Authors, Dictionaries, Journals, Memoirs, Transactions, Ephemerides, etc. in several languages.' The book was printed in two substantial folio volumes which is reflected in the double-volume format that we have adopted for our own work. Such is the scale of *Cyclopaedia* that no fewer than eighteen publishers were required to augment their resources in the publication of the first edition. By way of interest, these included Daniel Midwinter, who had assisted in the publication of John Harris's *Lexicon Technicum*, and Thomas Longman, founder of the famous publishing house, who nearly thirty years later was one of the publishers of Samuel Johnson's *A Dictionary of the English Language.*

In our study we trace the publication history of *Cyclopaedia* from the time of its announcement in Chambers' Prospectus. We reproduce in its entirety Chambers' flattering Dedication to George II and a substantial proportion of his scholarly Preface. A striking feature of the Preface is Chambers' *Division of Knowledge* in which he presents the principal categories or *Heads* of knowledge in the form of a *genealogical tree.* In view of the fact that copies of *Cyclopaedia* are very rare, we have incorporated Chambers' original diagrammatic representation of his division of knowledge. Within Chambers' schematization, Architecture is presented as a sub-set of Mechanics, i.e., applied mathematics. Chambers, as we show in Chapter 1, was considerably influenced in his understanding of the divisions of knowledge by the writings of Sir Francis Bacon. Chambers adopted Bacon's procedure of analysing the faculties of the mind and the objects of human knowledge for the intellectual basis of *Cyclopaedia.*

The first of the two great divisions of knowledge in Chambers' diagram is *Sensible and Rational Knowledge.* This set emanates directly from Bacon's view, as propounded in the Preface to his *The New Organon*, in which he argued the case for 'a new path for the mind to procede in, starting directly from simple sensuous perception'. The reference to 'sensuous perception' is consonant with John Locke's views as outlined in his *Essay Concerning Humane Understanding.* Bacon, Locke and Chambers demonstrate a mistrust of the concept of *innate* knowledge as prevalent in the seventeenth century. To Bacon, in particular, nature presented herself in the form of large classes of phenomena, the properties of which were concealed from mankind through their inherent complexity. The 'world of knowledge' was a labyrinth which the human mind must enter, observe and from which information was to be collected. The key to understanding was the philosopher's new method of induction: through analysing experience, facts and causes concealed from the senses could be comprehended and true conclusions could be established.

The second of the two great divisions of knowledge in Chambers' diagram, which has particular relevance to our survey, is *Artificial and Technical Knowledge*. This realm is dominated by quantities of bodies (mixed mathematics) which leads to optics (perspective), to phonics (music), and to mechanics (architecture) which in turn leads to trades and manufactures. Within Baconian and later Newtonian philosophy, the ordering principles of applied mathematics unified this seemingly heterogeneous grouping of subjects.

In France, Chambers' ideas were enthusiastically taken up by the eighteenth-century *philosophes*, the views of whom had in part been nurtured by René Descartes who had elaborated a perception of the entirety of human knowledge based upon a system of beliefs embracing mathematics, natural philosophy, medicine and morality, strengthened by a rigorous mental outlook and disposition to rationalism. We consider the general influence of Chambers' *Cyclopaedia* on the work of the French scholars in Chapter 1 and in more detail in our companion study *Architecture in the Encyclopédie of Diderot and d'Alembert*.

In the creation of *Cyclopaedia*, Chambers placed reliance upon learning, scholarship and the works of previous writers. Chambers was *par excellence* a systematiser and classifier. Moreover, in bringing to completion so large and complex a work virtually unaided, his energy and powers of concentration must have been prodigious.

John Harris had pioneered the concept of technical, encyclopaedic illustration ten years before Chambers had set to work, but Chambers went further and for its time *Cyclopaedia* was the most comprehensively illustrated work of its kind in the English Language. The broad sweep of Chambers' work far exceeds that of John Harris and his predecessors. The importance of *Cyclopaedia* was universally recognised: at home the Government passed a Bill authorising revisions and abroad the work was translated into Italian in the form of nine quarto volumes.

Chambers was alert to change and progress and in his Preface he confidently proclaimed:

> We have here the boundary that circumscribes our present prospect, and that separates the known from the unknown parts of the intelligible world . . . We promise ourselves this boundary will be removed, and the prospect extended far into the other hemisphere. Methinks I see trophies erecting in the yet undiscovered regions of science, and . . . inventions at present held impossible.

1 Ephraim Chambers and the *Cyclopaedia*

Ephraim Chambers: biographical note

Ephraim Chambers, the creator of the *Cyclopaedia*, was born about 1680 into a farming family. His father worked a small holding at Kendal, Westmorland where Chambers was educated at the local grammar school.[1] The family must have had the financial means and the determination to do the best for their son since he was sent to London to complete his education by being apprenticed to the cartographer and engraver John Senex. Senex is an interesting representative of the highly accomplished, eighteenth-century instrument maker whose skills contributed to the practice and development of the art and science of navigation. Chambers was highly fortunate to be engaged to so distinguished a practitioner and must have found himself surrounded in Senex's premises with maps and instruments of all kinds to stimulate his youthful and receptive mind. Senex engraved the plates for the London almanacs and Sir William Browne's *Account of Microscopes and Telescopes*. His reputation rests primarily on his work as a cartographer and globe maker. In 1710, with C. Price, he printed *Proposals for a New Sett of Correct Mapps* and later issued maps of America, Germany and 'Moscovy'. He presented ideas to Parliament on the subject of a new globular projection and for this, and his other works, he was elected to a Fellowship of the Royal Society in 1728.[2] Senex is known to have encouraged his young protégé in his desire to acquire knowledge and it was whilst working for his master that Chambers appears to have entertained the idea of creating an encyclopaedia modelled upon the *Lexicon Technicum* of John Harris. Harris had published the first volume of his pioneering work in 1704 under the full title:

LEXICON TECHNICUM
OR, AN
Universal English Dictionary
OF
ARTS
AND
SCIENCES:
Explaining not only
The TERMS of *ART*,
BUT THE
ARTS Themselves.

The second volume of *Lexicon Technicum*, in the form of a supplement to the first, was published in 1710. It is significant that John Senex was a subscriber to both of Harris's volumes which we can assume would be on his shelves for Chambers to consult at his leisure. The influence of the *Lexicon Technicum* on Chambers was considerable and led him, as we have intimated, to compile his own *Cyclopaedia* (see below).

Chambers contributed to the *Literary Magazine* which came out between 1735-37. This work must have been congenial to Chambers, the encyclopaedist, consisting as it did of writing reviews of recently published significant books. At this time he demonstrated his linguistic abilities by translating from the French, Jean Dubreuil's *Practice of Perspective* which was successful and ran to several editions. His collaboration with the botanist John Martin resulted in the publication of the five-volume survey, translated again from the French, of the *Philosophical History and Memoirs of the Royal Academy of Sciences at Paris* (1742). In his later years he spent time in France endeavouring to improve his health and whilst there declined an offer from the French publishing houses to bring out a translation of the *Cyclopaedia* dedicated to none other than Louis XV. Nevertheless, the indirect influence of the *Cyclopaedia* on French encyclopaedic writing was profound and led to the creation of the celebrated *Encyclopédie* (see later). Chambers died in 1740 just one year after *Cyclopaedia* had reached its third edition.

Chambers' Prospectus for the Cyclopaedia

The *Lexicon Technicum* of John Harris is significant in British encyclopaedic writing for being not only the first truly English encyclopaedia but also for the manner in which it established two important precedents. These were, firstly, the bringing together of a consortium of publishers to share the collective responsibilities for creating a large-scale publication and, secondly, the underwriting of the project by securing a list of subscribers. Harris names ten publishers on his Title Page and for Volume I of the *Lexicon Technicum* the names of subscribers run to eight pages, increasing to twelve pages of names for Volume II. Harris's lists would have provided Chambers with a valuable understanding of the encyclopaedia-purchasing public to which he intended to address his *Cyclopaedia*. As we have shown in our companion volume, Harris's subscribers included distinguished persons from the landed-gentry, academics, booksellers, merchants and the newly-emerging class of highly-skilled instrument makers such as Chambers' employer John Senex.

The key to securing a list of subscribers was the writing and dissemination of a Prospectus. This was therefore Chambers' first task and he duly announced on 17 January 1726 the following:

Proposals for Printing by Subscription,

Cyclopaedia:

or an Universal Dictionary of the Arts and Sciences;

containing The Definitions of the Terms, and the Description of the Things signify'd thereby, In the Several Arts, Both Liberal and Mechanical, and the several Sciences, Human and Divine: The Figures, Kinds, Properties, Productions, Preparations, and Use of Things Natural and Artificial; The Rise, Progress, and State of Things Ecclesiastical, Civil, Military, and Commercial; The various Phenomena of Nature and Discoveries of Men, and their Uses; With the several Systems, Sects, Opinions, etc. among Philosophers, Divines, Mathematicians, Physicians, Antiquaries, Criticks, etc. The Whole intended as a Course of Antient and Modern LEARNING. Compiled from the best Authors, Dictionaries, Journals, Memoirs, Transactions, Ephemerides, etc. in several Languages.[3]

In his Prospectus, Chambers promised potential subscribers that his book would constitute 'a course of antient and modern learning' and that it would be: 'Compiled from the best Authors, Dictionaries, Journals, Memoirs, Transactions, Ephemerides, etc. in Several Languages.' The Prospectus consists of two folio sheets printed in the style and format of the planned encyclopaedia to give would-be subscribers some indication of the form of the anticipated publication. Harris had adopted a folio format of 25cm x 31cm; for his Prospectus and the *Cyclopaedia* itself, Chambers used the slightly larger page size of 21cm x 40cm. From the outset, Chambers planned that his work should have the character of a dictionary but that additionally it was to be 'a system at the same time'. In this latter aspect he sought to reassure his subscribers that his work would have the virtues of order and methodology. He outlines his intentions as follows:

> It is a System taken asunder, and its Members re-compos'd in a different Order, to render 'em more obvious, and easy of access. Instead of the Natural Order, wherein the vast Assemblage would be unwieldy . . . they are here presented *singly*, and in the familiar Order of the Alphabet.

To tantalise the reader with the riches his book is to contain, Chambers offers the following prospect:

> For the *Matter*, it cannot here be insisted on. 'Tis a Scene of that Extent and Variety, that a Representation of it in little, wou'd confound the Eye rather than inform it. 'Tis the Result of Volumes without Number: the Sum of what the busy Mind of Man has been a-doing these many Ages; and what with grevious Toil, and great Expence of Time, and Books, we endeavour to arrive at in the long Course of *Education* and *Study*.

To further inform his readers and to reassure them as to the quality of his forthcoming work, Chambers incorporated within his Prospectus a specification which outlines in greater detail what the reader can expect from *Cyclopaedia* and the all important question of the financial outlay that was involved. Chambers' five-paragraph statement not only gives an anticipation of some of the qualities of *Cyclopaedia* but it also provides an insight into the sales-promotion procedures which were adopted at the launch of a major eighteenth-century publication. The following items are reproduced from the Prospectus of 1726:

> I. That the Book shall be printed in Two Volumes in Folio, on the same Paper and Letter with the Specimen annex'd.

To give his readers some idea of the quality of paper the printer was to use, Chambers had a sample page made up (his 'Specimen') with the articles for the letter '**A**' printed and other entries down to '**Abased**'. In this way readers could make an assessment of the layout of the text, the quality of the typography, and also gain some idea of how much information was provided under a particular headword.

II. The Whole Copy being finish'd, one Third of it already printed, and the rest in the Press at several Houses; 'tis proposed to deliver it to Subscribers by Midsummer next.

The claims made in this paragraph are remarkable when account is taken of the mass of material encompassed within the format of *Cyclopaedia's* two large folio volumes. From the time that he left his employer Senex to the date of publication of the Prospectus, Chambers must have worked with truly Herculean energy to compile and edit a text of such extent and complexity. The indefatigable Samuel Johnson, working with six copyists, informed his publishers that he would require three years to accumulate the materials for his *Dictionary*, and in the event required almost three times that long. That one third of Chambers' text was already at the printers illustrates how well advanced the work was and also the trust the publishers had in their youthful editor. The reference, 'at several Houses' is Chambers' allusion to his syndicate of no fewer than eighteen publishers who collaborated in the publication of the first edition of *Cyclopaedia*. These included: Daniel Midwinter, who had assisted in the publication of the *Lexicon Technicum*; Thomas Longman, who was then in partnership with John Osborne and who nearly thirty years later (1755) was to be one of the publishers of Johnson's *Dictionary*; and, of some interest, Chambers' former employer John Senex. The inclusion of Senex amongst the list of publishers is added confirmation of the high regard in which Ephraim Chambers was held in the mid 1720's even though he was still yet relatively unknown outside of his circle of London acquaintances.

III. The Work being computed to make upwards of Four Hundred Sheets, the Price to Subscribers will be after the Rate of Two-pence *per* sheet . . . One Guinea to be paid down, and the Remainder on Delivery of the Book.

The price being asked was a great deal when allowance is made for eighteenth-century monetary values but is typical of the sum required to secure two large, leather-bound folios of the period. By way of additional illustration, the cost of Johnson's *Dictionary* was £4. 10s. To buy such handsome books, a young schoolmaster who might expect to start his career on about £20 per year, would have to set aside upwards of one fifth of his annual earnings to have such folios on his shelves.

IV. Such as subscribe for Six Books, shall have a Seventh gratis.

Only two subscribers took up this offer in view of the very considerable financial outlay involved.

V. The names of the Subscribers will be printed before the Book.

By purchasing *Cyclopaedia*, the subscriber was securing for himself a small piece of immortality. The names of some 375 persons are now to be found, as Chambers' had promised, 'printed before the Book' (i.e. in the prefatory pages before the encyclopaedic entries). Judging from Chambers' list of subscribers, he appears to have circulated his Prospectus at a rather more exalted social level than the purchasers of Harris's *Lexicon Technicum*. Chambers' list is resplendent with Dukes, Viscounts, Earls, Lords, Baronets and Honourables. Intellectual notables include Colin MacLaurin, professor of mathematics at the University of Edinburgh. Less exalted but more typical subscribers included lawyers, clergymen, surgeons, merchants and booksellers, two of whom were those who took advantage of the offer of a free book by purchasing six. Among the booksellers subscribing to *Cyclopaedia* was Michael Johnson, 'bookseller at Lichfield' and father of Samuel Johnson. We know from Boswell's *Life of Samuel Johnson* that *Cyclopaedia* exerted an influence on Johnson when he was contemplating models for his *Dictionary*.

The above-quoted extracts are the principal parts of Chambers' Prospectus but one other aspect is worthy of mention. The text contains the following footnote:

> NB. The Figures necessary for explaining the several Matters, will be engraven on Copper; amounting to upwards of Thirty Folio Plates; for which nothing further will be required of the Subscribers.

John Harris had pioneered the concept of encyclopaedic illustration ten years before Chambers set to work. The *Lexicon Technicum* includes four whole-page plates and two fold-out format engravings. Chambers, as we see from above, was determined to do better and for its time the *Cyclopaedia* was the most comprehensively illustrated work of its kind in the English language.[4, 5]

Publication of the Cyclopaedia

Cyclopaedia was published in two folio volumes as the Prospectus had foretold in 1727, although the Title Page gives 1728. It was dedicated to the King. Subscribers opening their copies found the promise of the Prospectus fully upheld. The Title Page (see elsewhere) proclaims the splendour of the work and its attachment to both the arts and sciences whereas its great precursor, the *Lexicon Technicum*, had focused primarily on the sciences and what today we would describe as technology – even in Harris's day his work earned for him the epithet *Technical Harris*. *Cyclopaedia* immediately found favour with the literati and the importance of the work was soon widely recognised. Two years after publication it earned for its creator, Ephraim Chambers, an election to a Fellowship of the Royal Society. The folios retailed at four guineas for the set bringing revenues to the publishers of more than 1500 guineas, a not inconsiderable sum in the eighteenth century. It is said that the consortium of publishers was so pleased with the outcome that they donated £500 to Chambers as a mark of their appreciation.

Cyclopaedia had a long publication run. Chambers himself undertook to revise and expand the work and set out his ideas in a paper of *Considerations* which was probably seen by Samuel Johnson to the benefit of his *Plan* for *A Dictionary of the English Language*. *Cyclopaedia* even assumed national importance and Parliament introduced a Bill containing a clause requiring the publishers to issue a revised and improved edition. In 1738 a second edition was published with some improvements. The third edition appeared in 1739 and the fourth in 1741, one year after Chambers' death. A fifth edition came out in 1746 and between 1748-49 *Cyclopaedia* was published at Venice in Italian in nine quarto volumes.

The life-span of *Cyclopaedia* was considerably extended by Abraham Rees (1743-1825).[6] Rees was a resourceful improver of Chambers' text which he re-edited in 1778 and issued with new material as a four-volume work between 1781-86 of which a reprint followed in 1788-91. These considerable intellectual endeavours earned for Rees, like Chambers before him, a Fellowship of the Royal Society. Rees took *Cyclopaedia* into the nineteenth century in a greatly enlarged form when in January 1802 he published the first part of *The New Cyclopaedia, or a Universal Dictionary of Arts and Sciences . . . Biography, Geography, and History*. This monumental undertaking was finally completed in August 1820 in forty-five volumes, including six volumes of plates.

Since our observations on the work of Abraham Rees effectively bring us to the modern era of encyclopaedic publication, it is perhaps useful to note here that Chambers' *Cyclopaedia* and *Chambers's Encyclopaedia* though similar sounding are entirely different works. *Chambers's Encyclopaedia* was begun in Edinburgh by William and Robert Chambers who issued their new publication between 1859-68 in 520 parts (accumulated into ten volumes) which sold at three-half pence each. Through successive editors and having undergone many transformations it has survived to our own time to become a household word.

Chambers' Cyclopaedia of 1727

Cyclopaedia was published as we have already stated in 1727.[7] The Title Page (see later) informs the reader of the extent to which the book provides definitions and terms of subjects derived from many authors and their works. Chambers does not shrink from making scholarly demands of his readers: in his synthesising the etymology of words he acknowledges few linguistic boundaries and his compilations 'from the best Authors' make frequent use of Greek and Latin.

The Frontispiece is of double-folio page (fold-out) format placed opposite the Title Page for maximum impact. It is the most magnificent illustration in the entire work. A great academy is thronged with a multitude of scholars: 150 figures complete the composition, arranged in groups pursuing the various arts and sciences. They constellate into 47 *Heads* or sub-divisions of learning that are detailed in the Preface (see later). This scholarly fraternity is disposed about a large open concourse flanked by an array of classical columns rendered in the Corinthian order. The description *Cyclopaedia* appears on a scroll in the foreground to which a bearded sage points with his cane: this scene of idealised scholarship and learning consciously anticipates the majestic sweep of the *Cyclopaedia* and must have greatly impressed Diderot and his artist-engravers: we may even speculate that this art work perhaps exerted an influence on the creation of C.-N. Cochin's celebrated frontispiece for the *Encyclopédie*.

Chambers' Dedication, emboldened in one inch (25mm) capitals, is to George II. It is pregnant with obligatory flattery:

> The Arts and Sciences humbly crave audience of Your Majesty. The near concern they have in the happiness of a people, assures 'em of the favourable attention of a Prince who makes happiness his own. 'Tis by these, the parsimony of Nature is supplied, and life render'd easy and agreeable under its numerous infirmities. By these the mind is reclaim'd from its native wildness; and enriched with sentiments which lead to virtue and Glory.

Elsewhere, two passages catch the spirit of the Age of Enlightenment:

> The time seems at hand, when we are no longer to envy *Rome* her Augustus and Augustan age, but *Rome* in her turn shall envy ours.

And similarly:

> There is a time reserv'd in fate for every nation to arrive at its height; and the uppermost place on the terrestrial ball is held successively by several states. May not the numerous presages which usher in Your Majesty's Reign, give us room to expect that our turn is next; and that what *Greece* was under Alexander, and *Rome* under Augustus Caesar, *Britain* shall be under George and Caroline.

Chambers' final passage breathes the spirit of inquiry that is to herald a New Age:

> Methinks I see trophies erecting to Your Majesty in the Yet undiscovered regions of science; and Your Majesty's name inscribed to inventions at present held impossible.

Chambers' Preface is almost a book in itself. It amounts to thirty folio pages and served as a model for d'Alembert's *Discourse Préliminaire*. At the outset Chambers contrasts the circumstances of his one-man enterprise with the collaborative endeavours of European encyclopaedists:

> 'Tis not without some Concern that I put this Work in the Reader's Hands; a Work so disproportionate to a single Person's Experience, and which might have employ'd an Academy. What adds to my Jealousy, is the little measure of Time allow'd for a Performance to which a Man's whole Life scarce seems equal. The bare Vocabulary of the Academy *della Crusca* was above forty Years in compiling, and the Dictionary of the *French* Academy much longer; and yet the present Work is as much more extensive than either of them in its Nature and Subject, as it falls short of 'em in number of Years, or of Persons employ'd.

Notwithstanding these feelings Chambers acknowledges his debt to other encyclopaedists and gives intimation of his personal industry in transcribing from the original sources (for later editions he had assistance from amanuenses):

> I came like an Heir to a large Patrimony, gradually rais'd by the Industry, and Endeavours of a long Race of Ancestors. What the *French Academists*, the *Jesuits de Trevoux*, *Daviler*, *Chomel*, *Savary*, *Chauvin Harris*, *Wolfius*, and many more have done, has been subservient to my Purposes . . . Dictionaries . . . contributed their share . . . [and] Lexicons on almost every Subject . . . Recourse has been had to the originals themselves on the several Arts.

A noteworthy feature of the *Cyclopaedia* is Chambers' system of cross-referencing: he treats his subjects 'as so many Wholes, and as so many Parts of some greater Whole'. Chambers goes much further with this aspect of his methodology than John Harris. But the most striking feature of Chambers' Preface is his *Division of Knowledge*. This is a view of knowledge 'exhibiting only the grand, constituent Parts thereof' since, as Chambers adds, 'It . . . would be endless to pursue it into all its Members and Ramifications; which is the proper Business of the Book itself.' Chambers presents the principal categories or *Heads* of knowledge in the form of a *genealogical tree*. This serves, in part, as an outline of contents such that 'Heads of each Kind will here come in sight, and such as will naturally suggest, and lead to the rest; so that this will afford the Reader a sort of Summary of the Whole'. In Chambers' schematization of things, knowledge is either *Natural and Scientifical* or *Artificial and Technical*. These two categories branch into forty-seven subjects. Pre-eminent in the first category is Meteorology, which is held to include *aether*, *fire* and *vapour*. Lastly, in the second category, held to occupy the domain of the *symbolical*, is *Poetry*.

The influence of Chambers' systematisation of knowledge on the minds of the creators of the *Encyclopédie* was profound. They borrowed the idea and reinterpreted it in their own *Systeme Figuré Des Connoissances Humaines*.[8]

In Chambers' tree of knowledge, *Architecture* occupies position twenty-six in the hierarchy and is considered a derivative of *Mechanics* which in turn belongs to the realm of *Quantities of Bodies* (Mathematics). *Architecture* is used by Chambers to illustrate the complexities of language which is acknowledged to be 'in a perpetual flux.' *Names* are held to be 'Representatives of Ideas' and *Terms* to be 'Combinations of Ideas'. Chambers continues:

> A Man never knows when he is at the end of Terms, e.g. in Architecture. When he has got two or three Names, for some one Member, and thinks himself overstock'd, 'tis odds he has not half. 'Tis not enough he knows what it is named in the *English*; but he must likewise learn what the *French*, *Italians*, *Latins*, and *Greeks*, likewise call it, or frequently find himself at a stand. Thus it is in the Case of *Fillets*, *Lists*, *Listels*, *Reglets*, *Platbands*, *Bandeletts*, *Taenias*, and *Baguettes*; of *Chaplets*, *Astragals*, *Batoons* and *Tores*; of *Gulas*, *Gueules*, *Doucines*, *Cimas*, *Cymatiums*, *Ogees*, and *Talons*; *Ovums*, *Ovolos*, *Echinus's*, *Quarter-rounds*, *Boultins*, etc. between which, there is no known, allowed differences; but they are either used indiscriminately, or distinguish'd arbitrarily; one Person making this distinction, and the next another, or perhaps none at all. So that if we come strictly to Dictionaries, we should have a different one for every Author.

Chambers further illustrates the comprehensiveness of his method of subject compilation by, once more, citing architecture, this time as a listing of its several aspects and those of allied subjects. The reference is by no means inclusive of all the architectural terms given in the *Cyclopaedia* – these run to over seven hundred – but the catalogue as given in the Preface alerts the reader to the breadth of material the two folio volumes provide. The Preface may be quoted here as evidence also of Chambers' remarkable diligence and passion for classification:

> Architecture, including the Construction of Building; as House, Temple, Church, Hall, Palace, Theatre, etc. Ship, Galley, Galleon, Ark, Buccentaure, Boat, etc. Pyramid, Mausoleum, Pantheon, etc. Capitol, Seraglio, Escurial, etc. Arch, Vault, Bridge, Monument, Tomb, etc. Forms there of, Rotondo, Platform, Pinnacle etc. Plans, Design, Ichonography, Profile, etc. Door, Window, Stairs, Chimney, etc. Orders, as Tuscan, Doric, Corinthian, etc. Caryatides, Rustic, Gothic, etc. Column, Pilaster, Attic, etc. Parts there of, Entablature, Capital, Pedestal, etc. Cornice, Frieze, Base, etc. Volute, Pediment, Modillion, Console, etc. Mouldings, Ogee, Tore, Astragal, Sofia,

Abacus, Ovolo, etc. Materials, as Brick, Stone, Tyle, Slate, Shingle, etc. Timber, Wainscot, Glass, Lead, Plaster, etc. Beam, Rafter, Mortar, Nail, Hinge, Key, Lock, etc. Quarry, Masonry, etc.

As remarked, this extensive listing is but a microcosm of the full range of architectural subjects Chambers describes in the *Cyclopaedia*.

Commentary on Chambers' Preface would be incomplete without a brief look at what he has to say about the difference between art and science which he considers as 'two Words of familiar Use, and great importance; but I doubt little understood. The Philosophers have long labour'd to explain and ascertain their Notion and Difference.'

To science:

> . . . belong such Things as Men may discover by the use of Reasoning, and Sense . . . the Laws of Nature, the Affections of Bodies, the Rules and Canons of Right and Wrong, Truth and Error, the Properties of Lines and Numbers, etc.

To art:

> . . . belong such Things as mere general Reason would not have attained to: Things which lie out of the direct Path of Deduction, and which required a peculiar Cast, or Byass of Mind to see or arrive at.

Borrowing from Locke's *On Human Understanding*, Chambers sees the difference between art and science as amounting 'to much the same as between Wit and Humour'. Art is considered as 'a general Faculty of exciting agreeable and surprising Pictures in the Imagination'. Science is held to be 'particular' and tinged with something 'foreign and complexional'. The route a scholar must take to attainment in art and science is then laid forth:

> The first, every Man beside the Inventor must be taught; the latter, every Man must find. In effect, to attain to an Art, there is some previous Knowledge requir'd, which a Man's own Reason would never have suggested; whereas a Science requires no more than clear Ideas, and close Attention. With these Helps a Man may of himself go the whole length of a Science.

The discourse concludes with the view that art and science differ in essentials only as one is less and the other more pure:

> A *Science* is a System of Deductions made by Reason alone . . .

> An *Art*, on the contrary, requires a number of Data and Postulata to be furnish'd from without, and never goes any length, but at every turn it needs new ones.

Chambers' schematization of knowledge: the influence of Francis Bacon
Chambers' Preface is greatly influenced by Francis Bacon's genealogical division of knowledge: it is in essence a direct by-product of Bacon's text *Partitiones scientiarum*, the first book of the celebrated *Instauratio Magna* (*The Great Instauration*) with which Bacon endeavoured to reorganise the sciences and the restoration of man to a position of authority over nature which he considered to have been lost to mankind with the fall of Adam. Bacon was the herald of a new philosophy that anticipates the procedures of modern science. He represents the transition between mediaeval thought, rooted in myth, legend and superstition and enlightened thought based upon the rational interpretation of the best available evidence. Bacon's new plan, couched in a vivid, aphoristic style of writing, made the analytical thinker free of the reverential attitude to antiquity. It argued the case for a 'fresh examination of reality in all its detail', without rejecting as earlier theorists had done, 'certain topics as trivial or sordid', and asserted the virtues of 'observation and experiment' by means of which to gradually build up 'axioms or scientific laws of increasing generality and universality.' Bacon's Plan, as outlined in *The Great Instauration*, consists of six divisions as follows:

i. *The Divisions of the Sciences* (*Partitiones scientiarium*).
ii. *The New Organon*; *or Directions concerning the Interpretation of Nature* (*Interpretatio Naturae*).
iii. *The Phenomena of the Universe*; *or a Natural and Experimental History for the foundation of Philosophy* (*Historia Naturalis et Experimentalis*).
iv. *The Ladder of the Intellect* (*Scala Intellectus*).
v. *The Forerunners*; *or Anticipations of the New Philosophy* (*Prodromi*).
vi. *The New Philosophy*; *or Active Science.*

Bacon offers the reader the prospect of making a 'coasting voyage along the shores of the arts and sciences; not without importing into them some useful things on the way.' He embodies the spirit of these intentions in his famous classification of the attributes of the human mind and the properties of human understanding, in the text *The Advancement of Learning* (*De Augmentis Scientiarum*) which had a profound influence on the French *Encyclopédie* (see later). Bacon's treatise is divided into nine books of which the first sets out his division of all human learning into *History*, *Poesy* and *Philosophy* with reference to the three intellectual faculties *Memory*, *Imagination* and *Reason*. This remarkable text surveys the sciences, taken in their widest sense, and provides an inventory of all the possessions of the human mind.[9]

Chambers adopted Bacon's procedure of analysing the faculties of the mind and the objects of human knowledge for the intellectual basis of *Cyclopaedia* and in particular his diagram of the *Division of Knowledge* (see below). In this diagram, as we shall see, Chambers locates *Architecture* as a division of *Mechanicks* which itself is a branch of *Mathematics.* This intellectual construction can be traced back to Chapter VI, *Division of Mathematics into Pure and Mixed*, of Bacon's Book III of *The Advancement of Learning*. In this he states:

> Mathematic is either Pure or Mixed. To Pure Mathemtic belong those sciences which handle Quantity entirely severed from matter and from axioms of natural philosophy . . . Mixed Mathematic has for its subject some axioms and parts of natural philosophy, and considers quantity in so far as it assists to explain, demonstrate, and actuate these. For many parts of nature can neither be invented with sufficient subtlety, nor demonstrated with sufficient perspicuity, nor accommodated to use with sufficient dexterity, without the aid and intervention of mathematic: of which sort are Perspective, Music, Astronomy, Cosmography, Architecture, Machinery, and some others.

To further illustrate the influence of Bacon's writing on Chambers we conclude with a passage from Chapter II, *Division of the doctrine concerning the Body of Man into Medicine, Cosmetic, Athletic, and Voluptuary*, of Book IV. This text concerns the arts as related to the senses:

> Lastly I come to Arts of Pleasure Sensual, which are divided according to the senses themselves. The pleasure of the eye is chiefly Painting, with a number of other arts (pertaining to magnificence) which respect houses, gardens, vestments, vases, cups, gems and the like. The pleasure of the ears is Music, with its various apparatus of voices, wind, and strings; water instruments . . . Of all these arts those which belong to the eye and ear are esteemed the most liberal; for these two senses are the purest; and the sciences thereof are the most learned, as having mathematics like a handmaid in their train.

Chambers gives the reader 'the substance of what has been hitherto found in the several branches of knowledge, both *natural* and artificial.' Within his system, knowledge is constructed according to three faculties. First, is knowledge which appears to our *senses* 'either spontaneously, as in Natural History, or with the assistance of Art, as in Anatomy, Chymistry, medicine, Agriculture, etc.' Second, is knowledge which is revealed through our *imagination* 'as in Grammar, Rhetorick, Poetry, etc.' Third, is knowledge which is yielded by application of *reason* 'as in

Physicks, Metaphysicks, Logicks, and Mathematicks.'

Chambers elaborates his schematization according to whether knowledge is either *natural* and *scientific* or *artificial* and *technical.*

Natural and scientific knowledge is sub-divided into: *sensible* knowledge which consists in the perception of phenomena or external objects (meteorology, mineralogy, zoology); and *rational* knowledge which consists in the perception of the intrinsic characters or habitations of sensible objects according to their *powers* (natural philosophy), *abstracts* (ontology), *quantities* (arithmetic, geometry, statics), and *relations* (ethics, religion, theology).

Artificial and technical knowledge is either *internal* or *external.*

Internal knowledge is that employed in discussing *agreement, disagreement* and *relations* in respect of truth (logic).

External knowledge is either *real* or *symbolical.*

Real knowledge has: *powers* and *properties* (chemistry), *quantities* of *bodies* called mathematics (optics, hydrostatics, mechanics – from which arises architecture, sculpture, trades, military art); *structure* and *oeconomy* (anatomy); and *relations* to preservation and improvement (animals – from which medicine, vegetables – from which gardening).

Symbolical knowledge has *words* (grammar), *figures* (rhetoric) and *fables* (poetry).

Chambers presents his schematization or *Division of knowledge* in the form of a diagram in which he identifies forty-seven branches or 'Heads' of human understanding (see our reconstruction). Commenting upon this he observes:

> This is a View of Knowledge, as it were, *in semine*; exhibiting only the grand, constituent Parts thereof. It would be endless to pursue it into all its Members and Ramifications; which is the proper Business of the Book itself.

Opposite is an outline of Chambers' *Division of Knowledge* which is followed by our reconstruction of his original comprehensive schematization of the divisions of knowledge.

An Outline of Chambers' *Division of Knowledge*

KNOWLEDGE
is either

Natural and ***Scientifical***
which is either
Sensible Or, ***Rational***;

Or,

Artificial and ***Technical***
(consisting in the Application of Natural Notices to further Purposes) which is either

Internal
employ'd in discovering their Agreement and Disagreement; or their Relations in respect of Truth—call'd LOGICS
Or,
External,
which is either

Real,
Or,
Symbolical

KNOWLEDGE, is either

Natural and *Scientifical*, or, *Artificial* and *Technical*:

Natural and *Scientifical* knowledge, is either

- *Sensible*; consisting in the Perception of Phænomena, or External Objects—called PHYSIOLOGY, or NATURAL HISTORY; and which according to the different Kinds of such Objects, divides into
  - METEOROLOGY[1].
  - HYDROLOGY[2].
  - MINEROLOGY[3].
  - PHYTOLOGY[4].
  - ZOOLOGY[5].

O R

- *Rational*; consisting in the Perception of the intrinsick Characters or Habitudes of sensible Objects —either
  - Their *Powers* and *Properties*—called PHYSICKS, and NATURAL PHILOSOPHY[6].
  - *Abstracts* thereof—called METAPHYSICS[7] which subdivides into
    - ONTOLOGY.
    - PNEUMATOLOGY.
  - *Quantities* thereof, called MATHEMATICS—which divides, according to the Subject of the Quantity, into
    - ARITHMETIC[8]—whence
      - ANALYTICS[9].
      - ALGEBRA[10].
    - GEOMETRY[11] —whence
      - TRIGONOMETRY.
      - CONICS.
      - SPHERICS.
    - STATICS[12].
  - *Relations* thereof to our Happiness—called RELIGION, or the Doctrine of OFFICES, subdivides into—
    - ETHICS[13], or NATURAL RELIGION—whence
      - POLITICS[14].
      - LAW[15].
    - THEOLOGY[16], or REVELATION.

Artificial and *Technical* knowledge (consisting in the Application of Natural Notices to further Purposes) is either

- *Internal*; employ'd in discovering their Agreement and Disagreement; or their Relations in respect of Truth—call'd LOGICS[17].
- OR,
- *External*, which is either—
  - *Real*, employ'd in discovering and applying the—
    - Further *Powers* and *Properties* of Bodies—called CHYMISTRY[18]—whence
      - ALCHEMY.
      - NATURAL MAGIC, *etc.*
    - Quantities of Bodies—call'd MIX'D MATHEMATICS; which according to the different Subjects resolves into
      - OPTICS[19], CATOPTRICS, DIOPTRICS—whence
        - PERSPECTIVE[20].
        - PAINTING[21].
      - PHONICS—whence MUSICK[22].
      - HYDROSTATICS[23], HYDRAULICS.
      - PNEUMATICS[24].
      - MECHANICS[25]—whence
        - ARCHITECTURE[26].
        - SCULPTURE[27]. TRADES[28], and MANUFACTURES.
      - PYROTECHNIA[29]—whence
        - The MILITARY *Art*[30].
        - FORTIFICATION[31].
      - ASTRONOMY[32]—whence CHRONOLOGY[33]. DIALLING[34].
      - GEOGRAPHY[35], HYDROGRAPHY—whence
        - NAVIGATION[36].
        - COMMERCE[37].
    - OR,
    - *Structure* and *Oeconomy* of Organical Bodies—called ANATOMY[38].
    - Relations thereof to the Preservation and Improvement—either of
      - Animals—called MEDICINE[39]. PHARMACY[40].
      - Vegetables—called AGRICULTURE[41]. GARDENING[42].
      - Brutes—called
        - FARRYING[43], MANAGE—whence
          - HUNTING.
          - FALCONRY.
          - FISHING, etc.
  - *Symbolical*, employ'd in framing
    - *Words*, or Articulate Signs of Ideas—Called GRAMMAR[44].
    - *Figures*—called RHETORIC[45]—whence
      - The making of *Armories*, called HERALDRY[46].
    - *Fables*—called POETRY[47].

Chambers' diagram fulfils the office of a *Table of Contents* in order to 'afford the reader a sort of summary of the whole' such that in the book itself 'Heads', keywords and subject headings, 'will come into sight, and such as will naturally suggest and lead to the rest', this latter aspect being facilitated by Chambers' elaborate system of cross-referencing. From Chambers' lists of headwords we have compiled our inventory of letterpress articles on architectural subjects a summary of which follows.

Letterpress articles on architectural and allied subjects extracted from Ephraim Chambers' *Cyclopaedia*.

VOLUME I

ABACUS
ABBEY
ABBREUVOIR
ACANTHUS
ACOUSTICKS
ACROTERIA
ADDICE,
ALCOVE
ALLEY
ALTAR
AMPHIPROSTYLE
AMPHITHEATRE
ANNULETS
ANTECHAMBER
ANTEPAGMENTA
ANTIQUE
APARTMENT
APERTIONS
APOPHYGE
AQUÆDUCT
ARÆOSTYLE
ARC-BOUTANT
ARCH
ARCHITECT
ARCHITECTONICK
ARCHITECTURE
ARCHITRAVE
ARCHIVAULT
ART
ASH
ASHLAR
ASHLERING
ASPECT
ASSEMBLAGE
ASTRAGAL
ASYMMETRY
ATLAS
ATTIC
ATTITUDE
AUDITORIUM
AVENUE

BAGUETTE
BALCONY
BALDACHIN
BALLUSTRADE
BANQUETTE
BARBACAN
BASE
BASIL
BASILIC
BASTION
BATON
BATTLEMENTS
BEAD
BEAK
BEAM
BEARER
BEARING
BEAUTY
BED
BED, GARDENING
BED, MASONRY
BED, MOULDING
BED-CHAMBER
BERME
BEVEL
BITUMEN
BLINDS
BOLTS
BONNET
BOSSAGE
BOULDER-WALLS
BOULTINE
BOUTANT
BOYAU
BRACE
BRADS
BRANCHES
BREACH
BREST
BREST-SUMMERS
BRICK
BRIDGE
BUILDING
BULWARK
BUST
BUTMENTS
BUTTRESS

CABINET
CABLED-FLUTES
CALOTTE
CAMBER-BEAM
CANAL, LARMIER
CANAL, VOLUTE
CANTALIVER-CORNICE
CANTALIVERS
CANTONED
CAPITAL
CAPPONIERE
CARCASE

CARIATIDES
CARINA
CARNATION
CARPENTERS'-WORK
CARPENTRY
CARPET
CARTOUCH
CARYATIDES
CASCADE
CASEMATE
CASERNS
CASING
CASTING, FOUNDEY
CASTING, JOYNERY
CASTLE
CATACOMBS
CATACOUSTICS
CATAFALCO
CATHEDRAL
CATHETUS
CAULICOLES
CAVALIER
CAVETTO
CEILING
CELLS
CEMENT
CHAMBER
CHAMBRANLE
CHAMFER
CHANDELIERS
CHANNEL
CHANNELINGS
CHANTLATE
CHAPEL
CHAPLET
CHAPPEL
CHAPTREL
CHARGE
CHARNEL
CHANTELET
CHEMISE
CHILMINAR
CHIMNEY
CHOIR
CHURCH
CIMA
CINCTURE
CIRCUS
CITADEL
CLAIR-OBSCURE
CLAMPING
CLARO-OBSCURO
CLAYES
CLOISTER
COLARIN
COLLISEUM
COLONNADE
COLOSSUS
COLOUR
COLOURING
COLUMN
COMPARTIMENT
COMPARTITION
COMPOSITE
COMPOSITION
CONGE
CONSERVATORY
CONSOLE
CONTRAMURE
CONTRAST
CONTRAVALLATION
CONVENT
COPING
CORBEILS
CORBEL
CORINTHIAN
CORNICHE
CORONA
CORRIDOR
CORTIN
COTAGE
COUCH
COUNTER-APPROACHES
COUNTER-DRAWING
COUNTER-LIGHT
COUNTERFORTS
COURSE
COVERING
COVING
CRANE
CREUX
CROSSETTE
CROSS-GRAIN'D STUFF
CROWN
CROWN-POST
CROWN-WORK
CROWNING
CRYPTA
CUPOLA
CURLING
CURTAIN
CYMA
CYMATIUM

DADO
DECASTYLE
DECORATION
DECORUM
DEFILE
DEGREDATION
DEMI-GORGE
DEMI-LUNE
DENTICLES
DESIGN
DETACH'D
DIASTYLE
DIGLYPH
DILAPIDATION
DIMINUTION
DIPTERE
DITRIGLYPH
DOME
DONJON
DOOR
DORIC
DORMER
DORMITORY
DOUCINE
DOVE-TAIL
DOVETAILING
DRAPERY
DRAUGHT
DRIP
DRUM
DYE

EAGLE
EAVES
EAVES-LATH
ECHINUS
ECPHORA
EDGINGS
EDIFICE
ELEVATION
EMBOSSING
EMBRASURE
EMPASTING
ENAMEL
ENAMELLING
ENCAUSTICE
ENCHASING
ENFILADE
ENGRAVING
ENTABLATURE
ENTRY
ENVELOPE
EPAULE
EPAULEMENT
EPISTYLE
ESCALADE
ESCUTCHEON
ESPAULE
ESPLANADE
ESQUISSE
ESTRADE
ETCHING
EURYTHMY
EUSTYLE
EXCAVATION
EXPOSURE
EXPRESSION
EYE

FACE, FACADE
FACE, FACIA
FACE, PLACE
FACE, STONE
FACIA
FASCIA
FASCINES
FASTIGIUM
FEATHER
FENCE-WALLS
FIGURE
FILLET
FINISHING
FLANK
FLANKING
FLIGHT
FLOOR
FLOWER
FLUTES
FLYERS
FORT
FORTERESSE
FORTIFICATION
FORTIN
FOSS
FOUNDATION
FOUNTAIN
FRAISE
FRAME
FREEZE
FRESCO
FRET
FRIEZE
FRIZE
FRONT
FRONTAL
FRONTISPIECE
FRONTON
FURRING
FUSAROLE
FUST

GABIONS
GABLE
GAGE
GALLERY
GARDEN
GARDENING
GATE
GILDING
GIRDERS
GIRDING-BEAMS
GLACIS
GLADE
GLYPHE
GORGE
GOTHIC
GRADATION
GRASS-PLOTS
GRAVEL WALK
GREEK ORDERS
GREEN-HOUSES
GROTESQUE
GROTTO
GROUND
GROUND-SELL
GROUND-WORK
GROUP
GROVE
GUERITE
GUELE
GUILD-HALL
GULE
GUTTÆ
GUTTERS

HABITATION
HAIR
HALF-MOON
HALL
HANGINGS
HANGS-OVER
HEAD
HEALING
HEARTH, FOCUS
HEARTH-STONES
HEDGE
HERSE
HEWING
HIPPODROME
HOLLOW
HOLLOW TOWER
HORN
HORN WORK
HORTICULTURE
HOSPITAL
HOT-BED
HOUSE
HURDLES
HYPOCAUSTUM
HYPODROME

VOLUME II

ICONOGRAPHIA
ICHONOGRAPHY
ILLUMINATION
IMAGE
IMPOSTS
INCRUSTATION
INCRUSTED COLUMN
INSTRUMENTS
INSULATED
INTAGLIO'S
INTERCOLUMNS
IONIC
IRREGULAR COLUMN
ISLES

JAPANNING
JET D'EAU
JOINT
JOISTS
JONCTURE
JOYNERY
JUNCTURE

KARKRONI
KEY
KEY, ARCH

LACUNAR
LANDSKIP
LARMIER
LATH
LATHE
LAYMAN
LAZARETTO
LEAD
LEVEL
LEVELLING
LIBRARY
LICENCES
LIGHTS
LIME
LIME-STONE
LIMNING
LINE
LINTEL
LIST
LOCK
LODGMENT
LUNETTES
LUTHERN
LYCEUM

MAGAZINE
MANAGE
MANEQUIN
MANSE
MANSION
MANNER
MANTELETS
MANTLE
MANUFACTURE
MARBLE
MARBLED
MARQUETRY
MASON
MASONRY
MASQUE
MASSES
MASSIVE
MASTER-PIECE
MAUSOLEUM
MERLON
MESSUAGE
METOCHE
METOPE
METROPOLIS
MEZZANINE
MEZZO-TINTO
MILITARY
MILL
MINIATURE
MINUTE
MOAT
MODEL
MODERN
MODILLIONS
MODULE
MOINEAU
MONASTERY
MONUMENT
MORISCO
MORTAR
MORTOISE
MOSAIC
MOSQUE
MOULD
MOULDINGS
MOUND
MOYENAU
MUNIMENTS
MUNIONS
MURAL
MURAL ARCH
MURING
MUTILATION
MUTULE

NAILS
NAKED
NAVE
NERVES
NEWEL
NICHE
NUDITIES
NUEL
NURSERY

OBELISK
OBSERVATORY
OCTAGON
OCTOSTYLE
ODEUM
OFF-SETS
OFFICES
OGEE
OLITORY
OPENING FLANK
OPENING OF
TRENCHES
ORANGERY
ORCHESTRA
ORDER
ORDONNANCE
ORIGINAL
ORILLON
ORLE

ORNAMENTS
ORTEIL
ORTHOGRAPHY IN:
ARCHITECTURE,
GEOMETRY,
PERSPECTIVE,
FORTIFICATION
OUT-WORKS
OVA
OVICULUM
OVILIA
OVOLO

PAINTING
PALACE
PALÆSTRA
PALE
PALES
PALISSADE
PALLADIUM
PALLIER
PALLIFCATION
PANEL, JOYNERY
PANEL, MASONRY
PANNEL
PANNIER
PANTHEA
PANTHEON
PARAPET
PARASCENIUM
PARGETING
PARISH
PARK
PARTERRE
PASTIL
PAVEMENT
PAVILLION
PEDESTAL
PEDIMENT
PENÇIL
PENDENTIVE
PENTASTYLE
PERAMBULATOR
PERIPTERE
PERISTYLE
PERRON
PERSIAN
PERSPECTIVE
PHAROS
PIAZZA
PICKET
PICTURE
PIEDOUCH
PIEDROIT
PIGMENTS
PILA, PEER
PILASTER
PILE
PILLAGE
PILLAR
PINNACLE
PINNING
PIPE
PITCH
PITT
PLACARD
PLACE
PLAFOND
PLAISTER
PLAN
PLANCERE
PLANE
PLANTING
PLASTER
PLASTICE
PLAT-BAND IN:
ARCHITECTURE,
GARDENING
PLATFORM
PLINTH
PLINTH OF A WALL
PLOTT
PLOTTING
PLUMBERY
PORCH
PORTICO
POSTERN
POSTIQUE
POSTSCENIUM
POUDERINGS
PROFILE
PROJECTION
PROJECTURE
PROJECTURE,
ARCHITECTURE
PRONAOS
PROPERTY
PROPORTION
PROPYLÆUM
PROSCENIUM
PROSPECT
PROSTYLE
PROTHYRUM
PROTHYRIS,
PROTHYRIDE
PSEUDO-DIPTERE
PULVINATA,
PULVINATED
PUNCHIN
PUNCHION
PURBECK-STONE
PURLINS
PUTLOGS
PYCNOSTYLE
PYLING
PYRAMID

QUADRA
QUADRELS
QUARRY
QUARTERS
QUEUE D'ARONDE
QUINCUNX
QUIRK
QUOINS

RABBETING
RAFTERS
RAIL
RAISER
RAKING-TABLE
RAMMER
RAMPART
RANGING
RASANT
RAVELIN
RECTORY

REDENS, REDANT, REDAN
REDOUBT
REDUCT
REFECTORY
REGLET
REGULA
REJOYNTING
RELIEVO
REPOSE
REPOSITORY
RESEARCHING
RESSAUT
RETIRADE
RETRENCHMENT IN: ARCHITECTURE, WAR
REVERBERATION
REVERSED
RIDEAU
RIDGE
RONDEL
ROOD
ROOF
ROOM
ROSTRALIS
ROTONDO
ROUNDS
RUDENTURE
RUDERATION
RUINS
RULE, REGULA, RULER
RUSTIC

SAIGNER
SALLY
SALON
SANCTUARY
SAND
SAPHETA
SARCOPHAGUS
SAW
SAWING
SCABELLUM
SCAFFOLD
SCAMILLI IMPARES
SCANTLING
SCAPUS
SCARP
SCENE
SCENOGRAPHY
SCHOOL
SCIOGRAPHY
SCOTIA
SCRATCH-WORK
SCRIBING
SCROWLS
SCULPTURE
SCUTCHEON
SEALING
SEASONING
SECTION
SELL
SEPTIZON
SEPULCHER
SEPULCHRAL
SERPENTINE MARBLE
SEWERS
SGRAFIT
SHAFT
SHINGLES
SHOULDERING
SHOULDERING-PIECE
SILLON
SIMA
SIMATIUM
SLATE
SOCLE
SOFFITO
SPHINX
SPIRAL
SPIRAL STAIRS
SPIRE
SPITTAL-HOUSE
STADIUM
STAFF
STAGE
STAIRS
STANCHEONS
STAR-FORT
STATUARY
STATUE
STEEPLE
STEP
STEREOBATA
STILES
STONE
STONE-HENGE
STOVE
STRIAE
STRIGES
STRUCTURE
SUBSTRUCTION
SUMMER
SURVEYING
SURVEYOR
SUSPIRAL
SWALLOW-TAIL
SYMBOLICAL
SYMMETRY

TABERNACLE
TABLE
TABLE IN ARCHITECTURE
TÆNIA
TAIL
TAILLOIR
TALON
TALUS
TAMBOUR
TAPER
TAPESTRY
TARRASS
TASSELS
TAXIS
TECHNICAL
TEINTS
TEMPLE
TENAILLE
TENANCY
TENEMENT
TENON
TERM
TERRACE
TERRE-PLAIN

TESSELATED
TETRASTYLE
THEATER, THEATRE, THEATRUM
THEOLOGIUM
THERMÆ
THRONE
TIGE
TILE
TIMBER
TOILET
TOL-BOOTH, TOLL-BOOTH
TOMB
TONDIN
TORE
TOREUMATOGRAPHY
TOREUTICE
TOWER
TRABEATION
TRAJAN
TRANSOM
TRENCH
TRENCHING
TRIGLYPHS
TRIMMERS
TRINGLE
TUILERIE
TURNING
TUSCAN
TWIST
TYLE
TYLERY
TYMPAN

UNION
UPRIGHT

VAGINA
VAN-FOSSE
VASES
VAULT
VENEERING
VENTIDUCTS
VERNACULAR
VESTIBLE
VESTRY
VILLA
VILLAGE
VINEA
VISUAL
VIVO
VOLUTE
VOUSSOIR

WAINSCOT
WALKS IN GARDENING
WALL
WARDROBE
WASHING
WATER-SCAPE
WATER-TABLE
WELL-HOLE
WINDOW
WOOD
WORK-HOUSE

XYSTUS

YARD

ZETA, ZETECULA
ZOCCO, ZOCCOLO, ZOCLE, SOCLE
ZOOPHORUS

Illustrations of the letterpress articles
In the letterpress articles Chambers defines the chosen subjects with reference to the best authors, quoting from their texts across all aspects of the topic under consideration. We illustrate his style and methodology with paraphrases of the articles *Architecture* and *Building*.

Letterpress article Architecture
Architecture in the *Cyclopaedia* is allotted two and a half columns of text and a double plate.

An architect is:

A person skill'd in Architecture, or the Art of Buildings; who makes Plans and Designs of Edifices, conducts the Work, and directs the Motions and other Artificers employ'd therein.

Architecture is:

> The Art of *Building*; i.e., of erecting Edifices proper for Habitation, or Defence.

Civil architecture is:

> The Art of contriving and executing commodious Buildings for the Uses of Civil Life; as *Houses*, *Temples*, *Theatres*, *Halls*, *Bridges*, *Colleges*, *Portico's*, etc.

Chambers comments that architecture is 'scarce inferior to any of the Arts in Point of Antiquity'. He refers to the Grecian origins of classical architecture from which 'Three of the regular Orders or Manners of Building, are denominated . . . viz. *Corinthian*, *Ionic*, and *Doric*'.

Chambers traces the line of architectural succession from antiquity to continental Europe:

> In the last Centuries, the Architects of *Italy* and *France* were wholly bent upon retrieving the primitive simplicity and Beauty of antient Architecture, in which they did not fail of Success: in so much, that our Churches, Palaces, etc. are now wholly built after the Antique.

The article *Architecture* includes several references to writers on architecture. Pre-eminent among the ancients is *Vitruvius* whom Chambers considers to be 'the only entire Author'. Nonetheless, he chastises him for 'Want of Method, and Obscurity' and his mixture of Latin and Greek. Quoting *Leon Baptist Alberti*, Chambers observes 'he wrote Latin to the *Greeks*, and Greek to the *Latins*'. Of the authors who have endeavoured to interpret Vitruvius, Chambers lists *Perrault*, *Philander*, *Barbaro* and *Salmasius*. In similar vein, and worthy of quotation, is Chambers' commentary on 'The Authors upon *Architecture* since Vitruvius'. The most important of these are held to be:

> *Sebastiano Serlio*, who wrote seven books of Architecture, five of which concerning the five Orders, were made publick in 1602; throughout all which, he religiously keeps to Vitruvius's Rules.
>
> *Andrea Palladio*, who wrote four Books of Architecture, containing the fundamental Rules of the Art, with various Instances of all the Kinds of Works; published in Italian in 1575.
>
> *Philibert de Lorme*, who published nine Books of Architecture, in French, in 1567.

J. Barozzio de Vignola, who in 1631, made publick his Rules of the five Orders, in Italian.

Vincent Scamozzi, his Idea of *Universal Architecture*, published in 1615, in Italian.

C. D. Dieussart, in this *Theatre of Civil Architecture*, published in High-Dutch in 1697; wherein he not only delivers the Rules of *Architecture*, but explains and compares the five Orders as laid down by Palladio, Vignola, Scamozzi, etc. which same Design was also executed in French by R. Freart de Cambray, in a *Parralel of the antient Architecture with the Modern*, published in French in 1650, and since translated into *English* with Additions, by Mr. Evelyn.

François Blondel, Director of the Royal Academy of Painting, etc. in 1698, gave a *Course of Architecture*, in French; being a Collection from all the celebrated Writers upon the Subject of the Orders, etc.

Nicholas Goldman, in a *Treatise de Stylometris*, published in Latin and High-Dutch, in the Year 1661, has done good Service, in reducing the Rules and Orders of *Architecture* to a further Degree of Perfection, and shewing how they may be easily delineated by means of certain Instruments invented by him.

The last work cited by Chambers in his 'Authors upon Architecture' is the *Elements of Architecture* as laid down by:

Sir Henry Wotton – The same are reduced by Sturmis, and Wolfius, to certain Rules and Demonstrations; and thus is *Architecture* brought into Form of a Mathematical Art; by the first, in his *Mathesis Juvenil*, and the second, in his *Elementa Matheseos*, Tom. II. *An.* 1715.

Letterpress article Building

In the article *Architecture*, Chambers informs the reader: 'There are several Arts subservient to Architecture, as Carpentry, Masonry, Paving, Joynery, Smithery, Glazery, Plumbery, Plastering, Gilding, etc.' The article *Building* is a good illustration of Chambers' treatment of these. In its entirety the entry is two folio pages. Chambers firstly defines a building as 'any Place erected by Art, whether for Convenience, for Religion, or Magnificence'. In a frequently quoted passage, from Sir Henry Wotton, the entry continues: 'In *Building* there are three Things in view, viz. Commodity, Firmness, and Delight'. In interpreting these, the *Seat* or *Situation* of a residence is considered to be of paramount importance and the following guidance is offered:

> Regard must be had to the Quality, Temperature, and Salubrity of the Air; the Conveniency of Water, Fuel, Carriage, etc. and the Agreeableness of the Prospect.

Concerning the planning and orientation of rooms:

> The chief Rooms, Studies, Libraries, etc. to lie towards the East; Offices that require Heat, as Kitchens, Distillatories, Brew-Houses, etc. to the South: Those that require a cool fresh Air, as Cellars, Pantries, Granaries, etc. to the North: As also Galleries for Painting, Museums, etc. which require a steady Light.

For room planning, rectangular shapes are preferred 'as being a Medium between . . . Extremes': rule of thumb guidance takes the form: 'the Length don't exceed the Breadth by above one third'. Circular plan shapes are recognised as attractive but have practical disadvantages:

> The circular *Form* is very commodious, of the greatest Capacity of any; strong, durable beyond the rest, and very beautiful: but then it is found of all others the most chargeable; much room is lost in the bending of the Walls, when it comes to be divided; besides an ill Distribution of Light, except from the Centre of the Roof.

The article *Building* then acquaints the reader with several technical aspects of building construction, proceeding systematically from the foundations upwards:

Foundations:

> The Substruction, or Groundwork, to support the Walls . . . This to be level . . . close laid with Mortar, and the broader the better; at least, twice as broad as the Wall.

Walls:

> The great Laws of Muring, are that the Walls stand perpendicular to the Ground-work; the Right Angle being the Cause of all Stability: that the massiest and heaviest Materials be lowest, as fitter to bear than be borne . . . that certain Courses, or Ledges of more strength than the rest, be interlaid, like Bones, to sustain the Fabrick from total Ruin, if the Under-parts chance to decay.

Intermissions:

> Are either Columns or Pilasters; whereof there are five Orders, viz. Tuscan, Doric, Ionic, Corinthian, Composite.

The reader is then referred to separate articles on each of these, for which the double-folio plate on architecture is reserved.

Apertures:

Are either Doors, Windows, Staircases, Chimneys, or Conduits for the Suillage [sewage] . . . with regard to the last, it may be observ'd, that Art should imitate Nature in the ignoble Conveyances, and separate them from Sight, where a running water is wanting, into the most remote, lowest, and thickest Part of the Foundation; with secret Vents . . . for the Discharge of noisom Vapors.

Compartition, or Distribution of the Groundplot into Apartments:

The Architect [should] never fix his Fancy on a Paper Draught . . . much less on a mere Plan, without a Model, or Type of the whole Structure, and every Part there of.

A large Fabrick should have large Partitions, Entrances, Doors, Columns, and, in brief, all the Members large.

The Antients determin'd the Length of their Rooms, that were to be Oblongs, by double their Breadth; and their Height by half their Breadth and Length added together. When the Room was to be precisely Square, they made the Height half as much more as the Breadth: which Rules the Moderns take occasion to dispense with; sometimes squaring the Breadth, and making the Diagonal there of the Measure of the Height.

The *Italians* frequently distribute the Kitchen, Bake-house, Buttery, etc. under Ground, next above the Foundation, and sometimes level with the Floor of the Cellar; raising the first ascent into the House 15 Foot or more: which, beside the removing of Annoyances out of Sight . . . does, by elevating the Front, add a Majesty to the whole.

In the Distribution of Lodging Rooms, 'tis a popular and antient Fault, especially among the *Italians*, to cast the Partitions so, as when the Doors are all open, Man may see through the whole House; grounded on the Ambition of shewing a Stranger all the Furniture at once.

In the Compartition, the Architect will have occasion for frequent Shifts; thro which his own Sagacity, more than any Rules must conduct him. Thus he will be frequently put to struggle with Scarcity of Ground; sometimes to damn one room for the Benefit of the rest, as to hide a Buttery under a Staircase, etc. at other times, to make those the most beautiful which are most in Sight; and to leave the rest, like a Painter, in the Shadow.

Covering:

This is the last in the Execution, but the first in the Intention: for who would *build*, but to shelter? . . . The *Italians* are very curious in the Proportion and Gracefulness of the Pent or Slopeness; dividing the whole Breadth into nine Parts, where of two serve for the height of the highest Top or Ridge from the lowest: But in this Point, regard must be had to the Quality of the Region; for, as *Palladio* insinuates, those Climates which fear the falling of much Snow, ought to have more inclining Pentices than others.

Chambers concludes his long and informative article on *Building* with five aphoristic pieces of advice:

1st, *Let not the common Rooms be several nor the several Rooms be common*: i.e. The common Rooms not all to be private or retir'd, as the Hall, Galleries, etc. which are to be open, and the Chambers, etc. retir'd.

2nd, *A House had better be too little for a Day, than too big for a year*. Houses therefore to be proportion'd to ordinary occasions, not extraordinary.

3rd, *Country Houses must be Substantives, able to stand of themselves*. Not like City *Buildings*, supported and shelter'd on each Side by their Neighbours.

4th, *Let not the Front look asquint on a Stranger*; *but accost him right, at his Entrance*.

5th, *Let the Offices keep their due Distance from the Mansion House*; *those are too familiar, which are of the same Pile with it*.

Postscript: Chambers' achievement

As we have noted earlier, Chambers' *Cyclopaedia* was recognised from the outset as being a scholarly work of distinction and significance. Chambers was modest about his achievement. He even goes so far as to say, at the conclusion of his Preface, that his work is not designed for 'men of the first-rate erudition [who want to] study their authors at first hand.' He does though expect a high standard of learning from his readers particularly, as noted previously, a command of Greek and Latin and also mathematics. Above all else it is Chambers' sheer industry which commands our respect. In his role as editor-compiler, he likens himself to a bee which has 'rifled a thousand Flowers; prickly are many of 'em, to load your Hive.' The result is he concludes, changing the imagery, 'the mixed issue of a thousand Lions.'

Chambers saw books as 'Plans or Prospects of Ideas, [of] art fully arranged, and exhibited, not to the Eye, but to the Mind.' His own ordering mind saw virtue in a collection which 'has its advantages: where a number of things are thrown precariously together, we sometimes discover relations among 'em, we should never have thought of looking for.' Chambers saw himself as having made a path which he trusted others would 'enlarge, and widen, or even make it straighter and easier.' He certainly considered he had improved on his predecessors observing: 'Bayle's Dictionary has been called the Errata of *Morreri*; yet is not *Bayle* himself without his Errors. The most we can say, is, that we hope there will be few found in the present Work, in comparison of others of the like kind.'

Chambers was not as accomplished a mathematician as John Harris and the *Cyclopaedia* is not as strong as Harris's *Lexicon Technicum* on specifically mathematical subjects. Nevertheless, the broad sweep of Chambers' work far exceeds that of Harris and his predecessors. Chambers was alert to change and progress and saw these as being reflected in language, of which he remarked 'our Language is, and will continue in a perpetual flux'. Writing a quarter of a century later, Samuel Johnson expressed much the same idea in the Preface to his *A Dictionary Of The English Language* (1755) in which he likens words to buds, adding, 'some words are budding, and some falling away.'

The influence of Chambers' Cyclopaedia on the French Encyclopédie

We have stated that *Cyclopaedia* was well received in the United Kingdom, earning for its editor-compiler a Fellowship of the Royal Society; it ran to several editions, was translated into Italian and metamorphosed into a large-scale work of reference in the nineteenth century. However, perhaps the must enduring consequence of *Cyclopaedia* was the indirect influence it exerted on the creation of the celebrated *Encyclopédie*.

In France, at the time of creation of *Cyclopaedia*, the *philosphes* were imbued with the intellectual curiosity and reforming zeal that was so central to the ideals of the Age of Enlightenment. A century earlier Francis Bacon's great French contemporary René Descartes (1596-1650), the 'father of modern philosophy', had elaborated a perception of the entirety of human knowledge into a unified construction based upon a system of beliefs embracing mathematics, physics, medicine and morality, strengthened by a mental outlook founded on processes of rigorous rationalism. His influential works *Discours de la Méthode* (Discourse on Method, 1637), the *Meditationes de prima Philosphia* (Meditations of First Philosophy, 1641) and the *Principia Philosophiae* (Principles of Philosophy, 1644) had laid the foundations of the Cartesian doctrines enabling men to become 'masters and possessors of nature'. He shared Francis Bacon's view that 'the entire work of the understanding must be begun afresh and the mind itself be, from the start, not left to take its own course, but be guided step by step'. Francis Bacon's ideas for the advancement of learning had echoes in the works of certain of the French encyclopaedists such as Louis Moréri (*Grande dictionnaire historique*, 1674), Thomas Corneille (*Dictionnaire des arts et des sciences*, 1694) and Pierre Bayle (*Dictionnaire historique et critique*, 1696). The new generation of French intellectuals contemporary with Ephraim Chambers took up the ideas of their predecessors with enthusiasm. These included the philosopher and man of letters Denis Diderot (1713-1784) and the mathematician-philosopher Jean Le Rond d'Alembert (1717-1783).

It was natural, given the fame of *Cyclopaedia* and the intellectual climate then prevailing in France, for the idea to be mooted of bringing out a French translation of Chambers' work. Chambers, although conversant with the French language and a skilful translator, declined, as we have earlier noted, an invitation from the French publishing houses to produce such a work himself probably on the grounds of ill health. The translation was entrusted to the writer on agriculture John Mills (?-1784) and the German-born historian and translator Godefroy Sellius. Work on the translation commenced and the powerful publisher André-François Le Breton, who had obtained a licence to publish the work, issued a Prospectus for a French translation of a new edition to consist of four volumes of letterpress articles and one volume of 120 plates, under the title: *Encyclopédie, ou Dictionnaire universel des arts et des sciences . . . traduit de l'Anglois d'Ephraim Chambers.*[10]

The project for a translation into French of *Cyclopaedia* failed for a variety of reasons and the work was transferred into the editorial care of Denis Diderot and Jean Le Rond d'Alembert who, with the assistance of a veritable army of scholars, transformed the enterprise into an entirely different and more radical work, namely, the renowned *Encyclopédie.* Notwithstanding these transformations, the influence of *Cyclopaedia* on the minds of the French scholars can be traced in the preliminary pages of the *Encyclopédie*, notably in d'Alembert's celebrated *Preliminary Discourse*. D'Alembert, following, as he styles him, 'the immortal Chancellor of England, Francis Bacon', states the case for knowledge to be logically organised as a whole as in a *dictionnaire raisonné*, exemplifying the essential principles and applications of all arts and sciences. Citing Bacon's pioneering work, d'Alembert informs the reader:

> He divided . . . sciences into different branches, of which he made the most exact enumeration that was possible for him. He examined what was already known concerning each of these objects and made the immense catalogue of what remained to be discovered. This is the

> aim of his admirable book *The Advancement of Learning*. In his *Novum Organum*, he perfects the views that he had presented in the first book, carries them further and makes known the necessity of experimental physics, of which no one was yet aware.

In his contribution to the opening pages of the *Encyclopèdie*, Diderot further outlines the work of the English philosopher in his supplement to the *Preliminary Discourse*, namely in the text *Detailed Explanation of the System of Human Knowledge* and *Observations on Bacon's Division of the Sciences*. Accompanying these writings, in a style remarkably similar to that adopted by Chambers in the *Cyclopaedia*, is a representation of human understanding (*Entendement*) presented in the form of a chart titled *Systême Figuré Des Connoissances Humaines* (*Detailed System Of Human Knowledge*). Within the French schematization, three broad divisions of knowledge are identified: *History*, which is related to *Memory*; *Philosophy*, which emanates from *Reason*; and Poetry, which arises from *Imagination*.

The forty-seven divisions of knowledge identified by Chambers are elaborated in the *Encyclopédie* into several times that number. Chambers, we have earlier noted, locates *architecture* at heading twenty six under the division of *mechanics* which is a sub-set of *quantities of bodies*. In the *Encyclopédie*, by way of comparison, *architecture* appears in the genealogical table three times: under *Memory*, as *Practical Architecture* (a sub-set of *Working and Uses of Stone*, *Plaster*, *Slate*, *etc*.); under *Reason*, as *Military Architecture* (a sub-set of *Mathematics*); and under *Imagination*, as *Civil Architecture* (a sub-set of *Poetry*). To extend the comparison further, Chambers has a single entry (twenty eight) for *Trades and Manufactures*. In d'Alembert's *labyrinth*, as he was given to describing the tree of knowledge in the *Encyclopédie*, trades and manufactures are represented by many entries including, the working and uses of, gold, silver, precious stones, iron, glass, skins, stone, plaster, slate, silk and wool.

Whereas Chambers placed reliance on learning and scholarship to guide his creation of *Cyclopaedia*, the French philosophes emphasised the importance of visiting artists and craftsmen at their places of work in order to enlarge their personal experience of the arts, crafts and manufactures. This aspect of the *Encyclopédie* was particularly important to its chief editor, Denis Diderot, who, as the son of an artisan, understood the ways of workmen and took direct responsibility for assembling articles on the arts and crafts. He was somewhat critical of Chambers whom he considered 'added almost nothing to what he translated from our authors'. This led him to make his famous proclamation: 'Thus everything impelled us to go directly to the workers'. The result was the eleven folio volumes of plates published between 1762-72 which many consider to be one of the most important, and certainly the most beautiful, features of the *Encyclopédie*. The total number of plates greatly outnumber those given by Chambers in *Cyclopaedia* and are the best representations in any encyclopaedia of eighteenth-century arts and crafts.

In conclusion we may say that Chambers' influence on the French philosphes, as manifest through *Cyclopaedia*, is somewhat ambivalent. As we have noted, *Cyclopaedia* provided the initial inspiration for a French encyclopaedic work but was abandoned in favour of a more radical text which arose like a great Phoenix to become, as that eloquent commentator on the *Encyclopédie* Richard Schwab has observed 'one of the great enterprises in the history of ideas of Western civilisation'.

After completion of *Cyclopaedia* and his other literary works, Chambers lived frugally and died on 15th May 1740. He was honoured by burial in the cloisters of Westminster Abbey. He penned his own epitaph:

Multis pervulgatus,
paucis notus,
Qui vitam, interlucem, et umbram,
nec eruditas, nec idiota,
dicteris deditus, transegit, sed ut homo
Qui humani nihil a se alienum putat,
Victa simul et laboribus functus,
Hic requiescere voluit

EPHRAIM CHAMBERS
F. R. S.

Obijt 15 Maij MDCCXL.

2 *Cyclopaedia* Volume I

Title Page

Dedication

The Preface: Part One

CYCLOPÆDIA:

OR, AN

UNIVERSAL DICTIONARY

OF

ARTS and SCIENCES;

CONTAINING

The DEFINITIONS of the TERMS,

and ACCOUNTS of

The THINGS signify'd thereby,

In the several ARTS,

Both *LIBERAL* and *MECHANICAL*,

And the several SCIENCES,

HUMAN and *DIVINE*:

The Figures, Kinds, Properties, Productions, Preparations, and Uses, of Things NATURAL and ARTIFICIAL;

The Rise, Progress, and State of Things ECCLESIASTICAL, CIVIL, MILITARY, AND COMMERCIAL:

With the several Systems, Sects, Opinions, *etc.* among Philosophers, Divines, Mathematicians, Physicians, Antiquaries, Criticks, *etc.*

The Whole intended as a Course of Antient and Modern LEARNING

Compiled from the best authors, Dictionaries, Journals, Memoirs, Transactions, Ephemerides, *etc.* in several Languages.

In TWO VOLUMES.

By *E. CHAMBERS* Gent.

Floriferus ut apes in saltibus omnia libant,
Omnia nos——————————LUCRET.

VOLUME the FIRST.

LONDON:

Printed for *James* and *John Knapton, John Darby, Daniel Midwinter, Arthur Bettesworth, John Senex, Robert Gosling, John Pemberton, William* and *John Innys, John Osborn and Tho. Longman, Charles Rivington, John Hooke, Ranew Robinson, Francis Clay, Aaron Ward, Edward Symon, Daniel Browne, Andrew Johnston,* and *Thomas Osborn*, M.DCC.XXVIII.

TO THE

K I N G.

S I R,

THE ARTS and SCIENCES humbly crave Audience of Your Majesty. The near Concern they have in the Happiness of a People, assures 'em of the favourable Attention of a Prince who makes that Happiness his own. 'Tis by These, the Parsimony of Nature is supplied, and Life render'd easy and agreeable under its numerous Infirmities. By these the Mind is reclaim'd from its native Wildness; and enrich'd with Sentiments which lead to Virtue and Glory. 'Tis these, in fine, that make the Difference between your Majesty's Subjects, and the Savages of *Canada*, or the *Cape of Good Hope*.

THE Protection of ARTS has ever been esteemed the proper Province of the Great. 'Tis a Branch of the Regal Office; which a Prince, like Your Majesty, equal to the whole Charge of a Crown, will not suffer to be alienated into other Hands. From this, do the first and most distinguish'd Names in the List of Fame, derive a large Share of their Glory: and if there be any Age or Nation more conspicuous than the rest, and which is look'd on with Envy by our own; 'tis that wherein the Sovereigns have signaliz'd themselves most in this Quality.—Indeed, the Time seems at hand, when we are no longer to envy *Rome* her AUGUSTUS and AUGUSTAN AGE, but *Rome* in her turn shall envy ours.

SOMETHING extraordinary is apparently intended by Providence in calling such a Prince, to such a People: A Prince who feels a generous Impulse to devote his Cares and all his Toils to the Welfare of Mankind; and a People conspiring with unexampled Ardor and Unhumanity to all his glorious Views. Some of our best Princes have had their Hands ty'd down; check'd by reluctant Factions, who opposed every nobler Design: Your Majesty has found the happy Secret, to make even Contention do you Homage; and turn Opposition itself into Approbation, and Applause.

THERE is a Time reserv'd in Fate for every Nation to arrive at its Height; and the uppermost Place on the Terrestrial Ball is held successively by several States. May not the numerous Presages which usher in Your Majesty's Reign, give us room to expect that our Turn is next; and that what *Greece* was under ALEXANDER, and *Rome* under AUGUSTUS *Cæsar*, *Britain* shall be under GEORGE and CAROLINE?

BUT even this were to under-rate our Hopes, which are rais'd, by Your Majesty, to something still more truly glorious. Greatness, so fondly coveted, has already cost the World very dear; and, tho still pursued by unthinking Men under almost every Shape, is only desirable in a few. Of it self it is rather an Object of Terror and Alarm, than Delight; and at best only pleases, when join'd with something naturally amiable. From the Practice of Your majesty, Men may correct their Sentiments, and learn, that Greatness has no Charm except when founded in Goodness. To be Great, and a King, is but a small Matter with Your Majesty; 'tis a Quality many others enjoy in common with You, and to which some have even been doom'd, to their Infamy: 'tis what *Herod* was, and *Nebuchadnezzar* was; and *Nero*, and *Domitian* were. But, while other Princes chuse to be great in what is destructive, and others in things wholly indifferent; 'tis Your Majesty's Praise to be great in what is the Perfection of our Nature, and that whereby we approach nearest the Deity. Happy Choice! to use Power only as the Means of rendering your Beneficence more diffusive; and thus make Royality minister to the Happiness of Mankind, which it has too often invaded.

YOUR Majesty commands a People capable of every thing. Not more fitted to shine in Arms, or maintain an extended Commerce; than to succeed in the stiller Pursuits of Philosphy and Literature. And it will be Your Majesty's Glory, not to let any of their Talents lie unemploy'd.—If Your Majesty gives the Word, while some of 'em are busied in avenging Your Cause, by humbling some turbulent Monarch; some in extending your Dominions by new Settlements, and some in increasing your Peoples Wealth, by new Trades: Others will be employ'd in enlarging our Knowledge, by new Discoveries in nature, or new Contrivances of Art; others in refining our Language; others in improving our Morals; and others in recording the Glories of your Reign, in immortal Verse.

THE work I here presume to lay at Your Majesty's Feet, is an Attempt towards a Survey of the Republick of Learning, as it stands at the Beginning of Your Majesty's auspicious Reign. We have here the Boundary that circumscribes our present Prospect; and separates the known, from the unknown Parts of the Intelligible World. Under Your Majesty's Princely Influence and Encouragement, we promise our selves this Boundary will be removed, and the Prospect extended far into the other Hemisphere. Methinks I see Trophies erecting to Your Majesty in the yet undiscover'd Regions of Science; and Your Majesty's Name inscribed to Inventions at present held impossible!

I am, with all Sincerity and Devotion,

May it please Your MAJESTY,

Your Majesty's most Dutiful,

and Obedient Subject,

and Servant,

Ephraim Chambers.

THE

PREFACE.

'TIS not without some Concern that I put this Work in the Reader's Hands; a Work so disproportionate to a single Person's Experience, and which might have employ'd an Academy. What adds to my Jealousy, is the little measure of Time allow'd for a Performance to which a Man's whole Life scarce seems equal. The bare Vocabulary of the Academy *della Crusca* was above forty Years in compiling, and the Dictionary of the *French* Academy much longer; and yet the present Work is as much more extensive than either of them in its Nature and Subject, as it falls short of 'em in number of Years, or of Persons employ'd.

THE Reader might be here led to suspect something of Disingenuity; and think I first put a Book upon him, and then give him Reasons why I should not have done it.—But his Suspicions will cease, when he is appriz'd of the Advantages under which I engaged; which, in one Sense, are superior to what had been known in any former Work of the Kind; all that had been done in them accruing, of course, to the Benefit of this. I come like an Heir to a large Patrimony, gradually rais'd by the Industry, and Endeavours of a long Race of Ancestors. What the *French Academists*, the *Jesuits de Trevoux*, *Daviler*, *Chomel*, *Savary*, *Chauvin*, *Harris Wolfius*, and many more have done, has been subservient to my Purposes. To say nothing of a numerous Class of particular Dictionaries which contributed to their Share; Lexicons on almost every Subject, from Medicine and Law, down to Heraldry and the Manage.

Yet this is but a Part. I am far from having contented my self to take what was ready procured, but have augmented it with a large Accession from other Quarters. No part of the Commonwealth of Learning, but has been traffick'd to on this Occasion. Recourse has been had to the *Originals* themselves on the several Arts; and, not to mention what small Matters could be furnished *de proprio penu*, the Reader will here have Extracts and Accounts from a great Number of *Authors* of all Kinds, either overlook'd by former Dictionarists, or not then extant; and a Multitude of Improvements in the several Parts, especially of Natural

Knowledge, made in these last Years. I should produce Instances hereof; but I hope this would be needless, as it is endless; and that there are few Pages which will not afford several.

SUCH are the Sources from whence the Materials of the present Work were derived; which, it must be allowed, were rich enough not only to afford Plenty, but even Profusion: So to dispose such a Variety of Materials in such manner, as not to make a confused Heap of incongruous Parts, but one consistent Whole.—And here it must be confess'd there was no Assistance to be had; but I was forced to stand wholly on my own Bottom. Former Lexicographers have not attempted any thing like Structure in their Works; so seem to have been aware that a Dictionary was in some measure capable of the Advantages of a continued Discourse. Accordingly, we see nothing like a Whole in what they have done: And hence, such Materials as they did afford for the present Work, generally needed further Preparation, ere they became fit for our Purpose; which was as different from theirs, as a System from a Centro.

THIS we endeavoured to attain, by considering the several Matters not only absolutely and independently, as to what they are in themselves; but also relatively, or as they respect each other. They are both treated as so many Wholes, and as so many Parts of some greater Whole; their Connexion with which, is pointed out by a *Reference*. So that by a Course of References, from Generals to Particulars; from Premises to Conclusions; from Cause to Effect; and *vice versa*, *i.e.* in one word, from more to less complex, and from less to more: A Communication is opened between the several Parts of the Work; and the several Articles are in some measure replaced in their natural Order of Science, out of which the Technical or Alphabetical one had remov'd them.

FOR an Instance—The Article ANATOMY is not only consider'd as a *Whole*, *i.e.* as a particular Combination or System of Ideas; and accordingly divided into its Parts, *Humane* and *Comparative*: And Humane again subdivided into the Analysis of *Solids* and *Fluids*, (which are referr'd to in the several Places in the Book, where they themselves being treated of, refer to others still lower, and so on) but also as a *Part* of MEDICINE; which accordingly it refers to, and which it self refers to another higher, *etc.*—By which means a Chain is carried on from one End of an Art to the other, *i.e.* from the first or simplest Complication of Ideas appropriated to the Art, which we call the *Elements* or Principles thereof; to the most complex or general one, the *Name* or Term that denotes the whole Art.

NOR is the Pursuit dropt here: but as the Elements or Data in one Art, are ordinarily *quæsita* in some other subordinate one, and are furnished thereby; (as here for Instance, the Elements of *Anatomy* are furnished by *Natural History*, *Physicks*, and *Mechanicks*; and Anatomy may be considered as a Datum, or Element furnished to Medicine). We carry on the View farther, and refer out of one Art or Province into the adjoining ones, and thus lay the whole Land of Knowledge open: It appears indeed with the Face of a Wilderness; but 'tis a Wilderness thro' which the Reader may pursue his Journey as securely, tho not so expeditiously and easily, as thro' a regular Parterre.

IT may be even said, that if the System be an Improvement upon the Dictionary; the Dictionary is some Advantage to the System; and that this is perhaps the only Way wherein the whole Circle or Body of Knowledge can be deliver'd. In any other Form, many thousand Things must necessarily be hid and overlook'd: All the Pins, the Joints, the binding of the Fabrick must be invisible of course; all the lesser Parts, one might say all the Parts whatsoever, must be in some measure swallowed up in the Whole. The Imagination, stretch'd and amplified to take in so large a Structure, can have but a very general, indistinguishing Perception of any of the Parts.— Whereas the Parts are not less Matter of Knowledge when taken separately, than when put together. Nay, and in strictness, as our Ideas are all Singulars or Individuals, and as every Thing that exists is one; it seems more natural to consider Knowledge in its proper Parts, *i.e.* as divided into separate Articles denoted by different Terms; than to consider the whole Assemblage of it in its utmost Composition: which is a thing merely artificial and imaginary.

AND yet the latter Way must be allow'd to have many and real Advantages over the former; which in truth is only of use and significance as it partakes thereof: For this Reason, that all Writing is in its own Nature artificial; and that the Imagination is really the Faculty it immediately applies to. Hence it should follow, that the most advantageous way, is to make use of both Methods: To consider every Point both as a Part; to help the Imagination to the Whole: and as a Whole, to help it to every Part.— Which is the View in the present Work—so far as the many and great Difficulties we had to labour under would allow us to pursue it.

IN this View we have endeavoured to give the Substance of what has been hitherto found in the several Branches of Knowledge both natural and artificial; that is, of Nature, *first*, as she appears to our Senses; either spontaneously, as in Natural History; or with the Assistance of Art, as in Anatomy, Chymistry, Medicine, Agriculture, *etc. Secondly*, to our Imagination; as in Grammar, Rhetorick, Poetry, *etc. Thirdly*, to our Reason; as in Physicks, Metaphysicks, Logicks, and Mathematicks. With the several subordinate Arts arising from each; as Architecture, Painting, Sculpture, Trade, Manufactures, Policy, Law, *etc.* and numerous remote Particulars, not immediately reducible to any of these Heads; as Heraldry, Philology, Antiquities, Customs, *etc.*

THE *Plan* of the Work, then, I hope, will be allow'd to be good; whatever Exceptions may taken to the Execution of it. It wou'd look extravagant to say, That half the Men of Letters of an Age might be employ'd in it to advantage; and yet it will appear, that a Work accomplish'd as it ought to be, on the Footing of this, would answer all the Purposes of a Library, except Parade and Incumbance; and contribute more to the propagating of useful Knowledge thro' the Body of a People, than any, I had almost said all, the Books extant.—After this, let the Reader judge how far I may deserve Censure for engaging in it, even disadvantageously; and whether to have fail'd in so noble a Design, may not be some degree of Praise.

BUT, it will be here necessary to carry on the Division of Knowledge a little further; and make a precise Partition of the Body thereof, in the more formal Way of Analysis: The rather, as an Analysis, by shewing the Origin and Derivation of the several Parts, and the Relation in which they stand to their common Stock and to each other; will assist in restoring 'em to their proper Places, and connecting 'em together.

THIS is a View of Knowledge, as it were, *in semine*; exhibiting only the grand, constituent Parts thereof. It would be endless to pursue it into all its Members and Ramifications; which is the proper Business of the Book it self. It might here, therefore, seem sufficient to refer from the several Heads thus deduced, to the same in the Course of the Work; where their Division is carried on. And yet this would sometimes prove inconvenient for the Reader; who to find some particular Matter must go a long Circuit, and be bandied from one part of the Book to another: To say nothing of the Interruptions which may frequently happen in the Series of References. To obviate this we shall take a middle Course, and carry on the Distribution further, in a Note in the Margin; but this in a looser manner, to prevent the Embarrass of an Analysis so complex and diffusive as this must prove. Some of the principal Heads of each Kind will here come in sight, and such as will naturally suggest, and lead to the rest; so that this will afford the Reader a sort of Summary of the Whole: And at the same time will dispense a kind of auxiliary or succedaneous Order thro'out the Whole; the numerous articles omitted, all naturally enough ranging themselves to their proper Places among these. A Detail of this Kind is of the more Consequence, as it may not only supply the Office of a *Table of Contents*, by presenting the dispersed Materials of the Book in one View; but also that of a *Directory*, by indicating the Order they are most advantageously read in.—*Note*, then, That the *initial* Articles here, tally to the *final* ones of the Analysis; and that the several *Members* hereof, are so many *Heads* in the Book.

[20] PERSPECTIVE, or the PROJECTION of *Points*, *Lines*, *Planes*, *etc.* in *Scenography*, *Orthography*, *Ichnography*, *Anamorphosis*, etc.

[21] PAINTING, or the DESIGNING of Objects in *Clair-obscure*, *Proportion*, etc. with *Ordonnance*, *Expression*, etc. Circumstances hereof, *Attitude*, *Contrast*, *Group*, etc.

Kinds, *Limning*, *Miniature*, *Camieux*, *Fresco*, etc. *Enamelling*, *Mosaic*, etc.

26 ARCHITECTURE, including the Construction of BUILDINGS; as *House*, *Temple*, *Church*, *Hall*, *Palace*, *Theatre*, etc. *Ship*, *Galley*, *Galleon*, *Ark*, *Buccentaur*, *Boat*, etc. *Pyramid*, *Mausoleum*, *Pantheon*, etc. *Capitol*, *Seraglio*, *Escurial*, etc. *Arch*, *Vault*, *Bridge*, *Monument*, *Tomb*, etc. Forms thereof, *Rotondo*, *Platform*, *Pinnacle*, etc. Plans, *Design*, *Ichnography*, *Profile*, etc. Parts, as *Foundation*, *Wall*, *Roof*, etc. *Door*, *Window*, *Stairs*, *Chimney*, etc. Orders, as *Tuscan*, *Doric*, *Corinthian*, etc. *Caryatides*, *Rustic*, *Gothic*, etc. *Column*, *Pilaster*, *Attic*, etc. Parts thereof, *Entablature*, *Capital*, *Pedestal*, etc. *Cornice*, *Frieze*, *Base*, etc. *Volute*, *Pediment*, *Modillion*, *Console*, etc. Mouldings, *Ogee*, *Tore*, *Astragal*, *Scotia*. *Abacus*, *Ovolo*, etc. Materials, as *Brick*, *Stone*, *Tyle*, *Slate*, *Shingle*, etc. *Timber*, *Wainscot*, *Glass*, *Lead*, *Plaster*, etc. *Beam*, *Rafter*, *Mortar*, *Nail*, *Hinge*, *Key*, *Lock*, etc. *Quarry*, *Masonry*, etc.

27 SCULPTURE, or the framing of *Statue*, *Figure*, *Ornament*, etc. in *Relievo*, *Creux*, etc. as *Carving*, *Pottery*, *Porcelain*, etc. Engraving, *Seal*, *Dye*, etc. *Etching*, *Cutting*, *Mezzo Tinto*, etc. Foundery, of *Bell*, *Letter*, *Ordnance*, etc. Coining, *Money*, *Medal*, *Medallion*, *etc*. *Pile*, *Legend*, etc. *Lapidary*, *Turnery*, *Inlaying*, *Vaneering*, *Damasqueening*, *Enchasing*, etc.

28 TRADES and MANUFACTURES; as *Printing*, *Paper-making*, *Book-binding*, etc. *Gilding*, *Japanning*, *Glass-making*, *Grinding*, etc. *Plumbery*, *Glaziery*, *Forging*, *Hammering*, etc. *Weaving*, *Bleaching*, *Whitening*, etc. *Fulling*, *Dying*, *Pressing*, *Sheering*, *Calendring*, *Tabying*, *Freezing*, etc. *Woollen*, *Silk*, *Linum Incombustibile*, etc. *Cloth*, *Serge*, *Taffety*, *Stocking*, etc. *Velvet*, *Tapistry*, *Hat*, etc. *Tanning*, *Currying*, *Tawing*, etc. *Chamoising*, *Chagreen*, *Marroquin*, etc. Making *Parchment*, *Glue*. *Gun-Pouder*, *Smalt*, *Soap*, *Starch*, etc. *Candle*, *Taper*, *Torch*, etc. *Steel*, *Button*, *Pin*, *Needle*, *Pipe*, *Fan*, *Peruke*, etc.

31 FORTIFICATION, or the Construction of FORTRESSES; as *Citadel*, *Castle*, *Tower*, etc. *Fort*, *Star*, *Redoubt*, etc. Works, or Parts thereof, *Rampart*, *Bastion*, *Ditch*, *Counter-scarp*, *Curtain*, etc. *Ravelin*, *Horn-work*, *Crown-work*, etc. Approaches, *Trench*, *Sap*, *Mine*, etc. *Line*, *Parallel*, *Circumvallation*, *etc*. *Battery*, *Attack*, etc.

42 GARDENING, including the *Culture* of HERBS, *Flowers*, *Fruits*, etc. as *Dwarf*, *Standard*, *Stone*, *Wall*, *Espalier*, *Sallad*, etc. The Operations of *Planting*, *Transplanting*, *Replanting*, *Watering*. *Engrafting*, *Inoculating*, *Pruning*, *Pinching*, *Variegating*, etc. Preventing *Diseases*, *Blights*, *Gum*, etc. The Use and Ordering of a *Hot-bed*, *Green-house*, *Seminary*, *Nursery*, *Garden*, *Vineyard*, etc. Their *Exposure*, *Walls*, *Horizontal Shelter*, etc. *Walks*, *Grass-Plot*, *Terrace*, *Quincunx*, *Parterre*, etc.

3 *Cyclopaedia* Volume I

Architectural subjects portrayed in the letterpress articles

Our inventory of letterpress articles in the series *The Encyclopaedic-Dictionary in the Eighteenth Century*: *Architecture, Arts and Crafts* is composed of subjects derived from various branches of architecture and allied arts and crafts. The following short list of headwords is provided here to give the reader an outline impression of the range of material that is included in the letterpress articles:

Architectural terms, arch construction, building construction, building materials, cabinet work, carpentry, classical orders, decorative art, design principles including aesthetics, domes, drainage, fortification, furnishings, furniture, gardening, joinery, landscape terms, military art, mouldings, ornament, painting, perspective, sanitation, sculpture, surveying, terminology of the building trades and crafts, tools and equipment used in building construction, and vaults.

CYCLOPÆDIA:

OR, AN

UNIVERSAL DICTIONARY

OF

ARTS and SCIENCES.

VOLUME the FIRST

A

ABACUS, in Architecture, is the uppermost Member of the Capital of a Column; serving as a kind of Crowning, both to the Capital and the whole Column. See COLUMN.

Dr. *Harris*, and the rest of the Dictionary-Writers, make the *Abacus* to be the Capital it self; which is altogether as just, as to make the Crown of the Head the whole Head. See CAPITAL.

Vitruvius, and others after him, who give the History of the Orders, tell us, the *Abacus* was originally intended to represent a square Tile laid over an Urn, or rather over a Basket. An *Athenian* old Woman happening to place a Basket thus cover'd over the Root of an Acanthus; that Plant shooting up the following Spring, encompass'd the Basket all around, till meeting with the Tile, it curl'd back in a kind of Scrolls. An ingenious Sculptor passing by, took the Hint, and immediately executed a Capital on this Plan; representing the Brick by the *Abacus*, the Leaves by the Volutes, and the Basket by the Vase, or Body of the Capital. Such was the Rise of the first regular Order. See ORDER, ACANTHUS, VOLUTE, CORINTHIAN, *etc.*

There is some difference in the Form of the *Abacus* in different Orders. In the *Tuscan*, *Doric*, and antient *Ionic*, it is a flat, square Member, well enough representing its original Tile; whence the *French* call it *Tailloir*, Trencher. See TUSCAN, DORIC, and IONIC.

In the richer Orders it has lost its native Form; its four Sides, or Faces, being arch'd, or cut inwards; with some Ornament, as a Rose, or other Flower, or Fishes Tail in the middle of each Arch. See CORINTHIAN, and COMPOSITE.

But some Architects take other Liberties in the *Abacus*, both in respect of its Name, Place, and Office. Thus, in the *Tuscan* Order, where it is the largest and most massive, as taking up one third of the height of the whole Capital, it is sometimes call'd the *Die* of the Capital.—In the *Doric* 'tis not always the uppermost Member of the Capital; a Cymatium being frequently placed over it.—In the *Ionic*, some make it a perfect Ogee, and crown it with a Fillet. See CYMATIUM, OGEE, *etc*.

Add, that the *Abacus* is not constantly restrain'd to the Capital of the Column; *Scamozzi* using the Name for a concave Moulding in the Capital of the *Tuscan* Pedestal. See PEDESTAL.

ABBEY, or ABBY, a Monastery, or Religious House, govern'd by an Abbot, or Abbess.

ABBREUVOIR, or ABREVOIR, in Masonry, the Joint, or Juncture of two Stones; or the Interstice, or Space left between 'em to be fill'd up with Mortar. See STONE, MORTAR, MASONRY.

ACANTHUS, in Architecture, an Ornament in the *Corinthian* and *Composite* Orders; being the Representation of the Leaves of an Acanaceous Plant, in the Capitals thereof. See CAPITAL.

It takes its Name from the Name given the Plant among the *Greeks*, as being Prickly, or of the Thistle Kind. The *Latin* Botanists call it *Brancha Ursina*, Bearsfoot, from some supposed resemblance it bears thereto; or *Brancha Hircina*, by reason its Leaves bend and twist somewhat like a Goat's Horns.

There are two Kinds of the Plant *Acanthus*, one whereof grows wild, and is full of Prickles; the other grows in Gardens, and is by *Virgil* called *Mollis*, in regard it is soft, and without any Prickles.—The *Greek* Sculptors adorn'd their Works with the Figure of the latter; as the *Gothic* did with that of the former, which they represented not only in their Capitals, but also in other Ornaments.

The Garden *Acanthus*, is the most dented; bearing a good deal of resemblance to Parsley, or Smallage: And thus it is we find it represented in the *Composite* Capitals of *Titus*, and *Septimius Severus* at *Rome*.

These Leaves make the principal Character and Distinction of the two rich Orders from the Rest: and their different Number and Arrangement does also distinguish the two Orders from each other. See ORDER; see also CORINTHIAN, and COMPOSITE.

The Origin and Occasion of the Ornament, see under the Article ABACUS.

ACOUSTICKS, ACOUSTICA, the Doctrine or Theory of Sounds. *Acousticks* is what we otherwise call *Phonicks*.

ACROTERIA, or ACROTERS, in Architecture, little Pedestals, usually without Bases, placed in the Middle, and at the two Extremes of Frontispieces, or Pediments; and serving to support Statues. See PEDESTAL, and STATUE.

Those at the Extremes ought to be half the Height of the Tympanum; and that in the Middle, according to *Vitruvius*, one eighth Part more. See PEDIMENT, *etc.*

ACROTERIA sometimes also signify Figures, whether of Stone or Metal, placed as Ornaments, or Crownings, on the Tops of Temples, or other Buildings. See CROWNING.

Sometimes it also denotes those sharp Pinacles, or spiry Battlements, which stand in ranges about flat Buildings, with Rails and Ballusters. See PINNACLE, and BATTLEMENTS.

The Word, in its original *Greek*, signifies the Extremity of any Body; as the Tip of a Rock, *etc.*

ADDICE, or ADZE, a sharp Tool, of the Ax-kind, but different from the common Ax.—It is made crooked, and by that means more convenient for cutting the hollow side of any Board, or Timber; being what the coopers generally make use of for that Purpose.

ALCOVE, in Building, a part of a Chamber, separated by an Estrade, or Partition of Columns, and other corresponding Ornaments; in which is placed a Bed of State, and sometimes Seats, to entertain Company.

The word is deriv'd from the Spanish *Alcoba*; and that from the Arabick *Elcanf*, a Cabinet or sleeping Place; or from *Elcobat*, a TENT.

ALLEY, in Gardening, a strait parallel Walk, border'd or bounded on each hand with Trees, Shrubs, or the like. See GARDEN, WALKS in Gardening, EDGINGS, *etc.*

Alleys are usually laid either with Grass or Gravel.

An *Alley* is distinguish'd from a *Path*, in this; that in an Alley there must always be room enough for two Persons, at least, to walk a-breast: So that it must never be less than five Feet in Breadth; and there are some who hold that it never ought to be more than fifteen.

Counter-Alleys, are the little *Allies* by the Sides of the great ones. —A *Front-Alley*, is that which runs strait in the Face of a Building. —A *Transverse-Alley*, that which cuts the former at right Angles. —A *Diagonal-Alley*, that which cuts a Square, Thicket, Parterre, *etc.* from Angle to Angle.—A *Sloping-Alley*, is that which either by reason of the Lowness of the Point of Sight, or of the Ground, is neither parallel to the Front, nor to the Transverse *Alleys*.

An *Alley in Ziczac*, is that which has too great a Descent, and which, on that account, is liable to be damaged by Floods; to prevent the ill Effects whereof, it has Platbands of Turf running across it from Space to Space, which help to keep up the Gravel. This last Name is likewise given to an *Alley* in a Labyrinth, or Wilderness, form'd by several returns of Angles, in order to render it the more solitary and obscure, and to hide its Issue.

Alley in Perspective, is that which is larger at the Entrance than at the Issue; to give it a greater Appearance of length.

Alley of Compartiment, is that which separates the Squares of a Parterre.

The Word *Alley* is derived from the French Verb *Aller*, to go; the ordinary use of an *Alley* being for a Walk, Passage, or Thorow-fare from one Place to another.

ALTAR, ARA, ALTARE, a Place or Pile whereon to offer Sacrifice to some Deity.

The *Jews* had their Brazen *Altar*, for Burnt offerings, and a Golden *Altar*, or *Altar of Incense*. See TABERNACLE.

Among the *Romans*, the *Altar* was a kind of Pedestal, either square, round, or triangular; adorn'd with Sculpture, with Basso Relievo's, and Inscriptions, whereon were burnt the Victims sacrificed to Idols.

Those *Altars* set apart for the honour of the celestial Gods, and Gods of the higher Class, were placed on some pretty tall pile of Building; and for that reason were called *Altaria*, from the words *Alta* and *Ara*, a high elevated *Altar*. Those appointed for the terrestrial Gods, were laid on the Surface of the Earth and call'd *Arae*. And on the contrary, they dug into the Earth, and open'd a Hole for those of the infernal Gods, which they called *Scrobiculi*.

ALTAR, is sometimes also used among Christians, for a square Table, placed on the Eastern side of a Church, rais'd a little above the Floor, and set apart for the Celebration of the Eucharist.

Its Form is not borrow'd, either from that of the Heathen *Altars*, or even from that of the *Jews* in the Temple; but in regard the Eucharist was instituted by Jesus Christ at Supper, and upon a Table, the modern *Altar* is made in form of a Table; whence it is more usually, and even more significantly denominated *Communion* Table.

In the primitive Church, the *Altars* were only of Wood; as being frequently to be removed from place to place. The Council of *Paris*, in 1509, decreed, That no *Altar* should be built but of Stone.

At first there was but one *Altar* in each Church; but the number soon increas'd; and from the Writings of *Gregory* the Great, who lived in the VIth Century, we learn, that there were sometimes 12 or 13.

The *Altar* is sometimes sustained on a single Column, as in the Subterraneous Chapels of St. *Cecilia* at Rome, etc. sometimes, by four Columns, as the Altar of St. *Sebastian* of *Crypta Arenaria*; but by customary Form, is, to be a Massive of Stone-work, sustaining the *Altar-*

Table.

These *Altars* bear a resemblance to Tombs of the Martyrs, and celebrated the Mysteries of Religion thereon. For which Reason, it is a standing Rule to this Day in the church of *Rome*, never to build an *Altar*, without inclosing the Relicks of some Saints therein.

AMPHIPROSTYLE, in the antient Architecture, a kind of Temple, which had four Columns in Front, and as many behind. See TEMPLE.

The Word is derived from the *Greek about*, *before*, and *Column*.

AMPHITHEATRE, a spacious Building, either round or oval, having its Area or Arena incompassed with Rows of Seats, rising gradually one over another; with Portico's both within and without-side.

Among the Antients, the *Amphitheatre* was appointed for the exhibiting of Spectacles or Shews to the People: as, the Combats of Gladiators, and those of wild Beasts.

Their Theatre was built in form of a Semicircle, only exceeding a just Semicircle by one fourth part of the Diameter; and the *Amphitheatre* was nothing else but a double Theatre, or two Theatres join'd together; so that the longest Diameter of the *Amphitheatre*, was to the shortest, as 1½ to 1. See THEATRE.

There are *Amphitheatres* still standing at *Rome*, at *Nismes*, *etc*.—The *Amphitheatre* of *Vespasian*, called the *Coliseum*, and that at *Verona* in *Italy*, are the most celebrated now remaining of all Antiquity. See COLLISEUM.

Pliny mentions an *Amphitheatre* built by *Curio*, which turn'd on large Iron Pivots; so that of the same *Amphitheatre*, two several Threatres were occasionally made, whereon different Entertainments were presented at the same time.

The Word is deriv'd from the *Greek*, *about*, and *Theatre*; which comes from *specto*, *contemplator*; so that an *Amphitheatre*, strictly speaking, is a Place whence a Person may see all around him.

ANNULETS, in Architecture, are small square Members, in the Doric Capital; under the Quarter-round. See DORIC and CAPITAL.

They are also called *Fillets*, *Listels*, *etc*. See FILLET, *etc*.

Annulet is also a narrow flat Moulding, common to other Parts of a Column, *viz*. the Bases, *etc*. as well as the Capital; so called, because it encompasses the Column round.—In this Sense *Annulet* is frequently used for a *Baguette* or little Astragal. See BAGUETTE, and ASTRAGAL.

ANTECHAMBER, or ANTICHAMBER, an outer Chamber, before the principal Chamber of an Apartment; where the Servants wait, and Strangers stay, till the Person to be spoken withal is at leisure, *etc*. See APARTMENT.

The word is formed of the Latin *Ante Camera*. See CHAMBER.

ANTEPAGMENTA, or ANTIPAGMENTA, in the antient Architecture, the Jaumbs of a Door, or Lintels of a Window. See DOOR and WINDOW.

The Word is also used for the entire *Chambranle*, *i.e.* the Door-Case or Window-Frame. See CHAMBRANLE.

ANTIQUE, ANTIQUUS, something that is *antient*.

The Term is chiefly used by Architects, Sculptors, and Painters; who apply it to such pieces of Architecture, Sculpture, Painting, *etc.* as were made at the time when the Arts were in their greatest perfection, among the antient *Greeks* and *Romans*, *viz.* from the Age of *Alexander the Great* to the time of the Emperor *Phocas*, when *Italy* became over-run by the *Goths* and *Vandals*.

In this sense the Word stands opposed to *Modern*.

Thus we say, an *antique* Building, or a Building *after the Antique*; an *antique* Bust, or Bas Relievo; the *antique* Manner, Taste, *etc.*

ANTIQUE is sometimes even contradistinguish'd from *antient*, which denotes a lesser degree of Antiquity, when the Art was not in its utmost Purity: Thus, *antique* Architecture is frequently distinguish'd from *antient* Architecture. See ARCHITECTURE.

Some Writers use the Compound *Antiquo-modern*, in repsect of old *Gothic* Churches and other Buildings; to distinguish them from those of the *Greeks* and *Romans*.

APARTMENT, a Portion of a large House, wherein a Person may lodge separately; having all the Conveniencies requisite to make a compleat Habitation. See HOUSE, *etc.*

A compeat *Apartment* must consist at least of a Chamber, an Antichamber, and a Cabinet or Wardrobe. See CHAMBER, ANTECHAMBER, *etc.*

The word comes from the Latin *Apartimentum*, of the Verb *partiri*, to divide.

APERTIONS, in Architecture, are the Openings in a Building; as Doors, Windows, Stair-cases, Chimneys, Outlets and Inlets for Light, Smoke, *etc.* See BUILDING. See also DOOR, WINDOW, *etc.*

The *Apertions* should be as few as may be; it being a Rule that all Openings are Weakenings.

APOPHYGE, APOPHYGES, in Architecture, that part of a Column where it begins to spring out of its Base, and shoot upwards. See COLUMN and BASE.

The *Apophyge*, in its Original, was no more, than the Ring or Ferril heretofore fasten'd at the Extremities of wooden Pillars, to keep them from splitting; which afterwards was imitated in Stone-work. See ORDER.

The word in its original *Greek* signifies *Flight*; whence the *French* also call it *Escape*, *Conge*, *etc*. See CONGE.

AQUÆDUCT, AQUÆDUCTUS, q.d. *Ductus Aquæ*, a Conduit of Water; is a Construction of Stone, or Timber, made on an uneven Ground, to preserve the Level of the Water, and convey it, by a Canal, from one place to another.

There are *Aquæducts* under ground, and others rais'd above it, supported by Arches.

The *Romans* were very magnificent in their *Aquæducts*; they had some that extended an hundred Miles. *Frontinus*, who had the Direction of them, tells us of nine that emptied themselves thro' 13514 Pipes, of an Inch Diameter. *Blasius* has observ'd, that in the space of twenty-four Hours, *Rome* received, from these *Aquæducts*, no less than five hundred thousand Hogsheads of Water.—The *Aquæduct* built near *Maintenon* for carrying the River *Bure* to *Versailles*, is the greatest in the World. It is 7000 Fathoms long, and its Elevation 2560 Fathom; containing 242 Arcades.

ARÆOSTYLE, ARÆOSTYLOS, in the antient Architecture, a sort of Intercolumination; wherein the Columns were placed at the Distance of eight, or, as some say, ten Modules from one another. See INTERCOLUMNS.

In the *Aræostyle* the columns were the widest and openest they were ever planted at; whence the Name: from the Greek, *rarus*, and Column.

The *Aræostyle* is chiefly used in the *Tuscan* Order; at the Gates of great Cities and Fortresses.

ARC-BOUTANT, in Building, a flat Arch or part of an Arch abutting against the Reins of a Vault, to support, and prevent its giving Way. See VAULT.

Arc-Boutants are only arch'd Buttresses. See BUTTRESS.

The name is *French*; form'd of *arc* and *bouter*, to a-but.

ARCH, in Architecture, is a concave Building, rais'd with a Mould bent in form of the *Arch* of a Curve, and serving as the inward Support of any Superstructure. See BUILDING.

An *Arch*, says Sir *Henry Wotton*, is nothing but a narrow or contracted Vault; and a Vault, a dilated Arch. See VAULT.

Arches are used in large Intercolumnations of spacious Buildings; in *Porticos*, both within and without Temples; in publick Halls, as Ceilings, the Courts of Palaces, Cloisters, Theatres and Amphitheatres. See THEATRE, CEILING, *etc*.

They are also used as Buttresses and Counter-forts to support large

Walls laid deep in the Earth, for Foundations of Bridges and Aquæducts, for triumphal Arches, Gates, Windows, *etc.* See BUTTRESS, ARC-*boutant*, *etc.*

Arches are either *circular*, *elliptical*, or *strait*—

Circular ARCHES are of three kinds, *viz.*—1st, *Semicircular*, which make an exact Semicircle, and have their Centre in the middle of the Chord of the Arch.

2dly, *Scheme*, which are less than a Semicircle, and consequently are flatter Arches; containing some, 90 Degrees, others 70, and others only 60.

3dly, *Arches of the third and fourth Point*, as some of our Workmen call them; tho' the *Italians* call them *diterzo* and *quarto acuto*, because they always meet in an acute Angle at top—These consist of two *Arches* of a Circle ending in an Angle at the Top, and are drawn from the Division of a Chord into three of four Parts, at pleasure—Of this kind are many of the Arches in old *Gothick* Buildings; but on account, both of their Weakness and Unsightliness, they ought, according to Sir *Henry Wotton*, to be for ever excluded out of all Buildings.

Elliptical ARCHES consist of a Semi-Ellipsis; and were formerly much used instead of Mantle-trees in Chimneys—These have commonly a Keystone and Chaptrels or Imposts.

Strait ARCHES, are those whose upper and under Edges are strait; as in the others they are curved; and those two Edges also parallel, and the Ends and Joints all pointing towards a Center—These are principally used over Windows, Doors, *etc.*

The Doctrine and Use of *Arches* is well delivered by Sir *Henry Wotton*, in the following *Theorems.*—1st, All Matter, unless impeded, tends to the Center of the Earth in a perpendicular Line. See DESCENT, GRAVITY, CENTRE, *etc.*

2dly, All solid Materials, as Bricks, Stones, *etc.* in their ordinary rectangular Form, if laid in Numbers, one by the Side of another, in a level Row, and their Extreme ones sustained between two Supporters; those in the middle will necessarily sink, even by their own Gravity, much more if press'd down by any superincumbent Weight—To make them stand, therefore, either their Figure or their Position must be altered.

3dly, Stones, or other Materials being figured *Cuneatim*, *i.e.* Wedge-wise, broader above than below, and laid in a level Row, with their two Extremes supported as in the preceding Theorem; and pointing all to the same Center; none of them can sink, till the Supporters or Butments give way, because they want room in that Situation to descend perpendicularly. But this is but a weak Structure; in regard the Supporters are subject to too much Impulsion, especially where the Line is long; for which reason, the Form of strait *Arches* is seldom used, excepting over Doors and Windows, where the Line is short—In order to fortify the Work, therefore, we must not only change the Figure of the Materials, but also their Position.

4thly, If the Materials be shaped wedge-wise, and be disposed in form of a circular *Arch*, and pointing to some Center; in this Case, neither the pieces of the said *Arch* can sink downwards, for want of room to descend perpendicularly; nor can the Supporters or Butments suffer so much Violence as in the precedent flat Form: for the Convexity will always make the incumbent Weight rather rest upon the Supporters, than heave them outwards: whence this Corollary may be fairly deduced, that the securest of all the *Arches* above mention'd is the Semi-circular; and of all Vaults, the Hemispherical.

5thly, As Semi-circular Vaults, rais'd on the whole Diameter, are the strongest; so those are the most beautiful, which keeping to the same height, are yet distended, one fourteenth part longer than the said Diameter: which addition of width will contribute greatly to their Beauty, without diminishing any thing considerable of their Strength.

It is, however, to be observ'd, that according to Geometrical Strictness, to have the strongest *Arches*, they must not be Portions of Circles, but of another Curve, call'd the *Catenaria*, whose Nature is such, that a number of Spheres dispos'd in this form, will sustain each other, and form an *Arch*.

Dr. *Gregory* even shews, that Arches constructed in other Curves, only stand or sustain themselves by virtue of the Catenaria contain'd in their thickness; so that were they made infinitely slender or thin, they must tumble of course; whereas the Catenaria, tho' infinitely slender, must stand, in regard no one point thereof tends downward more than any other. *Philosoph. Transact.* No. 231.

See further of the Theory under the article VAULT.

Arches are sustain'd by Imposts. See IMPOSTS.

ARCH is particularly used for the Space between the two Peers of a Bridge. See PILA, PEER and BRIDGE.

The chief or *Master-Arch* is that in the middle; which is widest, and usually highest, and the Water under it deepest: being intended for the passage of Boats or other Vessels—We read of Bridges in the East, which consist of 300 Arches.

ARCH-*Stone*. See KEY, Arch.

Triumphal ARCH, is a Gate, or Passage into a City, magnificently adorned with Architecture, Sculpture, Inscriptions, *etc.* which being built of Stone or Marble, serves not only to adorn a Triumph, at the Return from a victorious Expedition, but also to preserve the Memory of the Conqueror to posterity.

The most celebrated Triumphal *Arches*, now remaining of Antiquity, are that of *Titus*, of *Septimus Severus*, and of *Constantine*, at *Rome*.

ARCHITECT, ARCHITECTUS, a Person skill'd in Architecture, or the Art of Buildings; who makes Plans and Designs of Edifices, conducts the Work, and directs the Masons and other Artificers employ'd therein. See ARCHITECTURE, PLAN, DESIGN, *etc.*

The Word is derived from the *Greek*, *Princeps*, and *Workman*, *q.d.* the *principal Workman.*

The most celebrated *Architects* are, *Vitruvius*, *Palladio*, *Scamozzi*, *Serlio*, *Vignola*, *Barbaro*, *Cataneo*, *Alberti*, *Viola*, *Bullant*, and *De Lorme*.

Vitruvius enumerates twelve Qualities requisite to an *Architect*; That he be Docil and Ingenious; Literate; skill'd in Designing; in Geometry; Opticks; Arithmetick; History; Philosophy; Musick; Medicine; Law, and Astrology. See BUILDING, *etc.*

ARCHITECTONICK, that which builds a thing up regularly, according to the Nature and Intentions thereof.

Thus, that plastick Power, Spirit, or whatever else it be which hatches the Ova of Females into living Creatures of the same Species, is by some called the *Architectonick* Spirit.

ARCHITECTURE, ARCHITECTURA, is the Art of *Building*; *i.e.* of erecting Edifices proper for Habitation, or Defence. See BUILDING, EDIFICE, *etc.*

Architecture is usually divided, with respect to its Objects, into three Branches, *Civil*, *Military*, and *Naval.*

Civil ARCHITECTURE, called also absolutely and by way of Eminence ARCHITECTURE, is the Art of Contriving and executing commodious Buildings for the Uses of Civil Life; as *Houses*, *Temples*, *Theatres*, *Halls*, *Bridges*, *Colleges*, *Portico's*, *etc.* See HOUSE, TEMPLE, THEATRE, *etc.*

Architecture is scarce inferior to any of the Arts in Point of Antiquity.—Nature and Necessity taught the first Inhabitants of the Earth to build themselves Hutts, Tents, and Cottages; from which , in course of Time, they gradually advanced to more regular and stately Habitations, with Variety of Ornaments, Proportions, *etc.* See *Vitruvius's Account of the Origin of* Arhictecture *under the Article* ORDER.

The antient Writers represent the *Tyrians* as the first among whom *Architecture* was carried to any tolerable Pitch; and hence it was that *Solomon* had Recourse thither for Workmen to build his Temple. *Villalpandus*, indeed, contends, that only Under-workmen were sent for from *Tyre*, Artificers in Gold, Silver, Brass, *etc.* and that the Rules of *Architecture* were delivered by God himself to *Solomon.* Hence, he adds, the *Tyrians* rather learnt their *Architecture* from *Solomon*; which they afterwards communicated to the *Egyptians*; these to the *Grecians*, and these again to the *Romans*—In Effect, the Author last cited, undertakes to prove, that all the Beauty and Advantages of the *Greek* and *Roman* Buildings, were borrow'd from this Fabrick. *Tom.* ii. *Part* ii. *Lib. Isagog.* iii. *C.* ix. x.

To confirm this, *Sturmius* produces several Passages in *Vitruvius*, where the Rules given by that Architect, *Lib.* vi. *C.* ii. and *Lib.* v. *C.* i quadrate exactly with what *Josephus* relates of the *Jewish* Temple, *Antiq. Jud.* Lib. vi. and viii. *etc.* See TEMPLE.

To what a Pitch of Magnificence the *Tyrians* and *Egyptians* carried *Architecture*, ere it came to the *Greeks*, may be learnt from *Isaiah* xxiii. 8. and *Vitruvius's* Account of the *Egyptian* Oecus; their Pyramids, Obelisks, *etc.* See OBELISK, PYRAMID, *etc.*

Yet, in the common Account, *Architecture* should be almost wholly of *Grecian* Original: Three of the regular Orders or Manners of Building, are denominated from them, *viz. Corinthian, Ionick,* and *Doric*: And scarce a Part, a single Member, or Moulding, but comes to us with a Greek Name. See CORINTHIAN, DORIC, IONIC; see also MOULDINGS, *etc.*

Be this as it will, 'tis certain the *Romans*, from whom we derive it, borrow'd what they had entirely from the *Greeks*; nor seem, till then, to have had any other Notion of the Grandeur and Beauty of Buildings, beside what arises from their Magnitude, Strength, *etc.*—Thus far they were unacquainted with any Order beside the *Tuscan*. See TUSCAN.

Under *Augustus*, *Architecture* arrived at its Glory: *Tiberius* neglected it, as well as the other polite Arts. *Nero*, amongst a Heap of horribe Vices, still retain'd an uncommon Passion for Building, but Luxury and Dissoluteness had a greater Share in it, than true Magnificence. —*Apollodorus*, excelled in *Architecture*, under the Emperor *Trajan*, by which he merited the Favour of that Prince; and it was he who raised the famous *Trajan* Column, subsisting to this Day. See TRAJAN.

After this, *Architecture* began to dwindle; and tho the Care and Magnificence of *Alexander Severus* supported it for some Time, yet it fell with the western Empire, and sunk into a Corruption, from whence it was not recovered for the Space of twelve Centuries.

The Ravages of the *Visigoths*, in the 5th Century, destroy'd all the most beautiful Monuments of Antiquity; and *Architecture* thence forwards, became so coarse and artless, that their profess'd Architects understood nothing at all of just Designing, wherein its whole Beauty consists: Hence a new Manner of Building took its Rise, called the *Gothic*. See GOTHIC.

Charlemange did his utmost to restore *Architecture*, and the *French* applied themselves to it with Success, under the Encouragement of *H. Capet*: His Son *Robert* succeeded him in this Design; till by Degrees the modern *Architecture* was run into as great an Excess of Delicacy, as the *Gothick* had before done into Massiveness. To these may be added, the *Arabesk* and *Morisk*, or *Moorish Architecture*; which were much of a-piece with the *Gothick*, only brought in from the South by the *Moors* and *Saracens*, as the former was from the North by the *Goths* and *Vandals*. See MORISCO, GROTESK, *etc.*

The Architects of the 13th, 14th, and 15th Century, who had some Knowledge of Sculpture, seem'd to make Perfection consist altogether in the Delicacy and Multitude of Ornaments, which they bestowed on their

Buildings, with a World of Care and Sollicitude; tho' frequently without any Conduct or Taste.

In the two last Centuries, the Architects of *Italy* and *France* were wholly bent upon retrieving the primitive Simplicity and Beauty of antient *Architecture*, in which they did not fail of Success: Insomuch, that our Churches, Palaces, *etc.* are now wholly built after the Antique.—

Civil Architecture may be distinguish'd, with regard to the several Periods or States thereof, into *Antique*, *Antient*, *Gothic*, *Modern*, etc. See ANTIQUE, ANTIENT, GOTHIC, and MODERN, *etc.*

Another Division of *Civil Architecture*, arises from the different Proportions which the different Kinds of Buildings render'd necessary, that we might have some proper for every Purpose, according to the Bulk, Strength, Delicacy, Richness, or Simplicity required.—

Hence arose five Orders or Manners of Building, all invented by the Antients at different Times, and on different Occasions, *viz. Tuscan*, *Doric*, *Ionic*, *Corinthian*, and *Composite*; the History, Characters, *etc.* of each whereof see under their respective Articles, TUSCAN, DORIC, IONIC, CORINTHIAN, and COMPOSITE.

What forms an Order, is the *Column* with its *Base* and *Capital*; surmounted by an Entablature, consisting of Architrave, Frieze, and Cornice; and sustain'd by a *Pedestal*. See ORDER; see also COLUMN, ENTABLATURE, PEDESTAL, *etc.* See also CAPITAL, *etc.*

For a general View of the Elements of Architecture, with the Rules which obtain with respect to the Matter, Form, Proportion, Situation, Foundation, Distribution, Covering, Apertures, etc. See the Article BUILDING.

For Particulars; see FOUNDATION, WALL, ROOF, WINDOW, DOOR, CEILING, *etc.* see also BEAUTY, *etc.*

There are several Arts subservient to *Architecture*, as Carpentry, Masonry, Paving, Joynery, Smithery, Glazery, Plumbery, Plastering, Gilding, *etc.* See CARPENTRY, MASONRY, PAVEMENT, JOYNERY, PLUMBERY, PLASTER, *etc.*—See also TIMBER, STONE, BRICK, TYLE, MORTAR, LEAD, *etc.*

We have no Greek Authors extant on *Architecture*—The first who wrote of it was *Agathareus* the *Athenian*, who was seconded by *Democritus* and *Theophrastus*—Among the *Latins*, the younger *Pliny* seems to speak the best; and indeed shews himself very knowing therein.

Of all the Antients, *Vitruvius* is the only entire Author; tho' *Vegetius* relates that there were 700 Architects at *Rome* in his Time—He lived under *Augustus*, and composed a compleat System of *Architecture*, in ten Books, which he dedicated to that Prince. There are two Things censured by the Moderns in this excellent Work, *viz.* Want of Method, and Obscurity. The Mixture of Latin, Greek, in *Vitruvius*, is such, that *Leon Baptist Alberti*, has observed, he wrote Latin to the *Greeks*, and Greek to the *Latins*: He adds, that the Work contains abundance of things superfluous and foreign to the Purpose.—For this Reason M. *Perrault* has extracted all the Rules out of *Vitruvius's* prolix Work, methodized and published them in a little Abridgment:—Several Authors have also endeavoured to explain the Text of *Vitruvius*, particularly *Philander*, *Barbaro*, and *Salmasius*, in Notes added to their several Latin Editions; *Rivius* and *Perrault* in the Notes to their *German* and *French* Versions; and *Baldus* in his *Lexicon Vitruvianum*:—The same M. *Perrault* has also composed an excellent Treatise *Of the five Orders*, which may be esteemed a Supplement to *Vitruvius*, who left the Doctrine of the Orders defective.

The Authors upon *Architecture* since *Vitruvius*, are —*Leòn. Baptist Alberti*, who in 1512, published ten Books of the Art of Building, in Latin, designed to outvie *Vitruvius*; in which, however, he has not succeeded: His Work has abundance of good Things, but is deficient in the Doctrine of the Orders—*Seb. Serlio*, who wrote seven Books of *Architecture*, five of which concerning the five Orders, were made publick in 1602; throughout all which, he religiously keeps to *Vitruvius's* Rules: The seventh was since publish'd in 1575; but the sixth, concerning private Buildings, has not yet appear'd.—*And. Palladio*, who wrote four Books of *Architecture*, containing the fundamental Rules of the Art, with various Instances of all the Kinds of Works; publish'd in Italian in 1575: The two first Books are render'd into High-Dutch, and enlarg'd with Annotations by *Boeckler—Phil. de Lorme*, who publish'd nine Books of *Architecture*, in *French*, in 1567—*J. Barozzio de Vignola*, who in 1631, made publick his Rules of the five Orders, in Italian; since translated, with large Additions, by *Daviler*, under the Title of *Cours d'Architecture*, etc. and since also into High-Dutch, with Notes.—

To these are to be added *Vincent. Scamozzi*, his *Idea of Universal Architecture*, publish'd in 1615, in *Italian*; *Car. Phil. Dieussart*, in his *Theatre of Civil Architecture*, publish'd in High dutch in 1697; wherein he not only delivers the Rules of *Architecture* but explains and compares the five Orders as laid down by *Palladio*, *Vignola*, *Scamozzi*, etc. which same Design was also executed in French by *R. Freart de Cambray*, in a *Parallel of the antient Architecture with the Modern*, publish'd in *French* in 1650, and since translated into *English* with Additions, by Mr. *Evelyln. Fr. Blondel* Director of the Royal Academy of Painting, *etc.* in 1698, gave a *Course of Architecture*, in *French*; being a Collection from all the celebrated Writers upon the Subject of the Orders, *etc.*—*Nich. Goldman*, in a Treatise *de Stylometris*, publish'd in Latin and High-Dutch, in the Year 1661, has done good Service, in reducing the Rules and Orders of *Architecture* to a further Degree of Perfection, and shewing how they may be easily delineated by means of certain Instruments invented by him.

Lastly, the *Elements of Architecture* are laid down by Sir *H. Wotton*—The same are reduced by *Sturmius*, and *Wolfius*, to certain Rules and Demonstrations; and thus is *Architecture* brought into the Form of a Mathematical Art; by the first, in his *Mathesis Juvenil.* and the second, in his *Elementa Matheseos*, Tom. II. *An.* 1715.

Military ARCHITECTURE, is the Art of strengthening and fortifying Places, to screen them from the Insults of Enemies, and the Violence of Arms. See FORTIFIED *Place.*

This we more usually call *Fortification*. See FORTIFICATION.

The Business of *Military Architecture*, is to erect *Forts*, *Castles*, and other *Fortresses*, with *Ramparts*, *Bastions*, etc. See FORTERESSE, RAMPART, BASTION, *etc*.

Naval ARCHITECTURE, or *Ship-Building*; is that which teaches the Construction of *Ships*, *Galleys*, and other floating Vessels for the Water; with Ports, Moles, Docks, *etc.* on the Shore.

ARCHITECTURE *in Perspective*, is a sort of Building, wherein the Members are of different Measures and Modules, and diminish in proportion to their Distance; to make the Work appear longer and larger to the View than really it is. See PERSPECTIVE.

Such is the celebrated Pontifical Stair-case of the *Vatican*, built under Pope *Alexander* VII. by the Cavalier *Bernino*.

Counterfeit ARCHITECTURE, is that which has its Projectures painted, either in Black or White, or coloured after the Manner of Marble; as is seen practised in the Facades and Palaces in *Italy*, and in the Pavilions of *Marly*.

This Painting is made in Fresco, upon plaister'd Walls; and in Oil, on Walls of Stone. See FRESCO.

Under the Name of *Counterfeit Architecture*, which we otherwise call *Scene-Work*, is likewise comprehended, that painted on slight Boards or Planks of Wood, whereon the Columns, Pilasters, and other Parts of Building, seem to stand out, with a *Relievo*; the whole being coloured in imitation of various Marbles, Metal, *etc.* and serving in the Decorations of Theaters, triumphal Arches, Publick Entries, Funeral Pomps, *etc*.

ARCHITRAVE, in Building, that part of a Column, or Order of Columns, which lies immediately upon the Capital. See ORDER and CAPITAL.

The *Architrave* is the lowest Member of the Frieze, and even of the whole Entablature. See FRIEZE and ENTABLATURE.

The *Architrave* is supposed to represent the principal Beam in Timber Buildings; whence the Name, which is form'd of the Greek, *chief*; and the Latin *Trabs*, Beam. See BEAM, *etc*.

The *Architrave* is different in the different Orders—In the *Tuscan* it only consists of a plain Face, crown'd with a Fillet; and is half a Module in Height. See FACE, TUSCAN, *etc*.

In the *Dorick* and *Composite*, it has two Faces, or Fasciæ; and three in the *Ionick* and *Composite*; in which last Order it is $^{10}/_{12}$ of a Module High, tho' but half a Module in the rest. See FASCIA, DORIC, CORINTHIAN, *etc*.

Architects, however, take a deal of Latitude in this Part; some using more Members than others; and many of them having two or three Forms of *Architraves*. What we give is after *Vignola*.

Architrave is sometimes also called the *Reason-piece*, or *Master-beam* in Timber-Buildings, as Porticos, Cloisters, *etc.* In Chimneys it is called the *Mantle-piece*; and over the Jambs of Doors, or Lintels of Windows, *Hyperthyron*. See MANTLE, *etc.*

The *Greek*, call it the *Epistyle*. See EPISTYLE.

ARCHITRAVE-*Doors*, are those which have an *Architrave* on the Jambs and over the Door; upon the Cap-piece, if strait, or on the Arch, if the Top be curved. See DOOR.

ARCHITRAVE-*Windows*, of Timber, are commonly an *Ogee* rais'd out of the solid Timber, with a List over it; tho' sometimes the Mouldings are struck and laid on; and sometimes are cut in Brick. See WINDOW.

The upper Fascia is called the *Header*, or *heading Architrave*; and the lower the *Jac*.

ARCHIVAULT, in Architecture, the inner Contour of an Arch; or a Band or Frame adorned with Mouldings, running over the Faces of the Arch-Stones, and bearing upon the Imposts. See ARCH, VAULT, IMPOSTS, *etc.*

It is different in the different Orders—In the *Tuscan*, it has only a single Face; two Faces crown'd, in the *Dorick* and *Ionick*; and the same Mouldings with the Architrave in the *Corinthian* and *Composite*.

The Word is *French*, *Archivolte*, where it signifies the same thing.

ART, ARS, is defined by the Schoolmen, a Habit of the Mind operative or effective according to right Reason; or, somewhat more intelligibly, a Habit of the Mind prescribing Rules for the due Production of certain Effects; or the introducing of Changes in Bodies from some Foreknowledge and Design in a Person endued with a Principle or Faculty of Acting.

On this Footing, *Arts* are divided into *active* and *factive*.—Such as Leave no external effect after their operation, as Dancing, Fiddling, *etc.* are called *active* or *practical Arts*: Those which do leave an Effect, as Painting, *etc.* are called *factive*, or *Effective* Arts.

ART is better defined, after my Lord *Bacon*, a proper Disposal or the Things of Nature by human Thought and Experience, so as to make them answer the Designs and Uses of Mankind.

Nature, according to that Philosopher, is sometimes free, and at her own Disposal; and then she manifests herself in a regular Order: as we see in the Heavens, Plants, Animals, *etc.*—Sometimes she is irregular, and disorderly, either thro' some uncommon Accident, or a Depravation in Matter, when the Resistance of some Impediment perverts her from her Course; as in the Production of Monsters.—At other Times she is subdued and fashion'd by human Industry, and made to serve the several Purposes of Mankind.

This last is what we call *Art*; in which Sense, *Art* stands opposed to Nature.

Hence, the Knowledge of Nature may be divided into the History of Generations, of Prætergenerations, and of *Arts*.—The first considers Nature at Liberty; the second, her Errors; and the third, her Restraints.—

ART is also used for Science or Knowledge reduced into Practice.

Several of the Schoolmen hold Logic and Ethicks to be *Arts*; inasmuch as they do not terminate in mere Theory, but tend to Practice.

In this Sense, some Branches of the Mathematicks are *Arts*, others Matters of Doctrine, or Science.

Staticks is wholly scientifical, as it takes up with the mere Contemplation of Motion: Mechanicks, on the contrary, is an *Art*, as it reduces the Doctrines of Staticks into Practice.

ART is principally used for a certain System or Collection of Rules, Precepts, and Inventions or Experiments, which being duly observ'd, make the Things a Man undertakes succeed, and render them advantageous and agreeable.

In this Sense, *Art* is opposed to *Science*, which is a Collection of speculative Principles and Conclusions.

The Nature and Origin of Art, *and its Distinction from Science, will be farther consider'd in the* Preface *to this Work.*

Arts, in this Sense, may be divided, with respect to their Scope and Object, into *human*, as Medicine; and *divine*, as Theology.

Human, again, may be subdivided into *Civil*; as Law, Politicks, *etc. Military*, as Fortification, *etc. Physical*, as Agriculture, Chymistry, Anatomy, *etc. Metaphysical*, as Logicks, pure Mathematicks, *etc. Philological*, as Grammar, Criticism, *etc. Mercantile*, to which belong the Mechanical Arts and Manufactures.

ARTS are more popularly divided into *Liberal* and *Mechanical*.

The *liberal Arts* are those that are noble, and ingenuous; or which are worthy of being cultivated without any regard to Lucre arising therefrom. —Such are *Poetry*, *Musick*, *Painting*, *Grammar*, *Rhetoric*, the *military Art*, *Architecture*, and *Navigation*.

Mechanical Arts, are those wherein the Hand, and Body are more concern'd than the Mind; and which are chiefly cultivated for the sake of the Profit they bring with them.—Of which kind are most of those which furnish us with the Necessaries of Life, and are popularly known by the Name of *Trades*.

The *mechanical Arts* take their Denomination from Machine; as being all practised by means of some Machine or Instrument.

With the *liberal Arts* it is otherwise; there being several of them which may be learnt and practiced without any Instrument at all: As Logic, Eloquence, Medicine properly so called, *etc*.

The *Arts* which relate to the Sight and Hearing, My Lord *Bacon* observes, are reputed liberal beyond those which regard the other Senses, which are chiefly employed in Matters of Luxury.

It has been well noted by some Philosophers, that during the Rise and Growth of States, the military *Arts* chiefly flourish; when arrived at their Height, the liberal *Arts*; and when on the declining hand, the voluptuary *Arts*.

There are also divers particular *Arts*, as the *Art* of Memory, the *Art* of Decyphering, *Art* of Flying, of Swimming, *Art* of Diving, *etc*.

Democritus maintain'd, that Men learnt all their *Arts* from Brutes; the Spider taught 'em Weaving, the Swallow Building, the Nightingale Music; and several Medicine.

ASH, in Building. See TIMBER.

ASHLAR, a Term among Builders; by which they mean common or free Stones, as they come out of the Quarry, of different Lengths and Thicknesses.

ASHLERING, among Builders, Quartering to tack to, in Garrets, about 2½, or 3 foot high, perpendicular to the Floor, up to the Underside of the Rafters. See BUILDING, ROOF, COVERING, *etc*.

ASPECT, ASPECTUS, in Perspective, Gardening, *etc*. See PROSPECT, EXPOSURE, *etc*.

ASSEMBLAGE, the joining or uniting of several Things together; or, the Things themselves so joined, or united.

The Carpenters and Joiners have various Kinds and Forms of *Assemblage*; as, with Mortoises and Tenons, with Dove-tails, *etc*. See MORTOISE, DOVE-TAIL, *etc*.

The *Europeans* admire the Carpentry of some *Indians*, where the *Assemblage* is made without either Nails or Pins. *Herrera*. See CARPENTRY, NAILS, *etc*.

ASTRAGAL, in Architecture, is a little round Member, in form of a Ring, or Bracelet; serving as an Ornament on the Tops and at the Bottoms of Columns. See MOULDING, COLUMN, *etc*.

The *Astragal* is sometimes also used to separate the *Fasciæ* of the Architrave; in which case it is wrought in Chaplets, or Beads and Berries.

It is also used both above and below the Lists, adjoining immediately to the Square, or Die of the Pedestal. See PEDESTAL.

ASYMMETRY, ASYMMETRIA, a want of Symmetry or Proportion. See SYMMETRY.

ATLAS, in Architecture, is a Name given to those Figures, or Half-Figures of Men, so much used instead of Columns, or Pilasters; to support any Member of Architecture, as a Balcony, or the like. See

COLUMN, *etc.*
These are otherwise call'd *Telamones.*

ATTIC is used in Architecture for a Kind of Building, wherein there is no Roof or Covering to be seen; thus called because usual in *Athens.*

ATTIC, or ATTIC-*Order*, is also a kind of little Order, rais'd upon another larger one; by way of crowning, or to finish the Building. See ORDER.

It is sometimes also used for the conveniency of having a Wardrobe, or the like; and instead of regular Columns, has only Pilasters of a particular Form. See COLUMN and PILASTER.

There are also *Attics* used for high Altars.—The Order takes its Name from *Athens*, where it was first practiced.

ATTIC *of a Roof*, is a Kind of Parapet to a Terrace, Platform, or the like. See TERRACE, *etc.*

ATTIC *Continued*, is that which encompasses the whole Pourtour of a Building, without any Interruption; following all the Jets, the Returns of the Pavillions, *etc.*

ATTIC *Interpos'd*, is that situate between two tall Stories, sometimes adorn'd with Columns, or Pilasters.

ATTIC *Base*, is a peculiar Kind of Base used by the antient Architects, in the *Ionick* Order; and by *Palladio*, and others of the Moderns, in the *Dorick*. See DORIC and IONIC.

The *Attic* is the most beautiful of all the Bases. See BASE.

ATTITUDE, in Painting and Sculpture, the Posture of a Figure, or Statue; or the Disposition of its Parts, by which we discover the Action it is engaged in, and the very Sentiment supposed to be in its Mind. See FIGURE, and STATUE.

The representing of these in a strong and lively manner, makes what they call a *good Expression*. See EXPRESSION.

The Word comes from the *Italian Attitudo*, which signifies the same thing.

AUDITORIUM, AUDITORY, in the antient Churches, was that part of the Church where the *Auditores* stood to hear, and be instructed. See CHURCH.

The *Auditorium* was that Part now call'd *Navis Ecclesiæ*. See NAVE.

AVENUE, in Gardening, is a Walk, planted on each side with Trees, and leading to some Place. See GROVE, GLADE.

All *Avenues*, *Mortimer* says, should lead to the Front of an House, Garden-Gate, Highway-Gate, or Wood, and terminate in a Prospect. In an *Avenue* to an House, whatever the Length of the Walk is, it ought to be as wide as the whole Breadth of the Front, and if wider, better.

B

BAGUETTE, in Architecture, a little round Moulding less than an *Astragal*, sometimes carv'd, and enrich'd with Foliages, Pearls, Ribbands, Laurels, *etc*. Tho according to M. *de Clerc*, when the *Baguette* is enrich'd with Ornaments, it changes its Name, and is called *Chapelet*.

BALCONY, in Architecture, a Projecture beyond the Naked of a Wall or Building, supported by Pillars, or Consoles, and encompassed with a Ballustrade. This Contrivance is not only made use of in Houses, but also in Ships. The Word comes from the *Italian*, *Balcone*, and that from the *Latin*, *Paleus*, or the *German*, *Palk*, a Beam. *Covaruvias* derives it from *jacere*; asserting that *Balconies* were originally little Turrets over the Gates of Citadels, whence Darts, *etc*. were thrown on the Enemy.

BALDACHIN, or *Baldaquin*, a Piece of Architecture in Form of a Canopy, supported with Columns, and serving as a Crown or Covering to an Altar. The Word comes from the *Italian*, *Baldacchino*.

BALLUSTRADE, in Architecture, an Assemblage of one or more Rows of little turn'd Pillars, high enough to rest the Elbows on, fixed upon a Terras, or the Top of a Building, or else to make a Separation between one Part of it and another. *Ducange* derives the Word from *Balaustrum*, or *Balaustrium*, a Place among the Antients where their *Baths* were rail'd in.

BANQUETTE, in Fortification, is a little Foot-Path or Elevation of Earth in Form of a Step, along a parapet, or that by which the Musquetters get up to discover the Counterscarpe, or to fire on the Enemies in the Moat, or in the Covert-Way: These are generally a Foot and half high, and almost three Foot broad.

BARBACAN, in Architecture, a Canal or Opening left in the Walls, for Water to come in and go out at, when Edifices are rais'd in Places liable to be overflow'd; or to drain off the Water from a Terras. The Word comes from the Italian *Barbacane*, which *Spelman* derives from the Arabick.

Barbacan, a military Term, signifying a Cleft or Aperture made in the Walls of a Castle or Fortress, to fire thro upon the Enemy. Some think the Word implies a Wooden Parapet; others any kind of defence or Safeguard against the Enemy: *Du Cange* calls it the *outer Defence*. 'Twas formerly a Fort at the Entrance of a Bridge, or without the City, having a double Wall, with Towers.

BASE, in Architecture, from the Greek *Rest*, or *Support*, is used to signify any Body which bears another; but particularly for the lower Parts of a Column and a Pedestal. The *Base* is sometimes also call'd *Spira*, from *Spiræ*, the Folds of a Serpent laid at Rest, which make a Figure not unlike it.

The *Base of a Column* is that part between the Shaft and the Pedestal, if there be any Pedestal; or, if there be none, between the Shaft and the Plinth, or Zocle. (See *Column* and *Pedestal.*) The *Base* is supposed to be the Foot of the Column; or, as some will have it, 'tis that to a Column which a Shoe is to a Man. The Members or Ornaments whereof it is composed, are supposed by others to have been originally intended to represent the Iron Circles wherewith the Feet of Trees and Posts which supported the antient Houses were girt, in order to strengthen them.

The Base is different in the different Orders.

The *Tuscan Base* is the most simple of all the Orders; consisting only of a single *Tore* besides the *Plinth.*

The *Dorick Base* has an *Astragal* more than the *Tuscan*; tho that was introduced by the Moderns.

The *Ionick Base* has a large *Tore* over two slender *Scotias*, separated by two *Astragals*; tho in the most antient Monuments of this Order there are no *Bases* at all; which the Architects are at a loss to account for.

The *Corinthian Base* has two *Tores*, two *Scotias*, and two *Astragals.*

The *Composite* has an *Astragal* less than the *Corinthian.*

The *Attick Base*, so call'd because first used by the *Athenians*, has two *Tores* and a *Scotia*, and is very proper for *Ionick* and *Composite* Columns.

BASIL, among Joiners, *etc.* the Angle to which the Edge of an Iron Tool is ground. To work on soft Wood they usually make their *Basil* 12 Degrees, for hard Wood 18; it being observ'd, that the more acute or thin the *Basil* is, the better and smoother it cuts; and the more obtuse, the stronger and fitter for Service.

BASILIC, or *Basilica*, in Architecture, from the Greek, *royal House*, or *Palace*; a Term antiently used for a large Hall, or publick Place, with Isles, Porticos, Galleries, *etc.* where the Princes administer'd Justice in Person: But the Word has been since transfer'd to signify any great Church, Court of Justice, or Exchange. *Azor* is of opinion, that *Basilics* were formerly such Churches as were not consecrated: but *Baronius* seems to have more Reason, when he says the Term was applied to the most magnificent Churches; such, which by their Grandeur as far surpass'd other Churches, as Princes Palaces do private Houses. Thus we still say, the *Basilic of St. Peter*, for the *Church of St. Peter*, etc. M. *Perrault* says, that *Basilics* differ'd from *Temples*, in that the Columns of *Temples* were without-side, and those of *Basilics* within. There were formerly four kinds of *Churches* at *Rome*, *viz. Patriarchal*, *Titular*,

Diaconial, and *Oratorial*; whereof the first were those particularly call'd *Basilics*. See CHURCH and TEMPLE.

BASTION, in the Modern Fortification, a huge Mass of Earth usually faced with Sods, sometimes with Brick, rarely with Stone, standing out from a Rampart, whereof it is a principal Part.

BATON, in Architecture, is a large Ring, or Moulding, in the Base of a Column, otherwise called TORE. See *Tore*.

BATTLEMENTS, Indentures or Notches on the top of a Wall, Parapet, or other Building, in form of Embrazures, for the sake of looking thro 'em, *etc*. These were much affected in the old Fortification.

BEAD, in Architecture, a round Moulding in the *Corinthian* and *Roman* Orders, carved in short Embossments, like *Beads* in Necklaces. See *Baguette*. A *Bead* is usually about ¼ of a Circle, and only differs from a *Boultine* in Bigness. When large, the Workmen call 'em *Boultines*; tho sometimes an Astragal is thus carved. A *Bead-plain* is also frequently set on the Edge of each Fascia of an Architrave. A *Bead* is often placed on the Lining-Board of a Door-Case, and on the upper Edges of Skirting-Boards.

BEAK, in Architecture, a little Fillet left on the Edge of a Larmier, which forms a Canal, and makes a kind of Pendant. *Chin-Beak* is also used for a Moulding, the same with the Quarter-round, except that its Situation is inverted. We find but few Examples hereof in the antient Buildings; but 'tis very frequent in the modern.

BEAM, in Building, the largest Piece of Wood in a Building, being laid a-cross the Walls, and serving to support the principal Rafters of the Roof. No House has less than two of these *Beams*, *viz*. one at each Head: Into these the Girders of the Garret-Floor are also framed; and, if the Building be Timber, the Teazle Tenons of the Posts. The Proportions of Beams near *London*, are fix'd by Statute, as follows: A Beam 15 Foot long, must be seven Inches on one side its Square, and five on the other; If it be 16 Foot long, one side must be eight Inches, the other six: If 17 Foot long, one side must be 10 Inches, the other six: In the Country they usually make 'em stronger. Sir *H. Wotton* advises these to be the strongest and most durable Timber. *Herrera* tells us, that in *Fer. Cortez's* Palace, in *Mexico*, there were 7000 Beams of Cedar: But he must certainly use the Word *Beam* in a greater Latitude than what we do. In effect, the *French*, under *Poutre*, *Beam*, take in not only the Pieces which support the Rafters, but also those which sustain the Joists for the Cielings. Some of their best Authors have consider'd the Force or Strength of *Beams*, and brought their Resistance to a Precise Calculation;

particularly M. *Varignon*, and M. *Parent*; the System of the latter is as follows.

When, in a *Beam* breaking Parallel to its Base, which is supposed to be a Parallelogram, two Plans of Fibres, which were before contiguous, are separated, there is nothing to be consider'd in those Fibres, but their Number, Bigness, Tension before they broke, and the Lever by which they act; all these together making the Resistance of the *Beam* to be broke. Suppose then another *Beam* of the same Wood, where the Base is likewise a Parallelogram, and of any Bigness, with regard to the other, at Pleasure; the Height of each of these, when laid Horizontal, being divided into an indefinite Number of equal Parts, and their Breadth into the same Number, in each of their Bases will be found an equal Number of little quadrangular Cells, proportional to the Bases whereof they are Parts: These then will represent little Bases, or which is the same thing, the Thicknesses of the Fibres to be separated for the Fracture of each *Beam*; and since the Number of Cells is equal in each, the Ratio of the Bases of both *Beams* will be that of the Resistance of their Fibres, both with regard to Number and Thickness. Now, the two *Beams* being supposed of the same wood, the Fibres, the most remote from the Points of Support, which are those which break the first, must be equally stretch'd when they break. Thus the Fibres, *v. g.* of the 10th Division, are equally stretch'd in each Case, when the first breaks; and in whatever Proportion the Tension be supposed, 'twill still be the same in both Cases; so that the Doctrine is entirely free, and unimbarrass'd with any System of Physicks. Lastly, 'tis evident the Levers whereby the Fibres of the two *Beams* act, are represented by the Height of their Bases; and of Consquence, the whole Resistance of each *Beam* is the Product of its Base by its Height multiplied by the Breadth: Which holds, not only in case of Parallelogrammatick, but also of Elliptick Bases.

Hence, if the Bases of two *Beams* be equal, tho both their Heights and Breadths be unequal, their Resistance will be as the Heights alone; and by Consequence, one and the same *Beam* laid on the smallest side of its Base, will resist more than when laid flat, in proportion as the first Situation gives it a greater Height than the second: and thus an Elliptick Base will resist more, when laid on its greatest Axis, than on its smallest.

Since in *Beams* equally long, 'tis the Bases that determine the Proportion of their Weights or Solidities; and since their Bases being equal, their Heights may be different, two *Beams* of the same Weight may have Resistances differing to Infinity: Thus, if in the one the Height of the Base be conceiv'd infinitely great, and the Breadth infinitely small, while in the other the Dimensions of the Base are finite; the Resistance of the first will be infinitely greater than that of the second, tho their Solidity and Weight be the same. If therefore all required in Architecture were to have *Beams*, capable of supporting vast Loads, and at the same time have the least Weights possible, 'tis plain they must be cut thin as *Laths*, and laid edge-wise.

If the Bases of the two *Beams* be suppos'd unequal, but the Sum of the Sides of the two Bases equal, *v.g.* if they be either 12 and 12, or 11 and 13, or 10 and 14, *etc.* so that they always make 24; and further, if they be suppos'd to be laid edge-wise; pursuing the Series, it will appear, that in the *Beam* of 12 and 12 the Resistance will be 1728, and the Solidity or Weight 144; and that in the last, or 1 and 23, the Resistance will be 529, and the Weight 23; The first therefore, which is square, will have less than half the Strength of the last, with regard to its Weight. Hence M. *Parent* remarks, that the common Practice of cutting the *Beams* out of Trees as square as possible, is ill Husbandry: He hence takes occasion to determine geometrically, what Dimensions the Base of a *Beam*, to be cut out of any Tree propos'd, shall have, in order to its having the greatest possible Resistance; or, which is the same thing, a Circular Base being given, he determines the Rectangle of the greatest Resistance that can be inscribed, and finds that the Sides must be nearly as 7 to 5, which agrees with Observation.

Hitherto the Length of the *Beams* has been supposed equal; if it be unequal, the Bases will resist so much the less, as the *Beams* are longer. To this it may be added, that a *Beam* sustained at each End, breaking by a Weight suspended from its middle, does not only break at the middle, but also at each Extreme; or, if it does not actually break there, at least immediately before the Moment of the Fracture, which is that of the Æquilibrium between the Resistance and the Weight, its Fibres are as much stretch'd at the Extremes as in the Middle: So that of the Weight sustain'd by the middle there is but one third Part which acts at the middle, to make the Fracture; the other two only acting to induce a Fracture in the two Extremes. A *Beam* may either be supposed loaden only with its own Weight, or with other foreign Weights applied at any distance, or only with those foreign Weights; since according to M. *Parent* the Weight of a *Beam* is not ordinarily above $1/70$ part of the Load given it to sustain, 'tis evident that in considering several Weights they must be all reduced by the common Rules to one common Centre of Gravity. M. *Parent* has calculated Tables of the Weights that will be sustained by the middle, in *Beams* of various Bases and Lengths, fitted at each End into Walls, on a Supposition that a Piece of Oak of an Inch square and a Foot long, retain'd horizontally by the two Extremes, will sustain 315 *lib.* in its middle before it breaks, which 'tis found by Experience it will.

BEARER, in Architecture, a Post, or Brick-Wall, trimmed up between the two Ends of a Piece of Timber to shorten its Bearing, or to prevent its bearing with the whole Weight at the Ends only.

BEARING *of a Piece of Timber*, in Carpentry, the Space either between the two fix'd Extremes thereof, when it has no other support; which is called *bearing at length*: or between one Extreme, and a Post,

Brick-Wall, *etc.* trimm'd up between the Ends to shorten its *Bearing*.

BEAUTY, a Term whereby we express a certain relation of some Object, either to an agreeable Sensation, or to an Idea of Approbation. When therefore I say a thing is *beautiful*, I either mean that I perceive something that I approve, or that something gives me Pleasure: Whence it appears, that the Idea affix'd to the Word *Beauty* is double; which renders the Word equivocal, and is the Source of all the Disputes on the subject of *Beauty*. We must therefore distinguish between Ideas and Sensations. Ideas take up the Mind; Sensations interest the Heart. Tho we see nothing in an Object to interest us, we may yet discover something in its Idea to merit our Approbation. Such an Object therefore pleases, and does not please, *i.e.* it pleases the Idea, and not the Sensation. On the contrary, there are some Objects whose Ideas don't offer anything laudable, which yet excite agreeable Sensations. There is therefore *Beauty* and *Beauty*. 'Tis exceeding hard to fix any general Characteristicks of *Beauty*: For, as the Ideas and Sensations of different Persons differ according to the Habitudes of the Body, and the turn of the Mind; so do the relations of Objects to those Ideas and Sensations vary, whence what we call *Beauty* results. Hence arise those different Opinions of a *beautiful Thought*, a *beautiful Woman*, a *beautiful Painting*, etc. M. *Perrault* distinguishes two Kinds of *Beauties* in Architecture, which chime in pretty aptly with the two Species of *Beauty* above. The one he calls *positive* and *convincing*, such as the Richness of the Materials, Grandeur of the Structure, Neatness of the Workmanship, Symmetry, *etc.* The others he calls *arbitrary*, which depend on the Will, and might have their Proportion changed without Deformity. These only please by the Connexion, or Association of their Ideas with others of a different Kind, and which please of themselves; and owe their Beauty to that Prepossession of the Mind, whereby a thing whose Value we do know, insinuates an Esteem for others which we do not. Thus he observes, there are many things in Architecture which Reason and good Sense would judge deformed, and which however, Custom has not only made tolerable, but even *beautiful*, by their being always joined with other *Beauties* that are positive. Thus, being at first pleased with viewing 'em in Company, and merely on account of their Company, at length we become pleased with 'em alone; and thus we frequently fall in love with Deformities, and grow fond of Faults. Hence the Use of Foils appears founded on an ill Philosophy. If a *Beauty* takes off from a *Deformity*, the *Deformity* in its turn takes off from the *Beauty*: 'Tis the Foil therefore is the Gainer, the Diamond loses. The Diamond pleases, we are pleased in some measure with every thing about it, particularly the Foil; the Foil displeases, we are in some measure displeased with every thing about it, particularly the Diamond: The Mind can't be well pleased and displeased at the same time. By viewing the Diamond and Foil together, a Man might in a long Course of Time find the one almost as *beautiful* as the other. The Disparity would be always diminishing, till they came near a Level: By removing 'em a-part, they would by degrees return to their original State, *i.e.* the Diamond would recover, and the Foil lose its Lustre.

BED. The old *Romans* had various kinds of *Beds* for Repose; as, their *Lectus cubicularis*, or *Chamber-Bed*, whereon they slept; their *Table-Bed*, or *Lectus discubitorius*, whereon they eat, (for they always eat lying) there being usually three Persons to one *Bed*, whereof the middle Place was accounted the most honourable, as well as the *middle Bed.* They had also their *Lectus lucubratorius*, whereon they studied; and a *Lectus funebris*, or *emortualis*, whereon the Dead were carried to the Pile.

BED, in Gardening, a Piece of made Ground, rais'd above the Level of the rest.

BED, in Masonry, a Course, or Range, of Stones; and the Joint of the *Bed* is the Mortar between two Stones placed over each other.

BED, or BEDDING-MOULDING, in Architecture, a Term used by the Workmen for those Members in a Cornice which are placed below the Coronet; and now a *Bed-Moulding* usually consists of these four Members, as O-G, a List, a Large Boultine, and another List under the Coronet.

BED-CHAMBER: with us, the *Gentlemen of the Bed-Chamber* are Persons of the first Rank, 11 in Number; whose Office is, each in his turn, to wait a Week in the King's *Bed-Chamber*, lying by the King on a *Pallat-Bed* all Night, and to wait on the King when he eats in private. The first of these is *Groom of the Stole*, or *Long-Robe*.

BERME, in Fortification, is a small Space of Ground four or five Foot wide, left without between the Foot of the Rampart and the Side of the Moat, to receive the Earth that rolls down from thence, and to prevent its falling into the Moat. Sometimes, for more Security, the *Berme* is pallisadoed.

BEVEL, in Masonry, and among Joiners, a kind of Square, one Leg whereof is frequently crooked, according to the Sweep of an Arch or Vault. 'Tis moveable on a Point or Centre, and so may be set to any Angle. Its Make and Use are pretty much the same as those of the common Square and Mitre, except that those are fix'd; the first at an Angle of 90 Degrees, and the second at 45; whereas the *Bevil* being moveable, may, in some measure, supply the Office of both, and yet, which 'tis chiefly intended for, supply the Deficiencies of both; serving to set off, or transfer, Angles either greater or less than 90 or 45 Degrees.

Hence a *Bevil* Angle is used to denote any other Angle but those of 90 or 45 Degrees.

BITUMEN, an inflammable Matter, fat and unctuous. Naturalists distinguish three Kinds of *Bitumens*, hard, soft, and liquid or oily; and each of these they subdivide into several others. Among the hard *Bitumens* are rank'd *yellow Amber*, (some add *Ambergreese*) *Jet*, *Asphaltum* or *Bitumen of Judea*, *Pisasphaltum*, *Pit-Coal*, *Black-Stone*, and *Sulphurs*. The soft are, *Maltha*, *Bitumen of Colao*, *of Sirnam*, and *Copal*. Lastly, the *Naptha of Italy*, and *Petroleum*, are rank'd among the liquid *Bitumens*.

Of these *Bitumens* some are fossil, others are found floating on the Surface of certain Lakes, and others spring from the Earth like Fountains. Some *Bitumens* are so hard, that they are used in Forges, instead of Coals; others so glutinous, that they serve instead of Cement, or Mortar in Buildings; of which kind it was, that the famous Walls of *Babylon* were built: and others so liquid, that they are burnt in Lamps instead of Oil. The *Bitumen* in most Esteem is that of *Judea*. See ASPHALTUM; and the other *Bitumens* under their proper Heads. *Strabo* derives the word from Pitch.

BLINDS, in Fortification, Defences made of Wood, or Branches interwoven and laid a-cross between two Rows of Stakes about the Height of a man, and four or five foot apart. They are used particularly at the heads of Trenches, when they are extended in Front towards the Glacis; serving to shelter the Workmen.

BOLTS of *Iron*, in Building, are distinguish'd into three Kinds, *Plate*, *Round*, and *Spring-Bolts*.

BOLTS, or Iron Pins in a Ship are of several sorts: As, *Ring-Bolts*, serving for the bringing to of the Planks, *etc*. *Drive-Bolts*, used to drive out others; *Set-Bolts*, employ'd for forcing the Planks and other Works, and bringing them close to one another; *Rag-Bolts*, on each side full of Jags or Barbs, to keep them from flying out of their Holes; *Clench-Bolts*, for the same end clench'd, or fasten'd at the Ends where they come thro; *Fore-Bolts*, made like Locks with an Eye at each end, whereinto a Forelock of Iron is driven to prevent starting-out; *Fender-Bolts*, made with long and thick Heads, struck into the uttermost Bends or Wales of a Ship, to save her Sides from Bruises and Hurts.

BONNET, in Fortification, a Work raised beyond the *Counterscarp*, having two Faces, which form a saliant Angle, and as it were a small Ravelin without any Trench: Its Height is about three Foot, and it is environ'd round with a double Row of Pallisadoes, ten or twelve Paces distant from each other; hath a Parapet three Foot high, and is like a little advanced *Corps du Guard*.

BONNET *a Prestre*, or *Priest's-Cap*, is an Outwork, having at the Head three saliant Angles, and two inwards: It differs from the *Double Tenaille* only in this, that its Sides, instead of being parallel, are made like a Swallow's Tail; that is, narrowing, or drawing close at the Gorge, and opening at the Head.

BOSSAGE, or *Boscage*, in Architecture, is us'd for any Stone that has a Projecture, and is laid in its Place in a Building uncut; to be afterwards carv'd into Mouldings, Capitals, Arms, *etc. Bossages*, are also what we otherwise call Rustick Work; consisting of Stones which seem to advance beyond the naked of a Building, by reason of Indentures, or Channels left in the Joinings; used chiefly in the Corners of Buildings, and thence call'd Rustick *Quoins*. The Cavity or Indenture is sometimes round, sometimes square, sometimes cham fraim'd, or bevel'd; sometimes in the Diamond Form: Sometimes it is inclos'd with a Cavetto, sometimes with a Listel.

BOULDER-WALLS, a kind of *Walls* built of round Flints or Pebbles, laid in a strong Mortar; used where the Sea has a Beach cast up, or where there are plenty of Flints.

BOULTINE, in Architecture, the Workman's Term for a Moulding, whose Convexity is just ¼ of a Circle; placed next below the Plinth, in the Tuscan and Dorick Capital. See ECHINUS.

BOUTANT, in Architecture. An *Arc-Boutant*, is an Arch or Buttress, serving to sustain a Vault; and which is it self sustained by some strong Wall or Massive-work. A Pillar *boutant*, is a large Chain or Jaumb of Stone, made to support a Wall, Terrass, or Vault. The Word is *French*, and comes from the Verb *Bouter*, *to but*, or *a-but*.

BOYAU, in Fortification, a Gut, or Branch of the Trenches: Or a Ditch cover'd with a Parapet, serving for a Communication between two Trenches. *Boyau* is also a Line, drawn winding about, in order to inclose several Tracts of Ground, or to attack some Works.

BRACE, in Architecture, a Piece of Timber, fram'd in with Bevil Joints. Its use is to keep the Building from swerving either way. When a *Brace* is fram'd into a King-piece, or principal Rafter, it is called by some a *Strut*.

BRADS, a kind of Nails used in Building, having no spreading Heads as other Nails have: Of these some are call'd *Joiners Brads*, and are for hard Wainscot; others *Batten Brads* for soft Wainscot; and some *Bill Brads* or *Quarter-Heads*, us'd when a Floor is laid in haste, or for shallow Joists subject to warp.

BRANCHES, in Architecture, the Arches of Gothic Vaults. These Arches, traversing from one Angle to another, Diagonal-wise; form a Cross between the other Arches which make the Sides of the Square, whereof the Arches are Diagonals. See VAULT, OGEE, *etc.*

BREACH, in Fortification, is the Ruins made in any Part of the Works of a Town, *etc.* either by playing Cannon, or springing Mines; in order to storm the Place, or take it by assault. They say, *Make good the Breach, Fortify the Breach, Make a Lodgment on the Breach, Clear the Breach*, etc.

BREST, in Architecture, a Term us'd by some, for that Member of a Column, otherwise called the *Thorus* or *Tore*. See TORE.

BREST-SUMMERS, in Timber-Building; are the Pieces in the outward Parts of the Building, and in the middle Floors, (not in Garrets or Ground-Floors) into which the Girders are fram'd. In the inner Parts of a Building, the Pieces into which the Girders are fram'd, are call'd *Summers*.

BRICK, a fat reddish Earth, form'd into long Squares, 4 Inches in Breadth, and 8 or 9 in Length, by means of a wooden Mould; and then bak'd or burnt in a Kiln, to serve for the Uses of Building.

Bricks appear to be of a very antient standing; the Tower of *Babel* being built thereof, as appears both from Sacred History, and from the Remains thereof still in being. Under the first Kings of *Rome*, they built with massive squar'd Stones, which they learnt from the *Tuscans*: Towards the latter Time of the Republick, they began to use *Brick*; borrowing the Practice from the *Greeks*: And the greatest, as well as the most durable Buildings of the succeeding Emperors, as the *Pantheon*, etc. were built therewith. In the Time of *Gallian*, the Buildings were compos'd of an Order of *Brick* and an Order of *Tofus*, a soft gritty Stone, alternately. After him, they laid aside the use of *Bricks*, and resumed Flints. In the *East* they bak'd their *Bricks* in the Sun: The *Romans* us'd 'em crude; only leaving 'em to dry in the Air a long space of Time, *viz.* four or five Years. The *Bricks* us'd by the *Greeks* were principally of three Kinds; the first call'd *of two Palms*: the second *of four Palms*; and the third *of five Palms*. They had other *Bricks*, just half each of these, which they join'd together to render their Works more solid, as well as more agreeable to the Eye, by the diversity of Figures and Sizes of the *Bricks*.

The *Bricks* among us are various; acquiring various Names, according to their various Forms, Dimensions, Uses, Method of making, Place, *etc.* The principal are, 1. *Compass Bricks*, of a circular Form, us'd in steyning of Walls. 2. *Concave*, or *hollow Bricks*, on one side flat, like a common *Brick*, on the other hollow'd: They are us'd to convey Water. 3. *Cogging*

Bricks, are us'd to make the indented Work, under the Coping of Walls built with great *Bricks*. 4. *Coping Bricks*, which are form'd on purpose for coping of Walls. 5. *Dutch*, or *Flemish Bricks*, us'd to pave Yards and Stables, and for Soapboilers Fatts, and Cisterns. 6. *Clinkers*, are such Bricks as are glazed by the heat of the Fire in making. 7. *Feather-edg'd Bricks*, are like the common *Statute Bricks*, only thinner on one Edge than on the other, and are us'd to pen up the Brick Pannels in Timber Buildings. 8. *Samel* or *Sandal Bricks*, are such as lie outmost in a Kiln or Clamp, and consequently are soft and useless; as not being thoroughly burnt. 9. *Great Bricks*, are 12 Inches long, 6 broad, and 3 thick: The Weight of one is about 15 Pounds; so that 100 will weigh 1500, and 1000 of them 15000 Pounds Weight: Their use is to build Fence-Walls; together with, 10. *Pilaster*, or *Buttress Bricks*, which are of the same Dimensions with them, only they have a Notch at one End, of half the Breadth of the Brick: Their use is to bind the Work at the Pilaster of Fence-Walls, which are built of *Great Bricks*. 11. *Paving Bricks*, or *Tiles*, are of several Sizes in several Countys and Places. 12. *Place Bricks*, are such as are made in a Place prepared on purpose for them, near the Building they are to be used in. 13. *Statute*, or *small Common Bricks*; these ought to be 9 Inches long, 4½ broad, and 2½ thick; 100 of these usually weigh about 550 Pounds, and 1000, 5500 Pounds; and about 407 in number, are a Tonn Weight: These are commonly us'd in paving Cellars, Hearths, Sinks, *etc*. 30 or 32, if true measure, will pave a Yard Square, and 330 will pave a Square of 100 Foot, laid flat; but if laid Edg-ways, there must be double the Number.

Barbaro, in his Comment on *Vitruvius*, recommends another Form of *Bricks*, *viz*. Triangular ones, every Side a Foot long; and some an Inch and half thick: These, he observes, wou'd have many Conveniences above the rest; as being more commodious in the Management, of less Expence, and of fairer Shew; adding much Beauty and strength to the Mural Angles, where they fall gracefully into an indented Work. And Sir. *H. Wotton* wonders they have never been taken into use; being recommended by so great Authority.

The Earth whereof *Bricks* are made, must not be sandy, which will make 'em both heavy and brittle; nor must it be too fat which will make 'em crack in drying. They should be made either in the Spring or Autumnal Season: when made, they must be shelter'd from the Sun if it be too hot, and yet be expos'd to the Air to dry. If they be made in frosty Weather, they are to be cover'd with Sand; if in hot Weather, with wet Straw. When they are well dried, they are burnt.

Bricks are burnt, either in a *Kiln* or *Clamp*: Those in the first, being set in it, and the Kiln cover'd with Pieces of Bricks, they put in Wood, to dry them with a gentle Fire; and this they continue till they are pretty dry, which is known, by the Smoak's turning from a whitish dark, to a black transparent Smoak. They then cease to put in Wood, and proceed to burn with Bush, Furze, Straw, Heath, Brake, or Fern Faggots; having first damm'd up the Mouth of the Kiln with pieces of *Bricks* pil'd upon one another, and clos'd with wet *Brick-Earth*, instead of Mortar: They then continue to put in more Faggots, till the Kiln and its Arches look white, and the Fire appear a-top of the Kiln. Then they slacken the Fire for an Hour, and let all cool by degrees: 48 Hours burning is generally sufficient. About *London*, they burn *Bricks* in Clamps, built of the *Bricks* themselves, after the manner of Arches in Kilns; with a Vacancy between each Brick's Breadth, for the Fire to play thro; but with this Difference, that instead of arching, they truss or span it over, by making the *Bricks* project one over another, on both sides the Place, for the Wood and Coals to lie in, till they meet, and are bounded by the *Bricks* at the Top, which close all up. The Place for the Fuel is carry'd up straight on both sides, till about three Foot high; then they fill it almost with Wood, and over that lay a covering of Sea-Coal; then over-span the Arch: But they strew Sea-Coal also over the Clamp, betwixt all the Rows of *Bricks*; then they fire the Wood, and that fires the Coal: and when it is burnt out, they conclude the *Bricks* burnt enough.

Goldman observes, that *Bricks* will have double the Strength, if, after one burning, they be steep'd in Water, and burnt afresh. If the Earth be too fat, it must be temper'd with Sand; and that trod out again, first by Cattle, then Men. *Bricks* made of common Earth, melt, nay vitrify, by too much Heat: For which Reason, the Kilns are made of Stones that will themselves calcine, that the Vehemence of the Fire may be broken by 'em: Besides which, they usually place other *Bricks*, made of an argillous Earth which would melt, next the Fire.

BRIDGE, a Work of Masonry, or Timber, built over a River, Canal, or the like, for the Convenience of crossing the same. Some learned Men make *Janus* to have been the first Inventor of *Bridges*, Ships, and Crowns; because in several *Greek*, *Sicilian*, and *Italian* Coins, there are represented on one side a *Janus* with two Faces, and on the other a *Bridge*, or a Crown, or a Ship. According to the Testimony given thereof by *Athenæus*, Lib. 15. one of the most notable Exploits of *Julius Cæsar*, was the expeditious making a *Bridge of Boats* over the Rhine. *Xerxes* is said to have made a *Bridge of Ships* over the *Hellespont*.

BUILDING, *D'aviler* defines, any Place erected by Art, whether for Convenience, for Religion, or Magnificence. See HOUSE, TEMPLE, THEATER.

A *Regular Building*, is that whose Plan is Square, its Opposite Sides equal, and the Parts dispos'd with Symmetry. An *Irregular Building*, is that, on the contrary, whose Plan is not contain'd within equal or parallel Lines, either by the Nature of its Situation, or the Artifice of the *Builder*; and whose Parts have not any just Relation to one another in the Elevation. An *Insolated Building*, is that which is not attach'd, join'd, or contiguous to any other; or is encompass'd with Streets, or any open

Square, or the like; as St. *Paul's*, the *Monument*, etc. A *Building* is said to be *engag'd*, when it is encompass'd with others, and has no Front towards any Street or publick Place, nor any Communication without, but by a back Passage. An *interr'd*, or *sunk Building*, is that whose Area is below the Level of the adjacent Street, Court, or Garden, *etc.* and whose lowest Courses of Stone are hid.

Felibien considers three Kinds of *Buildings* in Architecture; 1. *Sacred Buildings*; as the Temples of the Antients, and our Churches and Chappels. 2. *Publick Buildings*; as Basilicks, or Courts of Justice, Tombs, Theaters, Amphitheaters, Triumphal Arches, Gates, Bridges, Aqueducts, *etc.* 3. Palaces, and private Houses: Each of which see under their proper Heads, BASILIC, CHURCH, AMPHITHEATER, *etc.*

BUILDING is also us'd for the constructing or raising of an Edifice; in which Sense it comprehends, as well the Expenses, as the Invention and Execution of the Design thereof. See ARCHITECTURE.

In *Building* there are three Things in view, *viz.* Commodity, Firmness, and Delight: To attain these Ends, Sir *Henry Wotton* considers the whole Subject under two Heads, *viz.* the Seat or *Situation*, and the *Work*. For the Situation, either that of the *Whole* is to be consider'd, or that of its *Parts*.

For the first, regard must be had to the Quality, Temperature, and Salubrity of the Air; the Conveniency of Water, Fuel, Carriage, *etc.* and the Agreeableness of the Prospect. For the second, the chief Rooms, Studies, Libraries, *etc.* to lie towards the East; Offices that require Heat, as Kitchens, Distillatories, Brew-houses, *etc.* to the South: Those that require a cool fresh Air, as Cellars, Pantries, Granaries, *etc.* to the North: As also Galleries for Painting, Museums, *etc.* which require a steady Light. He adds, that the antient *Greeks* and *Romans*, generally situated the Front of their Houses to the South: But the modern *Italians* vary from this Rule. Indeed, in this Matter, regard must still be had to the Country; each being oblig'd to provide against its respective Inconveniences: So that a good Parlour in *Egypt*, might make a good Cellar in *England*.

For the second Head, the Situation being fix'd on, the next thing to be consider'd is the *Work*: under this come, first the principal Parts, then the Accessories, or Ornaments.

Under the Principals, are first, the *Materials*; then the *Form* or Disposition. The Materials are either Stone, as Marble, Free-stone, Brick, for the Walls, *etc.* or of Wood, as Firr, Cypress, Cedar, for Posts and Pillars of upright use; Oak, for Beams, Summers, Crop-work, or for Joining and Connection. See STONE, BRICK, WOOD, TIMBER, *etc.*

For the Form or Disposition of a *Building*, it must either be Simple or Mix'd: The simple Forms are either Circular or Angular; and the Circular ones either compleat, as just Spheres, or deficient, as Ovals.

The circular *Form* is very commodious, of the greatest Capacity of any; strong, durable beyond the rest, and very beautiful: but then it is found of all others the most chargeable; much room is lost in the bending of the Walls, when it comes to be divided; besides an ill Distribution of Light, except from the Centre of the Roof: On these Considerations it was, that the Antients only us'd it in Temples and Amphitheaters, which needed no Compartition. Oval Forms have the same Inconveniences, without the same Conveniences; being of less Capacity.

For Angular Figures, Sir *H. Wotton* observes, that *Building* neither loves many nor few Angles: The Triangle, *v. g.* is condemn'd above all others, as wanting Capacity and Firmness; as also, being irresolvable into any other Regular Figure in the inward Partitions, besides its own. For Figures of 5, 6, 7, or more Angles, they are fitter for Fortifications than Civil Buildings. There is, indeed, a celebrated *Building of Vignola*, at *Caprarole*, in Form of a Pentagon; but the Architect had prodigious Difficulties to grapple with, in disposing the Lights, and saving the Vacuities. Such *Buildings* then, seem rather for Curiosity than Conveniencey: and for this Reason, Rectangles are pitch'd on, as being a Medium between the two Extremes. But again, whether the Rectangle is to be a just Square or an Oblong, is disputed. Sir *Henry Wotton* prefers the latter, provided the Length don't exceed the Breadth by above one third.

Mix'd Figures, partly Circular and partly Angular, may be judg'd of from the rules of the Simple ones; only they have this particular Defect, that they offend against Uniformity. Indeed Uniformity and Variety may seem to be opposite to each other: but Sir *H. Wotton* observes, they may be reconcil'd; and for an Instance, mentions the Structure of the Human Body where both meet. Thus much for the first grand Division, *viz.* the *Whole* of a *Building*.

For the second Division, or the *Parts* of a Fabrick, *Baptista Alberti* comprises 'em under five Heads; *viz.* the Foundation, Walls, Apertures, Compartitions, and Cover.

For the *Foundation*, to examine its Firmness, *Vitruvius* orders the Ground to be dug up; an apparent Solidity not to be trusted to, unless the whole Mold cut thro be found solid: He does not indeed limit the Depth of the Digging; *Palladio* limits it to a sixth Part of the Height of the *Building*: This Sir *Henry Wotton* calls the Natural Foundation, whereon is to stand the Substruction, or Ground-work, to support the Walls, which he calls the Artificial Foundation: This then is to be level; its lowest Ledge, or Row of Stone only, close laid with Mortar, and the broader the better; at the least, twice as broad as the Wall: Lastly, some add, that the Materials below should be laid just as they grew in the Quarry; as supposing 'em to have the greatest Strength in their natural Posture. *De Lorme* enforces this, by observing, that the breaking or yeilding of a Stone in this Part, but the Breadth of the Back of a Knife, will make a Cleft of above half a Foot in the Fabrick above. For *Pallification*, or piling the Ground-Plot, so much commended by *Vitruvius*, we say nothing; that being requir'd only in a moist marshy Ground, which should never be chosen: Nor perhaps are there any Instances of this kind, where it was not Necessity that drove 'em to it. See FOUNDATION.

For the *Walls*, they are either entire and continued, or intermitted; and the Intermissions are either Columns or Pilasters. Entire, or continu'd Walls, are variously distinguish'd; by some, according to the Quality of the Materials, as they are either Stone, Brick, *etc.* others only consider the Position of the Materials; as when Brick, or square Stones are laid in their Lengths, with Sides and Heads together, or their Points conjoin'd, like a Network, *etc.* See MASONRY. The great Laws of Muring, are, that the Walls stand perpendicular to the Ground-work; the Right-Angle being the Cause of all Stability: that the massiest and Heaviest Materials be lowest, as fitter to bear than be born; that the Work diminish in Thickness, as it rises; both for ease of Weight and Expence: that certain Courses, or Ledges of more strength than the rest, be interlaid, like Bones, to sustain the Fabrick from total Ruin, if the Under-parts chance to decay: And lastly, that the Angles be firmly bound; these being the Nerves of the whole Fabrick, and commonly fortify'd, by the *Italians*, on each side the Corners, even in Brick *Buildings*, with squar'd Stones; which add both Beauty and Strength. See WALL.

The Intermissions, as before observ'd, are either Columns or Pilasters; whereof there are five Orders, *viz. Tuscan*, *Doric*, *Ionic*, *Corinthian*, *Composite*; each of which see distinctly consider'd under its respective Head. Columns and Pilasters are frequently, both for Beauty and Majesty, form'd Archwise; the Doctrine of which, see under ARCH.

For the *Apertures*, they are either Doors, Windows, Staircases, Chimneys, or Conduits for the Suillage, *etc.* which see under their Heads, DOOR, WINDOW, *etc.* Only with regard to the last, it may be observed, that Art should imitate Nature in these ignoble Conveyances, and separate them from Sight, where a running Water is wanting, into the most remote, lowest, and thickest Part of the Foundation; with secret Vents, passing up thro the Walls, like a Tunnel, to the open Air; which the *Italians* all commend for the Discharge of noisom Vapors.

For the *Compartition*, or Distribution of the Groundplot into Apartments, *etc.* Sir *H. Wotton* lays down these Preliminaries; That the Architect never fix his Fancy on a Paper Draught, how exactly soever set off in Perspective; much less on a mere Plan, without a Model, or Type of the whole Structure, and every Part thereof, in Pastboard or Wood; that this Model be as plain and unadorn'd as possible, to prevent the Eye's being impos'd on, and that the bigger this Model, the better. In the Compartition it self, there are two general Views, *viz.* the Gracefulness, and Usefulness of the Distribution, for Rooms of Office and Entertainment; as far as the Capacity thereof, and the Nature of the Country will allow. The Gracefulness will consist in a double Analogy or Correspondency; first, between the Parts and the Whole, whereby a large Fabrick should have large Partitions, Entrances, Doors, Columns, and, in brief, all the Members large: The second, between the Parts themselves, with regard to Length, Breadth, and Height. The Antients, determin'd the Length of their Rooms, that were to be Oblongs, by double their Breadth; and their Height by half their Breadth and Length added together. When the Room was to be precisely Square, they made the Height half as much more as the Breadth: which Rules, the Moderns take occasion to dispense with; sometimes squaring the Breadth, and making the Diagonal thereof the Measure of the Height; and sometimes more. This deviating from the Rules of the Antients, is ascrib'd to *M. Angelo*.

The second Consideration in the Compartition, is the Usefulness; which consists in the having a sufficient number of Rooms of all Kinds, with their proper Communications, and without Distraction. Here the chief Difficulty will lie in the Lights and Staircases. The Antients were pretty easy on both those Heads, having generally two cloister'd open Courts, one for the Women's Side, the other for the Men: Thus the Reception of Light into the Body of the *Building* was easy; which among us must be supply'd, either by the open Form of the *Building*, or by graceful Refuges or Breaks, by terrassing a Story in danger of Darkness, and by Abajours, or Sky-lights. For casting the Staircases, it may be observ'd, that the *Italians* frequently distribute the Kitchen, Bake-house, Buttery, *etc.* under Ground, next above the Foundation, and sometimes level with the Floor of the Cellar; raising the first ascent into the House 15 Foot, or more; which, beside the removing of Annoyances out of Sight, and gaining so much room above, does, by elevating the Front, add a Majesty to the whole. Indeed, Sir *H. Wotton* observes, that in *England* the natural Hospitality thereof, won't allow the Buttery to be so far out of sight; besides that, a more luminous Kitchen, and a shorter Distance between that and the Dining-Room, are requir'd, than that Compartition will well bear. In the Distribution of Lodging Rooms, 'tis a popular and antient Fault, especially among the *Italians*, to cast the Partitions so, as when the Doors are all open, a Man may see through the whole House; grounded on the Ambition of shewing a Stranger all the Furniture at once: an intolerable Hardship on all the Chambers, except the Inmost, where none can arrive but thro all the rest, unless the Walls be extreme thick for the Secret Passages: nor will this serve the Turn, without at least three Doors to each Chamber; a thing inexcusable, except in hot Countries: Besides its being a weakning to the *Building*, and the Necessity it occasions of making as many common great Rooms as there are Stories, which devours a great deal of room, better employ'd in Places of Retreat; and must likewise be dark, as running thro the Middle of the House. In the Compartition, the Architect will have occasion for frequent Shifts; thro which his own Sagacity, more than any Rules must conduct him. Thus he will be frequently put to struggle with Scarcity of Ground; sometimes to damn one room for the Benefit of the Rest, as to hide a Buttery under a Staircase, *etc.* at other times, to make those the most beautiful which are most in Sight; and to leave the rest, like a Painter, in the Shadow, *etc.*

For the *Covering* of the *Building*; this is the last in the Execution, but the first in the Intention: for who would *build*, but to shelter? In the covering, or Roof, there are two extremes to be avoided, the making it too heavy or too light: The first will press too much on the Underwork; the latter has a more secret Inconvenience; for the Cover is not only a bare Defence, but a Band or Ligature to the whole *Building*; and there requires a reasonable Weight. Indeed, of the two Extremes, a House top-heavy is the worst. Care is likewise to be taken, the Pressure be equal on each Side: and *Palladio* wishes, that the whole burden might not be laid on the outward Walls, but that the Inner likewise bear their Share. The *Italians* are very curious in the Proportion and Gracefulness of the Pent or Slopeness; dividing the whole Breadth into nine Parts, whereof two serve for the height of the highest Top or Ridge from the lowest; But in this Point, regard must be had to the Quality of the Region; for, as *Palladio* insinuates, those Climates which fear the falling of much Snow, ought to have more inclining Pentices than others. See COVERING.

Thus much for the principal or essential Parts of a *Building*: For the Accessories, or Ornaments, they are fetch'd from Painting and Sculpture. The chief things to be regarded in the first, are, that no Room have too much, which will occasion a Surfeit; except in Galleries, *etc.* that the best Pieces be placed where there are the fewest Lights: Rooms with several Windows are Enemies to Painters, nor can any Picture be seen in Perfection, unless illumin'd, like Nature, with a single Light: That in the Dispostion regard be had to the Posture of the Painter in working, which is the most natural for the Posture of the Spectator; and that they be accommodated to the Intentions of the Room they are us'd in: See PAINTING. For Sculpture, it must be observ'd, that it ben't too abundant; especially at the first approach of a *Building*, or at the Entrance; where a *Dorick* Ornament, is much preferable to a *Corinthian* one; that the Niches, if they contain Figures of white Stone, be not colour'd in their Concavity too black, but rather dusky; the Sight being displeas'd with too sudden Departments from one extreme to another. That fine Sculptures have the Advantage of nearness, and coarser of Distance; and that in placing of Figures aloft, they be reclin'd a little forwards: Because, the visual Ray extended to the Head of the Figure, is longer than that reaching to its Feet, which will of necessity make that Part appear further off; so that to reduce it to an erect Posture, it must be made to stoop a little forwards. M. *le Clerc*, however, won't allow of this Resupination, but will have every Part in its just Perpendicular. See STATUE.

To judge of a *Building*, Sir *H. Wotton* lays down the following Rules: 1. That before fixing any Judgement, a Person be inform'd of its Age; that, if the apparent Decays be found to exceed the Proportion of Time, it may be concluded, without further Inquisition, either that the Situation is naught, or the Materials or Workmanship too slight. If it be found to bear its Years well, let him run back, from the Ornaments and Things which strike the Eye first, to the more essential Members; till he be able to form a Conclusion, that the Work is commodious, firm, and delightful; the three Conditions, in a good *Building*, laid down at first, and agreed on by all Authors. This, he esteems the most Scientifical Way of judging. *Vassari* proposes another, *viz.* by passing a running Examination over the whole Edifice, according to the Structure of a well-made Man; as, whether the Walls stand upright on a clean Footing and Foundaton; whether the *Building* be of a beautiful Stature; whether, for the Breadth, it appear well burnish'd; whether the principal Entrance be on the middle Line of the Front, or Face, like our Mouths; the Windows, as our Eyes, set in equal Number and Distance on both Sides; the Offices, like the Veins, usefully distributed, *etc.* Lastly, *Vitruvius* gives a third Method of judging; summing up the whole Art under these six Heads: Ordination, or the settling the Model and Scale of the Work; Disposition, the just Expression of the first Design thereof; (which two, Sir *H. Wotton* thinks he might have spar'd, as belonging rather to the Artificer than the Censurer:) Eurithmy, the agreeable Harmony between the Length, Breadth, and Height of the several Rooms, *etc.* Symmetry, or the Agreement between the Parts and the Whole; Decor, the due Relation between the *Building* and the Inhabitant: Whence *Palladio* concludes, the principal Entrance ought never to be limited by any Rule, but the Dignity and Generosity of the Master. And lasty, Distribution, the useful casting of the several Rooms, for Office, Entertainment, or Pleasure. These last four are ever to be run over, e'er a Man pass any determinate Censure: And these alone, Sir *Henry* observes, are sufficient to condemn or acquit any *Building* whatever. See EURITHMY, SYMMETRY, *etc.*

Dr. *Fuller* gives us two or three good Aphorisms in *Building*; as,

1st, *Let not the common Rooms be several, nor the several Rooms common*: *i. e.* The Common Rooms not to be private, or retir'd, as the Hall, Galleries, *etc.* which are to be open, and the Chambers, *etc.* retir'd.

2d, *A House had better be too little for a Day, than too big for a Year.* Houses therefore to be proportion'd to ordinary Occasions, not extraordinary.

3d, *Country Houses must be Substantives, able to stand of themselves.* Not like City *Buildings*, supported and shelter'd on each Side by their Neighbours.

4th, *Let not the Front look asquint on a Stranger*; *but accost him right, at his Entrance.*

5th, *Let the Offices keep their due Distance from the Mansion House*; *those are too familiar, which are of the same Pile with it.*

BULWARK, the same, in the antient Fortification, with a *Bastion* in the Modern.

BUST, or BUSTO, in Sculpture, *etc.* a Term us'd for the Figure, or Portrait of a Person *in Relievo*; shewing only the Head, Shoulders, and

Stomach; the Arms being lopp'd off: ordinarily placed on a Pedestal, or *Console*. *Felibien* observes, the tho in Painting one may say a Figure appears *in Busto*; yet it is not properly call'd a *Bust*, that Word being confined to things *in Relievo*. The *Bust* is the same with what the Latins call'd Herma, from the Greek *Hermes*, *Mercury*; the Image of that God being frequently represented in this manner among the *Athenians*.

BUTMENTS, in Architecture, those Supporters or Props, on, or against which the Feet of Arches rest: Also little Places taken out of the Yard, or the Ground-plot of an House, for a Buttery, Scullery, *etc.* are sometimes called *Butments*. The word comes from the French *Bouter*, to abut or terminate on any thing.

BUTTRESS, a kind of Butment built archwise; or a Mass of Stone or Brick, serving to prop or support the Sides of a Building, Wall, *etc.* on the Outside, where it is either very high, or has any considerable Load to sustain on the other side, as a Bank of Earth, *etc.* The Theory and Rules of Buttresses, are one of the *Desiderata* in Architecture.

C

CABINET, the most retir'd Place in the finest Apartment of a Building; set apart for Writing, Studying, or preserving any thing very precious. A compleat Apartment consists of an Hall, Antichamber, Chamber, and Cabinet; with a Gallery on one Side. See APARTMENT. Hence a *Cabinet* of Paintings, of Curiousities, Mosaic *Cabinet*, *Cabinet* of a Garden, etc.

CABLED-FLUTES, in Architecture, such *Flutes* as are fill'd up with Pieces in form of *Cables*. See FLUTES.

CALOTTE, in Architecture, a round Cavity or Depressure, in form of a cup, or Cap, lathed and plaster'd; us'd to diminish the Rise or Elevation of a moderate Chapel, Cabinet, Alcove, *etc.* which, without such an Expedient, wou'd be too high for other pieces of the Apartment.

CAMBER-BEAM, in a Building, a piece of timber cut Archwise, or with an Obtuse Angle in the Middle, commonly us'd in Platforms; as Church Leads, and other Occasions, where long and strong Beams are requir'd. A *Camber-Beam* is much stronger than another of the Same Size; since being laid with the hollow Side downwards as they usually are, they represent a kind of Arch.

CANAL of a *Larmier*, the hollow'd Plafond or Soffit of a Cornice; which makes the Pendant Mouchette. See LARMIER and SOFFITO.

CANAL of the *Volute*, in the *Ionic* Capital, is the Face of the Circumvolutions, enclos'd by a Listel.

CANTALIVER-*Cornice*, is a *Cornice* with *Cantalivers* or Modillions under it; see CANTALIVERS.

CANTALIVERS, in Building, Pieces of Wood fram'd into the Front, or other Sides of a House, to sustain the Moulding and Eves over it: These seem, in effect, to be the same with Modillions, except that the former are plain, and the latter carv'd. They are both a kind of Cartouses, set at equal Distances under the Corona of the Cornice of a Building. See MODILLIONS; see also CORNICHE, and CORONA.

CANTONED is us'd in Architecture, when the Corner of a Building is adorn'd with a Pilaster, an angular Column, rustick Quoins, or anything that projects beyond the naked of the Wall.

CAPITAL, in Architecture, the uppermost Part of a Column or Pilaster, serving as the *Head* or Crowning thereof; plac'd immediately over the Shaft, and under the Entablature. See SHAFT, and ENTABLATURE.

CAPITAL *of a Column*, properly, is that whose Plan is round. See COLUMN.

CAPITAL *of a Pilaster*, is that whose Plan is square; or at least rectilineal. See PILASTER.

The *Capital* is a principal, and essential Part of an Order, of Column or Pilaster: It is made different in the different Orders; and is that which chiefly distinguishes and characterizes the Orders. See ORDER.

The Tuscan CAPITAL, is the most simple and unadorn'd: Its Members, or Parts, are but three, *viz.* an *Abacus*; under this an *Ovolo*, or quarter Round; and under that, a *Neck*, or *Colarino*. The Neck terminates in an *Astragal*, or Fillet, belonging to the *Fust*, or Shaft.

The Character of this *Capital*, whereby it is distinguish'd from the *Doric*, etc. is that the *Abacus* is square, and quite plain, and has no Ogee or other Mouldings; and that there are no Annulets under the *Ovolo*. Indeed, Authors vary a little as to the Character of the *Tuscan Capital*. *Vignola* gives the *Abacus* a Fillet, in lieu of an *Ovolo*. *Vitruvius* and *Scamozzi*, add an *Astragal* and Fillet, between the *Ovolo* and Neck: *Serlio* only a Fillet; *Philander* rounds the Corners of the *Abacus*. In the *Trajan* Column there is no Neck; but the *Astragal* of the Shaft, is confounded with that of the *Capital*. The Height of this *Capital* is the same with that of the Base, *viz.* one Module, or Semidiameter. Its Projecture is equal to that of the Cincture at the Bottom of the Column; *viz.* $\frac{8}{5}$ of the Module.

See TUSCAN.

The Doric CAPITAL, beside an Abacus, an Ovolo, and a Neck, in common with the *Tuscan*, has three Annulets, or little square Members, underneath the Ovolo, in lieu of the Astragal in the *Tuscan*; and a Talon, Cima, or Ogee, with a Fillet over the Abacus. Authors vary, too, as to the Characters of this *Capital*: *Palladio*, *Vignola*, etc. put Roses under the Corners of the *Abacus*, and in the Neck of the *Capital*. The Height of this *Capital*, *Vitruvius*, etc. makes one Module; and its Projecture 37 Minutes and an half. See DORIC.

The Ionic CAPITAL is compos'd of three Parts; an Abacus, consisting of an Ogee and a Fillet; under this a Rind, which produces the Volutes, or Scrolls, the most essential Part of this Capital; and at the Bottom, an Ovolo, or Quarter-Round: The Astragal, under that Ovolo, belongs to the Shaft; the middle Part is called a Rind, or Bark, from its suppos'd Resemblance to the Bark of a Tree laid on a Vase, whose Brim is represented by the Ovolo; and seeming to have been shrunk up in drying, and to have twisted into the Volutes. See VOLUTE, and SCROWLS. The Ovolo is adorn'd with *Eggs*, as they are sometimes call'd, from their oval Form: The *Greeks* call it the *Echinos*. See ECHINUS. The Height of this *Capital*, M. *Perrault* makes 18 Minutes; its Projecture one Module, seven Tenths.

The Differences in the Character of this *Capital*, flow mostly from the different Managements of the Volutes; and consist in this: That in the Antique, and some of the Modern, the Eye of the Volute answers not the Astragal of the Top of the Shaft, as *Vitruvius*, and most of the Moderns make it: That the Face of the Volutes, which usually makes a Flat, is sometimes curv'd and convex'd, so as the Circumvolutions go advancing outwards; as is frequent in the Antique. That the Border, or Rim of the Scroll in the Volute is sometimes not only a plain Swep, as ordinarily; but the Sweep is accompany'd with a Fillet. That the Leaves which invest the Balluster, are sometimes long and narrow; sometimes larger and broader: That the two Faces of the Volutes, are sometimes join'd at the outer corner; the Ballusters meeting in the inner, to make a Regularity between the Faces on the Front and Back of the Building, with those of the Sides. That among the Moderns, since *Scamozzi*, the *Ionic Capital* has been alter'd, and the four Faces made alike; by taking away the Balluster, and hollowing all the Faces of the Volutes inwards, as in the *Composite*: whereas, in the Antique, the Bark passes between the Ovolo and Abacus, quite straight, only twisting at its Extremities to form the Volute. And lastly, that of late Years, the Sculptors have added a kind of little Festoons, springing from the Flower, whose Stalk lies on the first Circumvolution of the Volute; and suppos'd to represent the Locks of Hair, hanging down on both sides of the Face. See IONIC.

The Corinthian CAPITAL is very much the richest: It has no Ovolo; and its Abacus is very different from those of the *Tuscan*, *Doric*, or *Ionic*; as having its Faces circular, hollow'd inward, with a Rose in the middle of each Sweep. Instead of an Ovolo, and Annulets, here is only a Brim of a Vase; and the Neck is much lengthen'd and enrich'd with a double Row, of eight Leaves in each, bending their Heads downwards; and between them, small Stalks arising: whence spring the Volutes; which don't resemble those of the *Ionic Capital*; and which, instead of the four in the *Ionic*, are here 16; four on each side; under the four Horns of the Abacus, where the Volutes meet in a small Leaf, which turns back towards the Corner of the Abacus. The Leaves are divided, each making three Ranges of lesser Leaves, whereof they are compos'd; each lesser Leaf is, again, generally parted into five, call'd *Olive Leaves*; sometimes into three, and call'd *Laurel Leaves*. The middle Leaf, which bends down, is parted into Eleven. In the Middle, over the Leaves, is a Flower, shooting out between the Stems and Volutes, like the Rose in the Abacus. The height of this *Capital* is two Modules ⅓, and its Projecture one ⅔.

The Differences in the Character of this *Capital*, are, that in *Vitruvius*, etc. the Leaves are in form of the *Acanthus*; whereas in the Antique, they are more ordinarily Olive Leaves: That their Leaves are usually unequal, the undermost being made commonly tallest, sometimes shortest; tho sometimes they are all equal. Sometimes the Leaves are raffled; sometimes quite plain: The first Row generaly belly out towards the Bottom, but are sometimes straight: sometimes the Horns of the Abacus are sharp at the Corners, which seems agreeable to the Rules of *Vitruvius*; but they are more commonly cut off. There is some difference too in the Form and Size of the Rose. Again, the Volutes are sometimes join'd to each other; sometimes wholly separate: Sometimes the spires of the Volutes continue twisting even to the End in the same Course; and sometimes turn back again near the Centre, in the form of an S. See CORINTHIAN.

The Composite CAPITAL, is so called, because compos'd of Members borrow'd from the *Capitals* of other Columns. It takes a quarter Round, or Ovolo, from the *Doric*; an Astragal under this, together with Volutes, or Scrolls, from the *Ionic*; and a double Row of Leaves from the *Corinthian*, which it resembles in most other things; consisting, generally, of the same Members and the same Proportions. In the Middle of the Abacus is a Flower; and under the Horns, Leaves which return upward; as in the *Corinthian*. Indeed, instead of Stalks in the *Corinthian*, the *Composite* has small Flowers, lying close to the Vase or Bell; twisting round towards the middle of the Face of the *Capital*, and terminating in the Rose. The Height of the *Composite Capital*, is two Modules ⅓, and its Projecture one Module ⅔, as in the *Corinthian*.

The Differences of the Character of this *Capital*, consist in this; that the Volutes, which ordinarily descend, and touch the Leaves, are in some Works of the Antique separated from them: That the Leaves, which are generally unequal in Height, the lower Rank being tallest, are sometimes equal: That the Volutes of the Moderns generally spring out of the Vase; whereas, in the Antique, they ordinarily run straight the Length of the

Abacus, over the Ovolo, without striking into the Vase: That the Volutes, whose Thickness is contracted in the Middle, and enlarg'd above and below in the Antique, have their Sides parallel in the Works of the Moderns: And lastly, that the Volutes, which have been hitherto, both by the Antients and Moderns, made as if solid, are now made much lighter and more airy; the Folds standing hollow, and at a distance from each other. See COMPOSITE.

For the Proportions of the several Members of the Capitals *of Columns*, see COLUMN. See also *each Member* under its proper Head; as, ABACUS, ACANTHUS, VOLUTE, ECHINUS, *etc.*

Some Architects distinguish the *Tuscan* and *Doric Capitals*, which have no Ornaments, by the Title of *Capitals of Mouldings*; and the three others, which have Leaves and Ornaments, they call *Capitals of Sculpture*.

Angular CAPITAL, is that which bears the Return of an Entablature, at the Corner of the Projecture of a Frontispiece.

CAPITAL *of a Balluster*, is that Part which crowns the *Balluster*; which sometimes bears a Resemblance to the *Capitals* of some Columns; particularly the *Ionic*. See BALLUSTRADE.

CAPITAL *of a Triglyph*, is the Plat-band over the *Triglyph*; called by *Vitruvius Tænia*. See TRIGLYPHS. 'Tis sometimes also a *Triglyph* which does the Office of a *Capital* to a *Doric* Pilaster.

CAPITAL *of a Niche*, is a kind of little Canopy over a shallow Niche, covering or crowning a Statue. See NICHE.

CAPITAL *of a Bastion*, in Fortification, is a Line drawn from the Angle of the Polygon, to the Point of the Bastion; or from the Point of the Bastion to the Middle of the Gorge. The *Capitals* are from 35 to 40 Fathom long, from the Point of the Bastion to the Point where the two Demigorges meet. See BASTION.

CAPPONIERE, in Fortification, a cover'd Lodgment, four or five Foot broad, encompass'd with a little Parapet about two Foot high, serving to support several Planks laden with Earth. The *Caponniere* is large enough to maintian 15 or 20 Soldiers; and is usually plac'd on the Extremity of the Counterscarp, having little Embrasures frequently made therein.

CARCASE, in Architecture, is the Shell, or Ribbs of a House; containing the Partitions, Floors, Rafters, *etc.* made by the Carpenter, *etc.* See HOUSE, BUILDING, and TIMBER. The *Carcase* is otherwise call'd the *Framing*, see FRAME.

CARIATIDES, or CARIATES, in Architecture; see CARYATIDES.

CARINA, is us'd in the antient Architecture. The *Romans* gave the Name *Carina* to all Buildings in form of a *Ship*; as we still call *Nave*, from *Navis* Ship, the middle or principal Vault of our Gothic Churches; because it has that Figure.

CARNATION, *Flesh-Colour*, in Painting, is understood of all the Parts of a Picture in general which represent Flesh; or which are naked and without Drapery. *Titian* and *Corregio*, in *Italy*, and *Rubens* and *Van Dyke* in *Flanders*, excell'd in *Carnations*. See Colouring.

It must be here observ'd, that the Word *Carnation* is not properly us'd for any particular Part of the Person painted; but for the whole Nudity of the Piece.

CARPENTERS-*Work*, in a Building, includes the *Framing*, *Flooring*, *Roofing*; the *Foundation*, *Carcass*, *Doors*, *Windows*, etc. See FRAME, FLOOR, ROOF, FOUNDATION, CARCASE, DOOR, WINDOW, *etc*.

CARPENTRY, the Art of cutting, framing, and joining large Pieces of Wood, for the Uses of Building; see BUILDING.

Carpentry is one of the Arts subservient to Architecture, and is divided into two Branches, *House Carpentry*, and *Ship Carpentry*: The first employ'd in Raising, Roofing, Flooring, *etc*. of Houses, *etc*. See HOUSE.

The second in the Construction of Vessels for Sea; as Ships, Barks, Barges, Boats, *etc*.

The Rules and Practices in *Carpentry*, as to *Planing*, *Sawing*, *Mortessing*, *Tenanting*, *Scribing*, *Paring*, *Moulding*, *etc*. are much the same as those in Joinery; so likewise are the *Tools*, or Instruments, and the *Stuff*, the same in both: All the Difference between the two Arts consisting in this, that Joinery is us'd in the smaller and more curious Works; see JOYNERY: and *Carpentry* in the larger, stronger, and coarser. See SAWING, SCRIBING, MOULDINGS, PLAN, *etc*. see also SAW, PLANE, *etc*.

Fr. Pyrard assures us, that the Art of Carpentry is in its greatest Perfection in the *Maldives* Islands: Their Works, there, he observes, are so artfully manag'd, that they will hold tight and firm without either Nails or Pins. He adds, they are so curiously put together, that no body can take 'em asunder, but those acquainted with the Mystery.

The Word is deriv'd from the Latin *Carpentum*, a Car, or Cart.

CARPET *Tapis*, a sort of Covering, work'd either with the Needle, or on a Loom; to be spread on a Table, Trunk, an Estrade, or even a Passage, or Floor. *Persian* and *Turky Carpets* are those most priz'd; especially the former. *Carpets* that had a Hair or Shag on one side only, were call'd by the Antients *Tapetes*; such as had a Shag on both sides, were call'd *Amphitapetes*.

CARTOUCH, an Ornament in Architecture, Sculpture, *etc.* representing a Scroll of Paper. It is usually a Table, or flat Member, with Wavings; whereon is some Inscription, or Devise, Ornament of Armory, Cypher, or the Like.

Cartouches are sometimes drawn on Paper, as in the Titles of Maps, *etc.* and sometimes made of Stone, Brick, Plaister, Wood, *etc.* for Buildings. The Word comes form the Italian *Cartoccio*, which signifies the same thing.

CARYATIDES, or CARIATES, in Architecture, a kind of Order of Columns, or Pilasters, under the Figures of Women, dress'd in long Robes. See ORDER *of Columns*.

The Origin of the *Caryatides* is related by *Vitruvius*: The *Greeks*, he observes, having taken the City of *Caria*, led away their Women Captives; and to perpetuate their Servitude, represented them in their Buildings, as charg'd with Burdens, such as those supported by Columns.

The *Caryatides*, M. *le Clerc* observes, are not now represented as among the Antients, *viz.* as Symbols of Slavery, with Hands ty'd before and behind; those Characters being suppos'd injurious to the Fair Sex. Among us, they are represented as Images of Justice, Prudence, Temperance, *etc.* Their Legs are always to be close to each other, and even across; their Arms laid flat to the Body, or to the Head, or at least as little spread as possible; that as they do the Office of Columns, they may have, as near as possible, the Figure thereof. Sometimes their Arms are cut off, for the greater Delicacy; as in the Hall of the *Swiss* Guards in the *Louvre*: but M. *le Clerc* does not approve of such Mutilations.

When insulated, they should never have any great Weight to support; and their Entablature and Pedestal are ordinarily to be *Ionic.*—When they join to a Wall, *etc.* 'tis advisable to put a Console over them, which may appear to sustain the Weight of their Entablature; otherwise, as they represent Women, they don't seem so proper to sustain great Loads. —When they are made in form of Angels, he would have them support the Entablature, which in that Case is to be *Corinthian*, with their Hands.

The Antients made the *Caryatides* frequently to support Baskets, or Corbels of Flowers; and these they call'd *Caniferæ*, or *Cistiferæ*.

CASCADE, a Cataract, or Fall of Water; either natural, as that of *Tivoli*, etc. or artificial, as those of *Versailles*, etc. and that, either falling with a gentle Descent, as those of the *Sceaux*; in form of a Buffet, as at *Trianon*; or by degrees, in form of a Perron, as at *St. Clou*; or *from* Bason to Bason, *etc.* The Word comes form the Italian *Cascata*, of *Cascare*; and that of *Cado*, to fall.

CASEMATE, or CASEMENT, in Building, is a hollow Moulding, which some Architects make 1/6 of a Circle; and others, 1/4. See MOULDINGS.

CASEMATE, or CASEMENT *of a Window*. See WINDOW.

CASEMATE, in Fortification, a kind of Vault of Masons-work, in that part of the Flank of a Bastion next the Curtain; serving as a Battery, to defend the Face of the opposite Bastion, and the Moat, or Ditch. See BASTION.

The *Casemate*, sometimes, consists of three Platforms, one above another; the highest being on the Rampart: But they commonly content themselves to withdraw the highest within the Bastion. The *Casemate* is also call'd the *low Place*, *low Flank*, and sometimes *retir'd Flank*; as being that Part of the Flank nearest the Courtine, and the Centre of the Bastion: It was formerly cover'd with an Epaulement, or a Massive Body of round or square Stone, which prevented those without from seeing within the Batteries. 'Tis now rarely us'd, by reason the Enemy's Batteries are apt to bury the Artillery of the *Casemate*, in the Ruins of the Vault.

The Name comes from a Vault, formerly made to separate the Platforms of the upper and lower Batteries; each of which was call'd in Italian *Casa Armata*, and in the Spanish *Casamata*. Tho others derive the Word from *Casa à Matti*, House of Tools: *Covarruvias*, from *Casa*, and *mata*, low House.

CASEMATE, is also us'd for a Well, with several subterraneous Branches, dug in the Passage of the Bastion, till the Miner is heard at work, and the Air given to the Mine.

CASERNS, or CAZERNS, in Fortification, erected between the Ramparts, and the Houses of fortify'd Towns, or even on the Ramparts themselves; to serve as Lodgings for the Soldiers of the Garison, to ease the Garison.

There are usually two Beds in each *Casern*, for six Soldiers to ly, who mount the Guard alternately; the third Part being always on Duty.

CASING *of Timber-work*, is the Plaistering a House all over on the Outside with Mortar; and striking it, yet wet, by a Ruler, with the Corner of a Trowel, or the like Instrument, to make it resemble the Joints of Freestone; by which means, the whole House appears as if built thereof.

CASTING, in Foundery, is the running of a melted Metal into a Mould prepar'd for that Purpose.

CASTING *of Lead on Cloth*, is the using a Frame or Mould cover'd with Woollen-Cloth and Linen over it, to cast the Lead into very fine Sheets.

CASTING *in Sand or Earth*, is the Running of a Metal between two Frames or Moulds fill'd with Sand or Earth, wherein the Figure the Metal is to take, has been impress'd, *in creux*, by means of the Pattern.

The Goldsmiths also use the Bone of the *Cuttle-Fish*, to mould and cast their lesser Works of Gold and Silver; that Bone, when dried, being reducible to a kind of a fine Pumice, very susceptible of all Impressions.

CASTING *in Stone* or *Plaister*, is the filling a Mould with fine liquid Plaister, that had been taken in pieces from off a Statue or other Piece of Sculpture, and run together again. There are two Things to be observ'd with regard to the Mould: The first, that it be well soak'd with Oil before the Plaister be run, to prevent its sticking: The second, that each Piece whereof it consists, have a Pack-thread, to draw it off the more easily when the Work is dry.

CASTING, in Joinery, *etc*. Wood is said to *cast* or *warp*, when either by its own Drought or Moisture, or the Drought or Moisture of the Air, or other Accident, it shoots or shrinks; in prejudice to its Flatness and Straightness.

CASTLE, a Place fortified by Nature or Art, either in City or Country, to keep the People in their Duty, or resist an Enemy. See FORTRESS, and FORTIFIED *Place*.

A CASTLE, is a little *Cittadel*; see CITTADEL.

CATACOMBS, Grottos, or subterraneous Places for the Burial of the Dead.

The Term is particularly us'd in *Italy*, for a vast Assemblage of subterraneous Sepulchres, three Leagues from *Rome*, in the *Via Appia*; suppos'd to be the Sepulchres of the Martyrs: Accordingly, they are visited out of Devotion, and Reliques thence taken, and dispers'd throughout the Catholick Countries; after having been first baptiz'd by the Pope, under the Name of some Saint.

These *Catacombs*, are said by many to be Caves, or Cells, wherein the Primitive Christians hid, and assembled themselves together; and where they interr'd such among them as were martyr'd. Each *Catacomb* is three Foot broad, and eight or ten high; running in form of an Alley, or Gallery, and communicating with each other: In many Places, thay extend within a League of *Rome*. There is no Masonry or Vaulting therein, but each supports it self: The two Sides, which we may look on as the *Parietes*, or Walls, were the Places where the Dead were deposited; which were laid lengthwise, three or four Rows over one another, in the same *Catacomb*, parallel to the Alley. They were commonly clos'd with large thick Tiles, and sometimes pieces of Marble, cemented in a manner inimitable by the Moderns. Sometimes, tho very rarely, the Name of the Deceased is found on the Tile: Frequently, a Palm is seen, painted, or engraven; or the Cypher Xp, which is commonly read *pro Christo*.

Some Authors will have the *Catacombs* to be the same with the *Puticuli* mention'd by *Festus Pompeius*; maintaining, that whereas it was the Practice of the antient *Romans* to burn their Dead, the Custom was, to avoid Expence, to throw the Bodies of their Slaves to rot in Holes of the Ground: and that the *Roman* Christians, observing, at length, the great

Veneration paid to Reliques, resolv'd to have a Stock of their own: Entering, therefore, the *Catacombs*, they added what Cyphers and Inscriptions they pleas'd, and then shut them up again, to be open'd on a favourable Occasion. Those in the Secret, add they, dying, or removing, the Contrivance was forgot, till Chance open'd them at last. But this Opinion has even less of probability than the former. Mr. *Monro*, in the *Philosophical Transactions*, gives it as his Opinion, that the *Catacombs* were the Burial Places of the first *Romans*, and dug in consequence of these two Opinions; that Shades hate the Light; and that they love to hover about the Places where the Bodies are laid.

Laying up the Bodies in Caves, is certainly the original way of disposing of the Dead; and appears to have been propagated by the *Phœnicians*, throughout the Countries to which they sent Colonies: The interring as we now do, in the open Air, or in Temples, was first introduc'd by the Christians. When an antient Hero dy'd, or was kill'd in a Foreign Expedition, as his Body was liable to Corruption, and for that Reason unfit to be transported intire, they fell on the Expedient of burning, in order to bring home the Ashes, to oblige the Manes to follow; that so his Country might not be destitute of the Benefit of his Tutelage. 'Twas thus Burning seems to have had its original; and by degrees it became common to all who could bear the Expences of it, and took place of the antient Burying: Thus *Catacombs* became disus'd among the *Romans*, after they had borrow'd the manner of Burning from the *Greeks*: and now, none but Slaves were laid in the Ground. See BURIAL.

Places thus prepar'd, might afford convenient Resortments for the primitive Christians; but could never be built by them. When the Empire became Christian, they were again disus'd; till the reading of I know not what Author, who mentions them, occasion'd 'em to be again look'd into. As to the fam'd Cypher Xp, 'tis observ'd to have been in use among the Antients, long before Christianity arose. The Abbot *Bencini* says, it was compos'd of the two *Greek* Letters XP, under which something mystical was comprehended; but no Author gives any account what that Mystery was.

Antiently, the Word *Catacomb* was only understood of the Tomb of St. *Peter* and St. *Paul*; and M. *Chastellan* observes, that among the more knowing of the People of *Rome*, the Word *Catacomb* is never apply'd to the subterraneous Burying-Places above mention'd, but only to a Chapel in St. *Sebastian*, one of the seven stational Churches; where the antient *Roman* Calendars say, the Body of St. *Peter* was deposited, under the Consulate of *Tuscus* and *Bassus*, in 258.

Some derive the Word *Catacomb* from the Places where Ships are laid up; which the modern *Latins* and *Greeks* call *Combes*. Others say, that *Cata* was us'd for *ad*, and *Catacumbas* for *ad-tumbas*: Accordingly, *Dadin* says, they antiently wrote *Catatumbas*.

CATACOUSTICS, call'd also *Cataphonics*, the Science of reflected

Sounds; or that Part of Acoustics which considers the Properties of *Echoes*.

CATAFALCO, an *Italian* Term, literally signifying *Scaffold*. It is chiefly us'd for a Decoration of Architecture, Sculpture, and Painting; rais'd on a Timber Scaffold, to shew a Coffin, or Tomb, in a Funeral Solemnity.

CATHEDRAL, a Church wherein is a Bishop's See, or Seat; see BISHOP.

The Word seems to take its rise from the manner of sitting in the antient Churches, or Assemblies of primitive Christians: In these, the Council, *i.e.* the Elders and Priests, was call'd *Presbyterium*; at their head was the *Bishop*, who held the Place of Chairman, *Cathedralis*; and the *Presbyters*, who sate on either side, were also call'd by the antient Fathers, *Assessores Episcoporum*. The Episcopal Authority did not reside in the Bishop alone, but in all the Presbyters, whereof the Bishop was President.

A *Cathedral* therefore, originally, was different from what it is now; the Christians, till the Time of *Constantine*, having no liberty to build any Temple: by their Churches they only meant their Assemblies; and by *Cathedrals*, nothing more than *Consistories*. Whence appears the Vanity of some Authors, especially the *Spaniards*, who pretend their *Cathedrals* to have been built in the Times of the Apostles.

CATHETUS, in Architecture, is a perpendicular Line, passing along through the middle of a Column: or that thro the Eye of the Volute, in the *Ionic* Capital; call'd also *Axis*. See VOLUTE.

CAULICOLES, CAULICULI, in Architecture, are eight lesser Branches, or Stalks, in the *Corinthian* Capital, springing out from the four greater or principal *Caules*, or Stalks. See *Corinthian* CAPITAL.

The eight Volutes of this Order, are sustain'd by four *Caules*, or Primary Branches of Leaves; and from which arise these *Caulicoles*, or lesser Foliages. See VOLUTE.

Some Authors confound these with the *Volutes* themselves; some with the *Helices* in the Middle, and some with the principal *Stalks* whence they arise.

The Word comes from the Latin *Caulis*, the Stalk, or Stem of a Plant, whence the Leaves and Branches arise.

CAVALIER, in Fortification, a Terrace, or Platform, commanding all around the Place. 'Tis rais'd 18, or 29 Foot high, atop of a Rampart, for placing the Cannon on, and securing the Country about. 'Tis sometimes also made in the Country, to play in upon the Town; and as much above the other Works, as a Man *on Horseback* is above a Footman; whence the

Denomination.

In form, the *Cavalier* is sometimes round, and sometimes square; having always a Parapet to cover the Cannon: Its Breadth depends on the Number of Pieces to be lodg'd on it; there being always allow'd an Interval of 10 or 12 Foot between each two, for the Conveniency of loading and firing.

The *Cavalier* is also sometimes call'd a *double Bastion.*

CAVETTO, in Architecture, a hollow Member, or Moulding, containing a Quadrant of a Circle, and having an Effect just contrary to that of a Quarter-round: It is us'd as an Ornament in Cornices.

M. *Felibien* observes, that the Workmen confound the *Cavetto* with a *Scotia*, but to ill Purpose; the *Cavetto* being in effect only half a *Scotia*: yet he himself is chargeable with the same Oversight. See SCOTIA.

When in its natural Situation, the Workmen frequently call it *Gula*, or *Gueule*; and when inverted, *Gorge*. See GULE, and GORGE.

CEILING, in Architecture, the upper Part, or Roof of a lower Room; or a Lay, or Covering of Plaister over Laths, nail'd on the Bottom of the Joists that bear the Floor of the upper Room; or on Joists for that purpose. See HOUSE, and ROOF.

The Word *Ceiling*, answers pretty accurately to the Latin *Lacunar*, every thing over Head.

Plaister'd *Ceilings* are much us'd in *England*, more than any other Country:Nor are they without their Advantages, as they make the Room Lightsome; are good in Case of Fire; stop the Passage of the Dust; lessen the Noise over Head; and, in Summer, make the Air cooler.

CELLS, *Cellæ*, *Cellulæ*, are little Houses, Apartments, or Chambers; particularly those wherein the antient Monks, Solitaries, and Hermits liv'd in Retirement.

The same Name is still retain'd in divers Monasteries. The Dormitory is frequently divided into so many *Cells*, or Lodges.

The *Carthusians* have each a several House, which serves 'em as *Cells*.

The Hall wherein the *Roman* Conclave is held, is divided by Partitions into divers *Cells*, for the several Cardinals to lodge in.

Some derive the Word from the Hebrew, a *Prison*, or Place where any thing is shut up.

CEMENT, CÆMENT,or CIMENT, in the general Sense of the Word, is any Composition of a glutinous or tenacious Nature, proper to bind, unite, or keep things in Cohesion.

In this Sense, Mortar, Solder, Glue, *etc.* are Cements. See MORTAR, *etc.*

The Bitumen brought from the *Levant*, is said to have been the

Cement us'd in the Walls of *Babylon*. See BITUMEN.

The Word is form'd of the Latin *Cæmentum*, of *cædo*, I beat. Tho M. *Felibien* observes, that what the antient Architects call'd *Cæmentum*, was a very different thing from our *Cement*: The Name *Cement*, with them, stood for a kind of Masonry, or manner of laying the Stones; and even for the Quality of the Stones; as when the Walls were fill'd up with little Pebbles, Rubbish, *etc*. See MASONRY.

CEMENT, is particularly us'd in Architecture, for a strong, binding sort of Mortar, us'd to bind, or unite Bricks or Stones together, for some kinds of Mouldings; or to make a block of Bricks, for the Carving of Scrolls, Capitals, *etc*.

It is of two Sorts: The *hot Cement*, which is the most common, is made of Rosin, Bees-Wax, Brickdust, and Chalk, boil'd together. The Bricks to be *cemented* are heated, and rubb'd one upon another, with *Cement* between.

The *cold Cement* is less us'd: 'tis made of Cheshire Cheese, Milk, Quick-lime, and Whites of Eggs.

CEMENT, is also us'd among Goldsmiths, Engravers, Jewellers, *etc*. for a Composition of fine Brickdust, well sifted, Rosin, and Bees-Wax, in use among those Artificers to keep the Metals to be engraven, or wrought on, firm to the Block, *etc*. as also to fill up what is to be chissel'd.

CHAMBER, in Building, a Member of a Lodging, or Piece of an Apartment, ordinarily intended for sleeping in; and call'd by the Latins *Cubiculum*.

A compleat Apartment is to consist of a *Hall*, *Antichamber*, *Chamber*, and *Cabinet*. See APARTMENT, *etc*.

The Word comes from the Latin *Camera*; and that, according to *Nicod*, from the *Greek*, *Vault*, or *Curve*; the Term *Chamber* being originally confin'd to Places arch'd over. The *Spaniards* call it *Camera*; whence *Camerade*.

Bed-CHAMBER. See BED-*Chamber*.

Privy-CHAMBER. The Gentlemen of the *Privy-Chamber* are Servants of the King, who are to wait and attend on him and the Queen at Court, in their Diversions, Progresses, *etc*.

Six of these are appointed by the Lord Chamberlain, together with a Peer, and the Master of the Ceremonies, to attend all Ambassadors from crown'd Heads in their publick Entries. Their Number is 48.

Their Institution is owing to King *Henry* VII. As a singular Mark of Favour, they are impower'd to execute the King's verbal Command, and without producing any written Order; their Person and Character being deem'd sufficient Authority.

CHAMBRANLE, in Architecture and Joinery, the Border, Frame, or Ornament of Stone or Wood, us'd in the three Sides round Chamber-Doors, large Windows, and Chimneys.

The *Chambranle* is different in the different Orders: when 'tis plain and without Mouldings, 'tis call'd simply and properly, *Band*, *Case*, or *Frame*.

The *Chambranle* consists of three Parts; the two Sides, called *Montants*, or *Ports*; and the Top, called the *Traverse*, or *Supercilium*.

The *Chambranle* of an ordinary Door, is frequently call'd *Door-Case*; of a Window, *Window-Frame*; of a Chimney, *Mantle-Tree*. See DOOR, WINDOW, MANTLE.

CHAMFER, or CHAMFRET, in Architecture, an ornament consisting of half a *Scotia*; being a kind of small Furrow, or Gutter on a Column, call'd also *Scapus*, *Stria*, *etc.* See STRIÆ.

Chamfering, or *Chamfraining*, is particularly us'd for the cutting of the Edge, or End of any thing a-slope, or bevel.

CHANDELIERS, in Fortification, are wooden Stakes six Foot high, and six or Seven apart, between which are plac'd Fascines, to cover the Workmen on Approaches. See FASCINES.

These are sometimes made to prevent the Enemy from seeing what passes within.

CHANNEL is particularly us'd in Architecture, for a Part of the *Ionic* Capital, a little hollow'd, in form of a *Canal*; lying under the Abacus, and running the whole length of the Circumvolution of the Volute; inclos'd by a Listel. See *Ionic* CAPITAL.

CHANNEL *of the Larmier*, is the Soffit of a Cornice; which makes the pendant *Mouchette*. See LARMIER, and SOFFITO.

CHANNEL *of the Volute*, in the *Ionic* Capital, is the Face of its Circumvolution; inclos'd by a Listel. See VOLUTE.

CHANNELINGS. See FLUTES.

CHANTLATE, in Building, a piece of Wood fasten'd near the Ends of the Rafters, and projecting beyond the Wall, to support two or three Rows of Tiles, so plac'd to prevent the Rain-Water from trickling down the Sides of the Wall.

CHAPEL, or CHAPPEL, a kind of little Church, serv'd by an Incumbent under the Denomination of a *Chaplain*.

There are two Kinds of *Chapels*, the one Consecrated, and held as *Benefices*; the other *Secular*, being of the Nature of *Oratories*.

The first are built apart and at a distance from the Parish-Church; being neither Parishes, Cathedrals, nor Priories, but subsisting of themselves.

These are call'd by the Canonists *Sub-Dio*, and by us *Chapels of Ease*; as being erected at a distance from the Mother Church, where the Parish is large and wide, for the Ease and Conveniency of some of the Parishioners who reside far off.

They are served by some inferior Curate, provided either by the Rector of the Parish, or by those whose Ease and Benefit they are intended.

The second kind are frequently built in, or adjoining to a Church, as a Part thereof; having only a Desk, *etc.* to read Prayers in; and, in the *Romish* Churches, an Altar, *etc.* to celebrate Mass on: but without any Baptistery, or Font.

These the Canonists call *Sub-Tecto.* They are generally erected by some considerable Person, for the Use of their own Families; *Ut ibidem Familiaria Sepulchra sibi constituant.*

Free CHAPELS, are those *Chapels of Ease* which have a settled Revenue for perpetual Maintenance of the Curate, *etc.* by charitable Donatives of Lands, or Rents bestow'd on 'em; so as not to be any Charge either to the Rector or the Parishioners.

There are several Collegiate Churches in *France*, which they call *Saintes Chapelles*, Holy Chapels; as those of *Paris*, *Dijon*, *Bourges*, *Bourbon*, etc. so call'd, by reason there are Relicks in them.

The Word *Chapel*, according to some, comes from the *Greek*, little Tents, or Booths, set up by Traders in Fairs to shelter them from the Weather. *Papias* fetches it both from the *Greek* and *Latin*, *quasi capiens*, or *populum vel laudem*: Others derive it from the *Chape*, or Cope which serv'd to cover the body: Others, *à pellibus Caprarum*; because these Places were antiently cover'd with Goatskins. *Rebuff* derives it *de Cappa*, *St. Martin's Cope*, which the Kings of France carry'd to War with them as their Standard, and preserv'd very carefully in particular Tents, thence call'd *Chapels.*

Hence, all those places where Relicks were preserv'd came to be call'd *Chapels*; and the Persons who had the Care of them, *Chaplains*.

CHAPLET, in Architecture, is a little Ornament cut or carv'd into round Beads, Pearls, Olives, and Pater-Nosters; as is frequently done in Baguettes.

A *Chaplet*, in effect, is little else but a *Baguette* enrich'd with sculpture. See BAGUETTE.

CHAPPEL, or CHAPEL. See CHAPEL

CHAPTREL. See IMPOSTS.

CHARGE, or *Overcharge*, in Painting, is the exaggerated Representation of any Person; wherein the Likeness is preserv'd, but withal, ridicul'd.

CHARNEL, a Portico, or Gallery, ordinarily about the Church-Yard;

over which were antiently dispos'd the Bones of the Dead, when the Flesh was consum'd.

The *Charnels*, or *Charnel-Houses*, are now usually contiguous to the Church.

CHATELET, antiently signify'd a little Castle, or Fortress, wherein the *Chatelain*, or Governour lodg'd. See CASTLE.

At present, the Term is us'd for certain Courts of Justice establish'd in several Cities in *France*: The *grand Chatelet* in *Paris*, *v. g.* is the Place where the Presidial, or ordinary Court of Justice of the Prevot of *Paris* is kept; consisting of a Presidial, a Civil Chamber, criminal Chamber, and a Chamber of Policy. The Term signifies the same at *Montpelier*, *Orleans*, etc.

The little *Chatelet* at *Paris*, is an antient Fort, now serving as a Prison.

The Word is a Diminutive of *Chateau*, form'd from *Castellum*, a Diminutive of *Castrum*; or from *Castelletum*, a Diminutive of *Castellum*.

CHEMISE, in Fortification, is a Wall wherewith a Bastion, or any other Bulwark of Earth, is lin'd; for its greater Support and Strength.

Or it is the Solidity of the Wall, from the *Talus* to the Row.

Fire CHEMISE, is a piece of Linen Cloth, steep'd in a Composition of Oil of Petreola, Camphor, and other combustible Matters; us'd at Sea, to set fire to the Enemy's Vessel.

CHILMINAR, CHELMINAR, or *Tchelminar*, the noblest and most beautiful Piece of Architecture remaining of all Antiquity; being the Ruins of the famous Palace of *Persepolis*, to which *Alexander* the Great, being drunk, set fire, at the Persuasion of the Courtesan *Thais*. See RUINS.

Authors and Travellers are exceedingly minute in their Descriptions of the *Chilminar*; particularly *Garcios de Silva Figueroa*, *Pietro de la Valle*, *Chardin*, and *Le Brun*.

A general Idea thereof, may be conceiv'd as follows.

There appear the Remains of near fourscore Columns; the Fragments whereof are at least six Foot high: but there are only nineteen that can be call'd entire; with a twentieth all alone, 150 Paces from the rest.

A Rock of black hard Marble, serves for the Foundation of the Edifice.

The first Plan of the Building is ascended to by fourscore and fifteen Steps cut in the Rock. The Gate of the Palace is twenty Foot wide; on the one side is the Figure of an Elephant, and on the other of a Rhinoceros, each thirty Foot high, and of Shining Marble. Near these Animals are two Columns; and not far off the Figure of a Pegasus.

After this Gate is pass'd, are found a great Number of Columns of white Marble; the Remains whereof shew the Magnificence of the Work: the smallest of these Columns is fifteen Cubits high, the largest eighteen;

each has forty Flutings, three large Inches broad; whence the highness of the whole may be guess'd at, with the other Proportions. Near the Gate is an Inscription on a square piece of Marble, smooth as Glass, containing about twelve Lines: The Characters are of a very extraordinary Figure, resembling Triangles and Pyramids.

These noble Ruins are now the Shelter of Beasts and Birds of Prey. Besides the Inscription above mention'd, there are others in *Arabic*, *Persian*, and *Greek*. Dr. *Hide* observes, that the Inscriptions are very rude and unartful; and that some, if not all of them, are in praise of *Alexander* the Great, and therefore later than that Conqueror.

M. *Le Brun* tells us, that he took his Voyage to the *East-Indies* merely for the sake of viewing the *Chilminar*.

The Word comes from the Persian *Tchehelminar*, forty Towers, or Columns; on account of the forty Columns of an enormous size seen there.

CHIMNEY, in Architecture, a part of an Apartment, wherein the Fire is made.

The Parts of a *Chimney* are the *Jambs*, or Sides, coming out perpendicularly, sometimes circularly, *etc.* from the Back; the *Mantle-Tree* rests on the Jambs: The *Tube*, or *Funnel*, which conveys away the Smoke; the *Chimney-piece*, or Moulding, on the foreside of the Jambs over the Mantle-Tree; and the *Hearth*, or Fireplace.

Palladio settles the Proportions of a Chamber *Chimney* thus; Breadth on the Inside, 5½, 6, or 7 Foot; Height to the Mantle-Tree, 4, or 4½; Depth, 2, or 2½ Foot.

According to *Wolfius*, the Breadth of the Aperture at bottom to be to the Height, as 3 to 2, to the Depth as 4 to 2. In small Apartments the Breadth is 3 Foot, in larger 5; in Bed-Chambers 4; in small Banquetting Rooms 5½, in large 6. But the Height never to exceed 2½, lest there being too much room for the Air and Wind, the Smoke be driven down into the Room. Nor must the Height be too little, lest the Smoke miss its way, and be check'd at first setting out. The same Author advises to have an Aperture, thro' which the external air may, on occasion, be let into the Flame, to drive up the Smoke; which the internal Air would otherwise be unable to do. See STOVE.

The Mouth of the Tube, or that part join'd to the *Chimney*-Back, *Felibien* orders to be a little narrower than the rest; that the Smoke coming to be repell'd downwards, meeting with this Obstacle, may be prevented from getting into the Room.

Some make the Funnel twisted, to prevent the Smoke's descending too easily: but the better Expedient is to make the Funnel narrower at bottom than at top; the Fire impelling it upward more easily when contracted at bottom; and in mounting it finds more space to disengage it self, and therefore has less occasion to return into the Chamber.

To prevent smoaking CHIMNEYS, M. *Lucar* advises two Holes, or two Pipes, one over the other, to be left in each side of the *Chimney*; one sloping upwards, the other downwards: Thro' one of these, says he, the Smoke will pass in any position.

De l'Orme orders a Brass Ball full of Water, with one small Aperture, to be hung up in the *Chimney*, at a Height a little above the greatest Flame: here, as the Water grows hot, 'twill rarefy and drive thro' the Aperture in a vapoury Stream, which will drive up the Smoke that wou'd otherwise linger in the Funnel.

Others place a kind of moveable Vane, or Weathercock a-top of the *Chimney*: so that what way soever the Wind comes, the Aperture of the *Chimney* will be skreen'd, and the Smoke have free egress. Indeed, the best prevention of a smoaking *Chimney*, seems to lie in the proper situation of the Doors of the Room, and the apt falling back of the Back, and convenient gathering of the Wings and Breast of the *Chimney*.

'Tis a Rule, in Building, that no Timber be laid within 12 Inches of the Foreside of the *Chimney* Jambs; that all Joists on the Back of the *Chimney* be laid with a Trimmer, at 6 Inches distance from the Back; and that no Timber be laid within the Funnel.

Chimneys are usually suppos'd a modern Invention; the Antients only making use of Stoves: But *Octavius Ferrareus* endeavours to Prove Chimneys in use among the Antients. To this End he cites the Authority of *Virgil*: *Et jam summa procul villarum culmina fumant*. And that of *Appian*, who says, 'That of those Persons proscrib'd by the Triumvirate, 'some hid themselves in Wells, and *Cloaca*, *Common-shores*; some in the 'Tops of Houses, and *Chimneys*:' for so he understands *Fumaria sub tecto posita*. Add, that *Aristophanes*, in one of his Comedies, introduces his old Man, *Polycleon*, shut up in a Chamber, whence he endeavours to make his Escape by the *Chimney*. However, the few instances remaining among the Antients; together with the Obscurity of the Rules of *Vitruvius* on this Head, make us rather conclude the Use of Stoves, whereof they had entire Apartments, made 'em neglect this Point of Building, which the Coldness of our Climates obliges us to have a principal regard to.

In the Year 1713, appear'd a *French* Book, entitled, *La Mechanique du Feu*, or The Art of augmenting the Effects, and diminishing the Expence of Fire, by M. *Gauger*; since publish'd in English, by Dr. *Desaguliers*: wherein the Author examines what Disposititon of *Chimneys* is most proper to augment the Heat; and proves Geometrically, that the Disposition of parallel Jambs, with the Back inclin'd, as in the common *Chimneys*, is less fitted for reflecting Heat into the Room, than parabolical Jambs, with the bottom of the Tablette horizontal.

He gives seven several Constructions of his new *Chimneys*, and the manner of executing them.

M. *Gauger*, however, does not appear to be the first Inventor of the *Chimney* he describes; the Description of a like Kind being found in a *German* Book, printed at *Leipsick*, in 1699.

The Word *Chimney* comes from the French *Cheminee*; and that from the Latin *Caminata*, a Chamber wherein is a *Chimney*: *Caminata*, again, comes from *Caminus*, and that from the *Greek*, *a Chimney*; of *uro*, I burn.

CHIMNEY-*Jambs*, are the Sides of a *Chimney*, usually standing out perpendicularly, sometimes circularly, from the Back; on the Extremities whereof the Mantle-Tree rests.

CHIMNEY-*Piece*, in Building, a Composition of certain Mouldings, of Wood or Stone, standing on the Foreside of the Jambs, and coming over the Mantle-Tree. See MANTLE.

CHIMNEY-*Money*, or *Hearth-Money*, a Tax impos'd by a Stat. 24 *Car*. II. expressing, that every Fire-Hearth and Stove of every Dwelling, and other Houses within *England* and *Wales*, except such as pay not to Church and Poor, shall be chargeable with two Shillings *per Annum*, payable at *Michaelmas* and *Lady-Day*, to the King and his Heirs. See CHIMNEY.

CHOIR, that part of a Church, Cathedral, *etc*. where the Priests, and *Choristers*, or Singers, are dispos'd. See CHURCH.

The *Choir* is distinguish'd from the *Chancel*, or *Sanctuary*, where the Communion is celebrated; as also from the *Nave*, or Body of the Church, where the People assist.

The Patron is said to be oblig'd to repair the *Choir* of a Church; and the Parishoners the Nave.

The *Choir* was not separated from the Nave, till the Time of *Constantine*: From that Time the *Choir* was rail'd in with a Ballustrade, with Curtains drawn over; not to be open'd till after the Consecration.

In the XIIth Century they began to enclose the Choir with Walls: But the antient Ballustrades have been since restor'd; out of a View to the Beauty of the Architecture. The Chantor is Master of the *Choir*.

In Nunneries, the *Choir* is a large Hall, adjoining to the Body of the Church; separated by a Grate, where the Religious sing the Office.

The Word, according to *Isidore*, is derived *à Coronis circumstantium*; because, antiently, the *Choristers* were dispos'd round the Altar to sing: which is still the manner of building Altars among the *Greeks*.

CHURCH, with regard to Architecture, *Daviler* defines a large Vessel extended in length, with Nave, Choir, Isles, Chapel, Belfry, *etc*. See each Part under its proper Head, NAVE, CHOIR, CHAPEL, *etc*.

A *simple Church*, he calls that which has only a Nave, and a Choir; a *Church with Isles*, that which has a row of Porticos, in form of vaulted Galleries with Chapels in its Pourtour.

Church in a Greek Cross, that where the Length of the Cross is equal to that of the Nave; so called, because most of the *Greek* Crosses are built in this form. *Church in a Latin Cross*, that whose Nave is longer than the Cross, as most of the *Gothic Churches*. *Church in Rotondo*, that whose Plan is a perfect Circle, in Imitation of the Pantheon. See ROTONDO, *etc*.

For the Form of the antient *Greek Churches*, when they had all their Parts, it was as follows: First was a Porch, or Portico, call'd the *vaunt Nave*, adorn'd with Columns on the outside, and on the inside surrounded with a Wall; in the middle whereof was a Door, thro' which they pass'd into a second Portico. The first of these Porticos was destined for the *Energumenes*, and Penitents in the first Stage of their Repentance; the second was much longer, destin'd for Penitents of the second Class, and the *Catechumens*, and hence call'd *Ferula*; because those plac'd in it began to be subject to the Discipline of the *Church*. These two Porticos took up about one third of the space of the whole *Church*.

From the second Portico, they pass'd into the *Nave*, which took up near another third of the *Church*. In the middle, or at one side of the Nave, was the *Ambo*, where the Deacons and Priests read the Gospel and preach'd. The Nave was destin'd for the Reception of the People, who here assisted at Prayers.

Near the Entrance of this was the *Baptistery*, or Font.

Beyond the Nave was the *Choir*, set with Seats, and round: the first Seat on the right, next the Sanctuary, being for the Chantor, or *Choragus*. See CHANTOR, *etc*.

From the Choir, they ascended by Steps to the *Sanctuary*, which was enter'd at three Doors. The Sanctuary had three *Apsides* in its length; a great one in the middle; under which was the *Altar*, crown'd with a *Baldaquin*, supported by four Columns: See SANCTUARY, BALDACHIN, *etc*. Under each of the small *Apsides*, was a kind of a Table, or Cupboard, in manner of a Buffet.

Tho, of the *Greek Churches* now remaining, few have all the Parts above describ'd; most of 'em having been reduc'd to Ruins, or converted into Mosques.

M. *Frezier*, Engineer to the *French* King, and F. *Cordemoy*, a Regular Canon, have disputed the Form of the antient and modern *Churches*, and the best manner of building them, with a good deal of Learning, in the *Journals de Trevoux*.

For the Form of the *Latin Churches*, tho it be various, yet may all the Variety be reduc'd to two Heads; *viz*. those in form of a Ship, and those of a Cross.

CIMA, or *Sima*, in Architecture, a Member or Moulding, call'd also *Gula*, and *Cymatium*. See GULE, and CYMATIUM.

CINCTURE, or CEINTURE, in Architecture, a Ring, List, or Orlo, at the top and bottom of the Shaft of a Column; dividing the Shaft, at one end, from the Base; and at the other from the Capital. See SHAFT, COLUMN, *etc*.

That at bottom is peculiarly call'd *Apophyges*; as if the Pillar took its height hence: and that at top *Colarin*, or Collar. See APOPHYGE, and COLARIN.

The *Cincture* is suppos'd to be an Imitation of the Girts, or Ferrils, antiently us'd to strengthen and preserve the primitive wooden Columns. See ORDER.

The Word, in its original *French*, signifies *Girdle*; of the Latin *cingo*, I gird.

CIRCUS, in Antiquity, a large Building, either round, or oval; us'd for the exhibiting of Shews to the People.

The *Roman Circus*, was a large Place, or Square, arch'd at one End; incompass'd with Porticos, and furnish'd with Rows of Seats, plac'd ascending over each other.

In the middle was a kind of *Banquette*, or Eminence, with Obelisks, Statues, and Posts at each End. This serv'd 'em for the Courses of their *Bigæ* and *Quadrigæ*.

There were no less than ten *Circus's* at Rome: the largest was that built by the Elder *Tarquin*, called *Circus Maximus*; between the *Aventine* and *Palatine* Mounts. *Pliny* says, it was enlarg'd by *J. Cæsar*, so as to take in no less than three *Stadia* in Length, and one in Width.

The most magnificent *Circus's* were those of *Augustus* and *Nero*. There are still some Remains of the *Circus's*, both at *Rome*, at *Nismes*, and other Places.

The *Romans* were excessively fond of the Games exhibited in the *Circus*, witness that Verse in *Juvenal*,

> *—Atque duas tantum res anxius optat,*
> *Panem et Circenses—*

Some derive the Word from *Circe*, to whom *Tertullian* attributes the Invention. *Cassiodorus* says, *Circus* comes *à circuitu*.

The *Romans*, *Servius* observes, at first, had no other *Circus* but that made by the *Tiber*, on one side, and a Pallisade of naked Swords on the other.

Hence, according to *Isidore*, came the Term *Ludi Circenses*, *quasi Cicum-Enses*. But *Scaliger* laughs at the Etymology.

The Games of the CIRCUS, which some call *Circensian Games*, were Combats celebrated in the *Circus*, in honour of *Consus* the God of Councils; thence also call'd *Consualia*.

They were also called *Roman Games*, *Ludi Romani*, either on account of their Antiquity, as being coeval with the *Roman* People, or because establish'd by the *Romans*: and the great Games *Ludi Magni*, because celebrated with more Expence and Magnificence than any others; and because held in honour of the great God *Neptune*, who was their *Consus*.

Those who say they were instituted in honour of the Sun, confound the *Pompa Circensis*, or Procession of the *Circus*, with the Games.

The Games of the *Circus* were instituted by *Evander*, and re-establish'd by *Romulus*: The Procession was only a Part of the Games, making the Prelude thereof; and consisting of a simple Cavalcade of Chariots.

Till the Time of the Elder *Tarquin*, they were held in an Island of the *Tiber*; and were called *Roman Games*: after that Prince had built the *Circus*, they took their Name therefrom; as being constantly held there.

There were six Kinds of Exercises in the *Circus*: the 1st Wrestling, fighting with Swords, with Staves, and with Pikes: the 2d was Racing: the 3d *Saltatio*, Dancing: the 4th *Disci*, Quoits, Arrows, and *Cestus*; all which were all on Foot: the 5th was Horsecoursing: the 6th Courses of Chariots, whether with two Horses, or with four.

In this last Exercise, the Combatants were at first divided into two *Quadrils*, then into four; each bearing the Names of the Colours they wore; *Factio Alba*, *Russea*, *etc.*

At first there was only white and red; then green was added, and blue. *Domitian* added two more Colours, but they did not hold.

It was *Oenomaus* who first invented this method of distinguishing the *Quadrils* by Colours. The green was for those who represented the Earth; the blue for the Sea, *etc.*

CITADEL, or CITTADEL, a Fort, or Place fortify'd with four, five, or six Bastions; built sometimes in the most eminent Part of a City, and sometimes only near the City.

In the first Case, the *Citadel* serves to defend the City against Enemies.

In the latter it serves to comand it, and to keep the Inhabitants in their Obedience: for which purpose the City is left unfortify'd on the Part toward the *Citadel*, but the *Citadel* fortify'd toward the City.

The most usual Form for *Citadels*, is that of a Pentagon: a Square being too weak, and a Hexagon too big. See FORTIFICATION.

There is always a large Esplanade between the *City* and *Citadel*. See ESPLANADE.

The Word is a Diminutive of the Italian *Citta*, City, *q.d.* little City.

CLAIR-OBSCURE, or CHIARO SCURO, in Painting, the Art of distributing, to advantage, the Lights and Shadows of a Piece; both with regard to the easing of the Eye, and the Effect of the whole Piece.

Thus, when a Painter gives his Figures a strong Relievo, loosens 'em free from each other, by the Management of his Lights and Shadows; he is said to understand the *Clair-obscure*.

The *Clair-obscure* makes one of the great Divisions, or Branches of Painting; the whole of a Picture being resolvable into *Light*, and *Shadow*.

The Doctrine of the *Clair-obscure*, will come under the following Rules.

Light may be either consider'd with regard to it self; to its Effects; the Place wherein 'tis diffus'd; or its Use.

For 1st, Light is either *Natural*, or *Artificial*.

Natural, either comes immediately from the Sun, which is brisk, and its Colour various according to the Time of the Day; or 'tis that of a clear Air thro' which Light is spread, and whose Colour is a little bluish; or a cloudy Air, which is darker, yet represents Objects in their genuine Colours with more ease to the Eye.

Artificial proceeds from Fire, or Flame, and tinges the Object with its own Colour: but the Light it projects, is very narrow and confin'd.

For the 2d, the *Effects* of Light are either *Principal*, as when the Rays fall perpendicularly on the Top of a Body, without any Interruption; or *Glancing*, as when it slides along Bodies; or *Secondary*, which is for things at a distance.

3. For the *Place*, 'tis either the open Campaign, which makes Objects appear with great Softness; or an inclos'd Place, where the Brightness is more vivid, its Diminution more hasty, and its Extremes more abrupt.

4. For the *Use*, or Application: The Light of the sun is always to be suppos'd without, and over against the Picture; that it may heighten the foremost Figures: the Luminaries themselves never appearing, in regard the best Colours can't express 'em. The chief Light to meet on the chief Group, and as much as possible on the chief Figure of the Subject. The Light to be pursu'd over the great Parts, without being cross'd, or interrupted with little Shadows. The full Force of the principal Light to be only in one Part of the Piece: taking care never to make two contrary Lights: Not to be scrupulously confin'd to one universal Light; but to suppose other accessory ones, as the opening of Clouds, *etc.* to loosen some Things, and produce other agreeable Effects. Lastly, the Light to be different, according to the Quality of Things whence it proceeds, and the Nature of the Subjects which receive it.

For *Shadows*, they are distinguish'd, 1st, into those form'd on the Bodies themselves, by their proper Relievos. 2d, Those made by adjacent Bodies; those that make Parts of any whole; and the different Effects, according to the difference of Places.

For the first, since the different Effects of Lights only appear by Shadows, their Degrees must be well manag'd. The Place which admits no Light, and where the Colours are lost, must be darker than any Part that has Relievo, and dispos'd in the Front. The Reflex, or return of the Light, brings with it a Colour borrow'd from the Subject that reflects it; and flies off at a greater or less Angle, according to the Situation of the reflecting Body, with regard to the luminous one: Hence, its Effects must be different in Colour, and in Force; according to the Dispositions of Bodies. Deepenings, which admit not of any Light, or Reflex, must never meet on the Relievo of any Member of any great elevated Part; but in the Cavities or Joints of Bodies, the Folds of Draperies, *etc.* And to find Occasions for introducing great Shadows, to serve for the Repose of the Sight, and the loosening of Things; instead of many little Shadows, which have a pittiful Effect.

For the 2d, The Shadows made by Bodies, are either in plain and smooth Places, or on the Earth; wherein they are deeper than the Bodies that occasion 'em, as receiving less reflex Light; yet still diminish as they depart further from their Cause; or on the neighbouring Bodies, where they are to follow the Form of the said Bodies, according to its Magnitude and its Position, with regard to the Light.

For the 3d, In Shadows that have Parts, the Painter must observe to take for a Light in a shadow'd Place, the Teint, or Lustre of the light Part; and, on the contrary, for the Shadow in the lighten'd Part, the Teint, or Lustre in the Shadow: To make an agreeable Assemblage of Colour, Shadow, and Reflex in the shadow'd Part; but without interrupting the great Masses of Shadows: To avoid forming little Things in the Shadow; as not being perceiv'd, unless closely look'd at; and to work, as it were, in the general, and at one sight: Never to set the strong Shadows against the Lights, without softening the harsh Contrast by the help of some intermediate Colour: tho the Mass of Light may be plac'd either before or behind that of the Shadow, yet ought it to be so dispos'd, as to illumine the principal Parts of the Subject.

For the 4th, The Effects of Shadows are different, as the Place is either wide and spacious; as in those coming immediately from the Sun, which are very sensible, and their Extremes pretty abrupt; from the serene Air, which are fainter and more sweet; from the dark Air, which appear more diffus'd, and almost imperceptible; and those from an artificial Light, which makes the Shadow deep, and their Edges abrupt: or as it is more narrow and confin'd, where the Lights coming from the same Place, make the Shadow more strong, and the reflex less sensible.

CLAIR-OBSCURE, CHIARO SCURO, is also us'd for a Design consisting only of two Colours; ordinarily black and white, sometimes black and yellow.

Or, it is a Design only wash'd with one Colour; the Shadows being of a dusky brown Colour, and the Lights heighten'd up with white.

The Word is also apply'd to Prints of two Colours, taken off at twice; whereof there are Volumes in the Cabinets of the Curious in Prints.

The Word *Clair-Obscure* is a compound of two others. *Clair* is us'd among the *French* for those Parts of a Painting which reflect the most Light; and comprehends not only the Lights themselves, but also those Colours that are luminous. By *Obscure*, is meant not only all the Shades, but also all the Colours that are dusky.

CLAMPING, in Joinery, *etc.* when a piece of Board is fitted with the Grain to the End of another piece of Board cross the Grain; the first Board is said to be *clamp'd*.

Thus the Ends of Tables are commonly *clamp'd*, to preserve them from warping.

CLARO-OBSCURO. See CLAIR-OBSCURE.

CLAYES, in Fortification, are Wattles made with Stakes, interwove with Oziers, *etc.* to cover Lodgments. See FASCINES.

CLOISTER, a Habitation surrounded with Walls, and inhabited by Canons and Religious.

In a more general Sense, *Cloister* is us'd for a Monastery of Religious of either Sex. See MONASTERY.

In a more restrain'd Sense, *Cloister* is us'd for the principal Part of a Regular Monastery, consisting of a Square built around; ordinarily, between the Church, the Chapter, and the Refectory; and over which is the Dormitory. See DORMITORY, *etc.*

The *Cloisters* serv'd for several Purposes in the antient Monasteries. 1st, *Peter de Blois* observes, that it was here the Monks held their Lectures: The Lecture of Morality at the North side, next the Church; the School on the West, and the Chapter on the East; Spiritual Meditation, *etc.* being reserv'd for the Church.

Du Cange concludes, that all these different Exercises were perform'd in the *Cloister* it self; but by mistake. The Church, the Chapter, and the School were not Parts of the *Cloister*, but Buildings adjoining to it.

Lancfranc observes, that the proper Use of the *Cloister* was for the Monks to meet in, and conserve together, at certain Hours of the Day.

The Form of the *Cloister* was square; and it had its Name, as being inclos'd on its four Sides with Buildings.

Hence, in Architecture, a Building is still said to be in form of a *Cloister*; when there are Buildings on each of the four Sides of the Court.

COLARIN, in Architecture, the little Frise of the Capital of the *Tuscan* and *Doric* Column; plac'd between the Astragal, and the Annulets: called by *Vitruvius*, *Hypotrachelium*. See FRIEZE, *etc.*

COLARIN, is also used for the Orlo, or Ring, a-top of the Shaft of the Column, next the Capital; called also *Cincture*. See CINCTURE.

COLLISEUM, or COLISEUM, in the antient Architecture, and oval Amphitheater, built at *Rome* by *Vespasian*, in the Place where stood the Pond of *Nero's* gilded House.

In this were seen Statues, representing all the Provinces of the Empire; in the middle whereof stood that of *Rome*, holding a golden Apple in her Hand. The same Term, *Colliseum*, is also given to another Amphitheatre of the Emperor *Severus*.

In these *Collisea* were represented Games, and Combats of Men and wild Beasts: There is now little remaining of either of them; Time and War having reduc'd 'em to Ruins. See AMPHITHEATRE.

The Word is form'd from *Colossæum*, on account of the *Colossus* of *Nero*, that stood near it: or, according to *Nardini*, from the Italian *Coliseo*.

COLONNADE, a Peristyle, of a circular Figure; or a Series of Columns, dispos'd in a Circle, and insulated within-side. See PERISTYLE.

Such is that of the little Park at *Versailles*, which consists of 32 *Ionic* Columns; all of solid Marble, and without Incrustation.

A *Polystyle* COLONNADE, is that whose number of Columns is too great to be taken in by the Eye at a single View. Such is the *Colonnade* of the Palace of St. *Peter* at Rome; which consists of 284 Columns of the *Doric* Order, each above four Foot and an half Diameter; all in *Tiburtine* Marble.

COLOSSUS, a Statue of a prodigious Size, representing a Giant. See STATUE.

The most eminent of this kind was the *Colossus of Rhodes*, a Statue of *Apollo*, so high, that Ships pass'd with full Sails betwixt its Legs. It was the Workmanship of *Chares*, a Disciple of *Lysippus*; who spent twelve Years in making it: 'Twas at length overthrown by an Earthquake, after having stood 1360 Years. Its Height was sixscore and six Foot: There were few People could fathom its Thumb, *etc.*

Some Criticks observe, that the *Colossus* of *Rhodes* gave its own Name to the People among whom it stood; and that many, at least among the antient Poets, call the *Rhodians*, *Colossians*: Hence they advance an Opinion, that the *Colossians* in Scripture, to whom St. *Paul* directs his Epistle, are, in reality, the Inhabitants of *Rhodes*. Of this Sentiment are *Suidas*, *Calepin*, *Munster*, etc.

When the *Saracens* became possess'd of the Island, the Statue was found laid along the ground: They sold it to a *Jew*, who loaded 900 Camels with the Brass.

The Basis that supported it was of a triangular Figure, its Extremities were sustain'd with 60 Pillars of Marble. There was a winding Stair-case to go up to the top of it; from whence one might discover *Syria*, and the Ships that went into *Egypt*, in a great Looking-Glass, that was hung about the Neck of the Statue.

Among the Antiquities of *Rome*, there are seven famous *Colossus's*: two of *Jupiter*, as many of *Apollo*; one of *Nero*, one of *Domitian*, and one of the Sun.

COLOUR, in Philosophy, a Property inherent in Light, whereby, according to the different Sizes, or Magnitudes of its Parts, it excites different Vibrations in the Fibres of the Optick Nerve; which propagated to the *Sensorium*, affects the Mind with different Sensations.

Or, *Colour* may be defin'd a Sensation of the Soul, excited by the Application of Light to the *Retina* of the Eye; and different, as that Light differs in the degree of its Refrangibility, and the magnitude of its component Parts.

In the former View, therefore, Light is the Subject of *Colour*; in the latter it is the Agent.

Various are the Opinions of antient and modern Authors, and of the several Sects of Philosophers, with regard to the Nature and Origin of the Phænomenon *Colour*.

COLOURING, in Painting, the manner of applying, and conducting the *Colours* of a Picture: Or, it is the Mixture of Lights and Shadows, form'd by the various *Colours*, employ'd in a Painting. See CLAIR-OBSCURE.

The *Colouring* is one of the principal Branches of Painting. M. *Felibien* divides the Painter's Art into three Parts; the Design, the Composition, and the *Colouring*. The *Colouring* strikes the most; but among Masters it always gives place to the Exactness of the Design.

De Piles observes, that the Word *Colouring*, in its confined Sense, is chiefly applicable to a History-piece; scarce at all to a Landskip. He adds, that the Term *Colouring* relates more immediately to the Carnations than any thing else.

The *Colouring*, in its general Sense, takes in what relates to the Nature and Union of *Colours*; their Agreement, or Antipathy; how to use them to advantage in Light and Shadow, so as to shew a Relievo in the Figures, and a sinking of the Ground: What relates to the aerial Perspective, *i.e.* the Diminution of *Colours*, by means of the Interposition of Air; the various Accidents and Circumstances of the Luminary and the Medium; the different Lights, both of the bodies illuminating and illuminated; their Reflections, Shadows, different Views, with regard either to the Position of the Eye, or the Object: What produces the Strength, Fierceness, Sweetness, *etc.* in Paintings well *colour'd*: The various Manners of *colouring*, both in Figures, Landskips, *etc.*

The Doctrine of COLOURING *is comprised under the following Rules.*

Colours are consider'd, either in respect of their Use, or their Oeconomy and Disposition.

1st, With regard to their *Use*. They are either in Oil, or Water: Those in Oil, again, are either consider'd with a View to their Preparation, or Application.

In the Preparation of Oil Colours, care must be taken that they be ground fine; that in putting them on the Pallet, those which won't dry themselves, be mix'd with Oil, or other Dryers; and that the ting'd *Colours* be mix'd in as small Quantities as possible.

For their *Application*, it is consider'd either with regard to the Kinds of Painting in Works of various *Colours*, or in those of one single *Colour*.

For the *first*: in the larger Pieces, the *Colours* are either laid on full, so as they may be *impasted*, or incorporated together, which makes them hold the more firmly.

Or else we mix those more agreeable ones, which dry too hard, and too hastily, with a little *Colour*, and the clearest of the Oil: But, in both

Cases, the Colours are to be laid on strong at first; it being easy to weaken those that are to be thrust back, and to heighten the others: The Touches to be bold, by the Conduct of a free and steady Pencil; that the Work may appear the most finish'd at a proper distance, and the Figures animated with Life and Spirit.

For glazed Colours, care must be taken that the Under-*Colour* be painted strong, and that it be a Body-*Colour*, and laid smooth.

In finish'd Works, which are to be view'd near at hand, they proceed, either by applying each *Colour* in its place; preserving their Purity, without fretting or tormenting them, but sweetly softening off their Extremities: Or by filling up all the great Parts with one single *Colour*; and laying the other *Colours*, which are to form the little Things, upon it: which is the more expeditious way, but more apt to decay.

For the *second*: the Kinds of Pictures in one *Colour* are two, *viz. Camieux*, where the Degradations of *Colours* of Objects afar off, are usually manag'd by Lights, as with Crayons; and *Basso-Relievo*, which is an Imitation of Sculpture, of whatsoever Matter and Colour: in both these, the *Colours* are wrought dry.

For Water-Colours, they are wrought various ways; *viz.* in *Distemper*, where the *Colours* are prepar'd in Size; which Method is used on all kinds of Matter: in *Fresco*, or Painting on fresh Mortar; where the Colouring must be quick, that the Matter dry not; and with much Care and Neatness, laying each Colour in its place, and intermingling them by Parcels: In *Agouache*, where the *Colours* are mix'd with Gum, and the Pencil dragg'd; as in Paint and Washings: In *Miniature*, for small and delicate Works; where the *Colours* are to be very fine and clean, mix'd with Gums and wrought in Dots or Points.

But in all the Kinds of Painting, both in Oil and Distemper, especially the latter, care must be taken that the Design be fix'd, and all the Parts mark'd out, ere any *Colours* be applied.

For the 2d Part of *Colouring*, or the *Oeconomy* and dispensing thereof in Paintings; regard is either had, first, to the Qualities of the *Colours*, to appropriate them according to their Value and Agreement: or, secondly, to their Effect, in the Union and Oeconomy of the Work.

For the first, it must be observ'd, that *white* represents Light, and gives the briskness and heightening; *black*, on the contrary, like Darkness, obscures and effaces the Objects: again, black sets off the light Parts; and by that they serve each other to loosen the Objects. A proper Choice to be made of *Colours*, and the too much charg'd manner to be avoided; both in *Carnations*, where red *Colours* are not to be affected, as rather resembling the Flesh when flead than the Skin; and all bright glowing *Colours*: the Skin, how delicate soever, being always of a Down-Colour. In the *Drapery*, where the Painter has his whole stock of *Colours* to chuse out of, to procure a good effect; and in the *Landskip*, to dispose those *Colours* near one another, which mutually assist and raise each other's Force and Briskness; as red and green, yellow and blue.

To manage 'em so, as that they be accommodated to the Effects of the great Parts of Light and Colour: that the strong *Colours* lead to the soft ones, and make 'em more look'd at; bringing 'em forwards, or keeping 'em back, according to the Situation, and the degree of Force requir'd.

For the *Effects of Colours*, they either regard the Union, or the Oeconomy: With respect to the first, care must be taken that they be laid so as to be sweetly united, under the briskness of some principal one; that they participate of the prevailing Light of the Piece; and that they partake of each other by the communication of Light, and the help of Reflection.

For the Oeconomy in managing their degrees, regard is to be had to the *Contrast*, or Opposition intervening in the Union of the *Colours*; that by a sweet Interruption, the brickness, which otherwise fades and palls, may be rais'd to the Harmony which makes the Variety of *Colours* agree; supplying and sustaining the Weakness of some by the Strength of others; neglecting some Places, on purpose to serve as a Basis or Repose to the Sight, and to enhance those which are to prevail thro' the Piece: To the *Degradation*; where, the better to proportion the *Colours* that fall behind, some of the same Kind are to be preserv'd in their purity, as a Standard, for those carried afar off to be compar'd by, in order to justify the Diminution: regard being always had to the quality of the Air, which, when loaded with Vapours, weakens the *Colours* more than when clear: To the *Situation* of the *Colours*; where care must be taken, that the purest and the strongest be placed before, or in the Front of the Piece; and that by their force, the compound ones, which are to appear at a distance, be kept back; particularly, the glazed *Colours* to be used in the first Rank: Lastly, to the *Expression* of the Subject, and the Nature of the Matters, or Stuffs; whether shining or dull, opake of transparent, polish'd or rough.

COLUMN, in Architecture, a round Pillar, made to support or adorn a Building. See PILLAR, BUILDING, *etc.*

The *Column* is the principal, or reigning Part of an Order. See ORDER.

The principal Laws and Properties of this eminent Member of Architecture are thus deduc'd:

Every *Fulcrum*, or support, is so much the more perfect as it is the firmer, or carries the greater appearance of firmness: and hence all *Columns*, or Pillars, ought to have their Base, or Foot, broader than themselves. See BASE.

Again, as a Cylinder and a quadrangular Prism are more easily remov'd out of their place than a truncated Cone, or Pyramid, on the same Base, and of the same Altitude: The Figure of *Columns* ought not to be cylindrical; nor, that of a Pilaster, pyramidical; but both the one and the other to be contracted or dimish'd, *i.e.* to grow less and less, like a truncated Cone and a truncated Pyramid. See DIMINUTION.

For the same Reasons, the lowest Parts of the *Columns* to be cylindrical, that of Pilasters pyramidical: Hence, again, as *Columns* are more firm, if their Diameter bears a greater proportion to their Height, than if it bore a less: The greater Ratio is to be chosen, where a large weight is to be sustain'd; and less where a less.

Further, as the Design of a *Column* is to support a weight, it must never be supposed without an Entablature: tho a *Column* rais'd on an eminent Place, so as to leave no room to fear its being thrust out of its place, needs no Pedestal. See ENTABLATURE, and PEDESTAL.

The entire *Column*, in each Order, is compos'd of three principal Parts, the *Base*, the *Shaft*, and the *Capital*. See the Proportions of each under its respective Head, BASE, SHAFT, *etc.*

Each of these Parts, again, is subdivided into a great Number of lesser, called *Members*, or Mouldings: some, whereof are essential, and found in all *Columns*; others are only accidental, and found in particular Orders. See MOULDINGS, ORNAMENTS, *etc.*

Columns are made different, according to the several *Orders* they are used in; and likewise, not only with regard to their *Order*, but to the *Matter*, *Construction*, *Form*, *Disposition* and *Use*.

COLUMNS, *with regard to* Order.

Tuscan COLUMN, is the shortest, and most simple of all the *Columns*. See TUSCAN.

Its height, according to *Vitruvius*, *Palladio*, and *Vignola*, is 7 Diameters, or 14 Modules; according to *Scamozzi*, 15 Modules; to *De Lorme*, 12; to *Trajan's Column*, 16. Its Diminution, according to *Vitruvius*, is one fourth of the Diameter; according to *Vignola*, a fifth; and according to *Trajan's Column*, a ninth.

Its several Parts, Mouldings, *etc.* see under their proper Heads, SHAFT, CAPITAL, *etc.* See it represented in Plate *Architecture*, Fig. 6.

Doric COLUMN, is somewhat more delicate. See DORIC.

Its Shaft is adorn'd with Flutings: its Height, according to *Vitruvius*, is from 14 to 15 Modules; to *Scamozzi*, 17; to *Vignloa*, 16; in the *Coliseum*, 19; in the *Theatre of Marcellus*, 15⅔. Its Diminution, according to the Theatre of *Marcellus*, 12 Minutes; to the *Coliseum*, 4 Min. and a half. See Plate *Architecture*, Fig. 7.

Ionic COLUMN, is more delicate still. See IONIC.

It is distinguish'd from the rest by the Volutes in its Capital; and by its Base. See VOLUTE.

Its Height, according to *Palladio*, is 17 Modules ⅓; according to *Vignola*, 18. Its Diminution, in the Temple of *Concord*, 10½; of *Fortuna Virilis*, 7½; *Coliseum*, 10 Min. See Plate *Architecture*, Fig. 8.

Corinthian COLUMN, is the richest and most delicate of all the *Columns*. See CORINTHIAN.

Its Capital is adorn'd with two rows of Leaves, and with Caulicoles; whence spring out little Volutes. Its Height, according to *Vitruvius*, and many Remains of antique Porticos, Temples, *etc.* is 19 Modules; according to *Serlio*, 18; to the *Coliseum*, 17.17; to the three *Columns* in the *Campo Vaccino*, 20; the Basilisk of *Antoninus*, 20. Its Diminution, according to the Temple of *Peace*, 6½ Minutes; the *Pantheon*, 6⅓; Temples of *Sibyl* and *Faustina*, 8; *Constantine's* Arch, 7; Portico of *Septimius*, 7½. See Plate *Architecture*, Fig. 9.

Composite COLUMN, has two rows of Leaves in its Capital, like the *Corinthian*; and angular Volutes, like the *Ionic*. See COMPOSITE.

Its Height, according to *Vignola*, and the Arch of *Titus*, is 20 Modules; to *Scamozzi*, and the Temple of *Bacchus*, 19½; the Arch of *Septimus*, 19.09. Its Diminution, according to *Titus's* and *Septimius* Arch, 7 Minutes; Baths of *Diocletion*, 11 Minutes ⅓; Temple of *Bacchus*, 6½. See Plate *Architecture*, Fig. 10.

It may be here observ'd, that there seems more of Caprice than Reason in that Diversity found in the Heights of *Columns* of the same Order, in different Authors; each of whom frequently takes the liberty of dispensing with his own Rules.

Vitruvius, for Instance, makes the *Doric Columns* of Temples, shorter than those of Porches behind Theatres; *Palladio* gives a greater height to *Columns* standing on Pedestals, than to those which have none; and *Serlio* makes his *Column* a third shorter, when insulate, or detach'd, than when contiguous to the Wall.

But notwithstanding the diversity of Height in *Columns* of the same Order in different Authors; they still bear a true Proportion in the several Orders compar'd with each other; by which they go increasing, as the Orders are less massive.

But this Augmentation is greater in some Ordonnances than in others; for in the Antique it is but of 5 Modules, or Semidiameters, for the five Orders: the shortest *Column*, *viz.* the *Tuscan*, being 15 Modules; and the longest, the *Composite*, 20. In *Vitruvius*, this Increase is also of 5 Modules; but commences from 14 Modules, and ends up at 19. The Moderns usually make it greater: *Scamozzi* makes it 5 Modules and a half; *Palladio* and *Serlio*, 6.

From the several Proportions of *Columns* assign'd by several Authors, M. *Perrault* has drawn a new one; which is a Mean, between the Extremes of the rest. Thus, he makes the *Tuscan Column* 14 Modules ⅔; which is a Mean between the *Tuscan* of *Vitruvius*, 14, and that of *Trajan's Columns*, 18: The height of the *Doric Column* he makes 16 Modules; which is a Mean between the 14 of *Vitruvius*, and the 19 of the *Coliseum*: The *Ionic* he makes 17 Modules ⅓; which is a Mean beween the 16 of *Serlio*, and the 19 of the *Coliseum*: The *Corinthian Column* he makes 18 Modules ⅔; as being a Medium between the 16 Modules 6 Minutes of the Temple of the *Sibyl*, and the 20 Modules 6 Minutes of the three Columns of the *Roman Forum*: Lastly, the *Composite Column*, by the same Rule, he makes 20 Modules; that height being a Mean between the Arch of *Titus*, and the Temple of *Bacchus*.

Indeed, the Rule he proceeds by seems very reasonable; *viz.* that the progressional Advance of each *Column* in the different Orders be equal:

so that having settled the whole Progression, from the *Tuscan* to the *Composite*, at 5 Modules 20 Minutes; this being a Mean between the 5 Modules of the Antique, and the 5½ of the Moderns; he divides this Sum, which is 160 Minutes, into four equal Parts, giving 40 Minutes to the Progression of each Order: This makes the *Tuscan Column* 14 Modules 20 Minuutes; the *Doric* becomes 16, the *Ionic* 17, 10 Min. the *Corinthian* 18, 20 Min. and the *Composite* 20 Modules. See PROPORTION.

For the distinguishing Characters of each Order of Columns; see ORDER.

COLUMNS, *with regard to their different* Matter.

Water-COLUMN, is a *Column* whose Shaft is form'd of a large *Jet d'Eau*; which spouting out Water violently from the Base, drives it within the Tambour of the Capital, which is made hollow; whence falling down again, it has the Effect of a liquid crystal *Column*. See FOUNTAIN.

An Instance of this we have at *Quinta d'Aveiro* near *Lisbon*.

Fusible COLUMN. Under this Term are comprehended, not only *Columns* of various Metals, and other fusible Matters, as Glass, *etc.* but also those of Stone, said to have been cast; the Secret of which, some will have us believe to have been known to the Antients.

Transparent COLUMN, any *Column* made of transparent Matter; as were those of Crystal in the Theater of *Scaurus*, mention'd by *Pliny*; and those of transparent Alabaster, in the Church of St. *Mark*, at *Venice*.

Hydraulic COLUMN, is that whose Shaft appears to be of Crystal, being form'd by a number of little Threads of Water, falling from Holes made in Girts of Metal, at equal distances, by means of a Pipe mounting thro' the middle thereof; as in the Gardens at *Versailles*.

Hydraulic COLUMN, is also a *Column* from whose top proceeds a *Jet d'Eau*, to which the Capital serves as a Bason; whence the Water descends by a little Pipe, which turns spirally around the Shaft.

Such are the *Ionic Columns* of the Cascade of the *Belvidera* at *Frescati*; and those of the Vineyard *Mathei* at *Rome*.

Molded COLUMN, is that made by Impastation, of Gravel and Flints of divers Colours, bound together with a Cement, which grows perfectly hard, and receives a polish like Marble.

The Secret of making these, it appears the Antients were Masters of, by the *Columns* lately discover'd near *Algiers*; which are doubtless the Ruins of the antient *Julia Cæsarea*: on these are found the very same Inscription in antique Characters, the Contours, Accents, and even Faults, being repeated in every Shaft: an incontestable Proof of their being molded.

COLUMNS *with regard to their* Construction.

COLUMN *of Joinery*, is made of strong timber boards, join'd, glu'd, and pinn'd together; is hollow, turn'd in the Lath, and usually fluted. Such are the *Columns* in most Altar-pieces.

Incrustated COLUMN, is made of several Ribs, or thin Shells of fine Marble, or other rare Stone, cemented upon a Mold of Stone, Brick, or the like. This is done with design both to save the precious Matter, as Oriental Jasper, Lapis, Agat, *etc.* or to represent pieces of such Matters of an extraordinary size, by the neatness and closeness of the Incrustation, which renders the Joints imperceptible.

Geminated COLUMN, that whose Shaft is form'd of three similar and equal Sides, or Ribs of Stone, fitted within one another; and fasten'd at bottom with Iron Pins, and a-top with Cramp-Irons. This is to be fluted, that the Joints may be the less discernible.

COLUMN *of Masonry*, is made of rough Stone, well laid and cover'd with Plaister; or of Bricks molded triangular wise, and cover'd with Stuc.

COLUMN *with Bands, or Tambours*; that whose Shaft is form'd of several Courses of Stone, or Blocks of Marble; less high than the Diameter of the *Column*: This is what *Ulpian* means by *Columna structilis*, or *adpacta*; which is opposite to the *Columna solida*, or *integra*, *i.e.* of one peice.

COLUMN *in Truncheons*, or *Pieces* consists of three, four, or five pieces of Stone, or Metal; differing from the Tambours, as being higher than the Diameter of the *Column*.

COLUMNS, *with regard to their* Form.

Fluted COLUMN, called also *channell'd*, and *striated Column*; that whose Shaft is adorn'd with Flutes, or Channelings; either from top to bottom, or only two thirds of its height. See FLUTES.

Cabled, or *Rudented* COLUMN, is a *Column* having Projectures in form of Cables, or Canes, in the Naked of the Shaft; each Cable having an Effect opposite to that of a Fluting; and accompany'd with a little List on each side. See CABLED-FLUTES.

Fluted and Cabled COLUMN, is that whose Flutes are fill'd up with Cables, Reeds, or Staves; beginning from the bottom of the Shaft, and reaching one third of its height.

Fluted COLUMN *enrich'd*, is that whose Flutings are fill'd up with Ornaments of Foliages, Rinds, Ribbons, *etc.* instead of Cables.

Twisted COLUMN, is that whose Shaft is twisted round, in manner of a Screw, with six Circumvolutions; being, ordinarily of the *Corinthian* Order: *Vignola* first found a Method of drawing it by Rule.

A twisted fluted COLUMN, is that whose Flutes follow the Contour of the Shaft, in a spiral Line throughout the whole length; whereof there are some antique ones of Porphyry, and hard Marble.

COLUMN *twisted and enrich'd*, is that, one third of whose Shaft is fluted, and the rest adorn'd with Branches and other Enrichments: and which being all of Marble, is enrich'd with Sculpture from bottom to top.

Sometimes, again, the *twisted Column* is form'd of two or three slender Shafts, twisted round, so as to leave a Cavity in the middle. Sometimes the Flutings are Spiral, yet the Shaft strait; which succeeds very well in the more delicate Orders.

Colossal COLUMN, a *Column* of enormous size, too large to enter any Ordonnance of Architecture; to be placed solitary, in the middle of a Square, *etc.* See COLOSSUS.

Such is the *Trajan Column*, whose Proportions are *Doric*, and its Profile *Tuscan*; 12 feet and ⅛ in Diameter, and 100 in Height, including the Base and the Capital: The Pedestal has 18 Feet, and the Crowning 16 and an half. It supports a brazen Statue of St. *Peter*, 13 Foot high: The whole making 147 antient *Roman* Feet.

It was built by *Apollodorus*; and consists of 34 Tambours, or Blocks, or Pieces of white Marble, taking in the Crowning.

The *Antonine Column*, which is also of white Marble, is inferior to the *Trajan* in beauty of Sculpture, but exceeds it in height; being 168 Feet to the Capital, besides the Feet of the Pedestal under ground.

Lastly, the *London Column*, or Monument, is of Stone, 15 Foot in Diameter, and 202 high; taking in the Pedestal and Crowning. See MONUMENT.

Carolitic COLUMN, that adorn'd with Foliages, or Leaves and Branches turn'd spirally around the Shaft; or in Crowns and Festoons.

These were used by the Antients for raising Statues on; which hence took the Denomination of *Carolitic*.

They are very suitable in triumphal Arches, and Decorations of Theaters.

Swell'd COLUMNS, are those which have a buldging, or swelling, in proportion to the height of the Shaft.

Authors are much divided on the Subject of this *Swelling*; Sir *H. Wotton* treats it as a most ridiculous Abuse: yet the Practice obtains among the modern Architects; who generally make their *Columns* a little bigger at one third of their height, than at the Base; *i.e.* they diminish the *Column* near the Base; which makes the Part appear big, and occasions a Swelling at about one third of the height.

This Swelling appears to have been unknown to the Antients. M. *le Clerc* observes, it ought not to exceed one Minute and a half at most. He thinks it ought never to be us'd, excepting where there is a particular occasion for it; as where *Columns* are to be placed over one another.

Diminish'd COLUMN, is that which has no swelling, but begins to taper, or diminish, from the Base; in imitation of Trees. See DIMINUTION.

Such are most of the Antique *Columns* of *Granate*; particularly the *Corinthian* ones in the Porch of the Pantheon.

Gothic COLUMN, a round Pillar, either too short for its bulk, or too slender for its height; as having, sometimes, 20 Diameters; and this without either Diminution, or Swelling: yet, its Ornaments, and the

Characters of its work, being as far from those of the Antique as its Proportions. See GOTHIC.

Hermetic COLUMN, a kind of Pilaster, in manner of a Terminus; having the Head of a Man, in lieu of a Captial. It had its Name from a Custom among the Antients, of placing *Mercury's* Head, whom the *Greeks* call *Hermes*, a-top of *Columns*.

Massive COLUMNS, are those too short for the Order whose Capital they bear; as the Pillars in *Gothic* Churches usually are. See MASSIVE.

Under the Term *Massive*, are, likewise, frequently comprehended *Rustic* and *Tuscan Columns*. See RUSTIC.

Cylindrical COLUMN, is that which has neither Swelling nor Diminution.

Oval COLUMN, that whose Shaft has a Flatness; its Plan being made oval, to reduce the Projecture.

Polygonous COLUMN, has several Sides, or Faces: The most Regular of these have eight Faces.

These three last, *Daviler* regards as Abuses in Architecture.

Pastoral COLUMN, that whose Shaft is form'd in imitation of the Trunk of a Tree, with Bark and Knots.

This Kind of *Column*, in the *Tuscan* Proportion, may be used in the Gate of Parks and Gardens; in the Decoration of Pastoral Scenes, *etc*.

Serpentine COLUMN, a *Column* form'd of three Serpents, twisted together; the Heads whereof serve as a Capital.

An Instance of this is at *Constantinople*, in the Square called *Atmeidan*, antiently the *Hippodrome*. *P. Gilles* calls this the *Delphic Column*; as imagining it antiently serv'd for the Tripod of *Apollo*, in the Temple at *Delphos*. It is now ordinarily called the *Talisman*, or *Enchanted Column*.

COLUMNS, *with regard to their* Disposition.

Insulated COLUMN, is that standing free, and detach'd, on all Sides, from any other Body.

COLUMN *inserted*, or *back'd*, is that attach'd to a Wall, by a third or fourth Part of its Diameter.

COLUMN *Nich'd*, is that whose Shaft enters, with half its Diameter, into a Wall, which is hollowed for its Reception; with its Plan parallel to the Projecture of the Tore.

Such is that in the Portal of St. *Peter* at *Rome*.

Angular COLUMN, is an insulated *Column*, placed in the Coin, or Corner of a Portico; or inserted into the Corner of a Building: Or, even a *Column* that flanks an Angle, either acute or obtuse, of a Figure of many Sides.

Attic COLUMN, according to *Pliny*, is a *Pilaster* insulated; having four equal Faces or Sides; and of the highest Proportion, *v. g. Corinthian*.

Flank'd COLUMN, according to M. *Blondel*, is a *Column* engag'd with one half, or at least, one third of its Diameter, between two Demi-

Pilasters.

Doubled COLUMN, is an Assemblage of two *Columns*; join'd in such a manner, as that the two Shafts penetrate each other with a third of their Diameter. Such are those of the four Angles in the Court of the *Louvre.*

Coupled COLUMNS, are those disposed by two and two; so as almost to touch each other at their Bases, and Capitals.

Canton'd COLUMNS, those engag'd in the four Corners of a square Pillar, to support four Springs of an Arch.

Grouped COLUMNS, those placed on the same Pedestal, or Socle; either by three and three, or by four and four.

Median COLUMNS. *Vitruvius* gives the Name *Columnæ Medianæ*, to the two *Columns* in the middle of a Porch, which have their Intercolumniation larger than the rest: So that if these last, for instance, be *Pycnostyles*; the *Medianes* are *Eustyle.*

The Term may also be applied to the middle Row of *Columns*, in a Frontispiece adorn'd with three Orders.

COLUMNS, *with regard to their* Use.

Astronomical COLUMN, is a kind of Observatory, in form of a very high Tower; built hollow, and with a spiral Ascent to an Armillary Sphere; placed a-top, for observing the Courses of the heavenly Bodies.

Such is that, of the *Doric* Order, erected at the *Hotel de Soissons* at *Paris*, by *Catherine de Medicis*, for the Observations of *Orontius Fineus*, a celebrated Astronomer of that Time.

Chronological COLUMN, that which bears some historical Inscription, digested according to the Order of Time: as by Lustra, Olympiads, Fasti, Epochas, Annals, *etc.*

At *Athens* there were *Columns* of this kind, whereon were inscrib'd the whole History of *Greece*, digested into Olympiads.

Hollow COLUMN, that which has a spiral Stair-case within-side, for the Convenience of ascending to the top: As the *Trajan Column*, the Stair-case whereof consists of 185 Steps, and is illumin'd by 43 little Windows: The *Antonine Column* has 198 Steps, with 56 Windows; each of these is Cut in Tambours of white Marble. The *Monument*, or *Fire-Column* at *London*, has also a Stair-case; but it does not reach to the top.

These kind of *Columns* are called also *Columnæ Cochlides*, or *Cochlidea*, *q.d.* a Stair-case in form of a Snail.

Funeral COLUMN, that which bears an Urn, wherein are supposed to be inclos'd the Ashes of some deceas'd Hero; and whose shaft is sometimes overspread with Tears, or Flames; which are Symbols of Sorrow, and of Immortality.

Gnomonic COLUMN, a Cylinder, whereon the Hour of the Day is represented by the Shadow of a Style.

Of these there are two Kinds: in the one, the Style is fix'd, and the Hour-Lines are no more than the Projection of a vertical Dial, on a cylindrical Surface.

In the other, the Style is movable; and the Hour-Lines are drawn to the different Heights of the Sun, in the different Seasons of the Year.

Historical COLUMN, is that whose Shaft is adorn'd with a Basso-Relievo, running in a spiral Line its whole length; and containing the History of some great Personage.

Such are the *Trajan* and *Antonine Columns*, at *Rome*.

Historical Columns may likewise be divided by Bands, or Tambours, into separate Basso-Relievos, containing distinct Subjects; by which means, the same *Columns* may likewise be made to answer the End of Chronological ones.

This Manner *Vignola* prefers to the former; which, he thinks, appears to much confused.

Indicative COLUMN, that which serves to shew the Tides, *etc*. along the Sea-Coasts. Of this kind, there is one at *Grand Cairo*, of Marble, whereon the Overflowings of the *Nile* are express'd: By this they form a Judgment of the succeeding Season: when the Water, for instance, ascends to 23 Foot, 'tis a Sign of great Fertility in *Egypt*.

Instructive COLUMN, that rais'd, according to *Josephus*, Lib. I. cap. 3. by the Sons of *Adam*, whereon were engraven the Principles of Arts and Sciences.

Baudelot tells us, that the Son of *Pisistratus* rais'd another of this kind, of Stone; containing the Rules and Precepts of Agriculture.

Itinerary COLUMN, a *Column* with several Faces, placed in the Cross-ways in large Roads; serving to shew the different Routs, by the Inscriptions thereupon.

Lactatory COLUMN, at *Rome*, according to *Festus*, was a *Column* erected in the Herb-Market, now the *Platea Montanara*; which had a Cavity in its Pedestal, wherein young Children, abandon'd by their Parents, out of Poverty or Inhumanity, were exposed, to be brought up at the publick Expence.

Legal COLUMN. Among the *Lacedemonians*, there were *Columns* rais'd in publick Places, whereon were engraven the fundamental Laws of the State.

Limitrophous, or *Boundary* COLUMN, that which shews the Limits of a Kingdom, or Country conquer'd. Such was that, which, *Pliny* says, *Alexander* the Great erected at the Extremities of the *Indies*.

As to those of *Hercules*, ordinarily call'd his *Columns*, or *Pillars*; they are only two very steep Mountains in the Straights of *Gades*, now *Gibraltar*.

Luminous COLUMN, a sort of *Column* form'd on a cylindrical Frame, mounted and cover'd over with oil'd Paper, or Gawze; so that Lights being dispos'd in ranks over each other, the whole appears as on fire.

This sort of *Column* is likewise made with rows of Lamps, or Torches, running round its Shaft; either in horizontal Belts, or Bands; or in a spiral Line, continued over a Festoon of Flowers.

Manubiary COLUMN, from the Latin *Manubiæ*, Spoils of the Enemy;

a *Column* adorn'd with Trophies, built in imitation of Trees, whereon the spoils of Enemies were antiently hung.

Memorial COLUMN, that rais'd on occasion of any remarkable Event; as the *Monument* in *London*, built to perpetuate the Memory of the Burning of that City, in 1666.

It is of the *Doric* Order, fluted, hollow, with a winding Stair-case; and terminated a-top with waving Flames.

There is, also, another of the like Kind, in form of an Obelisk, on the Banks of the *Rhine* in the *Palatinate*, in memory of the famous Passage of that River, by the great *Gustavus Adolphus* and his Army.

Menian COLUMN, any *Column* which supports a Balcony, or Meniane.

The Origin of this kind of *Column*, *Suetonius* and *Ascanius* refer to one *Menias*; who having sold his House to *Cato* and *Flaccus*, Consuls, to be converted into a publick Edifice; reserv'd to himself the Right of rasing a *Column* without side, to bear a Balcony; whence he might see the Shews.

Military COLUMN, among the *Romans*, a *Column* whereon was engraven a List of the Forces in the *Roman* Army, ranged by Legions, in their proper order; with design to preserve the Memory of the Number of Soldiers, and of the Order observ'd in any Military Expedition.

The *Romans* had another kind of *Military Column*, which they called *Columna Bellica*, standing before the Temple of *Janus*; at the Foot whereof the Consul declar'd War, by throwing a Javelin towards the Enemies Countries.

Miliary COLUMN, was a *Column* of Marble, rais'd by order of *Augustus*, in the middle of the *Roman Forum*; from whence, as a Centre, the Distances of the Several Cities, *etc.* of the Empire were reckon'd, by other *Miliary Columns*; dispos'd at equal distances, on all the grand Roads.

This *Column* was of white Marble; the same with that which is now seen on the Balustrade of the Perron of the Capital at *Rome*.

Its proportion is massive; being a short Cylinder, with a *Tuscan* Base and Capital, and a Brass Ball for a Crowning, the Symbol of the Globe of the Earth.

It was called *Miliarium Aureum*; as having been gilt, at least the Ball, by order of *Augustus*. It was restor'd by the Emperors *Vespasian* and *Adrian*; as appears by the Inscriptions.

Phosphorical COLUMN, a *Lighthouse*; or a hollow *Column*, built on a Rock, or the Tip of a Mole, or other Eminence, to serve as a Lanthorn to a Port.

Rostral COLUMN, that adorn'd with the Beaks, or Prows of Ships and Galleys, with Anchors and Grapplers; erected, either in memory of a Naval Victory; as the *Tuscan Column*, in the Capitol: or, in honour of some Admiral; as the *Doric* ones, at the Entrance of the Castle *de Richlieu*.

Sepulchral COLUMN, antiently, was a *Column* erected on a Tomb or Sepulchre; with an Inscription on its Base. See TOMB, *etc.*

Those over the Tombs of Persons of Distinction were very large; those for the common People small: these last are called *Stelæ* and *Cippi.*

Statuary COLUMN, that which supports a Statue. Such was that erected by Pope *Paul* V. on a Pedestal before the Church of St. *Maria major* at *Rome*; to support a Statue of the Virgin, which is of gilt Brass. See STATUE.

This *Column* was dug up in the Temple of *Peace*; its Shaft is a single Block of white Marble 49 Foot and an half high, and 5 Foot 8 Inches Diameter; of the *Corinthian* Order, fluted.

The Term *Statuary Column*, may likewise be apply'd to Caryatides, Persians, Termini, and other human Figures, which do the Office of *Columns*; and which *Vitruvius* calls *Telamones*, and *Atlantes.* See CARYATIDES, *etc.*

Symbolical COLUMN, is a *Column* representing some particular Country, by the Attributes proper thereto: As that of the *French* Order, set with Flower de Lis's, in the Frontispiece of the Jesuits Church at *Rouen*: or some memorable Action; as the *Corvinian Column*, on which was a Crow; erected to *Valerius Maximus*, Sirnam'd *Corvinus*, in memory of his Defeat of a Giant in the Army of the *Gauls*, by the Assistance of a Crow.

Under the Title of *Symbolic Columns*, may also be comprehended those which serve for Symbols. Such is that in a Medal of *Nero*, which expresses the Stability of the *Roman* Empire.

Triumphal COLUMN, a *Column* erected among the Antients in honour of an Hero; the Joints of the Stones, or Courses whereof, were cover'd with as many Crowns, as he had made different Military Expeditions.

Each Crown had its particular Name; as *Palissaris*, which was set with Spikes, in memory of having forced a Palisade. *Muralis*, adorned with little Turrets, or Battlements; for having mounted to an Assault. *Navalis*, of Prows and Beaks of Vessels; for having overcome at Sea. *Obsidionalis*, or *Graminalis*, of Grass; for having rais'd a Siege. *Ovans*, of Myrtle; which express'd an Ovation, or little Triumph: And *Triumphalis*, of Laurel; for a grand Triumph. See CROWN.

Procopius tells us of a *Column* of this Kind, erected in the place called *Augusteum*, before the Imperial Palace of *Constantinople*, supporting an *Equestrian* Statue of the Emperor *Justinian*.

Zophoric COLUMN, a kind of Statuary *Column*, whereon is placed a Figure of some Animal. Such is one of the two *Columns* of the Gate of *Venice*; whereon is the Lion of St. *Mark*, and the Arms of the Republick: or that at *Sienna*, which bears the Wolf that suckled *Romulus* and *Remus*.

COMPARTIMENT, or COMPARTMENT, a Design compos'd of several different Figures, dispos'd with Symmetry; to adorn a Parterre,

Plafond, Panes of Glass, or Pannels of Joinery, the Squares of a Cieling, *etc*.

A Compartiment of Tiles, is an Arrangement of white and red Tiles varnish'd, for the Decoration of the Covering of a Roof.

The Term *Compartiment* is also used in Painting: Thus, we say, all the *Turkish* and *Moorish* Paintings are only *Compartiments*: The fine Bindings of Books are in *Compartiments*, etc.

COMPARTITION, in Architecture, the useful and graceful *Distribution* of the whole Ground-Plot of an Edifice, into Rooms of Office, of Reception, or Entertainment. See HOUSE, *etc.*

Compartition makes one of the great Divisions of the Art of Building. See BUILDING.

COMPOSITE *Order*, in Architecture, the last of the five Orders of Columns; so called, because its Capital is *compos'd* out of those of the other Columns. See ORDER.

It borrows a Quarter-round from the *Tuscan*, and *Doric*; a double Row of Leaves, from the *Corinthian*; and Volutes from the *Ionic*: Its Cornice has simple Modillions or Dentils. See CAPITAL.

The *Composite* is also called the *Roman* and *Italic* Order; as having been invented by the *Romans*; conformably to the rest, which are denominated from the People among whom they had their rise. See its Figure in Tab. *Architecture.*

Most Authors rank this after the *Corinthian*; either as being the richest, or as the last that was invented: *Scamozzi* alone places it between the *Ionic* and *Corinthian*; out of a view to its delicacy and richness, which he esteems inferior to that of the *Corinthian*; and therefore makes no scruple to use it under the *Corinthian*: wherein he is follow'd by M. *le Clerc*. See CORINTHIAN.

The Proportions of this Order are not fix'd by *Vitruvius*; he only marks its general Character, by observing that its Capital is compos'd of several Parts taken from the *Doric*, *Ionic*, and *Corinthian*: He does not seem to regard it as a particular Order; nor does he vary it at all from the *Corinthian*, except in its Capital. In effect, it was *Serlio* who first added the *Composite* Order to the four of *Vitruvius*, forming it from the Remains of the Temple of *Bacchus*, the Arches of *Titus*, *Septimius*, and the Goldsmiths: Till then, this Order was esteem'd a Spécies of the *Corinthian*, only differing in its Capital.

The Order being thus left undetermin'd by the Antients, the Moderns have a kind of a Right to differ about its Proportions, *etc*. *Scamozzi*, and after him M. *le Clerc*, make its Column 19 Modules and an half; which is less by half a Module than that of the *Corinthian*: as, in effect, the Order is less delicate than the *Corinthian*. *Vignola* makes it 20; which is the same with that of his *Corinthian*: but *Serlio*, who first form'd it into an Order, by giving it a Proper Entablature and Base, and after him

M. *Perrault*, raise it still higher than the *Corinthian*.

This last does not think different Ornaments and Characters sufficient to constitute a different Order, unless it have a different Height too: Agreeably, therefore, to his Rule of augmenting the Heights of the several Columns by a Series of two Modules in each; he makes the *Composite* 20 Modules, and the *Corinthian* 18; which, it seems, is a Medium between the Porch of *Titus* and the Temple of *Bacchus*. See PROPORTION.

For the parts of the Order, see COLUMN, ENTABLATURE, CAPITAL, BASE, *etc*.

M. *Perrault*, in his *Vitruvius*, distinguishes between the *Composite* and *Composed* Order.

The latter, he says, is any Composition whose Parts and Ornaments are extraordinary and unusual; but have, withal, somewhat of Beauty; both on account of their Novelty, and in respect of the Manner or Genius of the Architect: So that a *Composed* Order is an arbitrary, humourous Composition, whether regular, or irregular.

The same Author adds, that the *Corinthian* Order is the first *Composite* Order, as being compos'd of the *Doric* and *Ionic*; which is the Observation of *Vitruvius* himself, *Lib*. 4. cap. 1.

COMPOSITION, in a Physical Sense, is the uniting or joining of several different Things, so as to form one whole, call'd a *Compound*.

The Schoolmen distinguish two Kinds of *Composition*; the one *Entitative*, which is between Things of the same Nature, *e.g*. two or more Drops of Water: the other *Essential*, when Things of different Kinds are join'd, and thus constitute new Things, or Essences, different from any of the Parts; and thus, say they, from the Matter and the Form of Wood, arises Wood; whose Essence is very different from either of those Ingredients taken separately.

COMPOSITION *of Ideas*, is an Operation of the Mind, whereby it combines several of its Simple Ideas into Complex ones.

Under the same Operation may likewise be reckon'd that of enlarging; whereby we put several Ideas together of the same Kind, as several Unites to make a Dozen.

COMPOSITION, in Painting, includes the Invention and Disposition of the Figures, the Choice of Attitudes, etc. *Composition*, therefore, consists of two Parts; one of which finds out, by means of History, proper Objects for a Picture; and the other disposes 'em to advantage.

CONGE, in Architecture, a Moulding either in form of a Quarter-Round, or of a Cavetto; which serves to separate two Members from one another.

Such is that which joins the Shaft of the Column to the Cincture, call'd also Apophyge; which, in Greek, signifies *Flight*; the Column seeming to arise hence: By the *Latins* it is called *Scapus*, the Shaft of the Column. See APOPHYGE.

CONSERVATORY, in Gardening. See GREEN-*Houses*.

CONSOLE, in Architecture, or Joinery, a Part or Member projecting in manner of a Bracket, or Shoulder-piece; serving to support a Corniche, Bust, Vase, Beam, little Vault, *etc*. See PROJECTURE.

These are also, upon occasion, call'd *Mutules*, *Modillions*, etc. according to their form. See MUTULE, *etc*.

Some of them are striated, or fluted; others in form of *Cartouches*; others have Drops, in the manner of Triglyphs.

Vitruvius calls all those of Gates *Prothyrides*; of *Thyra*, Gate. See PROTHYRIS.

Those made of the End of a Plank of Wood, cut triangular-wise, are called *Ancones*.

Consoles are frequently used as Keys of Arches, projecting out, to support a Vase, or other Ornament. See KEY.

The Word is deriv'd from the French *Consolider*, to re-unite, join, *etc*. agreeable to the Office of this Member.

CONTRAMURE, in Building, an Out-wall, built about the Wall of a City. See WALL, RAMPART, *etc*.

CONTRAST, in Painting and Sculpture, expresses an *Opposition*, or difference of Position, Attitude, *etc*. of two or more Figures; contriv'd to make Variety in a Painting. See DESIGN.

Thus when in a Group of three Figures, one is shewn before, another behind, and a third sidewise, there is said to be a *Contrast*. See GROUP.

M. *de Piles* defines *Contrast* an Opposition between the Lines which form the Objects; by means whereof they set off one another.

A *Contrast* well manag'd, is one of the greatest Beauties of a Painting.

The *Contrast* is not only to be observ'd in the Position of several Figures, but also in that of the several Members of the same Figure: Thus, if the right Arm advance the furthest, the right Leg is to be hindmost; if the Eye be directed one way, the right Arm to go the contrary way, *etc*. See FIGURE.

The *Contrast* must be pursued even into the Drapery. See DRAPERY.

To CONTRAST, in Architecture, is to avoid the Repetition of the same Thing, in order to please by Variety; as is done in the great Gallery of the *Louvre*, where the Pediments are, alternately, arch'd and angular.

The Word comes from the Italian *Contrastare*, to oppose, thwart; and that, according to M. *Huet*, from the Latin *Contrastatio*.

CONTRAVALLATION, or the *Line of* CONTRAVALLATION, in

Fortification, is a Trench guarded with a parapet; usually cut round about a Place by the Besiegers, to secure themselves on that side, and to stop the Sallies of the Garrison.

It is without Musket-shot of the Town; so that the Army forming a Siege, lies between the Lines of *Circumvallation* and *Contravallation*.

CONVENT, a Monastery of Religious, of the one or the other Sex. See MONASTERY.

The Word comes from the Latin *Conventus*, Meeting of *Convenire*.

COPING of a Wall, the Top or Cover of a Wall, made sloping to carry off the wet. See Wall.

COPING over, in Carpentry, a sort of hanging over, not square to its upright, but bevelling on its under Side till it end in an Edge.

CORBEILS, in Fortification, little Baskets, about a Foot and an half high, eight Inches broad at the bottom, and twelve at the top; which being fill'd with Earth, are frequently set one against another upon the Parapet, or elsewhere; leaving certain Port-holes, from whence to fire upon the Enemy under covert, without being seen by them.

CORBEL, in Architecture, the Representation of a Basket, sometimes seen on the Heads of Cariatides. See CARYATIDES.

The Word is also used for the Vase or Tambour of the *Corinthian* Column; so call'd from its resemblance of a Basket; or because it was first form'd on the Model of a Basket. See CORINTHIAN *Order*.

CORBELL, CORBIL, or CORBEL, is also used in Building for a short Piece of Timber placed in a Wall, with its End sticking out six or eight Inches, as occasion serves, in manner of a Shouldering-piece.

The under Part of the End thus sticking out, is sometimes cut into the form of a Boultin; sometimes of an Ogive, and sometimes of a Face, *etc.* according to the Workman's Fancy; the upper side being plain and flat.

These *Corbels* are usually placed for Strength immediately under the Semi-girders of a Platform, and sometimes under the Ends of Chamber-Beams: in which latter Case, they are commonly placed a Foot or two below the Beam, and have a piece of Timber standing upright close to the Wall from the *Corbel* to the Beam.

CORBEL is also used by some Architects for the hollow Niches, or hollows left in Walls for Images, Figures, or Statues to stand in. See NICHE.

CORINTHIAN *Order*, the fourth, or as *Scamozzi*, and M. *le Clerc* make it the fifth and last of the Orders of Architecture; being the noblest, richest, and most delicate of all others. See ORDER.

The Invention of this Order most of the Moderms, after *Vitruvius*, ascribe to *Callimachus* a *Corinthian* Sculptor, who passing by the Tomb of a young Lady, over which her Nurse had placed a Basket with some of her Play-things, and cover'd it up from the Weather with a Tile; the whole having been plac'd on a Root of Acanthus; as it sprung up, the Branches encompass'd the Basket, and bending down atop under the Corners of the Tile, form'd a kind of Volutes.

Hence *Callimachus* took his hint: The Basket he imitated in the Vase of his Column; the Leaves in the Volutes; and the Tile in the Abacus of his Order. See ABACUS, ACANTHUS, *etc.*

Villalpandus treats this Story of *Callimachus* as a Fable; and will have the *Corinthian* Capital to have taken its Origin from an Order in *Solomon's* Temple, the Leaves whereof were those of the Palm-Tree.

The *Corinthian* Order has several Characters whereby it is distinguish'd from the rest: Its Capital is adorn'd with two Rows of Leaves, between which rise little Stalks, or Caulicoles, whereof the Volutes are form'd, which support the Abacus, and which are in Number 16.

It has no Ovolo, nor even Abacus properly speaking; for the Member which goes by that Name is quite different from the Abacus in the other Orders, being cut with a Sweep, in the middle of which is carved a Rose, or other Ornament. See CAPITAL, VOLUTE, ABACUS, CAULICOLES, *etc.*

CORNICHE, or CORNICE, in Architecture, the uppermost Member of the Entablature of a Column; or that which crowns the Order. See ENTABLATURE, CROWNING, *etc.*

The *Cornice* is the third grand Division of the Trabeation, commencing from the Frize and ending with the Cymatium.

The Word is form'd from the Latin *Coronis*, a Crowning.

The *Cornice* is different in the different Orders: In the *Tuscan* Order it is the most plain. *Vignola* makes it to consist of an Ovum or Quarter-round, and Astragal or Baguette, the Reglet or Fillet, the Larmier, and the Talon. *See* TUSCAN.

In the *Doric*, he makes Capitals of the Triglyphs of the Frize with their Bandilettes, a Talon, Mutules or Dentils, a Larmier with its *Guttæ* underneath, a Talon, Fillet, Cavetto, and Reglet. See DORIC.

In the *Ionic*, the Members are in most respects the same as in the *Doric*; except that they are frequently enrich'd with carving, and there are always Dentils. See IONIC.

In the *Composite* there are Dentils; its Mouldings are carv'd, and there are Channels under the Soffit. See COMPOSITE.

The *Corinthian Cornice* is the richest; and is distinguish'd by having both Modillions and Dentils: contrary to the Opinion of *Vitruvius*, who looks on those two Ornaments as incompatible; and of M. *le Clerc*, who regards the Dentils as peculiar to the *Ionic*. See MODILLIONS, DENTICLES, *etc.* see also ORDER.

For the Heights and Projectures of the Cornices in the several Orders;

Goldman makes the Height of the *Tuscan* $1\frac{1}{3}$, and its Projecture, $2\frac{2}{5}$ Modules: The Height of the *Doric* $1\frac{1}{3}$, its Projecture $2\frac{2}{5}$; Height of the *Ionic* $1\frac{3}{5}$, its Projecture $2\frac{2}{5}$; Height of the *Composite* $1\frac{3}{5}$, Projecture $2\frac{13}{30}$; Height of the *Corinthian* $1\frac{2}{5}$, Projecture $2\frac{13}{30}$.

CORNICE is used in the general for all little Projectures of Masonry or Joinery; even where there are no Columns.

Thus, we say, the *Cornice* of a Chimney, a Buffet, *etc*.

Architrave CORNICE, is that immediately contiguous to the Architrave; the Frize being retrench'd.

Mutilated CORNICE, is that whose Projecture is cut or interrupted, to the right of the Larmier, or reduced into a Platband with a Cimaise.

Cantaliver CORNICE, a Term used by the Workmen for a *Cornice* that has Cantalivers underneath it. See CANTALIVERS.

Modillion CORNICE, a *Cornice* with Modillions under it. See MODILLIONS.

Coving CORNICE, a *Cornice* which has a great Casement or hollow in it; ordinarily lathed and plaister'd upon Compass, Sprockets, or Brackets.

CORNICE is also used for the Crownings of Pedestals. See PEDESTAL.

This *Cornice*, too, is different in the different Orders: In the *Tuscan*, according to M. *Perrault*, it has a Platband which serves as a *Corona*, and a Cavetto with its Fillet: In the *Doric*, it has a Cavetto with a Fillet, which bears a Drip crown'd with a Square: In the *Ionic*, a Cavetto with its Fillet above, and a Drip or hanging Square crown'd with an Ogee and its Fillet: In the *Corinthian*, an Ogee with its Fillet, a Cymatium under the *Corona*, which it hollows to make a Drip, a Corona, and an Ogee with its Fillet: Lastly, in the *Composite*, a Fillet with a Sweep over the Die, an Astragal, Cyma with its Fillet, *Corona*, and Ogee with its Fillet.

CORONA, *Crown*, or *Crowning*, in Architecture, a large flat strong Member of the Cornice; so called, because it crowns not only the Cornice, but the Entablature, and the whole Order. See ENTABLATURE.

The *French* call it the *Larmier*, our Workmen the *Drip*, as serving, by its great Projecture, to skreen the rest of the Building from the Rain. See LARMIER.

Some *Latin* Authors call it *Supercilium*; but, as it shou'd seem, by mistake for *Stillicidium*. Some *French* Authors call it *Mouchette*; some *Latin* ones *Mentum*, Chin; from its keeping off the Weather from the Parts underneath, as the Chin does the Sweat, *etc*. out of the Neck.

Some call it absolutely the *Cornice*, as being the principal Member thereof. *Vitruvius* frequently uses the Word *Corona* for the whole Corniche. See CORNICHE.

The *Corona* is it self *crown'd* or finish'd with a Reglet, or Fillet.

There are sometimes two *Corona's* in a Cornice; as in the *Corinthian* of the *Rotunda*.

CORRIDOR, in Fortification, a Road or Way along the Edge of the Ditch, withoutside; incompassing the whole Fortification.

It is also called the *Covert way*; because cover'd with a Glacis, or Esplanade, serving it as a Parapet. See COVERT Way.

The *Corridor* is ordinarily about 20 Yards broad.

The Word comes from the Italian *Coridore*, or the Spanish *Coridor*.

CORRIDOR is also used in Architecture for a Gallery, or long Isle, round a Building, leading to several Chambers at a distance from each other.

CORTIN, in Fortification, signifies the Wall, or Distance, between the Flanks of two Bastions.

COTAGE, or COTTAGE, a House without Land belonging to it, Stat. 4 Edw. I. By a later Statute, 31 Eliz. no Man may build a House without laying four Acres of Land to it: So that, properly, a *Cotage* now is a House without four Acres of Land to it.

COUCH, in Painting, is used for each Lay or Impression of Colour, whether in Oil or Water, wherewith the Painter covers his Canvas, Wall, Wainscot, or other Matter to be painted.

Thus, we say, A Painting has had its last *Couch*, or Lay; a Ceiling has had two *Couches*. There is a great deal of Art in *couching* the Colours well, one after another.

The Word is also us'd for a Lay or Impression on any thing, to make it more firm and consistent, or to skreen it from the Weather.

Thus, Paintings are cover'd with a *Couch* of Varnish: A Canvas to be painted, must first have two *Couches* of Size ere the Colours be laid. Two or three *Couches* of white Lead, are laid on Wood ere the Couch of Gold be applied: The Leather-gilders lay a *Couch* of Water and Whites of Eggs on the Leather, ere they apply the Gold or Silver Leaf.

The Gold Wire-drawers use the Word *Couch* for the Gold or Silver Leaf wherewith they cover the Mass to be gilded or silver'd, before they draw it thro' the Iron that is to give it its proper thickness. See WIRE-DRAWING.

The Gilders use *Couch* for the Quantity of Gold or Silver Leaves applied on the Metals in gilding or silvering. Each *Couch* of Gold is but one Leaf, or two at most, and each of Silver three; to gild: If the Gilding be hach'd, there are requir'd from eight to twelve *Couches*, and only three or four if it be without haching. To Silver there are requir'd from four to ten *Couches*, according to the Beauty of the Work.

COUNTER-*Approaches*, in Fortification, Lines or Works made by the Besieg'd, when their Trenches meet the Lines of Attack of the Besiegers.

COUNTER-*Battery*, a Battery on the Side opposite to that of the Enemy; especially that which aims to dismount the Enemy's Cannon.

COUNTER-GUARD, in Fortification, a triangular Work, in form of a large Parapet, placed beyond the Ditch, before the Point and Faces of the Bastion.

It differs from the Demi-lune, in that it covers the whole Bastion. The Engineers ordinarily call it the *Enveloppe*.

'Tis chiefly us'd when the Bastion is on an Eminence; and 'tis by means hereof that Bastions may be doubled and tripled. *Counter-guards* are very fit for covering the weak Parts of a Place.

COUNTER-DRAWING, in Painting, *etc.* the copying a Design or Painting, by means of a fine linen Cloth, an oil'd Paper, or other transparent Matter; whereon the Strokes appearing thro', are follow'd and traced with a Pencil, with or without Colour.

Sometimes they *counterdraw* on Glass, and with Frames or Nets divided into Squares, with Silk or with Thread; and also by means of Instruments invented for the Purpose; as the Parallelogram.

COUNTER-LIGHT, or *Counter-jour*, a Window or Light opposite to any thing, which makes it appear to a disadvantage. A single *Counter-light* is sufficient to take away all the Beauty of a fine Painting.

COUNTERFORTS, *Buttresses*, *Spurs*, are Pillars of Masonry, serving to prop or sustain Walls or Terrasses, subject to buldge, or be thrown down. See BUTTRESS.

These Works are usually bent into Arches, and placed at a distance from each other.

When any thing is built on the Descent of a Mountain, it must be strengthen'd with *Counterforts* well bound to the Wall, and at the Distance of about twelve Yards from each other.

COURSE, in *Architecture*, a continued Range of Stones, level, or of the same height, throughout the whole length of the Building; without being interrupted by an Aperture. See BUILDING, WALL, MASONRY, *etc.*

A *Course of Plinths*, is the continuity of a Plinth of Stone or Plaister, in the Face of a Building; to mark the Separation of the Stones. See PLINTH.

COVERING, in Architecture, one of the principal Parts of a Building. See ROOF.

COVING, in Building. When Houses are built projecting over the

Ground-plot, and the turn'd Projecture arch'd with Timber, lath'd and plaster'd; the Work is call'd *Coving*. See ALCOVE.

COVING *Corniche*, is a Corniche with a great Casement, or Hollow therein. See CORNICHE.

CRANE, a Machine used in Building, for the raising large Stones and other Weights.

M. *Perrault*, in his *Notes on* Vitruvius, makes the *Crane* the same with the *Corvus*, *Crow* of the Antients.

The modern *Crane* consists of several Members, or Pieces, the principal whereof is a strong perpendicular Beam, or Arbor, firmly fix'd in the Ground, and sustain'd by eight Arms, coming from the Extremities of four Pieces of Wood laid across, thro' the Middle whereof the Foot of the Beam passes. About the middle of the Arbor the Arms meet, and are mortais'd into it: its Top ends in an Iron Pivot, whereon is bore a transverse Piece, advancing out to a good distance in manner of a *Crane's* Neck; whence the Name.

The middle and Extremity of this is again sustain'd by Arms from the middle of the Arbor; and over it comes a Rope, or Cable, to one end whereof the Weight is fix'd; the other is wound round the Spindle of a Wheel, which turn'd, draws the Rope, and that heaves up the Weight; to be afterwards applied to any Side or Quarter, by the mobility of the transverse Piece on the Pivot.

CREUX, a Term in Sculpture, much us'd by the *French*; tho not yet, that we know of, naturaliz'd among us: But the want of a Word of equal import in *English*, as it has frequently put us under a necessity of using this in the Course of the present Work; so it pleads strongly for its Admission into our Language.

The Word originally signifies a *Hollow*, *Cavity*, or *Pit*, out of which something has been scoop'd, or dug: The Term is used particularly for that kind of Sculpture and Graving, where the Lines and Figures are cut, and form'd within the Face, or Plan of the Plate or Matter engraven: In which Sense, it stands in opposition to *Relievo*; where the Lines and Figures are imboss'd, and rise prominent above the Face of the Matter.

CROSSETTE, in Architecture, the Returns in the Corners of Chambranles, or Door-Cases, or Window-Frames; call'd also *Ears*, *Elbows*, *Ancones*, *Prothyrides*.

Crossette of a Luthern, the Plaister or covering near a Luthern.

CROSS-GRAIN'D *Stuff*, in Joinery. Wood is said to be *cross-grain'd*, when a Bough or Branch has shot out of it: for, the Grain of the Branch shooting forward, runs athwart that of the Trunk. In Wood well grown this Defect is scarce perceivable, except in working; but in Deal-Boards these Boughs make Knots. If the Bough grew up with the young

Trunk, instead of a Knot is found a *curling in the Stuff*; very sensible under the Plane.

CROWN, in Architecture, the uppermost Member of the Cornice; call'd also *Corona* and *Larmier*. See CORONA, and LARMIER.

CROWN-POST, in Architecture, a Post which in some Buildings stands upright in the middle, between two principal Rafters.

CROWN-WORK, or CROWNING, in Fortification, Outworks running in the Campagne, to keep off the Enemy, to gain some Hill, or advantageous Post, and to cover the other Works of the Place. See OUT-WORKS.

It consists of two Demi-Bastions at the Extremes, and an entire Bastion in the middle, with Courtines.

CROWNING, is understood, in the general, of any thing that terminates, or finishes a Decoration of Architecture: Thus, a Cornice, a Pediment, Acroteria, *etc.* are call'd *Crownings*. See ACROTERIA, *etc.*

Thus, also, the Abacus is said to *crown* the Capital; and thus any Member or Moulding is said to be *crown'd*, when it has a Fillet over; and a Niche is *crown'd*, when it is cover'd with a Capital. See MOULDING, NICHE, *etc.*

CRYPTA, a subterraneous Place, or a Vault; especially under a Church, for the Interment of particular Families, or Persons. See TOMB.

S. *Ciampini*, describing the Outside of the *Vatican*, speaks of the *Cryptæ* of St. *Andrew*, St. *Paul*, etc. See CATACOMBS.

Vitruvius uses the Word for a Part of a Building, answering nearly to our Cellar; *Juvenal*, for a *Cloaca*.

Hence, *Crypto-Portico*, a subterraneous Place, arch'd, or vaulted; us'd as an Underwork, or Passage in old Walls. The same is also us'd for the Decoration at the Entry of a Grotto. See GROTTO.

CRYPTA, is also used by some of our antient Writers, for a Chapel, or Oratory under ground: *Egresso toto conventu, accepta absconsa si nox est vadit per* cryptam. *Du Cange*.

The Word is form'd from the *Greek*, *abscondo*, I hide; whence *Crypte*.

CUPOLA, in Architecture, a spherical Vault; or the round Top of the Dome of a Church, in form of a *Cup* inverted. See DOME.

Some call it *Lanthorn*.

The Word is *Italian*, form'd of the barbarous Latin *Cuppola*; otherwise call'd *Thola*, and *Fornix*.

CURLING Stuff, in Joinery. See CROSS-GRAIN'D *Stuff*.

CURTAIN, or COURTINE, in Fortification, that Part of a Wall, or Rampart, which is between two Bastions; or which joins the Flanks thereof.

It is usually border'd with a parapet five Foot high; behind which the Soldiers stand to fire upon the Covert Way, and into the Moat.

Besiegers seldom carry on their Attacks against the *Curtain*; because it is the best flanked of any Part. See FLANK.

Du Cange derives the Word *Curtain* from the Latin *Cortina*, *quasi minor Cortis*, a little Country Court, inclos'd with Walls: He says, it was in Imitation hereof, that they gave this Name to the Walls and Parapets of Cities, which inclose 'em like *Courts*: He adds, that the *Curtains* of Beds take their name from the same Origin; that *Cortis* was the Name of the General's, or Prince's Tent; and that those who guarded it were call'd *Cortinarii* and *Curtisani.*

CYMA, in Architecture. See CIMA, SIMA, and CYMATIUM.

CYMATIUM, CIMATIUM, or CIMA, in Architecture, (from the Greek *undula*;) a Member, or Moulding of the Cornice, whose Profile is waved, *i.e.* concave a-top, and convex at bottom; frequently also call'd *Doucine*, *Gorge*, or *Gula Recta*; especially by the *French*: by the Italians *Goletta*, i.e. *parva Gula*; but more usually, *Cymatium*, among us; as being the last, or uppermost Member, *q.d.* the *Cyma*, or Summet of the Corniche. See CORNICHE.

Some write it *Simaise*, from *Simus*, *Camus*, flat-nos'd; but this Etymology is unlucky: the Beauty of the Moulding consisting in its having its Projecture equal to its Height.

M. *Felibien*, indeed, will not allow this Etymology; contending, that the Moulding is not so denominated from its being the uppermost Member of the Corniche; but, according to the Sentiment of *Vitruvius*, from its being waved.

This is certain, that *Vitruvius* sometimes uses the Word *Unda* for *Cymatium*, and sometimes *Lysis*, i.e. *Solution*, *Separation*; in regard, Corniches, where the *Cymaises* are found, separate one piece of Architecture from another; as the Pedestal from the Column, and the Frieze from the Corniche.

But withal, it must be observed, that he does not confine *Cymatium* to the Corniche; but uses it indifferently for any similar Moulding, wheree'er he meets with it: in which he differs from the most accurate among the Moderns.

Felibien makes two Kinds of *Cymatiums*; the one *right*, the other *inverted*: In the first, that Part which projects the furthest is concave, and is otherwise call'd also *Gula Recta*, and *Doucine*. See DOUCINE.

In the other, the Part that projects furthest is convex, called *Gula Inversa*, or *Talon*. See TALON.

Our Architects don't use to give the Name *Cymatium* to these

Mouldings, except when found on the Tops of Cornices; but the Workmen apply the Name indifferently, wherever they find 'em.

Palladio distinguishes the *Cymatium* of the Corniche by the Name *Intavolatum*.

Tuscan CYMATIUM, consists of an Ovolo, or Quarter-Round. *Philander* makes two *Doric Cymatiums*, whereof this is one: *Baldus* calls this the *Lesbian Astragal*.

Doric CYMATIUM, is a Cavetto, or a Cavity less than a Semicircle, having its Projecture subduple its Height.

Lesbian CYMATIUM, according to *Vitruvius*, is what we otherwise call *Talon*, *viz*. a concavo-convex Member, having its Projecture subduple its Height. See TALON.

D

DADO, in Architecture, is by some Writers used for the Dye; which is the Part in the Middle of the Pedestal of a Column, between its Base and Cornice: It is of a Cubick Form, whence the Name of Dye. See DYE.

DECASTYLE, in the ancient Architecture, a Building with an Ordonnance of ten Columns in Front. The Temple of *Jupiter Olympius* was *Decastyle*.

DECORATION, in Architecture, an Ornament in a Church, or other public Place: Or, what adorns, and enriches a Building, Triumphal Arch, *etc.* either without Side, or within. See ORNAMENTS.

The Orders of Architecture contribute a great deal to the *Decoration*; but then the several Parts of those Orders must have their just Proportions, Characters, and Ornaments; Otherwise the finest Order will bring Confusion, rather than Richness. See ORDER.

Decorations in Churches, are Paintings, Vases, Festoons, *etc.* occasionally applied to the Walls; and with so much Conduct and Discretion, as not to take off any Thing from the Form of the Architecture: As is much practised in *Italy*, at the solemn Feasts.

DECORATION is particularly applied to the *Scenes* of Theatres. See SCENE.

In Opera's, and other Theatrical Performances, the *Decorations* must be frequently changed, in Conformity to the Subject. The Ancients had two Kinds of *Decorations* for their Theatres: The first, call'd *Versatilis*, having three Sides, or Faces which were turned successively to the Spectators: The other, call'd *Ductilis*, shewing a new *Decoration* by drawing, or sliding another before it. This latter Sort of *Decoration* is still used, and apparently with much greater Success than among the Ancients, who were oblig'd to draw a Curtain whenever they made a

Change in the *Decoration*; whereas on our Stage the Change is made in a Moment, and almost without being perceiv'd. See THEATRE.

DECORUM, DECOR, *Decency*, is particularly used in Architecture, for the Suitableness of a Building, and the several Parts, and Ornaments thereof, to the Station, and Occasion. *Vitruvius* is very severe in this Point; and gives express Rules for the Appropriating, or Suiting the several Orders to their natural Characters. So, *e.g.* as a *Corinthian* Column should not be set at the Entrance of a Prison, or Gate-House; Nor a *Tuscan*, in the Portico of a Church; As some have done among us, in breach of *Decorum*. Even in the Disposition of the Offices of our ordinary Houses, we find grievous Offences against *Decorum*: The Kitchin being often where the Parlour should be; and that in the first and best Story, which should have been damned to the lowest and worst.

DEFILE, in Fortification, is a streight narrow Lane, or Passage, through which a Company of Horse, or Foot, can pass only in File, by making a small Front, so that the Enemy may take an Opportunity to stop their March, and to charge them with so much more Advantage, in Regard that the Front and Rear cannot reciprocally come to the Relief of one another.

The word is form'd from the French *deffiler*, to unthread, or unstring.

DEGREDATION, in Painting, expresses the lessening, and rendring dim and confused, the Appearance of distant Objects in a Landskip; so as they shall appear there, as they would do to an Eye placed at that Distance from them.

DEMI-*Gorge*, in Fortification, is half the Gorge, or Entrance into the Bastion; not taken directly from Angle to Angle, where the Bastion joins to the Courtine, but from the Angle of the Flank, to the Centre of the Bastion, or the Angle of the two Courtins would make, were they thus protected to meet in the Bastion.

DEMI-*Lune*, *Half-Moon*, in Fortification, consisting of Faces and Flanks; frequently built before the Angle of a Bastion, and sometimes also before the Courtine, tho' now much disused.

DENTICLES, or DENTILS, in Architecture, an Ornament in Corniches, bearing some Resemblance to Teeth; particularly affected in the *Ionic*, and *Corinthian* Orders.

They are cut on a little square Member, properly call'd *Denticulus*; and the Notches, or Ornaments themselves, *Dentes*, from *Dens*; as having the Appearance of a Set of Teeth.

Anciently, *Dentils* were never used but in the *Ionic* Cornich: Yet we find them in the Remains of the Theatre of *Marcellus*; which is an

Argument with some, that *Vitruvius* had not the Direction of that Building. *Vitruvius* prescribes the Breadth of each Dentil, or Tooth, to be half its Height; and the Indenture, or Interval between each two, he orders to be ⅔ of the Breadth of the *Dentil.*

The same Author, *C*.2. of his IVth Book, observes, that the *Greeks* never used *Dentils* underneath Modillions; by reason Modillions represent Forces; and *Dentils* represent Ends of Rafters, which can never be plac'd underneath Forces. See MODILLIONS.

The *Romans* were not so scrupulous as to this *Decorum*; excepting in the *Pantheon*, where there are no *Dentils* under the Modillions, neither in the *Portico*, nor in the Inside of the Building. The *Italians* call them *Dentelli*, and *Denticoli.*

DESIGN, or *Draught*, with regard to the Arts and Sciences, signifies the Thought, Plan, Geometrical Representation, Distribution, and Construction of a Painting, Poem Book, or Building.

This Painter has shewn the first *Design* of his Piece, in which the Figures are well disposed. The *Design* of that Poem or Book is artfully laid.

Claudian never sees his whole *Design* together: When he composes a Part, he thinks of nothing else; and works up every Member, as if it were separate from all the rest.

In Building, one may use the Term *Ichnography*, when by *Design* is only meant the Plan of a Building, or a flat Figure drawn on Paper.

Orthography, when some Face, or Side of the Building is rais'd from the Ground.

And *Scenography*, when both Front and Sides are seen in Perspective.

DESIGN, in the Manufactories, the Figures wherewith the Workman enriches his Stuff, or Silk; and which he copies after some Painter.

In undertaking such Kinds of figured Stuffs, 'tis necessary, that before the first Stroak of the Shuttle, the whole *Design* be represented on the Threads of the Warp; we don't mean in Colours, but with an infinite Number of little Pack-threads, which being disposed so as to raise the Threads of the Warp, let the Painter see from Time to Time what Kind of Silk is to be put in the Eye of the Shuttle, for Woof.

This Method of preparing the Work, is call'd *reading the Design*, or reading the Figure; which is perform'd after the following Manner.

A Paper is provided considerably broader than the Stuff, and of a Length proportionate to what is intended to be represented thereon. This they divide length-wise by as many black Lines as there are intended to be Threads in the Warp; and cross these Lines, by others drawn breadth-wise; which with the former make little equal Squares. On the Paper thus squared, the Painter *designs* his Figures, and heightens them with Colours as he sees fit. When the *Design* is finish'd, a Workman reads it, while another lays it on the Simblot.

Now, to read the *Design*, is to tell the Person, who manages the Loom, the Number of Squares, or Threads comprized in the Space he is reading; intimating at the same Time, whether it is Ground, or Figure.

To put what is read on the Simblot, is to fasten little Strings to the several Pack-Threads, which are to raise the Threads named. And thus they continue to do, till the whole *Design* is read.

Every Piece being composed of several Repetitions of the same *Design*; when the whole *Design* is drawn, the Drawer, to re-begin, as I may say, the *Design* a-fresh, has nothing to do, but raise the little Strings with Slip-knots to the Top of the Simblot, which he had let down to the Bottom. And this he is to repeat as often as is necessary, till the Whole be manufactured.

The Ribband-Weavers have likewise a *Design*, but much more simple than that now described. 'Tis drawn on Paper, with Lines, and Squares, representing the Threads of the Warp, and Woof. But in lieu of Lines, whereof the Figures of the former consist, these only consist of Points, or Dotts, placed in certain of the little Squares, form'd by the Intersection of the Lines. These Points mark the Threads of the Warp that are to be raised; and the Spaces left blank, *design* the Threads that are to keep their Situation. The rest is manag'd as in the former.

DESIGN, is particularly used in Painting, for the first Idea of a large Work drawn roughly, and in little; with Intention to be executed, and finish'd in large.

In this simple Sense, the *Design* is the simple *Contour*, or *Outline* of the Figures, or Things intended to be represented; or the Lines that terminate, and circumscribe them.

Such *Design* is sometimes drawn in Crayons, or Ink, without any Shadows at all: Sometimes it is *hatch'd*, that is, the Shadows are express'd by sensible Lines, usually drawn a-cross each other with the Pen, Crayon, or Graver. Sometimes, again, the Shadows are done with the Crayon rubb'd, so as there does not appear any Lines: Sometimes, the Grain, or Strokes of the Crayon appear, as not being rubb'd: Sometimes the *Design* is *wash'd*, that is, the Shadows are done with a Pencil, in *Indian* Ink, or some other Liquor: And sometimes the *Design* is *colour'd*, that is, Colours are laid on, much like those intended for the Grand-Work.

The Qualities, or Parts required in a *Design*, are *Correctness*, *Good Taste*, *Elegance*, *Character*, *Diversity*, *Expression*, and *Perspective*.

Correctness depends principally on the Justness of the Proportions, and a Knowledge of Anatomy. *Taste* is an Idea, or Manner of *Designing*, which arises either from the Complexion and natural Disposition, or from Education, the Masters, Studies, *etc.* *Elegance* gives the Figures a Kind of Delicacy, which strikes People of Judgment, and a certain Agreeableness, which pleases every body. The *Character* is what is peculiar to each Thing; In which there must be a *Diversity*; in as much as every Thing has its particular Character, to distinguish it. The *Expression* is the Representation of an Object according to its Character, and the several Circumstances it is supposed to be in. The *Perspective* is the

Representation of the Parts of a Painting, or Figure, according to the Situation they are in, with respect to the Point of Sight.

The *Design*, or *Draught*, is a Part of the greatest Import, and Extent in Painting. 'Tis acquired chiefly by Habit, and Application; Rules being of less Avail here than in any of the other Branches of the Art, as *Colouring*, *Clair-Obscure*, *Expression*, etc.

The principal Rules that regard the Design, are: That Novices accustom themselves to copy good Originals at first Sight: Not to use Squares in drawing; for Fear of Stinting and confining their Judgment: To stay till they can *design* well after the Life 'ere they begin the Practice of Perspective Rules: In *Designing* after the Life, to learn to adjust the Bigness of their Figures to the visual Angle, and the Distance of the Eye, from the Model or Object: To mark out all the Parts of their *Design*, before they begin to shadow: To make their Contours in great Pieces, without taking Notice of the little Muscles, and other Breaks: To make themselves Masters of the Rules of Perspective: To observe every Stroak as to its perpendicular Parallel, and Distance; and particularly so to compare, and oppose the Parts that meet upon, and traverse the Perpendicular, so to form a Kind of Square, in the Mind; which is the great, and almost the only Rule of *Designing* justly: To have a regard not only to the Model, but also to the Part already *design'd*; there being no such Thing as *designing* with strict Justness, but by comparing, and proportioning every Part to the first.

The rest relates to *Perspective*: As, that those Objects are seen at one View, whose Rays meet in a Point: That the Eye and Object be always conceiv'd as immoveable: That the Space, or *Medium* between them, be conceiv'd transparent: And that the Eye, Object, and Picture, be at a just Distance; which is usually double the Bigness of the Subject, or Picture.

DETACH'D *Pieces*, in Fortification, are Demi-Lunes, Ravelins, Horn- and Crown-Works, and even Bastions, when separated, or at a Distance from the Body of the Place.

In Painting, the Figures are said to be well *detach'd*, or loosen'd, when they stand free, and disengag'd from each other, are no where confounded together, but stand out with a strong *Relievo* from the Ground, and each other.

DIASTYLE, in the ancient Architecture, an Edifice, where the Columns stand at such Distance one from another, that eight Modules, or four Diameters, are allowed for the Intercolumniation. See INTER-COLUMNS.

DIGLYPH, a Kind of imperfect Triglyph, Console, or the like, with only two Channels, or Engravings. See TRIGLYPH.

DILAPIDATION, is a wastful Destroying, or letting Buildings run to

Ruin and Decay, for Want of necessary Reparation: 13 *Eliz. Cap. 13.*

The Money recover'd for *Dilapidations*, by *14. Eliz. 11.* shall be employed in the Repair of the same Houses.

DIMINUTION, in Architecture, a Contraction of the upper Part of a Column, whereby its Diameter is made less than that of the lower Part. See COLUMN.

To attain those two important Points in Architecture, Strength, and the Appearance of Strength, all Architects have made their Columns lesser above than below; which is call'd their *Diminution*. Some have likewise made them a little bigger towards the middle than at the Bottom, which is call'd their *Swelling*. The *Gothic* Architects, indeed, observe neither *Diminution*, nor *Swelling*: Their Columns are perfectly Cylindrical; for which reason they are properly call'd *Pillars*, in Contra-distinction to *Columns*. See PILLAR.

The *Diminution* generally commences from one Third of the Height of the Column. Some make it begin from the very Basis, and hold tapering to the Capital; but this has not so good an Effect. *Vitruvius* would have the *Diminution* of Columns differ according to their Height, and not according to their Diameter. Thus, a Column 15 Foot high, he *diminishes* a sixth Part of its Diameter, and another of 50 Foot, only an 8th Part; but we don't find this Rule observ'd in the Antique.

The Difference of Orders, Mr. *Perrault* observes, does not infer a Difference of *Diminution*; there being small and great *Diminutions* in different Works of the same Order: Excepting, however, the *Tuscan*, which *Vitruvius* diminishes by a 4th Part; tho' *Vignola* only makes it a 5th, and the *Trajan* Column a 9th. The *Diminutions* are very diferently adjusted in the different antique Buildings, as well as in the different modern Authors.

DIPTERE, or DIPTERON, in the ancient Architecture, a Temple surrounded with two Rows of Columns, which form a Sort of Portico's, call'd *Wings*, or Isles. See TEMPLE.

DITRIGLYPH, in Architecture, the Space between two Triglyphs. See TRIGLYPH.

DOME, in Architecture, a spherical Roof; or a Work of a Spherical Form rais'd over the Middle of a Building, as Church, Hall, Pavillion, Vestible, Stair-Case, *etc.* by Way of Crowning, or Acroter. See ROOF, and CROWNING.

Domes are the same with what the *Italians* call *Couppolas*; and we frequently *Cupola's*: The *Latins*, according to *Vitruvius*, *Tholi.* See CUPOLA.

They are usually made round; tho' we have Instances of square Ones; as those of the *Louvre*: and others, that are Polygons, as that in the

Jesuites Church in the *Rue* St. *Anthoine* at *Paris*. They have usually Columns rang'd around their Out-sides, both by Way of Ornament, and to support the Vault.

A *Flat* DOME is that whose Sweep, or Contour, is less than a Hemisphere.

The Word is form'd from the Latin *Doma*, which among the Ancients signified a *Roof*, or *Open Porch*; as is observed by *Papias*. 'Tis frequent in the corrupt *Latin* Authors, who borrowed it from the *Greeks*, among whom *Trullus*, or *Trullum*, is a common Name for any round Building, such as the Palace of *Constantinople*, wherein was held the Council; thence call'd *in Trullo*.

DONJON, in Fortification, is generally taken for a large Tower, or Redoubt of a Fortress, where the Garrison may retreat in Case of Necessity, and capitulate with greater Advantage.

DOOR, in Architecture, an Aperture in a Wall, to give Entrance, and Exit in, and out of the building, or an Apartment thereof. See HOUSE, BUILDING, *etc*.

It ought to be a Rule, that the *Doors* of a House be as few in Number, and as moderate in Dimensions, as possible: For, in a Word, all *Openings are Weakenings*. 2. That they do not approach too near the Angles of the Walls; It being a most glaring Sollecism to weaken that Part, which must weaken all the rest: A Precept, well recorded, but ill practised by the *Italians* themselves, particularly at *Venice*. 3. That the *Doors*, if possible, be right over one another; that void may be over void, and full over full. 4. That, if possible, they be opposite to each other, so as that one may see from one End of the House to another; which will not only be graceful, but also convenient; as it affords a Means of cooling the House in Summer, by letting in Air; and of keeping out the Wind in Winter, which Way soever it sit. 5. 'Tis not only Ornamental, but secure, to turn Arches over *Doors*; as it discharges 'em in great Measure of the superincumbent Weight.

The Proportions of *Doors* are adjusted by those of a Man: In larger Buildings they must always be larger than in smaller; but in none should they be less than 6 Foot high, to admit a Man of just Stature, erect; And as the Breadth of a Man, with his Arm placed a-kimbo, is nearly subduple his Height, the Width should never be less than 3 Foot.

Some Architects give us these Dimensions: In small Buildings the Breadth of the *Door* 4 Foot, or 4 and a half; in middling Buildings, 5 or 6; in large Ones, 7 or 8: In Chambers of the first, 3 and a half, $3\frac{3}{4}$, or 4; of the second, 4, or 4 and a half; and of the third, 5, or 6; in Churches, 7 or 8; in Gates, 9, 10, or 12: hence their Height is easy determin'd; except for the Gates of Cities, which should only be $\frac{4}{5}$ of their Breadth.

'Tis an Observation of the excellent *Palladio*, that the principal *Door*, or Entrance of a House, must never be regulated by any certain

Dimensions, but by the Dignity of the Person that is to live in it: yet, to exceed rather in the more, than the less, is a Mark of Generosity; and may be excused with some noble Emblem, or Inscription, as that of the Conte *di Bevilacqua* over his large Gate at *Verona*, where had been committed a like Disproportion, *Patet Janua, Cor magis.*

DORIC, in Architecture, is the second of the five Orders; being that between the *Tuscan* and *Ionic.* See ORDER.

The *Doric* Order seems the most natural, and best proportioned of all the Orders; all its Parts being founded on the natural Posititon of solid Bodies. Accordingly, the *Doric* is the first, and most ancient of the Orders of Architecture, and is what gave the first Idea of regular Building. See ARCHITECTURE.

At its first Invention it was more simple than at present; and when in After-times they came to adorn, and enrich it more, the Appellation *Doric* was restrain'd to this richer Manner, and the primitive simple Manner they call'd by a new Name the *Tuscan* Order. See TUSCAN.

The Tradition is, that *Dorus*, King of *Achaia*, having first built a Temple of this Order at *Argos*, which he dedicated to *Juno*, occasion'd it to be call'd *Doric*: Tho' others derive its Name, I know not how, from its being invented, or us'd by the *Dorians*.

Some time after its Invention they reduced it to the Proportions, Strength, and Beauty of the Body of a Man. Hence, as the Foot of a Man was judg'd the 6th Part of his Height, they made the *Doric* Column, including the Capital, 6 Diameters high, *i.e.* six times as high as thick. Afterwards they added another Diameter to the Height, and made it 7 Diameters; with which Augmentation it might be said to be nearer the Proportion of a Man: The human Foot, at least in our Days, not being a 6th, but nearly a 7th Part of the Body. See COLUMN.

The Characters of the *Doric* Order, as now manag'd, are, the Height of its Column, which is 8 Diameters; its Frize, which is enrich'd with Triglyphs, Drops, and Metopes; its Capital, which is without Volutes; and its admitting of Cymatiums. See TRIGLYPH, CYMATIUM, *etc.*

The Ancients, we have already observ'd, had two *Doric* Orders: The first, which was the more Simple and Massive, they us'd chiefly in Temples; The second, which was the more light and delicate, serv'd in Portico's, and Theatres.

Vitruvius, indeed, complains of the *Doric* as very troublesome and perplexing, on Account of the Triglyphs, and Metopes, so as scarce to be capable of being us'd, except in the Pycnostyle, by putting a Triglyph between each two Columns, or in the Aræostyle, by putting 3 Triglyphs between each two Columns. See INTERCOLUMNS.

The Moderns, on Account of its Solidity, use it in large strong Buildings; as in the Gates of Cities, and Citadels, the Outsides of Churches, and other massy Works, where Delicacy of Ornaments would be unsuitable.

The most considerable ancient Monument of this Order, is the Theatre of *Marcellus*, at *Rome*, wherein the Capital, the Height of the Frize, and its Projecture, are much smaller than in the modern Architecture.

Vignola adjusts the Proportions of the *Doric* Order thus: The whole Height of the Order, without Pedestal, he divides into 20 Parts, or Models; one of which he allows the Base, 14 to the Shaft, or Fust, one to the Capital, and 4 to the Entablature.

The particular Proportions, *etc.* of the several Parts, and Members, see under their respective Articles, SHAFT, CAPITAL, ENTABLATURE, PEDESTAL, BASE, ARCHITRAVE, FRIZE, and CORNICHE.

DORIC *Cymatium.* See CYMATIUM.

DORIC, in Masic. The *Doric Mode* is the first of the authentic Modes of the Ancients.

Its Character is to be severe, temper'd with Gravity, and Joy; And it is proper for Occasions of Religion, and War. It begins with *D, Sol, Re.*

Plato admires the Music of the Doric Mode; and judges it proper, to preserve good Manners, as being Masculine. And on this Account allows of it in his Common-wealth.

The Ancients had likewise their *Subdoric Mode*, which was one of their Plagal Modes. It began with *G Ut*, a Diatessaron lower than the *Doric* Mode.

DORMER, or DORMANT, in Architecture, is a Window made in the Roof of an House, or above the Entablature; being rais'd upon the Rafters. The *Latins* call it *Lucerna*, whence the French *Lucerne*, and our *Luthern.*

There are divers Kinds; Square, Round, *etc.* See LUTHERN.

DORMITORY, a Gallery in Convents, or Religious Houses, divided into several Cells, wherein the Religious lodge, or lye, *etc.* See CONVENT.

'Tis deem'd a Crime in a Religious to lye out of the *Dormitory.* By Chap. XXII. of *The Rule* of St. *Benedict*, it appears, that the ancient *Dormitories* were not divided into Cells, but were a Kind of large open Halls, fill'd with Beds, as in our Hospitals. See CELLS.

The Word is form'd from the Latin *Dormitorium*, of *dormire*, to sleep.

DOUCINE, in Architecture, a Moulding, or Ornament on the highest Part of the Cornice, in form of a Wave, half convex, and half concave. See CORNICHE.

The *Doucine* is the same with a *Cymatium*, or *Gula.* See CYMATIUM, and GULE *Recta*, and *Inversa.*

DOVE-*Tail*, in Carpentry, the strongest of the Kinds of Jointings, or Assemblages; wherein the *Tenon*, or Piece of Wood that enters the other, goes widening to the Extreme, and so cannot be drawn out again, by reason the Tip, or Extreme is bigger than the Hole.

It has its Denomination from the Resemblance the Tenon bears to a *Dove's* Tail.

The French call it *Queue d'Aronde*, Swallow's Tail; which Name the *English* themselves retain in Fortification. See QUEUE *d'Aronde*.

DOVETAILING, in Architecture, is a Way of fastening Boards, or Timbers together, by letting one Piece into another indentedly, with a *Dove-tail* Joint, or with a Joint in the Form of a *Dove's* Tail.

DRAPERY, in Painting, and Sculpture, the Representation of the Garments, or Cloathing of human Figures.

In the general Sense, *Drapery* includes not only the Garments, but also the Tapistry, Linings, and most other Things that are not *Carnations*, nor Landskips.

The Art of *Drapery* consists chiefly in three Points, *viz.* The Order of the Folds, or Plates: The different Quality of the Stuffs: And the Variety of their Colours.

As to the *Folds*, they should be so manag'd, that you may easily perceive what it is that they cover, and distinguish it from any Thing else. For Instance, that you see 'tis an Arm that is under the *Drapery*, and not a Leg, *etc.* The Folds, again, must be large, as breaking, and dividing the Sight the less. There should likewise be a Contreast between 'em, otherwise the *Drapery* will be stiff.

The *Quality of the Stuffs* should likewise be well consider'd; some making their Folds abrupt, and harsh, and others more soft, and easy. The Surface of some, again, has a Lustre, others are dead: Some are fine, and transparent, others firm, and solid.

The Variety of *Colours*, when well manag'd, makes the great Beauty of a Painting; all being not equally amicable, and friendly, with respect to each other; and some never to be placed near certain others.

Mr. *de Piles* gives Abundance of good Observations, as to *Drapery*. Their first Effect, he observes, and that which the Painter ought to have principally in View, is, that they express the Thing they are suposed to cover: They must never be made to adhere, and stick to the Parts of the Body: A great Lightness, and Motion of the *Drapery*, are only proper for Figures in great Agitation, or exposed to the Mind. The Nudities of the Figures should always be design'd, 'ere the Painter proceed to the *Draperies*. *Paolo Veronese* excell'd in *Draperies*.

The Word is French, *Drapperie*, form'd from *Drap*, Cloth.

DRAUGHT, or, as 'tis pronounced, DRAFT, in Architecture, the Figure of an intended Building described on Paper; wherein is laid down, by Scale, and Compass, the several Divisions, and Partitions of the Apartments, Rooms, Doors, Passages, Conveniencies, *etc.* in their due Proportion. See BUILDING.

'Tis usual, and even exceedingly convenient, 'ere a Building is begun to be rais'd, to have *Draughts* of the *Ichnography*, or Ground-plot of each Floor, or Story; As also of the Form and Fashion of each Front, with the Windows, Doors, Ornaments, *etc.* in an *Orthography*, or Upright. See ICHNOGRAPHIA, and ORTHOGRAPHY.

Sometimes the several Fronts, *etc.* are taken, and represented in the same *Draught*, to shew the Effect of the whole Building, call'd a *Scenography*, or *Perspective*. See SCENOGRAPHY.

DRIP, in Architecture. See LARMIER.

DRIPS, are also us'd in Building for a Kind of Steps, on flat Roofs, to walk upon. This way of Building is much us'd in *Italy*: where the Roof is not made quite flat, but a little raised in the Middle; with *Drips*, or Steps, lying a little inclining to the Horizon. See ROOF.

DRUM, in Architecture. See TAMBOUR.

DYE, in Architecture, is the Middle of the Pedestal; or that Part which lies between the Base and the Cornice: being so nam'd, because it is frequently made in the Form of a Cube, or *Dye*. It is also call'd *Dado*, by the *Italians*; and by *Vitruvius*, *Truncus*. See PEDESTAL.

DYE, is also us'd for a Cube of Stone, placed under the Feet of a Statue, and over its Pedestal; to raise it, and shew it the more.

E

EAGLE, in Architecture, a Figure of that Bird antiently used, as an Attribute, or Cognizence of *Jupiter*, in the Capitals and Friezes of the Columns of Temples, consecrated to that God.

EAVES, the Margin, or Edge, of the Roof of a House; being the lowest Tiles, Slate, or the like, which hang over the Walls, to throw off the Waters to a Distance from the Wall. See ROOF.

EAVES-*lath*, is a thick feather-edge Board, generally nail'd round the *Eaves* of a House for the lowermost Tiles, Slate, or Shingles to rest upon.

ECHINUS, in Architecture, a Member, or Ornament, near the Bottom of the *Ionic*, *Corinthian*, and *Composite* Capitals; which, from its circular Form of Contour, is called by the *French*, *Quart de Rond*, and by the *English*, *Quarter round*, or *Boultin*; and from its being usually carved, or cut with Figures of Eggs, *etc.* is called also by the *Latins*, *Ovum*, by the *Italians*, *Ovolo*, the *French*, *Ove*, and the *English*, *Eggs* and *Anchors*.

See QUARTERS and OVOLO.

Lastly, The Eggs being encompassed with a Cover, and thus bearing some Resemblance to a Chessnut cut open; The *Greeks* have call'd it *Echinus*, that is, *Chessnut*. See BOULTINE, *etc*.

ECHINUS, among Botanists, is used for the prickly Head, or Top of any Plant; thus called from its Likeness to an Hedge-hog or Chessnut.

ECPHORA, *Projecture*, in Architecture, is usually the Line, or Distance, between the Extremity of a Member, or Moulding, and the Naked of the Column, or other Part it projects from.

Some Authors, however, account the *Ecphora* from the Axis of the Column; and define it to be the right Line intercepted between the Axis, and the outermost Surface of a Member or Moulding. See PROJECTURE.

EDGINGS, in Gardening, Rows of Shrubs, or Plants, placed by Way of Borders, around Beds, Compartiments, *etc*.

For the *Edgings* of Compartiments, Box seems the most proper. But they are frequently made of Aromatic Plants, as Sage, Sweet Marjoram, Thyme, Lavender, Hyssop, *etc*.

EDIFICE, a *Building*. See BUILDING. The Word is form'd of the *Latin*, *Ædes*.

ELEVATION, in Architecture, a Draught, or Description, of the principal Face, or Side of a Building; called also an *Upright*, or *Orthography*. See ORTHOGRAPHY.

ELEVATION, we also find us'd in some Writers of Perspective, for the *Scenography*, or perspective Representation of the whole Body or Building. See SCENOGRAPHY.

EMBOSSING, or IMBOSSING, the act of forming, or fashioning Works in Relievo, whether they be cast, or moulded, or cut with the Chissel, *etc*. See RELIEVO, STATUE.

Embossing is one great Part of Sculpture; being that which has to do with Figures rais'd, or prominent from the Plane or Ground: The other Part, which makes Figures, *etc*. that are indented, or cut in below the Ground, is call'd *Engraving*. See SCULPTURE and ENGRAVING.

EMBRASURE, in Architecture, the Enlargement made of the Gap, or Aperture of a Door, or Window, on the inside the Wall. See DOOR.

Its Use is to give the greater Play, for the opening of the Door, Wicket, Casement, *etc*. or to take in the more Light.

The *Embrasure* coming sloping inwards, makes the inner Angles obtuse. When the Wall is very thick, they sometimes makes *Embrasures* on the out-side.

EMBRASURES, in Fortification, are the Holes, or Apertures, thro' which the Canons are pointed; whether in Casemattes, in Batteries not cover'd with Gabions, or in the Parapets of Walls.

The *Embrasures* are generally about 12 Foot a-part, from six to nine Foot wide without, and two or three within. Their Height from the Platform is usually three Foot on the inside, and a Foot and half on the outside: That so the Muzzle, on Occasion, may be sunk, and the Piece brought to shoot low.

They are particularly called *Cannonieres*, when big enough to pass the Mouth of a Canon through; and *Meurtrieres*, when only big enough for Muskets.

When the Parapet is so low, that the Canon may shoot without *Embrasures*, it is said to shoot *en barbe*.

EMPASTING, a Term used in Painting, for the laying on of Colours, thick and bold; or applying several Lays of Colours, so as they may appear thick.

A Painting is said to be well *Empasted* with Colours, when the Colours are bestow'd plentifully, or it is well fed and saturated with Colours.

The Term is also used, when the Colours are laid distinct and asunder; and not soften'd, or lost in each other. This Head is not painted: It is only *Empasted.*

It is form'd of the French *Empaster*, which has the same Signification, of *Paste*, or *Pâte*, Paste.

ENAMEL, popularly AMEL, a Kind of coloured Glass, by the *Latins* call'd *Encaustum*, and by the French *Esmail*; used in *Enamelling*, and Painting in *Enamel.* See ENAMELLING, and *Painting in Enamel.*

The Basis of *Enamel* is the finest Crystal Glass, made of the best Kali from *Alicant*, and Sand vitrified together.

To this Glass are added Tin and Lead in equal Quantities, calcined by a reverbatory Fire. Such is the fundamental Composition of *Enamel*; to which are added other metallic or mineral Matters, to give it the Colour required: As *Æs Ustum* for Green, *Crocus Martis* for Yellow, *etc.*

We may distinguish three Kinds of *Enamel*: The first intended for the conterfeiting and imitating of pretious Stones: The second, for *Painting in Enamel*: And the third used by the *Enamellers*, Jewellers, and Goldsmiths, on Gold, Silver, and other Metals; with which last Kind, particularly the White, it is that the Makers of the *Dutch* Ware, give the Lustre and Polish to their Works.

The *Enamel* used in imitating pretious Stones, and that for Painting, is prepared by the Workmen themselves, employ'd in those Arts: The rest is brought from *Venice* and *Holland.* The Composition is the same, in the main, in all the three Kinds: All the Difference consists in giving it the Colour, or Transparency.

ENAMELS *for Painting.*

The *white Enamel*, or Colour, used by the Painters in *Enamel*, is the same with the common *Enamel*, used by *Enamellers*: Only, 'tis to be prepared by grinding, and cleansing it with *Aqua fortis*. After which, washing it well in fair Water, 'tis ground, or pounded a-fresh in a Flint or Agat Mortar. The *ruddy brown*, is made with Feces of Vitriol, and Salt-Petre, or with Iron Rust, well ground on an Agat, with Oil of Aspic. *Black*, is made of Perigueux well calcined, and ground with Oil of Aspic; to which is added an equal Quantity of the Goldsmiths, or *Enamellers* black. *Yellow*, is the same with the Goldsmiths Yellow, the Composition whereof will be given hereafter. *Blue* is made of the *Azure*, or *Lapis*, used by the Painters in Oil, well purified and prepared with Brandy, and exposed in a Bottle five or six Days to the Rays of the Sun. *Vermillion Red*, is made with Vitriol calcined between two Crucibles luted together; then wash'd in *Aqua fortis*, and afterwards in fair Water: The Fire to be moderate, and to remain about half an Hour. *Lake Red*, is composed of fine Gold, dissolved in *Aqua Regia*, with *Sal Armoniac*, or common Salt. The Dissolution being compleated, 'tis put in a Cucurbit with Spring Water, and Mercury, o'er hot Sand for 24 Hours. The Powder remaining at the Bottom of the Curcurbit, when the Water is pour'd off, is ground up with double its Weight of Flower of Sulphur, and put in a Crucible over a gentle Fire. And when the Sulphur, which takes Fire, is exhaled, the red Powder remaining, is ground up with *Rocaille*. Lastly, White Copperas calcined, makes a Colour much like the amber Colour used by the Limners.

These seven or eight Colours, or *Enamels*, serve for the Composition of all the rest, by a discrete Mixture and Combination thereof. Thus, blue and yellow make green; blue and red, Violet; and so of the rest.

ENAMELS *used by the Jewellers, Goldsmiths, and Enamellers.*

These Kinds of *Enamel*, we have observed, come chiefly from *Venice* and *Holland*: They are in little thin Cakes of different Sizes; usually four Inches in Diameter, and ½ an Inch thick. Every Cake has the Maker's Mark struck on it with a Punchion. The most usual Marks are the Name *Jesus*, a Syren, Monkey, Sun, *etc.*

Those bought from *Venice* are chiefly white, Slate-Colour, sky Blue, Carnation, Yellow, Green, and a deep Blue, call'd a false *Lapis*. These seven are the Principles of all the others, which arise out of the Mixture of these. And the white, in particular, is, as it were, the Basis of all the other six principal Colours.

White is made, as already observed, of Crystal Glass, Tin, and Lead calcined by a reverberatory Fire. And this *Enamel* is used not only by the Jewellers and *Enamellers* on Metal, but also by the Painters in *Enamel*, the Makers of *Dutch* Ware, *etc*. By adding Azures thereto, it becomes a Slate Colour; by adding Copper, and *Cyprus* Vitriol, it makes a sky Blue.

By Perigueux, a flesh Colour; and by Iron-Rust, a yellow. To make a green, they add Copper-filings, *etc.*

ENAMELLING, by the Antients call'd *Encaustice*, the Art, or Act of applying *Enamels* of various Colours, on Metals, *etc.* either in the Way of Painting, or by the Lamp.

Painting in ENAMEL, is a Method of Painting with *Enamels*, or Metal Colours, ground, reduc'd to Powder, and used, like other Colours, with a pencil; then fused, baked again, and vitrified by Force of Fire.

The Art of *Painting in Enamel* is very antient; and appears to have been first practised on Earthen, or Potters-Ware. As early as the Age of *Porsenna* King of *Tuscany*, we hear of beautiful Vases made in his Territories, *Enamell'd* with various Figures; though far short of those afterwards made at *Faenza*, and *Castel Durante*, in the Dutchy of *Urbino*, in the Time of *Raphael*, and *Michael Angelo*. There are still some of those Vases extant in the Cabinets of Antiquaries; in all which the Design, or Drawing of the Figures, is much better than the Colouring. For they were, at that Time, only acquainted with two Colours, *viz.* black and white; either for Earthen, or Metal Works: Excepting a faint Kind of Carnation in the Faces and other Parts.

In the Time of *Francis* I. of *France*, the Art was retriev'd in that Country, particularly at *Limoges*; where there were produced Abundance of very valuable Pieces, in the Manner of the Antients, that is, well perform'd as to the Draught, and the Clairobscure, chiefly in two Colours.

There are two Ways of *Painting in Enamel*; the one with clear and transparent, and the other with thick and opake Colours.

To use the first, they are only ground up with Water: The second are ground with Oil of Aspic.

The first are laid on the Metal flat, and border'd, or edg'd, with a Rim of the Metal, to keep the Colours asunder. Tho', we have seen Pieces laid on contiguous, and without any Partition; which is very difficult to practise, by Reason the transparent Colours, in Melting, are apt to run in each other; especially in the little Works.

The Invention of opake Colours, is much later, and greatly preferable to that of transparent ones.

All Metals, however, will not equally admit both Kinds. Copper, for Instance, which bears all the opake Colours, will not bear the transparent ones: But to employ these latter upon Copper, they are forced, first, to cover it with a Lay, or Couch, of black *Enamel*, over which they lay a silver Leaf, and on this apply the other suitable Colours, that is, the Colours or *Enamels* proper for Silver, which it self does not allow of all Kinds.

Those which suit it best, are Purple, Green, Azure, and Ague marin. But Gold receives all the Kinds, and Colours, both opake and transparent, perfectly well. It must be added, however, that only the finest Gold must be used herein. For the transparent Colours being laid on a base Gold, grows dim and livid; there being a Kind of Smoak that settles on it not unlike black Lead.

Of transparent *Enamels*, the hardest are the best; tho' there is a Difference even among these; some losing their Colour in the Fire, and others retaining it. As to the reds, they are only red by Accident, being only yellow when made and applied on the Gold; and becoming red in the Furnace. The best transparent Reds are those made of calcin'd Copper, Iron Rust, Orpiment, and calcin'd Gold, melted with the due Proportions of Glass.

But 'tis the Method of Painting with opake, or thick *Enamel*, to which we owe all our fine modern Pieces of *Enamel*; particularly those curious ones on Gold, representing Portraits to as much Perfection as the best Painting in Oil; and even some History Pieces: With this great Advantage, that their Beauty and Lustre never decays.

This Art we are indebted for to the *French*: Nothing of the Kind having been attempted before the Year 1630; when *Jean Toutin*, a Goldsmith of *Chasteaudun*, and a great Master in the common Way of Painting with transparent *Enamel*, first applied himself to find a Way to use thick Colours of different Teints, which should melt with Fire, yet retain their Lustre, Purity, *etc.*

Toutin succeeded in his Attempt, and having got the Secret, communicated it to his Fellow Artists; who, in their Turns, contributed to the bringing it to Perfection: The first who distinguish'd himself was *Dubie*, a Goldsmith, who wrought in the Galleries of the *Louvre*. After him came *Mortiere*, a Native of *Orleans*, who applied himself chiefly to the Painting on Rings and Watch Cases. His Disciple, *Robert Vauquer* of *Blois*, exceeded them all both in his Designs and his Colours. After him *Pierre Chartier* of *Blois*, took himself to the Painting of Flowers, wherein he succeeded to Admiration.

By this Time, the *English* were fallen into the Way; who, as is allowed by Foreigners themselves, seem to have been the first that applied it with Success to the Painting of Portraits, which was now become mightily in Vogue, in Lieu of those in Miniature.

M. *Felibien* observes, that the first, and most finish'd Portraits, and those in the finest Colours, were brought into *France* by *Petitot*, and *Bordier* from *England*: This occasion'd *Louis Hance*, and *Louis du Guernier*, two good Painters in Miniature, to attempt the like; in which the latter succeeded beyond every Body. He likewise invented several new Teints for the Carnations; and had he lived, had probably merited the Glory of carrying the Art to its last Perfection.

This Kind of Painting, to be in Perfection, must be on Plates of Gold: For Copper, beside that it emits a Fume which tarnishes the Colours, is apt to scale and crackle; and Silver turns the Whites, Yellow.

These Plates are made a little Hollow on one Side, and rais'd on the other, either in a circular or oval Manner, to prevent the Gold's fretting by the Fire, and making the Colours crack and fly: Nor must they be made too thick. 'Tis sufficient they can bear the Colours; tho' 'tis usual to strengthen them all around with a Circle somewhat thicker.

The Plate being hammer'd very evenly throughout, they apply a white *Enamel* on both Sides, tho' the Design be only to paint on one. The Intent of this is to prevent any swelling and warping by the Fire: For otherwise, in large Pieces, and especially if the Colours be laid on any Thing unequally, they are apt to rise up in Puffs or Blisters. Now, this first Lay, which is white, remaining smooth and uniform, serves as a Ground for all the other Colours. The Composition of the white *Enamel*, with the other opake Colours, is already delivered under the Article ENAMEL.

The Gold Plate thus *Enamell'd* in *white*; the Draught, or Design, to be painted, must be chalked thereon; and, afterwards, the whole accurately drawn out in a ruddy brown. The Draught, or Out-line, thus finish'd, the Piece is set to the Fire, and then painted with the Colours above prescribed.

The white Ground they paint on, serves all the Colours for white. The Method being to spare the Ground from first to last, in the Places where the Lights are to be, after the same Manner as in Miniature: Tho' they have another white, to lay over the other Colours, when there is Occasion to raise them.

Add, that as the Painters in Oil re-touch their Paintings several Times, and let them dry; so in this Sort of Painting, they touch the Piece as often as they please, setting it each Time to a reverbatory Fire, and taking it away again, as soon as they perceive the *Enamel* has got its full polish.

The reverberatory Fire is made in a little Furnace, wherein there is Fire both a-top and all around; only a void Place in the middle, where the Piece is to be put for the *Enamels* to melt. The Colours are laid on with the Tip or Point of the Pencil, as in Mignature; with this only Difference, that they use Oil of Aspic to dilute them instead of Gum Water.

Method of ENAMELLING *by the Lamp.*

The Works of this Kind are all perform'd by the Fire of a Lamp; wherein, in Lieu of Oil, is put Horses Grease, call'd Caballin Oil.

The Lamp is of Copper, or Tin, and consists of two Parts, the Box and the Lamp: In the latter of these, which is a Kind of flat Oval, is put the Oil, and out of this rises the Wick. All the Use of the Box is to receive the Oil, which the Ebullition, occasion'd by the intense Heat, might throw abroad.

This Lamp, or even where two or three Artists work together, two or three more Lamps, are placed on a Table of a proper Height, *etc.* under which, about the middle of its Height, is a double Bellows like those of an Organ, which one of the Workmen raises and falls with his Foot, to increase and quicken the Flame of the Lamps, which is by such Means rais'd to a Degree of Vivacity almost incredible.

The Wind of the Bellows is convey'd to the several Lamps, tho' never so many, by Means of Grooves cut along the Thickness of the Table, and cover'd with Parchment, extending from the Bellows, to a Tube or Pipe placed before each Lamp. These Tubes are of Glass, and that the *Enamellers* may not be incommoded with the Heat of the Lamp, each Tube is cover'd at about six Inches Distance with a Piece of Tin, call'd a Fan, fixed in a Hole of the Table. In Works that don't take up much Time, they content themselves with a Glass Blow-Pipe, to heighten the Flame of the Lamp.

Applying their Cake of *Enamel* to the Flame of this Lamp, they draw it out into Threads inconceivably fine. Those made use of in artificial Plumes of Feathers are so very slender, that they may be wound on a Reel, like Silk or Thread.

The factitious Jeats of divers Colours, sometimes used in Embroideries, are also made of *Enamel*; and that with so much Art, that each Piece has its Hole to pass the Silk thro', wherewith it is sew'd. These Holes are made by blowing them in long Pieces, which are afterwards cut off with a proper Tool.

'Tis rare that the *Dutch*, or *Venetian* Enamels, are used pure: The common Way is to melt them in an Iron Ladle, with an equal Quantity of Glass or Crystal. And when the two Matters are in perfect Fusion, they draw it out into Threads of different Sizes, as Occasion requires, by taking a Quantity between two Pipe-Stopples, held in the two Hands, and removing them asunder as far as the Arms will reach. If the Thread is to be longer than the Workman can stretch, a second Person takes one of the Ends, and continues to draw it out while the first continues to keep the *Enamel* to the Flame. All the Threads drawn after this Manner are round; so that if the Nature of the Work require them to be flat, they are afterwards drawn thro' a Pair of Pinchers while yet hot. They have another Iron Instrument, in Manner of Plyers, to draw out the *Enamel* by the Lamp, when it is to be work'd, or disposed in Figures, or otherwise. Lastly, they have glass Tubes, of various Sizes, serving to blow the *Enamel* into various Figures, and preserve the necessary Vacancies therein, as also to spare the Stuff, and form the Contours.

When the *Enameller* goes to Work, he seats himself before his Lamp, with his Foot on the Treddle; and holding in his left hand the Piece of Work he has to *Enamel*, or the Brass or Iron Wiars his Figures are to be form'd with; with his right Hand, he draws out the Thread from the *Enamel* held to the Lamp: And this with a Dexterity and Patience equally surprizing.

There is nothing but may be thus represented in *Enamel*; and there are Figures so finely finish'd of this Kind, that one would take them to have come out of the Hands of the ablest Sculptors.

ENCAUSTICE, the art of *Enamelling*. See ENAMELLING.

ENCHASING, INCHASING, or CHASING, the Art of *Enriching*, and beautifying Gold, Silver, and other Metal Works, by some Design, or Figures represented thereon, in low Relievo.

ENFILADE, a *French* Term, sometimes used in *English.* It signifies a Series, or Continuation, of several Things, disposed, as it were, in the same *Thread* or Line.

As an *Enfilade* of Rooms, of Doors, of Buildings, *etc.*

It is also used figuratively; as a long *Enfilade* of Stories, Examples, Narrations, *etc.*

The Word is form'd of the *French* Verb *Enfiler*, to *string* a Thing, which is compounded of *en*, in, and *fil*, or *filum*, Thread: *q.d.* a *Thread*, or *String*, of any Thing.

ENGRAVING, or GRAVING, the Art, or Act of cutting Metals and precious Stones, and representing Figures, Letters, and other Matters thereon.

The Word is a Compound of the Preposition *in*, and the *Greek, I write.* Tho' *Menage*, after *Salmasius*, derives it from *cavare*, to hollow; others from the *Latin*, *graphiare*; and others from the *German*, *graben*, *q.d. Effodere*, to dig out. In *Latin* it is called *Sculptura* and *Cœlatura.*

Engraving is properly a Branch of Sculpture; tho', with us, it generally stands contra-distinguished from Sculpture; As the latter, popularly call'd *Carving*, expresses the forming of Figures, *etc.* Emboss'd, or in Relievo; and the latter, those indented, or *in Creux.*

Among the *French*, the Term *Engraving*, *Graveure*, is of more Extent. It includes all Sculpture, both that perform'd with the Chissel and Knife, on Marbles and Woods, in Relievo; and that in Metals, Stones, *etc.* with the Graver, indented.

Engraving is divided into several Branches, according to the matters it is practised on, and the Manner of performing it. The Original Way of *Engraving* on Wood is now distinguished by the Name of *Cutting in Wood*; that on Metals with *Aqua Fortis*, is called *Etching*; that by the Knife, Burnisher, Punch, and Scraper, *Mezzotinto*; that on Stones, for Tombs, *etc. Carving* or *Stone cutting*; and lastly, that perform'd with the Graver on metals, as Copper, Brass, Steel, Silver, *etc.* as also on precious Stones, Crystals, *etc.* retain alone the primitive Denomination of *Engraving.*

Engraving is an Art, for the greatest Part, of Modern Invention; having its Rise no earlier than the 16*th* Century. The Antients, 'tis true, practised *Engraving* on pretious Stones and Crystals, with very good Success; and there are still some of their Works remaining, equal to any Production of the later Ages. But the Art of *Engraving* on plates, or Blocks of Wood, to afford Prints, or Impresssions, was not known till after the Invention of painting in Oil.

The Discovery is ascribed to *Maso Finiguerra*, a Goldsmith of *Florence*; who having used to take Impressions of every thing he cut, in

Clay; and to cast melted Sulphur in this Mould; at length hit on a Way of taking the Impressions thereof on Paper, by smearing this Figure of Sulphur with Oil and Lamp-black.

The Secret soon got Abroad, and coming to the Hands of *Albert Durer* and *Lucas*, they greatly improved it, and began to *Engrave* on Wood and Copper, in which they succeeded to Admiration.

ENGRAVING *in Copper*, is employ'd in representing divers Subjects, as Portraits, Histories, Landskips, Foliages, Figures, Buildings, *etc.* either after Paintings, or Designs made for the Purpose.

It is perform'd either with the *Graver*, or with *Aqua Fortis*.

For the first, there needs but little *Apparatus*, and few Instruments. The Plate you work on being well polish'd, is cover'd over with a thin Skin of Virgins Wax, and on this, the Draught, or Design, done in black Lead, red Chalk, or other ungum'd Matter is laid; and rubb'd down for Wax to take off. The Design thus transferr'd upon the Wax, is traced thro' on the Copper, with a Point or Needle: Then heating the Plate, and taking off the Wax, the Strokes remain; to be follow'd, heighten'd, *etc.* according to the Tenor of the Design, with the *Graver*, which is to be very sharp, and well temper'd.

In the Conduct of the *Graver*, consists all the Art; for which there are no Rules to be given; all depending on the Habitude, Disposition, and Genius of the Artist.

The other Instruments are, a *Cushion*, or Sand-Bag, to lay the Plate on, to give it the necessary Turns and Motions: A *Burnisher*, round at one End, and usually flattish at the other, to rub out Slips and Failures, soften the Stroaks, *etc.* A *Scraper*, to pare off the Surface, on Occasion: And a *Rubber* of black Cloth, or Hat, to fill up the Strokes, that it may appear how the Work proceeds.

ENGRAVING *in Copper with* Aqua Fortis. See ETCHING.

ENGRAVING *on precious Stones*, consists in representing Figures or Devices, either in Relievo, or Indented, on divers Kinds of hard, polish'd Stones.

The Art of *Engraving* on precious Stones, is one of those wherein the Ancients excell'd; there being divers antique Agats, Cornelians, and Onyx's, which surpass any Thing of that Kind the Moderns have produced. *Pyrgoteles* among the *Greeks*, and *Dioscorides* under the first Emperors of *Rome*, are the most eminent Engravers we read of: The former was so esteem'd by *Alexander*, that he forbad any body else to *Engrave* his Head: And *Augustus's* head, *Engraven* by the latter, was found so beautiful, that the succeeding Emperors chose it for their Seal.

All the polite Arts having been buried under the Ruins of the *Roman* Empire, the Art of *Engraving* on Stones met the same Fate. It was retriev'd in *Italy*, at the Beginning of the 15*th* Century; when one *John* of *Florence*, and after him *Dominic* of *Milan*, perform'd Works of this Kind no Way to be despised.

From that Time, such Sculptures became common enough in *Europe*, and particularly in *Germany*, whence great Numbers were sent into other Countries; but they came short of the Beauty of those of the Ancients; especially those on precious Stones; for as to those on Crystal, the *Germans*, and after their Example, the *French*, etc. have succeeded well enough.

In this Branch of *Engraving*, they make Use either of the Diamond, or Emery. The Diamond, which is the hardest, and most perfect of all precious Stones, is only cut by it self, or with its own Matter.

The first Thing, is to cement two rough Diamonds to the Ends of two Sticks, big enough to hold them steady by, in the Hands; and to rub or grind 'em against each other, till they be brought to the Figure desired. The Dust, or Powder, rub'd off, serves afterwards to polish 'em; which is perform'd with a Kind of Mill, that turns a Wheel of soft Iron.

The Diamond is fix'd in a Brass Dish, and thus apply'd to the Wheel, which is cover'd with Diamond Dust, mix'd up with Oil of Olives: And when the Diamond is to be cut facet-wise, they apply first one Face, then another, to the Wheel.

Rubies, Saphirs, and Topazes, are cut and form'd the same Way, on a Copper Wheel; and polish'd with *Tripoli*, diluted in Water.

As to Rubies, Emeralds, Hyacinths, Amethists, Grenats, Agats, and other of the softer Stones, they are cut on a Leaden Wheel, moisten'd with Emery and Water, and polish'd with *Tripoli*, on a Pewter-Wheel. *Lapis*, Opal, *etc*. are polish'd on a Wooden Wheel.

To fashion and *Engrave* Vases of Agat, Crystal, *Lapis*, or the like, they have a Kind of Lathe, like that of the Pewterers; excepting, that whereas the latter is to hold the Vessels, which are to be wrought with proper Tools; the former generally holds the Tools, which are turn'd by a Wheel, and the Vessels held to 'em to be cut and *Engraved*, either in Relievo, or otherwise: Remembring from Time to Time, to moisten the Tools with Diamond Dust and Oil; or, at least, Emery and Water.

To *Engrave* the smaller Works of any of these Stones, when polish'd; such as Medals, or Seals, *etc*. they use a little Iron Wheel, the two Ends of whose Axes are receiv'd within two Pieces of Iron placed upright, as in the Turners Lathe; to be brought closer, or set further apart, at Pleasure. At one End of one of the Axes, are fitted the proper Tools, being kept tight by a Skrew. Lastly, the Wheel is turn'd by the Foot; and the Stone applied by the Hand to the Tool; and thus shifted and conducted as Occasion requires.

The Tools are generally of Iron, sometimes of Brass. As to their Form, it is various, but generally bears some Resemblance to Chissels, Gouges, *etc*. Some have small round Heads like Buttons; others, like Ferrels, to take the Pieces out; others, flat, *etc*. These Tools are not applied directly against the Stone; but, as it were, sidewise; thus wearing, and, as it were, grinding off the Substance. And still, whether it be Figures, or Letters, or Characters, the Manner of Application is the same. The Tools, as above observ'd, are to be frequently moisten'd with

Diamond Dust, and Oil of Olives. When the Stone is *Engraven*, they polish it on Wheels of Brushes, made of Hogs Bristles and *Tripoli*. For the larger, and less delicate Works, they have Copper, or Pewter Tools, on purpose to polish the Ground, or plain Parts, with *Tripoli, etc.* which they apply after the same Manner, as those wherewith the *Graving* is perform'd.

ENGRAVING *on Steel*, is chiefly employ'd in cutting Punches, Matrices, and Dies, proper to strike Coins, Medals, and Counters.

The Method of *Engraving*, with the Instruments, *etc.* are the same for Coins, as for Medals and Counters: All the Difference consists in their greater, or less Relievo; the Relievo of Coins being much less considerable than that of Medals; and that of Counters still less than that of Coins.

The *Engraver* in Steel usually begins with Punches, which are in Relievo, and serve for making the Creux, or Cavities of the Matrices, and Dies. Tho', sometimes, he begins immediately with the Creux; but 'tis only when the intended Work is to be cut very shallow. The first Thing is to design his Figures; then he moulds 'em in white Wax, of the Size and Depth required: And from this Wax he graves his Punch.

This Punch, is a Piece of Steel, or at least of Iron and Steel mix'd; on which, ere they temper, or harden it, the intended Figure, whether a Head, or a Reverse, is cut or chissel'd in Relievo.

ENTABLATURE, or ENTABLEMENT, in Architecture, by *Vitruvius* and *Vignola* call'd *Ornament*, is that Part of an Order of Column which is over the Capital; comprehending the *Architrave*, *Frize*, and *Corniche*. See ARCHITRAVE, FRIZE, and CORNICHE.

The *Entablature* is also called the *Trabeation*: It is different in the different Orders: Indeed it consists of the three grand Parts, or Divisions above-mentioned, in all; But those Parts consist of a greater, or less Number of particular Members, or Sub-divisions, as the Orders are more or less rich. See ORDER.

Vignola makes the *Entablature* a Quarter of the Height of the whole Column, in all the Orders. See COLUMN.

In the *Tuscan* and *Doric*, the *Architrave*, *Frize*, and *Corniche*, are all of the same Height. See TUSCAN and DORIC.

In the *Ionic*, *Corinthian*, and *Composite*, the whole *Entablature* being 15 Parts; 5 of them are allow'd for the *Architrave*, 4 for the *Frize*, and 6 for the *Corniche*. See IONIC, CORINTHIAN, *etc.*

ENTABLATURE, or ENTABLEMENT, is also used for the last Row of Stones on the Top of the Wall of a Building, whereon the Timber and the Covering rest.

The Word is form'd of the Latin *Tabulatum*, or *Intabulamentum.*

As it is frequently made to project beyond the Naked of the Wall to carry off the Rain, some Authors call it in Latin *Stillicidium*, or Drip.

Such an *Entablement* does not bear far enough out: It lets the Water fall on the Foot of the Wall.

ENTRY, or ENTRANCE, in its general Sense, a Door, Gate, Passage, *etc.* thro' which to *Enter*, or arrive within a Place. See DOOR, GATE, *etc.*

ENVELOPE, in Fortification, is a Mount of Earth, sometimes raised in the Ditch of a Place, and sometimes beyond it; being either in Form of a simple Parapet, or of a small Rampart border'd with a Parapet.

These *Envelopes* are made, where one would only cover the weak Places with single Lines; without any Design of advancing towards the Field, which cannot be done but by Works which require a great deal of Breadth; such as Horn-Works, Half-Moons, *etc.*

These *Envelopes* are sometimes called *Sillons*, *Contregards*, *Conserves*, *Lunettes*, etc.

EPAULE, or ESPAULE, in Fortification, the *Shoulder* of the Bastion; or the Angle made by the Face and Flank; whence that Angle is often called the *Angle of the Epaule*. See BASTION and SHOULDERING.

The Word is pure *French*, and literally signifies *Shoulder*.

EPAULEMENT, in Fortification, a Side-Work hastily thrown up, to cover the Canon, or the Men.

It is made either of Earth thrown up, of Bags of Earth, Gabions, or of Fascines and Earth, of which latter make, the *Epaulements* of the Places of Arms, for the Cavalry, behind the Trenches are.

EPAULEMENT, is also used for a Demi-Bastion, consisting of a Face and Flank, placed at the Point of a Horn- or Crown-Work. Also, for a little Flank, added to the Sides of Horn-Work, to defend them when too long. Also for the Redoubts made on a right Line, to fortifie it. And, lastly, for a Square Orillon, which is a Mass of Earth almost square, faced and lined with a Wall, and designed to cover the Canon of a Casement.

EPISTYLE, in the antient Architecture, a Term used by the *Greeks*, for what we call *Architrave*, *viz*. a Massive of Stone, or Piece of Wood, laid immediately over the Capital of a Column. See ARCHITRAVE.

The *Epistyle* is the first, or lowest Member of the Entablature. See ENTABLATURE.

ESCALADE, or SCALADE, a furious Attack upon a Wall, or a Rampart; carried on with Ladders, to mount by without proceeding in Form, breaking Ground, or carrying on Works to Secure the Men.

ESPAULE, ESPAULEMENT, *etc.* in Fortification. See EPAULE, EPAULEMENT, *etc.*

ESPLANADE, in Fortification, call'd also *Glacis*, a Part serving the Counterscarp, or cover'd Way, for a Parapet; being a Declivity, or Slope of Earth, commencing from the Top of the Counterscarp, and losing it self insensibly in the Level of the Campaign.

ESPLANADE, also signifies the Ground which has been levell'd from the *Glacis* of the Counterscarp, to the first Houses; or the vacant Space between the Works and the Houses of the Town.

The Term is also applied in the general for any Piece of Ground render'd flat or level, which before had some Eminence that incommoded the Place.

ESQUISSE, in the *French* Painting, a Term signifying the first slight sketch, or draught of a Picture; the first Thought of a Design drawn hastily with a Crayon, or in Colours on Paper, Canvas, or the like; in Order to be finish'd, and painted or engraven afterwards. See DESIGN.

He had not the Trouble of making a finish'd, and correct Design; but went to work upon the *Esquisse.*

The Word is form'd of the Italian *Schizzo*, a Splash; by Reason an *Esquisse* of a Painting only represents, as it were, Splashes, or Dabs of Colours.

ESTRADE, a *French* Term, literally signifying a public Road, or High-way. Hence the Military Phrase, *Battre l'Estrade*, to *Beat the Estrade*, that is, to send Scouts, or Horsemen, to get Intelligence, to learn the Dispositions of the Enemy, and inform the General of every Thing like to fall in the Way. An Army never marches, without sending *Batteurs d'Estrade* on every Side.

The Word is form'd of the Italian *Strada*, Street, or Road, which is derived from the Latin *Strata*, a paved Street. Some derive it from *Estradiots*, who were Cavaliers anciently employ'd in *beating the Estrade*.

ESTRADE is also us'd, for a little Elevation in the Floor of a Room, frequently encompass'd with an Alcove, or Rail, for the placing a Bed in; and sometimes, as in *Turkey*, only cover'd with fine Carpets, to receive Visitors of Distinction in.

ETCHING, a Method of Engraving on Copper; wherein the Lines, or Strokes, instead of being cut with a Tool, or Graver, are eat with Aqua fortis.

Etching was invented much about the same Time with *Engraving* on Copper, properly so call'd; by *Alb. Durer*, and *Lucas*. It has several Advantages over that Art; as, that it is done with more Ease and Expedition; that it requires fewer Instruments; and even, that is represents divers Kinds of Subjects better, and more agreeably to Nature, as Landskips, Ruins, Grounds, and all small, faint, loose, remote Objects, Buildings, *etc.*

The Method of *Etching* is thus: The Plate being well polish'd, is heated over the Fire; and when hot, cover'd over with a peculiar *Ground*, or Varnish. When cold again, the Ground is blacken'd with the Smoak of a Candle; and on this Ground, thus blacken'd, the Back of the Design, or Draught, is laid.

This done, the Design remains to be calk'd, or transferr'd upon the Plate; which is more easily effected, than in the common Graving; for the Back of the Design having been before rubb'd over with red Chalk, nothing remains but to trace over all the Lines and Strokes of the Draught with a Needle or Point; which pressing the Paper close down to the Ground, occasions the Wax therein to lay hold of the Chalk, and so bring off the Marks of the several Lines: So as at length, to shew a Copy of the whole Design in all its Correctness.

The Draught thus calk'd, the Artist proceeds to draw the several Lines, and Contours with a Point, thro' the Ground, upon the Copper.

To finish his Work, he makes use of Points of divers Sizes, or Bignesses; and presses on them sometimes more strongly, and sometimes more lightly, according as the several Parts of the Figures, *etc.* require more or less Strength or Boldness: Some of the Points being as fine as Needles, for the tender, Hair Stroaks, as the remoter, fainter Objects; and others again, as big as Bodkins, made oval-wise, for the deeper Shadows, and the Figures in the Front of the Work.

Things thus prepared, a Rim, or Border of Wax, is rais'd round the Circumference of the Plate, and *Aqua fortis* pour'd on; which, by the said Border, is kept from running off at the Edges.

The Ground being impenetrable to that corrosive Water, the Plate is defended from it every where but in Lines, or Hatches, cut thro' it with the Points; which, lying open, the Water passes thro' them to the Copper, and eats into the same, to the Depth required: Which done, it is pour'd off again.

Of *Etching* Grounds, it must be observed, there are two Kinds, the one *soft*, and the other *hard*. There are also two Kinds of *Aqua Fortis*: The one *White*, which is only used with the soft Ground, and is applied as above directed: The other *Green*, made of Vinegar, Common Salt, *Sal Ammoniac* and Verdegrease. This is used indifferently with either Kind of Ground: Its Application is somewhat different from that of the *White*.

Without making any Border, they pour it on the Plate, which is placed for that Purpose a little inclined; and as the Water runs off, it is received in a Vessel placed underneath. This they repeat, pouring it again and again, till it has eaten deep enough.

Add, that the *Aqua fortis*, of which Kind soever it be, most not continue equally long, or be pour'd equally often, on all the Parts of the Design. The remote parts must be bitten more slightly, than those nearer to the View.

To manage this, they have a Composition of Oil and Grease, wherewith they cover the Parts that are to be bitten no further. Or else they lay the Composition on as a defensitive at first, and take it off again when they find proper. In Effect, they are every now and then covering and uncovering this or that Part of the Design, as Occasion requires; the Conduct of the *Aqua fortis* being one of the principal Concerns in the whole Art, and that on which the Effect of the whole very much depends. The Operator is also to be very attentive to the Ground, that it don't fail, or give Way, in any part, to the Water; and where it does, to stop up the Place with the Composition aforesaid.

Lastly, it is to be remember'd, that a fresh Dip of *Aqua Fortis* be never Given, without first washing out the Plate in Fair Water, and drying it at the Fire.

The *Aqua fortis* having done its Part, the Ground is taken off, and the Plate wash'd and dried; after which nothing remains but for the Artist to examine it with his Graver in his Hand, to touch it up, and heighten it, where the *Aqua fortis*, etc. has miss'd.

EURYTHMY, EURYTHMIA, in Architecture, Painting, and Sculpture; a certain Majesty, Elegance, and Easiness appearing in the Composition of divers Members, or Parts of a Body, Building, or Painting; and resulting from the fine Proportions thereof. See PROPORTION.

The Word is *Greek*, and signifies literally a Consonance, or fine Agreement, or as we may call it, a Harmony of all the Parts; being compounded of *well*, and *Rythmus*, a Cadence, or Agreement of Numbers, Sounds, or the like Things.

Vitruvius ranks the *Eurythmia* among the essential Parts of Architecture: He describes it as consisting in the Beauty of the Construction, or Assemblage of the several Parts of the Work, which render its Aspect, or whole Appearance graceful: *E. gr.* when the Height corresponds to the Breadth, and the Breadth to the Length, *etc.*

'From these three Ideas, (or *Designs*, *viz. Orthography*, *Scenography*, 'and *Profile*) it is, that the same *Eurythmy*, *Majestica*, and *Venusta* '*Species Edificii*, does Result; which creates that agreeable Harmony 'between the several Dimensions; so as nothing seems disproportionate, 'too long for this, or too broad for that, but corresponds in a just and 'regular Symmetry, and Consent of all the Parts with the whole.' *Evelyn's Account of Archit. etc.* See SYMMETRY.

EUSTYLE, in Architecture, a Kind of Edifice, where the Pillars are placed at a most convenient Distance one from another; the Intercolumniations being all just two Diameters and a Quarter of the Column; except those in the middle of the Face before and behind, which are three Diameters distant. See INTERCOLUMNS.

The *Eustyle* is a Medium between the *Pycnostyle* and *Aræostyle*. See PYCNOSTYLE, *etc.*

The Word is *Greek*, being form'd of *bene*, well, and *Column*.

Vitruvius, L. III. *C*. 2. observes, that the *Eustyle* is the most approved of all the Manners of Intercolumniation; and that it surpasses all the rest, in Conveniency, Beauty, and Strength.

EXCAVATION, the Act of hollowing, or digging a Cavity; particularly in the Ground.

The *Excavation* of the Foundations of a Building, by the *Italians* call'd *Cavasione*, is settled by *Palladio* at a sixth Part of the Height of the whole Building. Unless there be Cellars under Ground, in which Case he would have it somewhat more. See FOUNDATION.

The Word is Latin, *Excavatio*, form'd of *ex* and *cavus*, hollow, or *Cavea*, a Pit, *etc*.

EXPOSURE, or EXPOSITION, in Gardening, the Aspect, or Situation of a Garden, Wall, Building, or the like, with respect to the Sun, Winds, *etc*. See GARDEN.

There are four regular Kinds of *Exposures*, viz. *East*, *West*, *North*, and *South*. But it must be observed, that among Gardeners, these Terms signifie just the contrary to what they do among Geographers.

The Gardeners, in Effect, do not give the Names East, West, *etc*. to the Places where the Sun is; but to those whereon he shines; and they consider the Manner wherein he shines, whether as to the whole Garden, or some one of its Sides.

If they find, that the Sun is Rising, and during the first half of the Day, continues to shine on one Side of a Garden or Wall; they call that an *Eastern Exposure*, or *East Wall*, *etc*. And if the Sun begin to shine later, or end sooner, 'tis not a proper *Eastern Exposure*.

For the same Reason, they call the *West*, the Side the Sun shines on in the latter half of the Day, *i.e.* from Noon to Night. And accordingly, the *South*, or *South Exposure*, is the Place whereon he shines, from about Nine a Clock in the Morning, till Night; or which, in the general, he shines longest on in the whole Day; and the Part he shines least on, is the *North*, or *Northern Exposure*; at what Hour soever it begin or end, being usually from 11 a Clock to 1.

Such is the Gardeners Language with Regard to the *Exposures*, and particularly those of Walls; by which we are let into the Signification of this or the like Expressions, usual among them.—My *East* Wall proves, hits, or thrives better, than my *West*. My *East* Fruit Trees have had fewer Showers than my *West*, etc.

The *Eastern* and *Southern Exposures* are, by common Consent of all Gardeners, the two principal; and have a considerable Advantage above the rest. A *West Exposure* is not much Amiss; at least, 'tis better than a *Northern* one, which is the worst of all. Each has its Inconveniencies.

The *Eastern*, commencing differently at different Seasons of the Year, and ending about Noon, subjects the Trees, *etc*. to the *N. E.* Winds, which wither the Leaves and new Shoots, blow down the Fruit, *etc*. beside that it has little Benefit of Rains, which come mostly from the *West*. Yet does the Reverend Mr. *Lawrence* judge, the *East* better than the *West* Wall, for all Kinds of Fruit: Not that it has more Hours of Sun, or that there are any peculiar Virtues in the *Eastern* Rays, but because the early Rays of the Sun do sooner take off the cold chilly Dews of the Night.

M. *Gentil* recommends the *Eastern Exposure*, as best for all Kinds of Peaches; adding, that they ripen soonest, grow bigger, are better colour'd, and of a finer Tast than in any other: But Mr. *Carpenter* restrains the Rule to the early and middle Sorts: For the backward, he rather chuses a *Southern*, or *South East Exposure*; which is best for all late Fruits; because the Influence of the Sun is strongest, and continues longest therein.

The *Western*, accounted from half an Hour past 11, till Sun-set, is backwarder than an *Eastern* one by 8 or 10 Days; but it has this Advantage, that it receives little Damage from the Frosts, which melt before the Sun comes to shine upon the Fruit, and fall off like Dew, without doing any Prejudice: So that it may bear Apricocks, Peaches, Pears, and Plums. But it is incommoded with *North West* Winds in the Spring, as also the Autumn Winds, which blow down a deal of Fruit.

The *Northern Exposure* has less Sun than the West; yet is it not without its Advantages. In the Northerly Parts of *England*, it bears little else but Pears, Cherries, and Plums. But in the warmer Parts is serves for Apricocks, which have the Advantage of continuing later, than in any other *Exposure*, besides being free from Insects.

The *Southern Exposure*, accounted from about 9 till 4, is recommended for Peaches, Pears, Grapes, and Plums.

EXPRESSION, in Painting, the natural and lively Representation of the Subject, or of the several Objects intended to be shewn.

The *Expression* consists principally in representing the human Body, and all its Parts, in the Action suitable to it; in exhibiting in the Face, the several Passions proper to the Figures; and observing the Motions they impress on the other external Parts.

The Term *Expression*, is ordinarily confounded with that of Passion: But they differ in this, that *Expression* is a general Term, implying a Representation of an Object, agreeably to its Nature and Character, and the Use, or Office it is to have in the Work; and Passion, in Painting, is a Motion of the Body, accompanied with certain Dispositions, or Airs of the Face, which mark an Agitation in the Soul. So that every Passion is an *Expression*; but not every *Expression* a Passion.

The Laws, or Rules of Expression *in Painting.*

Expression, we have said, is a Representation of Things according to their Character; and may be consider'd either with respect to the *Subject* in general; or to the *Passions* peculiar thereto.

I. With regard to the *Subject*, 'tis to be observed, 1°. That all the

Parts of the Composition are to be transform'd, or reduced, to the Character of the Subject; so as they may conspire to impress the same Sentiment, Passion, or Idea: Thus, *e. gr.* in a Representation of Joy and Peace, every Thing is to appear calm and agreeable; of War, turbulent, and full of Terror, *etc.*

2°. In Order to this, if any Circumstance occur in History, or Description, that would invert, or take from the Idea; it must be suppressed; unless essential to the Subject.

3°. To this End, the History, or Fable, is to be well studied in the Authors who describe it, in Order to conceive its Nature and Character truly, and impress it strongly on the Imagination; that it may be diffused and carried thro' all the Parts of the Subject.

4°. A Liberty to be taken of chusing favourable Incidents, in order to diversifie the *Expression*; provided they be not contrary to the principal Image of the Subject, or the Truth of History.

5°. The Harmony of the *tout Ensemble* to be particularly regarded, both with regard to the Actions, and the Light and Colour.

6°. The Modes and Customs to be observed; and every Thing made conformable to Time, Place, and Quality.

7°. The three Unities of Time, Place and Action, to be observed: That is, nothing to be represented in the same Picture, but what passes at the same Time, and may be seen at the same View.

II. With regard to the Particular *Passions* and *Affections* of the Subject; the Rules are, 1°. That the Passions of Brutes are few and simple, and have almost all an immediate Respect, either to Self-Preservation, or the Propagation of the Kind: But, in Men, there is more Variety; and accordingly, more Marks and *Expressions* thereof. Hence, man can move his Eye-brows, which, in Brutes, are immoveable: And can likewise move the Pupil every way, which Brutes cannot.

2°. Children, having not the Use of Reason, act much on the Footing of Brutes; and *express* the Motions of their Passions directly, and without Fear or Disguise.

3°. Tho' the Passions of the Soul, may be *express'd* by the Actions of the Body; 'tis in the Face they are principally shewn; and particularly in the Turn of the Eye, and the Motion of the Eye-brows.

4°. There are two Ways of lifting up the Eye-brows, the one at the middle, which likewise draws up the Corners of the Mouth, and argues pleasant Motions: The other, at the Point next the Nose, which draws up the middle of the Mouth, and is the Effect of Grief and Sadness.

5°. The Passions are all reducible to Joy and Sadness; each of which is either simple, or mix'd and passionate.

6°. Simple Joy causes a Dilatation of all the Parts: The Eye-brows rise in the middle, the Eyes half open, and smiling, the Pupil sparkling and moist, the Nostrils a little open, the Cheeks full, the Corners of the Mouth drawn a little upwards, the Lips red, the Complexion lively, the Forehead serene.

7°. Passionate Joy proceeding from Love, shews the Forehead smooth and even, the Eye-brows a little elevated on the Side the Pupil is turn'd to, the Eyes sparkling and open, the Head inclined towards the Object, the Air of the Face smiling, and the Complexion ruddy:—That proceeding from Desire, shews it self by the Body, the Arms extended towards the Object, in uncertain and unquiet Motions.

8°. Simple Sadness, is *Express'd* by the Body being cast down, the Head carelessly hanging aside, the Fore-head Wrinkled, the Eye-brows rais'd to the midst of the Fore-head, the Eyes half shut, the Mouth a little open, the Corners downwards, the under-Lip pointing and drawn back, the Nostrils swell'd, and drawn downwards.—That mix'd with *Fear*, causes the Parts to contract and palpitate, the Members to tremble and fold up, the Visage to be pale and livid, the Point of the Nostrils elevated, the Pupil in the middle of the Eye, the Mouth openest at the Sides, and the under-Lip drawn back.—In that mix'd with *Anger*, the Motions are more violent, the Parts all agitated, the Muscles swell'd, the Pupil wild and sparkling, the Point of the Eye-brows fix'd towards the Nose, the Nostrils open, the Lips big and press'd down, the Corners of the Mouth a little open and foaming, the Veins swell'd, and the Hair erect.—That with *Despair*, resembles the last, only more excessive and disorder'd.

9°. The Hand has a great Share in the *Expression* of our Sentiments and Passions: The raising of the Hands conjoyn'd—towards Heaven, *expresses* Devotion: Wringing the Hands, Grief: Throwing them towards Heaven, Admiration: Fainting, and dejected Hands, Amazement and Despair: Holding the Hands, Idleness: Holding the Fingers indented, Musing: Holding forth the Hands together, Yielding and Submission: Lifting up the Hand and Eye to Heaven, calling God to Witness: Waving the hand from us, Prohibition: Extending the right hand to any one, Pity, Peace, and Safety: Scratching the head, Thoughtfulness and Care: Laying the Hand on the Heart, solemn Affirmation: Holding up the Thumb, Approbation: Laying the Fore-finger on the Mouth, bidding Silence: Giving with the Finger and Thumb, *parce dare*: And the Fore-finger put forth, the rest contracted, *Monstrari et dicier hic est.*

10. The *Sex* of the Figure to be regarded; and Man, as he is of a more vigorous and resolute Nature, to appear in all his Actions freer and bolder than Women, who are to be more reserved and tender.

11. So also the *Age*, the different Stages whereof incline to different Motions both of Body and Mind.

12. The *Condition*, or Honours, a Person is invested withal, renders their Actions more reserved, and their Motions more grave; contrary to the Populace, who observe little Conduct or Restraint; giving themselves, for the most part, up to their Passions; whence their external Motions become rude and disorderly.

Lastly, in *Spirits* we must retrench all those corruptible Things, which serve only for the Preservation of Life, as Veins, Arteries, *etc.* only retaining what may serve for the Form and Beauty of the Body.—In

Angels, particularly, as being symbolical Figures, we are to mark out their Offices and Virtues, without any Draught of sensual Passions; only appropriating their Characters to their Functions of Power, Activity, and Contemplation.

EYE, in Architecture is used for any round Window made in a Pediment, an Attic, the Reins of a Vault, or the like. See WINDOW.

BULLOCKS-EYE, or *Oeuil de Bœuf*, is a little Sky-Light in the Covering, or Roof, intended to illumine a Granary, or the like. The same Term is applied to the little Lutherns in a Dome, as in that at St. *Peters* at *Rome*, which has 48 in three Rows. See LUTHERN.

EYE *of a Dome*, is an Aperture at the Top of the Dome; as that of the *Pantheon* at *Rome*, or of St. *Pauls* at *London*. It is usually cover'd with a Lanthorn. See DOME.

EYE *of the Volute*, in Architecture, is the Centre of the Volute, or that Point wherein the Helix, or Spiral, whereof it is form'd, commences: Or it is the little Circle in the middle of the Volutes, wherein are found the thirteen Centres for the describing the Circumvolutions thereof. See VOLUTE.

F

FACE, or *Facade*, is sometimes used for the *Front* of a Building; or the Side on which the chief Entrance is: as also for the Side it presents to a Street, Garden, Court, *etc*. And sometimes for any Side opposite to the Eye. See FRONT.

FACE, *Facia*, or *Fascia*, in Architecture, is a flat Member, having a considerable Breadth, and but a small Projecture: such are the Bands of an Archdrave, Larmier, *etc*. See FASCIA.

FACE *of a Place*, includes the Curtain, together with the two Flanks raised above it, and the two *Faces* of the Bastion, which look towards one another, and flank the Angle of the Tenaille.

FACE *of a Bastion*, or *of a Bulwark*, is the most advanced Part of a Bastion towards the Field; or the Distance comprehended between the Angle of the Shoulder, and the Flank'd Angle. See BASTION.

The *Face*, or Flat of a Bastion, Ravelin, Demi-bastion, Horn-work, Crown-work, *etc*. is more usually understood of that part of the Work terminated between the Angle of the Epaule, and the flank'd Angle; or between the Point of the Bastion, and the Flank.

FACE *prolonged*, in Fortification, is that Part of a Line of Defence rasant, which is between the Angle of the Epaule or Shoulder of a

Bastion, and the Curtain; or the Line of a Defence rasant cut by the length of the *Face*.

FACE *of a Stone*, is the Surface or Plain of Stone, which is to lie in the Front of the Work. The *Face* is easily known when the Stone is scapled, as being always opposite to the Back; and the Back going rough as it comes from the Quarry.

But rough Stones, the Workmen generally chuse to make one of those Sides the *Face*, which when in the Quarry, lay perpendicular to the Horizon; and consequently the Breaking, and not the Cleaving way of the Stone. See STONE.

FACIA, in Architecture. See FASCIA.

FASCIA, in Architecture, by the Workmen call'd *Facia*, *Facio*, or *Face*, a Broad-list, Fillet, or Band; particularly used in Architraves, and Pedestals. See ARCHITRAVE, and PEDESTAL.

The Architrave consists of three *Fasciæ*, or Bands; thus call'd by *Vitruvius*, as resembling *Swaths*, call'd in Latin *Fasciæ*.

That Author admits no *Fasciæ*, in the *Tuscan*, and *Doric* Architrave, *i.e.* he makes it all plain, with out any Division, or cantoning into Parts or *Fasciæ*: But the modern Architects take Liberty to differ from him herein. See TUSCAN, DORIC, *etc*.

In Brick-buildings, the Juttings out of the Bricks, beyond the Windows, in the several Stories, except the highest, are call'd *Facia's*, or *Fasciæ*.

These are sometimes plain, and sometimes moulded. But the Moulding is only a *Cima reversa*, or an O G, at the Bottom, with two plain Courses of Bricks over it; then an *Astragal*, and lastly a *Boultine*.

FASCINES, or *Faggots*, in Fortification, are small Branches of Trees, or Bavins, bound up in Bundles; which being mix'd with Earth, serve to fill up Ditches, to screen the Men, make the Parapets of Trenches, *etc*.

Some of them are dipp'd in melted Pitch or Tar; and being set on Fire, serve to burn the Enemies Lodgements, or other Works.

A pitch'd *Fascine* is a Foot and half about: A *Fascine* for Defence, two or three Foot.

In the corrupt *Latin*, they use *Fascenina*, *Fascennia*, and *Fascinata*, to signify the Pales, *Fascines*, etc. used to enclose the antient Castles, *etc*.

FASTIGIUM, in Architecture, the same with *Pediment*. See PEDIMENT.

FEATHER *Edg'd Boards* and *Planks*, are those which are thicker on one side, than on the other.

FENCE-Walls, are Walls of Brick, or Stone, made round Gardens, *etc.* See WALL.

FIGURE, in Architecture, and Sculpture, is used for Representations of Things, made in solid Matters; such as Statues, *etc.*

In this Sense we say, *Figures* of Brass, of Marble, of Stuck, of Plaster, *etc.* But in this Sense too, the Term is more usually applied to human Representations, than other Things. Thus we say, an Equestrian *Figure*, for a Man on Horseback. See STATUE.

Daviler, however, observes, that those, either represented sitting, as Popes, *etc.* or kneeling, as those on Monuments, *etc.* or laid all along, as Rivers, *etc.* are more properly called *Figures*, than Statues.

FIGURE, in Painting, and Designing, is the Lines and Colours, that form the Representation of a Man, or other Animal.

Thus we say, There are above an hundred *Figures* in this piece: Such a *Figure* is lame, *etc.*

But the Term *Figure* is in a more immediate and peculiar manner understood of human Personages: Thus, a Painting is said to be full of *Figures*, when there are abundance of Representations of Men: And a Landskip is without *Figures*, when there is nothing but Trees, Plants, Mountains, *etc.*

FIGURE, in Fortification, is the Plan of the fortified Place; or the interior Polygon.

When the Sides and Angles are equal, 'tis called a *Regular*; when unequal, an *Irregular Figure*.

FILLET, or FILET, in Architecture, a little square Member, or Ornament, used in divers Places, and on divers Occasions; but generally as a sort of Corona, over a greater Moulding.

The *Fillet* is the same with what the *French* call *Reglet*; the *Italians*, *Lista*, or *Listella*; and others, *Band*, or *Bandelette*. See REGLET, LIST, *etc.*

The Word is *French*, Fillet, of Fil, Thread.

FILLET is also used among Painters, Gilders, *etc.* for a little Rule, or Reglet of Leaf Gold, drawn over cerain Mouldings; or on the Edges of Frames, Pannels, *etc.* especially when painted white; by way of Enrichment.

FINISHING, in Architecture, *etc.* is frequently applied to a Crowning, Acroter, *etc.* rais'd over a Piece of Building, to terminate, and *finish*, or compleat it. See CROWNING.

FLANK, in Fortification, is a Line, drawn from the Extremity of the Face, towards the Inside of the Work.

Such is the Line B A Tab. Fortification. Fig. 1.

Or, *Flank* is that Part of the Bastion, which reaches from the Courtin to the Face, and *defends* the opposite Face, the *Flank*, and the Courtin. See BASTION.

There is also an *Oblique*, or *Second Flank*, which is that Part of the Courtin, where they can see to scour the Face of the opposite Bastion, and is the Distance between the Lines Razant, and Fitchant.

The *Low*, *Cover'd*, or *Retired Flank*, is the Platform of the Casemate, which lies hid in the Bastion.

Simple FLANKS, are Lines, going from the Angle of the Shoulder, to the Courtin; whose chief Office is for Defence of the Moat, and Place.

FLANK *of the Courtin*, or *Second Flank*, is that Part of the Courtin between the *Flank* and the Point, where the Fitchant Line of Defence terminates.

FLANK *Fitchant* is that from whence a Canon playing, *fireth* its Bullets directly in the Face of the opposite Bastion.

FLANK *Razant* is the Point, from whence the Line of Defence begins, from the Conjunction of which with the Courtin, the Shot only raseth the Face of the next Bastion; which happens when the Face cannot be discover'd but from the *Flank* alone.

Retired FLANK, or the *Lower* and *Covert Flank*, is that exterior Part thereof, which advanceth to secure the innermost; which advanced Part, if it be rounded, is called the Orillon; so that this *Flank* is only the Platform of the Casemate, which lies hid in the Bastion.

FLANK'D *Angle*, is the Angle form'd by the two Faces of the Bastion, and which of course forms the Point of the Bastion.

FLANK'D *Tenaille*, call'd also *double Tenaille*.

FLANKING, in the general, is the Act of discovering and firing upon the Side of a Place, Body, Batalion, *etc.*

To *flank* a Place, is to dispose a Bastion, or other Work in such manner, as that there shall be no Part of the Place, but what may be defended, *i.e.* may be play'd on, both in Front, and Rear.

To *flank* a Wall with Towers. This Bastion is *flank'd* by the opposite Flank, and a Half-Moon. This Horn-Work is *flank'd* by the Courtin.

Any Fortification that has no Defence but just right forwards, is faulty: And to render it compleat, one Part ought to be made to *flank* the other. Hence the Courtin is always the strongest Part of any Place, because 'tis *flank'd* at each End.

Batalions also are said to be *flank'd* by the Wings of the Cavalry: And a House is sometimes said to be *flank'd* with two Pavillions, or two Galleries; meaning it has a Gallery, *etc.* on each Side.

FLIGHT of a *Stair-Case*, see STAIRS.

FLOOR, in Building, the Underside of a Room, or that Part we walk on.

Floors are of divers sorts; some of Earth, some of Brick, others of Stone, *etc.*

Carpenters, by the Word *Floor*, understand as well the framed Work of Timber, as the Boarding over it.

Earthen Floors are commonly made of Lome, and sometimes, especially to make Malt on, of Lime, and Brook Sand, and Gun Dust, or Anvil Dust from the Forge.

Ox-blood, and fine Clay, temper'd together, Sir *Hugh Plat* says, makes the finest *Floor* in the World.

For *Brick* and *Stone Floors*, see PAVEMENT.

For boarded *Floors*, it is observable, that the Carpenters never *floor* their Rooms with Boards, till the Carcass is set up, and also inclos'd with Walls, lest the Weather should wrong the *Flooring*; yet they generally rough-plane their Boards for the *Flooring*, before they begin any thing else about the Building, that they may set them by to dry and season; which is done in the most careful manner.

FLOOR *of a Ship*, strictly taken, is only so much of her Bottom, as she rests on, when a-ground; so that such Ships as have long, and withal broad *Floors*, lye on the Ground with most Security, and are not apt to feel, or tilt on one Side; whereas others, which are narrow in the *Floor*, or in the Sea-Phrase, *Cranck'd by the Ground*, cannot be grounded without Danger of being overthrown.

FLOWER of the *Capital*, is an Ornament of Sculpture, in form of a Rose, in the middle of the Sweep, of the *Corinthian* Abacus: In that of the Composite, 'tis an imaginary kind of Flower. See CAPITAL.

FLUTES, or *Flutings*, in Architecture, are perpendicular Channels, or Cavities, cut along the Shaft of a Column or Pilaster. See COLUMN, and PILASTER.

They are supposed to have been first introduced in Imitation of the Plaits of Womens Robes; and are therefore called by the *Latins*, *Striges*, and *Rugæ*. See STRIGES, *etc.*

The *French* call them *Cannelures*, as being Excavations; and we, *Flutes*, or *Flutings*, as bearing some Resemblance to the musical Instrument so called.

They are chiefly affected in the *Ionic* Order, where they had their first Rise; tho' they are also us'd in all the richer Orders, as the *Corinthian* and *Composite*; but rarely in the *Doric*; scarce ever in the *Tuscan*. See IONIC, CORINTHIAN, COMPOSITE, DORIC, *etc.*

Their Number is 24, tho' in the *Doric* only, 20.

Each *Flute* is hollow'd in, exactly a Quadrant of a Circle.

Between the *Flutes* are little Spaces that separate them, called by *Vitruvius*, *Striæ*, and by us, *Lists*; tho' in the *Doric*, the *Flutes* are frequently made to join each other, without any intermediate Space at all;

the List being sharpen'd off to a thin Edge, which forms a Part of each *Flute*.

In some Buildings we see Columns with *Flutes* that go winding round the Shaft, spirally; but this is rather look'd on as an Abuse.

Vignola determines the Depth of the *Flutes* by taking the Angle of an equilateral Triangle for the Center.

Vitruvius describes it from the Middle of the Square, whose Side is the Breadth of the *Flute*: Which latter Method makes them deep.

The *Flutes*, or *Striges*, are frequently fill'd up with a prominent, or swelling Ornament, sometimes plain, in form of a Staff or Reed; and sometimes a little carved, or enrich'd, in imitation of a Rope, or otherwise, and therefore called a *Rudenture*, or *Cabling*; and the Columns thus enrich'd, *Cabled Columns*. See COLUMN.

This is most frequent in the *Corinthian* Order. The Cablings, or Fillings up commence from about one Third of the Height of the Column, reckoning from the Base; and are continued to the Capital: That is, they begin and end with the Diminution of the Column. See DIMINUTION.

FLYERS, in Architecture, such Stairs, as go strait and do not wind round, nor its Steps made tapering; but the fore and back part of each Stair and the Ends respectively parallel to one another.

So that if one *Flight* do not carry you to your design'd Height, there is a broad half Space; and then you *fly* again, with Steps everywhere of the same Breadth and Length as before. See STAIRS.

FORT, a little Castle, or Fortress; or a Place of small Extent, fortify'd by Art, or Nature, or both.

A *Fort* is a Work encompass'd round with a Moat, Rampart, and Parapet; to secure some high Ground, or Passage of a River; to make good an advantageous Post; to fortify the Lines and Quarters of a Siege, *etc*.

Royal FORT is a *Fort* whose Line of Defence is at least 26 Fathoms long.

Star FORT, is a Sconce, or Redoubt, constituted by re-entering and salient Angles; having commonly from five, to eight Points; and the Sides flanking each other.

FORTERESSE, or FORTRESSE, a general Name for all fortify'd Places, whether so fortify'd by Nature, or Art: Such are fortify'd *Towns*, *Castles*, *Cittadels*, *Forts*, *Towers*, *Redoubts*, *Sconces*, etc.

FORTIFICATION, call'd also *Military Architecture*, is the Art of *fortifying*, or strengthening a Place; by making Works around the same, to render it capable of being defended by a small Force, against the Attacks of a more numerous Enemy.

Some Authors go back to the Beginning of the World, for the Author and Origin of Military Architecture.

According to them, GOD himself was the first Engeneer; and Paradice, or the Garden of *Eden*, the first *Forteresse*. *Cain* improv'd on the Hint, in building the first City, *Gen*. iv. 17. After him came *Nimrod*, *Gen*. x. 17. Then *Semiramis*, as *Policenus* relates, *Stratagem. L. VIII.* c. 27. The *Canaanites*, *Numb*. xiii. 19. *Deut* i. 28. *David*, 2 *Kings* v. 9. *Solomon*, 2 *Chron*. iii. 5 *Rehoboam* his Son, 2 *Chron*. xi. 5, 10. and the other Kings of *Juda* and *Israel*, and at length the *Greeks* and *Romans*. *Vitruvius L. X.* last Chap. and *L. I.* c. 5.

Such is the Series of those who fortify'd Places: To which might be added *Pharoah*, the Persecutor of the *Israelites*, who built the Cities of *Pithom* and *Raamses*. *Exod*. i. 11.

But how antient soever the surrounding of Cities with Walls, Towers, *etc*. may be; the Name *Fortification*, and the Art now understood thereby, are of no very old Standing.—

They had their Rise since the Invention of Cannons; the terrible Effects whereof render'd it necessary to change the Structure of the antient Walls, and add so many Things thereto, that those Changes were thought enough to constitute a new Art, which was called *Fortification*, by reason of the Strength it afforded those in Cities, to defend them against an Enemy.

The first Authors who have wrote of *Fortification*, consider'd as a particular form'd Art, are *Ramelli*, and *Cataneo*, *Italians*. After them *Errard*, Engineer to *Henry* the Great of *France*; *Stevinus*, Engineer to the Prince of *Orange*, *Marolois*, the Chevalier *de Ville*, *Lorini*, the Count *de Pagan*, and the Mareshal *de Vauban*: Which two last Noble Authors contributed very greatly to the Perfection of the Art.

From the Idea and Office of *Fortification*, some General Fundamental Rules or Axioms may be drawn, as

1°. That the Manner of fortifying should be accommodated to that of Attacking: So that no one Manner can be assured will always hold, unless it be assured the Manner of Besieging be incapable of being alter'd; and to judge of the Perfection of a *Fortification*, the Method of Besieging at the Time when it was built, must be consider'd.

2°. All the Parts of a *Fortification* should be able to resist the most forcible Machines us'd in Besieging.

3°. A *Fortification* should be so contriv'd, as that it may be defended with as few Men as possible; which Consideration, when well attended to, saves a world of Expence.

4°. That the Defendents may be in the better Condition, they must not be expos'd to the Enemy's Guns and Mortars; but the Aggressors be expos'd to theirs.

Hence, 5°. All the Parts of a *Fortification* should be so dispos'd, as that they may defend each other; in order to this, every Part thereof is to be *flank'd*, *i.e*. capable of being seen and defended from some other; so that there be no Place were an Enemy can lodge himself, either unseen, or under Shelter.

6°. All the Campaign around must lie open to the Defendents; so that no Hills or Eminences must be allow'd, behind which the Enemy might shelter himself from the Guns of the *Fortification*; or from which he might annoy them with his own.

The Fortress, then, is to command all the Place round about; consequently the Outworks must all be lower than the Body of the Place.

7°. No Line of Defence to be above Point blank Musket Shot, which is about 120 Fathom.

8°. The acuter the Angle at the Centre; the stronger is the Place, as consisting of the more Sides, and consequently more defensible.

Such are the General Laws and Views of *Fortification*: The Particular ones, respecting each several Work or member thereof, will be deliver'd under their proper Articles.—

FORTIFICATION is also us'd for the *Place fortified*; or the several Works rais'd to defend and flank it, and keep off the Enemy.

All *Fortifications* consist of Lines and Angles, which have various Names, according to their various Offices.

The principal Angles are those of the CENTRE, the FLANKING *Angle*, FLANK'D *Angle*, *Angle of the* EPAULE.

The principal Lines are, those of CIRCUMVALLATION, of CONTRAVALLATION, of the CAPITAL, *etc.*

Fortifications are divided into *Regular*, and *Irregular*; and again into *Durable*, and *Temporary*.

Regular FORTIFICATION, is that wherein the Bastions are all equal; or that built in a regular Polygon; the Sides and Angles whereof are generally about a Musket Shot from each other.

In a Regular *Fortification*, the Parts being all equal, have the Advantage of being equally defensible; so that there are no weak Places.

Irregular FORTIFICATION, is that wherein the Bastions are unequal, and unlike; or the Sides, and Angles not all equal, and equidistant.

In an Irregular *Fortification*, the Defence and Strength being unequal; there is a Necessity for reducing the Irregular Figure, as near as may be, to a Regular one.

And, as the Irregularity of a Figure depends on the Quantity of Angles and Sides; the Irregularity of a *Fortification* arises either from the Angles being too small, or the Sides being too long, or too short.

Consequently, an Irregular Figure being propos'd to be fortify'd; all the Angles, with the Quantity of the Sides, must be found, to be able to judge how it is to be fortify'd.

Durable FORTIFICATION, is that built with design to remain a standing Shelter for Ages. Such are the usual *Fortifications* of Cities, Frontier Places, *etc.*

Temporary FORTIFICATION, is that erected on some emergent Occasion, and for little time.

Such are Works cast up for the seizing and maintaining a Post, or

Passage; those about Camps, *etc.* as Circumvallations, Contravallations, Redoubts, Trenches, Batteries, *etc.*

FORTIFIED *Place*, a *Forteress*, or *Fortification*; *i.e.* a Place well flank'd, and shelter'd with Works.

Places fortified, after the modern way, consist chiefly of *Bastions*, and *Curtains*, and sometimes of *Demi-Bastions*; according to the Situation of the Ground, of *Cavaliers*, *Ramparts*, *Fausse-brayes*, *Ditches*, *Counter-scarps*, *Cover'd Ways*, *Half Moons*, *Ravelins*, *Hornworks*, *Crownworks*, *Outworks*, *Esplanades*, *Redents*, and *Tenailles*.

There are other moveable and additional Parts, as BERMES, PARAPETS, BANQUETTES, EMBRASURES, CORDONS, BOYAUX, MOULINETS, CHEVAUX *de Frise*, CHAUSSETRAPES, GALLERIES, MANTELETS, BATTERIES, SAPS, MINES, BLINDS, GABIONS, and PALISSADES.

Several of these Works, again, consist of divers Parts, which have different Denominations: Thus a Bastion consists of FACES, FLANKS, CASEMATES, ORILLONS, GORGE, *etc.*

The Methods of *Fortifying*, that have been invented, are various; and new Methods continue still to be propos'd. The principal, and those which chiefly obtain thro' *Europe*, are those of *Coehorn*, *Pagan*, *Vauban*, and *Scheiter*, from which all the rest are easily conceiv'd.—

FORTIFICATION, *according to the* Dutch *Method*, is that, which making the Flank perpendicular to the Curtains makes the Flank subduple of the Face, and subtriple of the Curtain; and the Angle of the Bastion, equal, either to two thirds of the Angle of the Polygon; or to half thereof, with the Addition of 15 or 20 Degrees, to make it equal to a right Angle.

Freitach determines the Quantity of the Flank in a square to be six; in a Pentagon, seven; in a Hexagon, eight; in a Heptagon, nine; in a Enneagon, ten; in a Decagon, eleven and in all other Figures, twelve *Rhinland* Perches: Consequently, the Face 24, and the Curtain 36 Perches.

To draw the Profile of a Fortification *after the* Dutch *manner*; Draw the inner Polygon G H *Tab. Fortification. Fig.* 6 into five equal Parts: The Demigorge A B is to contain one of them; and the Capital A B, two. Then divide the Curtain E A into four Parts; and make the Flank A B one.

This is *Freitach's* Method: But the *French* make both the Demigorge, and Flank sixth Part of the Polygon; and in Squares, and Pentagons, setting aside the second Flanks, describe a Semicircle upon B S, that the Angle of the Bastion C may be a right Angle.

The *Italians* giving the same Dimensions to the Gorge and Flank, make the secondary Flank half the Curtain, in Polygons exceeding a Hexagon, and a third of the Curtain in the rest.

Lastly, the *Spaniards* giving the same Dimensions to the Flanks, omit the secundary Flanks.

But Flanks perpendicular to the Curtain, with secundary Flanks, are in Disrepute, on account of the Obliquity of the Defence; so that the *Dutch*

Method of fortifying, with the other antient ones, founded on the same Principles, are now out of Doors.—

FORTIFICATION, *according to the* COUNT *de* PAGAN'S *Method*; supposes in the larger Fortifications, the external Polygon A B, *Fig*. 7 to be 100, the Face A G, 30 Perches; in the smaller, the first, 80, and the second, 25; and in the middle Size, the first 90, and the second 27 ½: The Perpendicular C D, 15, and the Flanks G F and H E perpendicular to the Lines of Defence A F and B E, cover'd with an Orillon, and tripled; and a Ravelin, Curtain and Bastions built on the Outside the Rampart.

This Method was receiv'd with great Applause, and must be allow'd greatly preferable to the *Dutch* Method.

But it has its Defects; for besides that the *fortifying* of Places by it is very expensive; its triple Flanks are too close, so as to lie expos'd to the Violence of the Bombs; the Orillon is so large as to prejudice the Length of the Flanks; the outer Rampart of the Bastion is too big, *etc*.

FORTIFICATION, *according to Mons*. Blondel's *Method*; has a great Affinity with that of the Count *de Pagan*, only that the Quantity of the Lines and Angles are differently determined.

Thus, a Right Angle being subtracted from the Angle of the Polygon, and to a third part of the Remainder, 15 added; the Sum gives the diminish'd Angle. In the greater Fortifications, the outer Polygon is 100, in the smaller 85. The external Polygon being divided into ten Parts, seven of them give the Lines of Defence; and the Faces are half those of the Tenaille.

This Method is very well calculated for the Purposes of Architecture; only, being somewhat expensive, 'tis but little us'd.—

FORTIFICATION, *according to Mons*. Vauban's *Method*, supposes the outer Polygon in the larger Fortifications to be 100; in the smaller, 80; and in the middle Size, 90: The Faces are made 2/7 of the same; the Perpendicular, in a Square, 1/8, in a Pentagon 1/7, in the other Polygons, 1/6 of the same. The Complement of the Face to the Line of Defence, he makes equal to the Distance of the Epaule: He uses re-entering and crooked Flanks; and places a low Tenaille before the Curtain; and the *Dutch* Ravelin with Lunettes.

This Method of *fortifying* pleases most People; both as it increases the Strength, without much Expence; and as it is perfectly well accomodated to the Principles of military Architecture, above laid down.

Yet it seems defective in this, that the Faces are too much expos'd to the View of the Enemy; and that the Lunettes are a little too long to be defended.

Mons. Vauban's *newer Method of* FORTIFYING, is that which builds large Bastions before small ones, and covers the Courtin with a double Ravelin; drawing a low Tenaille before the same Curtain.

Detached Bastions are perfectly suitable to what we have deliver'd in the former Method; excepting that the Flanks are rectilinear, and destitute of an Orillon.

FORTIFICATION, *according to the Method of* Scheiter, supposes the external Polygon, in the larger Fortifications to be 100 Perches; in the lesser 80; in the middle Size 90: The Flanks perpendicular to the Lines of Defence; and the Lines of Defence in the greater Fortifications 70 Perches, in the lesser 60, and the middle Size 65.

It detaches Bastions from the Curtain, and forms a kind of inner Recess behind the Curtain: It assumes the Angle of the Bastion, in a Square, to be 64 Degrees; to this adding 8, the Product is the Angle in a Pentagon; to which adding six Degrees, the Sum is the Angle of the Bastion in a Hexagon; and adding 5 to this, the Sum is the same Angle in a Heptagon.

FORTIN, a Diminutive of the Word *Fort*, importing a little *Fort*, or Sconce, call'd also *Field Fort*, built in Haste, for defence of a Pass, or Post; but particularly to defend a Camp, in the time of a Siege, where the principal Quarters are join'd, or made to communicate with each other by Lines defended by *Fortins* and Redoubts.

Fortins being very small, their flank'd Angles are generally 120 Fathom distant from each other; but their Figure and Extent is various, according to the Place and Occasion; some having whole Bastions, and others only half Bastions.

Star-FORTIN, is that whose Sides flank each other, *etc.*

FOSS, in Fortification, *etc.* a *Ditch*, or *Moat*.

The Word is form'd of the Particle *Fossum*, of the Verb *fodio*, I dig.—

Foss, in Anatomy, a kind of Cavity in a Bone, with a large Aperture, but no Exit, or Perforation.

When the Aperture is very narrow, it is called a SINUS.

In the Cranium there are six internal, and fourteen external *Fosses*. The Cavity of the Orbit, which contains the Eye, is a *Foss*.

Foss is particularly us'd for a Cavity, or Denture in the Back Part of the Neck.

Foss-*Way*, was antiently one of the four great Highways of *England*; so call'd, because in some Places it was never perfected, but left as a great Ditch.

FOUNDATION, that Part of a Building under Ground: Or, the Mass of Stone, *etc.* which supports a Building; or upon which the Walls of a Superstructure are rais'd: Or, it is the Coffer, or Bed, dug below the Level of the Ground, to raise a Building upon; in which Sense the *Foundation* either goes the whole Area and Extent of the Building, as when there are to be Vaults, Cellars, or the like; or it is drawn in Cuts, or Trenches, as when only Walls are to be rais'd. See BUILDING.

The *Foundation* is properly so much of the Masonry, as reaches as high as the Surface of the Ground; and is always to be proportion'd to the Load, or Weight of the Building it is to bear.

Sometimes it is massive, and continued under the whole Building; as in the antique Arches, and Aqueducts, and some Amphitheatres: More usually it is only in Spaces, or Intervals; either to avoid Expence, or because the Vacuities are at too great a distance, in which latter case they make use of insolated Pillars, bound together by Arches.—

That we may found our Habitation firmly, says Sir. *H. Wotton*, we must first examine the Bed of Earth, upon which we are to build; and then the Under-cielings, or Substruction, as the Antients call'd it. For the former, we have a general Precept in *Vitruvius*, *Subtructionis* Fundationes *fodiantur, si queant inveniri ad solidum et in solido*: By which he recommends not only a diligent, but even jealous Examination of what the Soil will bear; advising us not to rest upon any appearing Solidity, unless the whole Mould thro' which we cut, have likewise been solid. But how deep we should go in this Search, he has no where determined, as perhaps depending more on Discretion, than Regularity, according to the Weight of the Work: Yet *Palladio* has ventur'd to reduce it to a Rule; allowing for the *Cavasione*, *i.e.* hollowing or underdigging, a sixth Part of the Height of the whole Fabrick, unless there be Cellars under Ground, in which case he would have it somewhat lower. Sir *H. Wotton's Elem. of Architecture.*—

The Foundations of Buildings are either *Natural*, or *Artificial.*

Natural, as when we build on a Rock, or a very solid Earth; in which case we need not seek for any further Strengthening.

But, if the Ground be sandy, or marshy, or have lately been dug; in such case recourse must be had to *Art*. In the former Case, the Architect must adjust the Depth of the *Foundation* by the Height, Weight, *etc.* of the Building: A sixth Part of the whole Height is look'd on as a Medium: And as to Thickness, double that of the Width of the Wall, is a good Rule.

Where the Natural *Foundation* may not be trusted, they either fortify the Ground by *pallifying* it, *i.e.* driving it full of Piles; see PALLIFICATION; Or else lay large wooden Planks at the Bottom of the Trenches, dug for the *Foundation*.

In some Places they *found* the Peers of Bridges, and other Buildings near the Water, on Sacks of Wool, laid like Matrasses; which being well press'd, and greasy, will never give way, nor rot in Water.

FOUNDATION is also us'd for the Establishment of a City, or Empire.

The *Romans* reckon'd their Years from the *Foundation* of *Rome*, *Ab Urbe Condita*; which we sometimes express by *ab U. C.* Chronologers make 479 Years from the passing out of *Egypt*, to the *Foundation* of *Rome*.

FOUNDATION is also a Donation, or Legacy, either in Money, or Lands, for the Maintenance and Support of some Community, Hospital, School, Lecture, or other work of Piety. See SCHOOL, HOSPITAL, *etc.*

Among the Order of *Augustins*, there is a *Foundation* for the marrying of poor Maids; and another for the furnishing of Trusses to poor People, who have Ruptures, or Hernias.—

FOUNTAIN, with regard to Architecture, is an Assemblage of Masonry, Sculpture, *etc.* either for the Decoration, or Conveniency of a City, Garden, or the like.

Fountains acquire various Denominations, according to their Form, and Situation; as,

Spring FOUNTAIN, a kind of plain Spout, or Stream of Water, issuing out of a Stone, or Hole in a Wall, without any Decoration.

Such is the *Fountain* of *Trevi* at *Rome.*

Cover'd FOUNTAIN, a kind of Pavillion, built of Stone; either Insulate, and square; or round; or Multilateral; or back'd: And that, either with a Projecture, or Indenture: Inclosing a Reservoir, and spouting, or darting forth the Water thereof thro' one or more Cocks; in the Middle of a Street, Square, Garden, Court, or the like. Such is that in *Lincolns-Inn, London.*

Open FOUNTAIN, is any spouting *Fountain*, with a Bason, Cup, and other Ornaments; all open: As is frequent in our Gardens, and the Vineyards at *Rome.*

Spouting FOUNTAIN, or *Jet d'Eau*, is any *Fountain*, whose Water is darted forth impetuously thro' one, or more Jets, or Adjutages, and returns in Rain, Nets, Folds, or the like. See JET *d'Eau.*

Bason FOUNTAIN, is that with only a simple Bason, of any Figure whatever; in the Middle whereof is a Jet, or Spout, or perhaps a Statue; or even a Group of Figures: As in the Court of *Buckingham* House.

Cup FOUNTAIN, is that which beside a Bason, has likewise a Cup of one single piece of Stone, or Marble, supported on a Shaft, or Pedestal, and receiving a Jet, or Spout, rising out of the Middle thereof. As the *Fountain* in the Court of the *Vatican*; the cup whereof is of Granate, and antique; being brought from the Bath of *Titus* at *Rome.*

Pyramidal FOUNTAIN, that form'd of several Basons, or Cups, rang'd in Stories over each other, and diminishing all the way; being supported by a hollow Stem, or Shaft.

As the *Fountain* of *Monte-Dragone* at *Frescati*: Or else supported by Figures, Fishes, or Consoles; the Water whereof in its Fall makes Nets in divers Stories, and represents a kind of Water Pyramid: As that at the Head of the Cascades at *Versailles.*

Statuary FOUNTAIN, that which being open and insulated, or even back'd, is adorn'd with one or more Statues, by way of finishing, or crowning: As the *Fountain* of *Latona* at *Versailles*; and that of the Shepherd at *Caprarola.* There are also Statues, which spout forth Water at some of their Parts; or at Sea Shells, Vases, Urns, and other Attributes of the Sea.

Rustic FOUNTAIN, is that form'd, or enrich'd with Rock-work, Shell-work, Petrifactions, *etc.*

Satyric FOUNTAIN, a kind of Rustic *Fountain*, in manner of a Grotto; adorn'd with Termini, Mascaroons, Fauns, Sylvans, Bacchantes, and other Satyrical Figures, serving for Ornaments as well as Jets d'Eau.

Marine FOUNTAIN, that compos'd of Aquatic Figures, as Sea Divinities, Naiads, Tritions, Rivers, Dolphins, and other Fishes, and Shells. As the *Fountain* of the Place *Palestrina* at *Rome*; where a Shell supported by four Dolphins, serves as a Cup, and supports a Triton, that spouts Water out of his *Concha Marina*.

Naval FOUNTAIN, is that made in form of a Vessel. Such is that of the Place *d'Espagne* at *Paris*, representing a *Bark*: That at *Monte-Cavallo*, representing a *Galley*; or that before the Vineyard *Mattei* at *Rome*, resembling a Boat.

Symbolical FOUNTAIN, that where the principal Ornaments are the Attributes, Arms, or Cognizances of the Person who erected it. As a *Fountain* of St. *Peter* in *Montorio*, resembling a Castle flank'd with Towers; and *Doujons*, representing the Arms of *Castile*: And some other *Fountains* at *Rome*; among which are the *Flower-de-Lis*, and the *Dove*, the Bearings of the Family of Pope *Innocent* X.

Arch'd FOUNTAIN, that whose Bason and Jet are plac'd perpendicularly under an Arch. Such are the *Fountains* of the *Colonnade*, and Triumphal Water Arch at *Versailles*.—

FRAISE, in Fortification, a kind of Defence, consisting of pointed Stakes, driven parallel to the Horizon, into the Retrenchments of a Camp, a Half Moon, or the like; to send off and prevent any Approach or Scalade.

Fraises differ from Palissades chiefly in this, that the latter stand perpendicular to the Horizon; and the former jet out parallel to the Horizon.

Fraises are chiefly us'd in Retrenchments, and other Works, thrown up of Earth: Sometimes they are found under the Parapet of a Rampart; serving instead of the Cordon of Stone, us'd in Stone Works.

He fortified all the weak Places of his Camp with *Fraises* and Palissades.—All the Outworks of the Place were *frais'd* and palissaded.

FRAISING *of a Batallion*, is the lining of the Musqueteers all around with Pikes; in case of their being charg'd by a Body of Horse.

FRAME, in Joinery, *etc.* a kind of Case, wherein a thing is set, or inclos'd, or even supported; as a Window *Frame*, *Frame* of a Picture, of a Table, *etc.* See WINDOW, *etc.*

The *Frame* is also a Machine, us'd in divers Arts. The Printers *Frame* is more usually call'd *Chase*.

The Founders *Frame* is a kind of Ledge, inclosing a Board; which being fill'd with Sand, serves as a Mould to cast their Work in. See MOULD.

FRAME, among Painters, *etc.* is a kind of *Chassy*, or Square, compos'd of four long Pieces, or Slips of Wood, join'd together; the intermediate Space whereof is divided into a great Number of little Squares, like the Mashes of a Net; and for that reason sometimes call'd *Reticula*.

Its Use is in the reducing of Figures from great to small; or from small to great. See REDUCT.

FRAMING *of a House*, is all the Timber Work therein; *viz.* the *Carcase*, *Flooring*, *Partitioning*, *Roofing*, *Cieling Beams*, *Ashlering*, *etc.* See CARCASE, ROOF, FLOOR, *etc.*

FREEZE, or FRIZE, in Architecture, that Part of the Entablature of Columns, between the Architrave and Corniche. See ENTABLATURE.

The *Freeze* is properly a large, flat Face, or Member separating the Architrave from the Corniche. See CORNICHE.

The Antients call'd it *Zophoros*, by reason it was usually enrich'd with Figures of Animals; and our Denomination *Freeze* has a like Origin, being form'd of the Latin *Phrygio*, an Embroiderer, because commonly adorn'd with Sculptures in Basso Relievo, imitating Embroidery.

The *Freeze* is suppos'd to be intended to represent the Heads of the transverse Beams that sustain the Roof or Covering.

In the *Tuscan* Order it is quite plain: In the *Doric*, enrich'd with Triglyphs: In the *Ionic*, 'tis sometimes made arch'd, or swelling; particularly in which case 'tis call'd by *Vitruvius*, *Pulvinatus*, *q. d.* pillow'd: In the *Corinthian*, and *Composite*, it is frequently join'd to the Architrave, by a little Sweep; and sometimes to the Corniche. And in these richer Orders it is usually adorn'd with Sculpture, Figures, Compartiments, Histories, Foliages, Festoons, *etc.* See TUSCAN, DORIC, IONIC, *etc.*

As to the Height of the *Freeze*, 'tis in the general much the same with that of the Architrave.

The *Tuscan* FREEZE, *Vitruvius* makes 30 Minutes; *Vignola*, 35; *Palladio*, who makes it swelling, gives it but 26; and *Scamozzi*, 42. The *Doric*, in *Vitruvius* and *Vignola*, is 30 or 40 Min: In *Palladio*, *etc.* 45. The *Ionic*, *Vitruvius* makes flat, adorned with *Acanthus* Leaves, Lions, *etc.* and makes it 30 Minutes high: *Vignola*, who also makes it flat, gives it 45 Minutes: And *Palladio*, who makes it convex, or swelling, 27 Minutes; and *Scamozzi* 28. The *Corinthian*, *Vitruvius* enriches with *Acanthus* Leaves, human Figures, *etc.* and makes its Height 37 Minutes: *Vignola*, 45; *Palladio*, 28; and *Scamozzi* 31¾. Lastly, the *Composite*, which in *Vitruvius* is set with Cartouses, and carv'd between them, is 52½ Min; *Vignola*, who makes it like *Vitruvius*, only gives it 45 Min. *Palladio* who makes it swelling, only 30: And *Scamozzi*, 32.

From the Variety of Enrichments practic'd on the *Freezes*, they become variously denominated.

Convex, or *Pulvinated* FREEZES, are those whose Profile is a Curve; the best Proportion whereof, is, when drawn on the Base of an equilateral

Triangle.

In some the Swelling is only a-top, as in a Console: In others at Bottom, as in a Balluster.

Rustic FREEZES are those whose Courses are rusticated, or imboss'd; as in the *Tuscan Freeze* of *Palladio*.—

Flourish'd FREEZES are those enrich'd with Rinds of imaginary Foliages; as the *Corinthian Freeze* of the Frontispiece of *Nero*: Or with Natural Leaves, either in Clusters; or Garlands: Or Continued, as in the *Ionic* of the Gallery of *Apollo* in the *Louvre*.

Marine FREEZES are those representing Sea Horses, Tritons, and other Attributes of the Sea; or Shells, Baths, Grotto's, *etc.*—

Symbolical FREEZES are those adorn'd with the Attributes of Religion; as the *Corinthian*, of the Temple behind the Capitol at *Rome*, whereon are represented the Instruments and Apparatus of Sacrifice.

FRESCO, a kind of Painting; perform'd on fresh Plaister; or on a Wall laid with Mortar not yet dry; and with Water Colours.

This sort of Painting has a great Advantage: By its incorporating with the Mortar, and drying along with it, it is render'd extremely durable; and never fails or falls, but along with it. The *Italians,* from whom we borrow the Term, call it *a Fresco*: *Vitruvius*, *lib*. 7. *c*. 3. calls it *Udo Sectorio*.

Painting in *Fresco* is very antient; having been practiced in the earliest Ages of *Greece* and *Rome*.

It is chiefly perform'd on Walls and Vaults, newly plaister'd with Lime and Sand: But the Plaister is only to be laid, in proportion as the Painting goes on; no more being to be done at once, than the Painter can dispatch in a Day, while it is dry.

Before he begins to paint, a Cartoon or Design is usually made on Paper; to be calk'd, and transferr'd to the Wall, about half an Hour after the Plaister is applied.

The Antients painted on Stuck; and we may remark in *Vitruvius*, what infinite Care they took in making the Incrustation or Plaistering of their Buildings to render them beautiful and lasting: Tho' the modern Painters find a Plaister made of Lime and Sand preferable thereto; both as it does not dry so hastily; and as being a little brownish, it is fitter to lay Colours on, than a Ground so white as Stuck.

In this kind of Painting, all the Compound and Artificial Colours, and almost all the Minerals are set aside; and scarce any thing us'd but Earths; which are capable of preserving their Colour, defending it from the burning of the Lime, and resisting its Salt, which *Vitruvius* calls its Bitterness.

For the Work to come out in all its Beauty; the Colours must be laid on quick, while the Plaister is yet moist: Nor must they ever be retouch'd, dry, with Colours mix'd up with the White of an Egg, or Size, or Gum, as some Workmen do; by reason such Colours grow blackish: Nor do any preserve themselves, but such as were laid on hastily at first.

The Colours us'd, are *White* made of Lime slack'd long ago, and white Marble Dust; *Oker*, both red, and yellow; Violet *Red*; Verditer; *Lapis Lazuli*; Smalt; Earth Black, *etc*. All which are only ground, and work'd up with Water; and most of them grow brighter and brighter, as the *Fresco* dries.

FRET, or FRETTE, in Architecture, a kind of Knot, or Ornament, consisting of two Lists, or small Fillets variously interlac'd, or wove; and running at parallel Distances, equal to their Breadth.

A necessary Condition of these *Frets* is, that every Return, and Intersection be at Right Angles. This is so indispensible, that they have no Beauty without it; but become perfectly *Gothic*.

Sometimes the *Fret* consists but of a single Fillet; which, if well managed, may be made to fill its Space exceedingly well.

The Antients made great use of these *Frets*: The Places they were chiefly applied on, were even, flat Members, or Parts of Building; as the Faces of the Corona, and Eves of Cornices; under the Roofs, Soffits, *etc.* on the Plinths of Bases, *etc*.

The Appellation was occasion'd hence, that the word *Frette* literally signify'd the Timber Work of a Roof, which consists chiefly, of Beams, Rafters, *etc*. laid a-cross each other, and as it were, *fretted*.

FRET-*work*, an Enrichment of *Frets*; or a Place adorn'd with something in manner thereof.

Fret-work is sometimes used among us, to fill up, and enrich flat, empty Spaces; but 'tis principally practiced in Roofs, which are *fretted* over with Plaister Work.

The *Italians* also apply it to the Mantlings of Chimneys with great Figures: A cheap Piece of Magnificence, and as durable almost within Doors, as harder Matters in the Weather.

FRIEZE, or FRIZE, or FREEZE, in Architecture, a Member, or Division of the Entablature of Columns, by the Antients call'd *Zophoros*. See FREEZE.

FRIZE, or FRIEZE, in Architecture, a Part of the Entablature of Columns, more usually wrote, and pronounced *Freeze*. See FREEZE.

FRONT, in Architecture, the principal Face, or Side of a Building: Or that presented to the chief Aspect, or View. See FACE.

The Palace consists of a large Pile in *Front*, flank'd with two Wings. See FRONTISPIECE.

FRONT, in Perspective, a Projection, or Representation of the Face or Fore-part of an Object; or of that Part directly opposite to the Eye; call'd also, and more usually, *Orthography*. See ORTHOGRAPHY.

FRONTAL, in Architecture, a little *Fronton*, or Pediment, sometimes placed over a little Door, or Window. See PEDIMENT.

FRONTISPIECE, in Architecture, the *Portail*, or principal Face of a fine Building.

The *Frontispiece* of the *Louvre* is the finest Piece of Architecture in *France*.

The Word is form'd of the Latin *Frontispicium*, q. d. *Frontis hominis inspectio*. Hence, by a Figure, we say the *Frontispiece* of a Book, meaning the Ornament with an engraven Title on the first Page.

FRONTON, in Architecture, an Ornament among us, more usually called *Pediment*. See PEDIMENT.

FURRING, in Architecture, the making good the Rafters Feet in the Cornice. See RAFTERS.

Thus, when Rafters are cut with a Knee, these *Furrings* are Pieces which go strait along with the Rafter, from the Top of the Knee, to the Cornice.

Also, when Rafters are rotten, or sunk hollow in the Middle, there are Pieces cut thickest in the Middle, and tapering towards each End, which are nailed upon them to make them strait. Such Pieces are called *Furrs*; and the Putting them on, *Furring the Rafters*.

FUSAROLE, in Architecture, a Moulding, or Ornament, placed immediately under the Echinus, in the *Doric*, *Ionic*, and *Composite* Capitals.

The *Fusarole* is a round Member, carved, in manner of a Collar or Chaplet, with oval Beads. The *Fusarole* should always answer exactly under the Eye of the Volute in the *Ionic* Capital.

The *Italians* call it *Fusciolo*; and the *French*, from whom we borrow it, *Fusarole*.

FUST, in Architecture, the Shaft of a Column; or that Part comprehended between the Base, and the Capital; called also the *Naked*. See SHAFT.

The *Fust* is that cylindrical Part, which makes, as it were, the Body, or Trunk of the Column, exclusive of the Head and Foot. See COLUMN.

The Word is *French*, and literally signifies a Cask.

G

GABIONS, in Fortification, *etc*. large Baskets, made of Osier Twigs, wove of a cylindrical Form, six Foot high, and four wide; which being fill'd with Earth, serve as a Defence, or Shelter from the Enemy's Fire.

They are commonly used in Batteries, to screen the Engineers, *etc.* In order to which one is placed on either Side each Gun, only leaving room for the Muzzle to appear thro'.

There are also a smaller sort of *Gabions*, us'd on Parapets, in Trenches, *etc.* to cover the Musqueteers; being placed so near as that a Musquet can but just peep thro'.

They also serve as a Parapet on Lines, Lodgments, *etc.* where the Ground proves too hard to dig into.

To render the *Gabions* useless, they endeavour to set them on fire, by throwing pitched Faggots among them.

GABLE, or GABLE-*End* of a House, is the upright triangular End; from the Cornice, or Eaves, to the Top of its Roof. See HOUSE.

GAGE, in Joinery, is an Instrument made to strike a Line truly parallel to the streight Side of any Board or Piece of Stuff.

GALLERY, in Architecture, a cover'd Place in a House, much longer than broad, and which is usually on the Wings of the Building; serving to walk in.

The *Galleries* of the *Louvre* are magnificent: A *Gallery* of Painting: a compleat Apartment is to consist of a Hall, Antichamber, Chamber, Cabinet, and *Gallery*. See APARTMENT.

Savol, in his Architecture, derives the Word *Gallery* from *Gaul*, as supposing the antient *Gauls* to have been the first who us'd them: *Nicod* fetches it form the *French aller*, to *go*, *q. d. Allerie*. Others bring it from *Galere*, Gally, by reason it bears some Resemblance thereto in respect of Length. In the corrupt *Latin* we meet with *Galilæa*, for the *Gallery* of a Monastery.

GALLERY, is also a little Isle, or Walk, serving as a common Passage to several Rooms, placed in a Line, or Row.

The *Gallery of a Church* is a kind of continued Tribune, with a Balustrade; built along the Sides or lower End of a Church, to hold the more People: And, in the *Greek* Churches, to seperate the Women from the Men.—*Gallery of a Theatre*, see CHURCH, THEATRE, *etc.*

GALLERY, in Fortification, is a cover'd Walk, made of Timber; serving for the safe Passage of a Ditch.

The Sides of the *Gallery* are to be Musket proof, and consist of a double Row of Planks, lined with Plates of Iron; and the Top is sometimes cover'd with Earth, or Turf, to hinder the Effect of the Stones, artificial Fires, *etc.* of the Enemy.

Galleries are chiefly used to secure and facilitate the Miners Approach to the Face of the Bastion, over the Moat, which is already supposed fill'd up with Faggots and Bavins, and the Artillery of the opposite Flank dismounted. Sometimes it is called a *Traverse*.

GARDEN, an Inclosure, or Plot of Ground, curiously cultivated, and furnish'd with Variety of Plants, Flowers, Fruits, *etc*. See GARDENING.

Gardens are distinguish'd into *Flower Gardens*, *Fruit Gardens*, and *Kitchen Gardens*: the first for Pleasure, and Ornament; and therefore placed in the most conspicuous Parts: The two latter, for Service; and therefore made in By-places. See FLOWER.

In a *Garden*, the principal Things to be consider'd, are, the *Form*, *Soil*, *Situation*, and *Aspect* or *Exposure*.

For the *first* Point, the *Form*: A Square, or rather Oblong, is most eligible; leading from the Middle of the House, with a Gravel Walk in the Midst; narrow Grass-borders on each Side; and on either Side of these, Rows of Variety of Winter Greens. If the Ground be irregular, it may be made uniform so as to afford a Prospect nothing inferior to the most regular; strait Lines will reduce any Figure to Order. A Triangle has its Beauty, as well as a Square; and the most irregular Spots may be brought by Borders, and Walks to those two Figures.

Indeed, an Irregularity is easily hid in a large *Garden*, by long Walks, and tall Hedges, interrupting a distant View: And the little Corners, and Triangular Spaces, may be agreeably fill'd up with Borders of Flowers, Dwarf Trees, flowring Shrubs, or Ever-Greens. Nor is it prudent, to be solicitous to throw the whole *Garden* into a single View; as Irregularities, and Unevennesses, afford many uncommon, pretty Devices, *etc*. See WALKS.

For the *second* Point, or the *Soil*: A deep, rich, black Mould is best for Plants; sandy Land is warm and forward, and good for Flowers; chalky Land is cold and backward. But both are easily corrected by Compost's, or Materials of opposite Kinds.

For the *Situation*: If the *Garden* be too high, it will be expos'd to the Winds, which are highly prejudicial to Trees; if too low, the Dampness will be injurious, beside the breeding of Vermine: A Flat, therefore, or the Side of a Hill, are the happiest Situation; especially the latter, as it is usually well water'd, and shelter'd from the Extremes of Weather; beside that the Water descending from on high, will supply Fountains, Cascades, and other Ornaments of a *Garden*. See FOUNTAIN, *etc*.

For the *Aspect*, or *Exposure*, we have already consider'd what relates thereto, under the Article *Exposure*: See EXPOSURE.

Of all things, the Ground, or Soil of the Garden is the most important: Unless this be rich, and fertile, all the other Advantages will be vain. To judge of the Quality of the Soil, some direct us to look, whether there be any Heath, Thistles, or other Weeds growing spontaneously therein, as a certain Sign of a poor Ground: The Growth of the Trees, too, thereabout, is to be consider'd; if they grow crooked, ill shap'd, and grubby, of a faded Green, and full of Moss; the Place is to be immediately rejected. If the Contrary be found; you must proceed to examine the Depth of the Soil, by digging Holes under Ground. The Soil should be three Foot deep; but less than two, is not sufficient.

The chief Furniture of Pleasure Gardens are, *Parterres*, *Vista's*, *Glades*, *Groves*, *Compartiments*, *Quincunces*, *Verdant Halls*, *Arbour Work*, *Mazes*, *Labyrinths*, *Fountains*, *Cabinets*, *Cascades*, *Canals*, *Terrasses*, etc. See each under its proper Article, CASCADE, FOUNTAIN, *etc.*

In the *planting of a Fruit or Kitchen Garden*, if the Soil be a hungry Gravel, or Sand, Mr. *Switzer* directs the Holes where the Trees are to be planted, to be dug two Foot deep, and three or four over, and filled with rotten Horses or Cows Dung, mix'd with rich Mould: If it be Marle or stiff Clay, a Compost of Rubbish, Lime, Pieces of Brick, Ashes, Sand, *etc.* will be best to mix with Dung and Mould; tho' he is of opinion, that untried Earth, dug from a Waste or Common where Cattle has been fed, would prove the best Soil for young Trees.

The Trees being now taken out of the Nursery, the biggest Roots are to be shorten'd to about six Inches; all the small Fibres taken off; and the Head to be pruned, so as not to leave above two Branches; and those not above six Inches long. See PLANTING.

The *Wall Trees*, to be placed as far from the Wall, as possible; that there may be the more Room for the Roots to spread. Then, filling up the Hole with Mould, there remains nothing but to secure the Roots from the Winter's Frost, by covering the Spot with Straw, Fern, Dung, *etc.* And in Summer, from the Sun; by Sand, and Pebble Stone.

For Trees planted in Borders, the common Practice is to make a Trench by the Wall-side, two Foot broad, and as many deep. This Trench they fill with old Dung, mix'd with Earth, lightly laid, near as high as the Borders are intended to be: and then trodden down to half the Height in the Places where the Trees are intended to be. 'Tis prudent, to plant the Trees shallow, and to raise the Earth about them; especially in a wet, clayey Soil.—See PLANTING.

'Tis an Observation of some Importance, that Wall Trees, and Fruit thrive best, when the Walks that run parallel to them, are Gravel; more of the Rays of the Sun being thereby reflected to them, than if they were Grass. Add, that no sort of tall Trees are to be suffer'd to grow in any of the opposite Borders, or intermediate Spaces, so, as their Shade might reach to the South East, or South West Walls. The Places near the Walls are most advantageously filled with Dwarfs.

GARDENING, HORTICULTURE, the Art of cultivating a *Garden.* See GARDEN.

Gardening has in all Ages been esteem'd an Employment, worthy the greatest Hero's, and Philosophers: The Emperor *Diocletian*, at *Salona*; and *Epicurus*, and *Metrodorus*, at *Athens*, have ennobled the Art beyond all Encomium.

Goetzius, Superintendant of *Lubeck*, printed a Dissertation in 1706, entitled *seu De Eruditis Hortorum Cultoribus*, Of the Learned Men who have loved, and cultivated Gardens; among which Number he ranks *Adam*, *Gregory Nazianzen*, St. *Augustin*, *Pliny*, and *Cicero*: Nor should he have forgot the elder *Cato*, *Democritus*, *Plato*, and his Academy.

The Sieur *le Nostre* first carried *Gardening* to any thing of Perfection: M. *la Quintinie* has gone yet further; nor have our own Countrymen of late been wanting to its Improvement; as *London*, and *Wise*, *Bradley*, *Laurence*, *Fairchild*, etc.

M. *Fatio* has lately applied Mathematical Reasoning to *Gardening*, and shewn how to make the best use of the Sun's Rays in Gardens. See WALL.

The principal Operations in *Gardening*, are *Planting*, *Transplanting*, *Engrafting*, *Inoculating*, *Pruning*, *Sowing*, etc. Which see under their proper Articles, PLANTING, TRANSPLANTING, ENGRAFTING, PRUNING, *etc.*

GATE, a large Door, leading, or giving Entrance, into a City, Town, Castle, Palace, or other considerable Building. See DOOR.

Thebes, in *Egypt*, was antiently known by the Appellation, *With an hundred Gates*. *Fez*, in *Africa*, has 31 *Gates*. In Antient *Rome* there was a *Triumphal Gate*, *Porta Triumphalis*. In the Modern *Rome* there is the *Jubilee Gate*, which is only open'd the Year of a Grand Jubilee.

The *Gates* of *London* are many of them converted into Gaols, or Prisons; as *Ludgate*, *Newgate*, etc. The lesser, or By-gates, are call'd *Posterns*.

Gates thro' which Coaches, *etc*. are to pass, should not be less than seven Foot broad; nor more than twelve: The Height to be 1½ the Breadth.

GILDING, or GUILDING, the Art, or Act of spreading or covering a Thing over with Gold, either in Leaf or Liquid.

The Art of *gilding* was not unknown among the Antients, tho' it never arrived among them at the Perfection to which the Moderns have carried it.

Pliny assures us, that the first *gilding*, seen at *Rome*, was after the Destruction of *Carthage*, under the Censorship of *Lucius Mummius*; when they began to *gild* the Ceilings of their Temples, and Palaces; the Capitol being the first Place on which the Enrichment was bestow'd. But he adds, that Luxury advanced on them so hastily, that in a little time you might see all, even private and poor Persons, *gild* the very Walls, Vaults, *etc*. of their Houses.

We need not doubt but they had the same method with us, of beating Gold, and reducing it into Leaves; though it should seem they did not carry it to the same Height; if it is true which *Pliny* relates, that they only made five-hundred Leaves four Fingers square of a whole Ounce. Indeed he adds that they could make more; that the thickest were called *Bracteæ Prænestinæ*, by reason of a Statue of that Goddess at *Præneste* gilt with such Leaves; and that the thinner sort were called *Bracteæ Questoriæ*.

The modern Gilders do also make use of Gold Leaves of divers Thicknesses; but there are some so fine, that a thousand don't weigh above four or five Drachms. The thickest are used for *gilding* on Iron, and other Metals; and the thinnest, on Wood.—

But we have another Advantage over the Antients, in the manner of using, or applying the Gold: The Secret of painting in Oil, lately discover'd, furnishes us with means of *gilding* Works that shall endure all the Injuries of Time and Weather, which to the Antients was impracticable. They had no way to lay the Gold on Bodies that would not endure the Fire but with Whites of Eggs, or Size; neither of which will endure the Water: So that they could only *gild* such Places as were shelter'd from the Moisture of the Weather.—

The *Greeks* called the Composition which they used for *gilding* on Wood, *Leucopheum*, or *Leucophorum*; which is described as a sort of glutinous, compound Earth, serving, in all Probability, to make the Gold stick, and bear polishing. But the Particulars of this Earth, its Colour, Ingredients, *etc.* the Antiquaries and Naturalists are not agreed upon.

There are several Methods of *Gilding* in use among us; *viz. Gilding in Oil*; *Water Gilding*; *Gilding by the Fire*, which is peculiar to Metals; *Gilding of Books*, etc.

Method of GILDING *in Oil.*

The Basis, or Matter whereon the Gold is applied, in this Method, is the Remains of Colours found settled to the Bottom of the Pots wherein the Painters wash their Pencils. This Matter, which is very viscid, they first grind; then pass it thro' a linen Cloth; and thus lay it with a Pencil on the Matter to be gilt, after having first wash'd it once or twice over with Size; and if it be Wood, with some white Paint.

When almost dry, but while yet unctuous enough to catch and retain the Gold, they spread their Gold Leaves thereon, either whole, or cut in pieces: To take up and apply the Leaves, they make use of a Piece of fine, soft, well carded Cotton; or of a Palet for the purpose; or barely the Knife, wherewith the Leaves were cut, according to the Parts of the Work they are to *gild*, or the Breadth of the Gold to be applied.

In proportion as the Gold is laid, they pass over it a coarse, stiff Pencil or Brush, to make it stick, and, as it were, incorporate, with the Ground: With the same Pencil, or a smaller one, they mend any Cracks that may happen therein, after the same manner as will be hereafter shewn in *Water* GILDING.

This sort of *Gilding* is chiefly used for Domes, and Roofs of Churches, Courts, Banquetting Houses, *etc.* and for Figures of Plaster, Lead, *etc.* that are to stand exposed to the Weather.

Method of GILDING *in Water.*

Water *Gilding* is not perform'd without more Apparatus; nor can it be used so ordinarily, nor on such large Works, as the former: Wooden Works, and those of Stuc, are almost the only ones *gilt* in this way: which, beside, must be shelter'd from the Weather.

The Size made use of for *gilding*, is to be made of Shreds, *etc.* of Parchment, or Gloves, boil'd in Water, to the Consistence of a Gelly. See GLUE.

If it be Wood that is to be *gilt*, they first give it a *Wash* of this Size, boiling hot; and when this is dry, another of white Paint, mix'd up with the same Size. For this White, some use Plaster of Paris, well beaten and sifted; others Spanish White, *etc.* It is laid on with a stiff Brush, and oftener, or seldomer repeated, according to the Nature of the Work: For Pieces of Sculpture, seven or eight Lays suffice; for flat, or smooth Works, ten, or twelve. In the latter Case they are applied by drawing the Brush over the Work; in the former by dabbing it.

When the whole is dry, they moisten it with fair Water, and rub it over with several Pieces of coarse Linen, if it be on flat; otherwise, they beat or switch it with several Silps of the same Linen tied to little Sticks, to make it follow and enter all the Cavities and Depressures thereof.

The White thus finish'd, they proceed to yellow it; observing, that if it be a Piece of Sculpture in Relievo, they first touch up, and repare the several Parts which the white Ground may have disfigur'd, with little iron Instruments, as Gouges, Chissels, *etc.*

The yellow, which they use, is only common Oker well ground and sifted, and thus mix'd up with the Size us'd for the White, only weaker by one half. This Colour is laid on hot; and in Works of Sculpture supplies the Place of Gold, which frequently cannot be carried into all the Cavities and Dentings of Foliages and other Ornaments.

Over this yellow is applied a Lay, which is to serve for the Ground whereon the Gold is to be immediately laid: It is usually composed of the *Armenian* Bole, Blood-stone, Black-Lead, and a little Fat, to which some add Soap and Oil of Olives; other,s burnt Bread, Bistre, Antimony, Glass of Tin, Butter, and Sugar Candy. These Ingredients being all ground together, with hot Size, they apply three Lays of the Composition upon the yellow, each after the other is dried; taking Care not to put any in the small Cavities of the Work, to hide the yellow. The Brush used in this Application, is to be soft; and when the Matter is well dried, they go over it with a stronger Brush, to rub it down, and take off the little prominent Grains, and thus facilitate the burnishing of the Gold.

In order to proceed to *gild*, they must have three sorts of Pencils; one, to wet; another to touch up, and amend; and a third to flatten: There is also required a Cushion, to spread the Gold Leaves on, when taken out of the Book; a Knife to cut them; and a Squirrel's Tail fitted with a Handle; or else a piece of fine Stuff on a Stick, to take them up, direct, and apply them.—

They begin with the wetting Pencils, by which they moisten the Lay last laid on with Water, that it may the better receive and retain the Gold: The Leaves are then laid on the Cushion; and taken up, if whole, with the Squirrel's Tail; if in pieces, with the other Instrument, or even the Knife they are cut withal, and laid, and spread gently on the Parts of the Work before moisten'd.

When the Leaves happen to crack, or break in laying on, they make up the Breaches with little Pieces of Leaf taken up on the repairing Pencil; and with the same Pencil, or another somewhat bigger, they smooth the whole; pressing the Gold into the Dents, where it could not so well be carried with the Squirrel's Tail.

The Work thus far *gilt*, when dry, remains either to be burnish'd, or flatted.—

To *burnish* it, is to smooth, and polish with a Burnisher, which is usually a Dog's, or Wolf's Tooth, or a Blood-stone fitted in a Handle for the purpose. See BURNISHING.

To *flat*, is to give it a light Lick in the Places not burnish'd, with a Pencil dipt in Size, wherein a little Vermillion sometimes has been mix'd. This helps to preserve and prevent its flawing, when handled.

The last thing is to apply the Vermeil in all the little Lines, and Cavities; and to stop and amend any little Faults with Shell Gold.

The Composition here call'd *Vermeil*, is made of Gum Gutta, Vermillion, and a little of some ruddy, brown Colour, ground together, with *Venice* Varnish, and Oil of Turpentine. Some *Gilders*, in lieu hereof, content themselves with fine Lacca, or Dragons Blood, with Gum Water.

Sometimes, instead of burnishing the Gold, they burnish the Ground or Composition laid on last before it; and content themselves afterwards to wash the Part over with the Size. This Method is chiefly practiced for the Hands, Face, and other Nudities in Relievo; which, by this means, don't appear so very brilliant as the Parts burnish'd; though much more so than the Parts perfectly flat.

To *gild* a Work, and yet preserve white Grounds; they use to apply a Lay of Spanish White mix'd with a weak Fish Glue, on all the Parts of the Ground whereon the yellow, or the last Lay might run.—

Method of GILDING *by the Fire.*

There are two ways of *Gilding by Fire*, *viz.* that with Liquid Gold, and that with Leaf Gold.

Thd first is perform'd with Gold, reduc'd to a Calx, and amalgamated with Mercury, in the Proportion of about an Ounce of Mercury to a Drachm of Gold.

In order to the Operation, they heat a Crucible red hot, then put the Gold and Mercury in it, and stir them gently about, till the Gold be found melted and incorporated into a Mass with the Mercury. This done, they cast them into Water, to wash and purify; and thence into other Waters, where the Amalgama which is almost as liquid as if there were nothing but Quicksilver in it, may be preserved a long time for Use.

Before they proceed to apply this amalgamated Gold, they first *roughen* the Metal it is to be applied on; which is done by washing it over with *Aqua fortis*, or *Aqua secunda*: Then rincing the Metal in fair Water, and scouring it a little with small Sand, it is ready for the Gold.—

They now cover over the Metal with the Mixture of Gold, and Mercury, taking it up with a Slip of Copper, or a Brush made of brass Wire, and spreading it as equably thereon as possible; in order to which they wet the Brush from time to time in fair Water.—

GIRDERS, in Architecture, the largest Pieces of Timber in a Floor—Their Ends are usually fasten'd into the Summers, or Brest-Summers, and the Joists are framed in at one End to the *Girders*. See JOISTS.

By the Statute *for rebuilding London*, no Girder is to lie less than ten Inches into the Wall; and their Ends to be always laid in Lome, *etc.*

<table>
<tr><td rowspan="7">Girders
and
Summers
in
Length.</td><td>From</td><td>to</td><td colspan="2">must be in</td></tr>
<tr><td>F. In.</td><td>F. In.</td><td>Breadth Inches.</td><td>Depth Inches.</td></tr>
<tr><td>10 0</td><td>15 0</td><td>11</td><td>8</td></tr>
<tr><td>15 0</td><td>18 0</td><td>13</td><td>9</td></tr>
<tr><td>18 0</td><td>21 0</td><td>14</td><td>10</td></tr>
<tr><td>21 0</td><td>24 0</td><td>16</td><td>12</td></tr>
<tr><td>24 0</td><td>26 0</td><td>17</td><td>14</td></tr>
</table>

GIRDING-*Beams*, in Building, the same as *Girders*. See GIRDERS.

GLACIS, in Building, *etc.* is an easy, insensible Slope, or Declivity.

The Descent, or Inclination of the *Glacis* is less steep than that of the *Talut*. See TALUS.

In Gardening, a Descent sometimes begins in *Talut*, and ends in *Glacis*.

The GLACIS *of the Cornich* is an easy, imperceptible Slope in the Cymaise of a Corchich, to promote the Descent and draining off of the Rain-Water.

GLACIS, in Fortification, is particularly used for that of the Counterscarp.

The *Glacis* of the Counterscarp, or cover'd Way, is the Sloping Bank that reaches from the Parapet of the Corridor, or Counterscarp, to the level Side of the Field.

The *Glacis* is about six Foot high, and loses itself by an insensible Diminution in the space of ten Fathoms. See ESPLANADE.

GLADE, in Agriculture, Gardening, *etc.* a *Vista*, or open, and light

Passage, made thro' a thick Wood, Grove, or the like, by lopping off the Branches of Trees along the Way.

GLYPHE, or GLYPHIS, in Architecture, a general Name for any Cavity, or Canal, whether round, or terminating in an Angle; us'd as an Ornament, in any Part.

The Word is *Greek*, which literally signifies *graving*.

[The word] is properly a Nitch, or Indenture made in graving; or, more properly, the Notch in the Head of an Arrow, over which the String goes. See TRIGLYPH.

GORGE, in Architecture, a sort of concave Moulding, wider, but not so deep as a Scotia; used chiefly in Frames, Chambranles, *etc*.

GORGE *of a Chimney*, is the Part between the Chambranle, and the Crowning of the Mantle.—Of this there are divers Forms, straight, perpendicular, in form of a Bell, *etc*.

GORGE is sometimes also used for a Moulding that is concave in the upper Part, and convex at Bottom; more properly called *Gula*, and *Cymatium*.

GORGE is also used for the Neck of a Column; more properly called *Collarino*, and *Gorgerin*.

GORGE, in Fortification, the Entrance of a Bastion; or of a Ravelin, or other Out-work.

The *Gorge* is what remains of the Sides of the Polygon of a Place, after retrenching the Curtains: In which case it makes an Angle in the Centre of the Bastion—Such is A H D *Tab. Fortification*, *Fig*. 1.

In flat Bastions, it is a Right Line on the Curtain by which we enter either the Bastion, or the Out-works. It is also the Aperture thro' which we enter, or the Interval between, the two Wings, or Flanks on the Side of the Ditch.

All the *Gorges* are to be destitute of Parapets; otherwise, the Besiegers having taken possession of a Work, might make use thereof to defend themselves from the Shot of the Place: So that they are only fortified with Palisado's, to prevent a Surprise.

Half the Gorge, *Demi* GORGE, or That Part of the Polygon between the Flank, and the Centre of the Bastion, as A H, is call'd the *Demi Gorge*.

GOTHIC *Architecture* is frequently very solid, heavy, and massive; and sometimes, on the contrary, exceedingly light, delicate, and rich. The Abundance of little, whimsical, impertinent Ornaments, are its most usual Character.

Authors distinguish two Kinds of *Gothic Architecture*; the one Antient, the other Modern.—The *Antient*, is that which the *Goths* brought with them from the North, in the Vth Century: The Edifices built in this manner, were exceeding massive, heavy, and coarse.

Those of the *Modern Gothic* run into the other Extreme, being light, delicate, and rich to a Fault: Witness *Westminster* Abby, the Cathedral of *Lichfield*, the Cross of *Coventry*, etc.

The last kind continued long in use, especially in *Italy*, *viz*. from the XIIIth Century, to the Restauration of the antique Building in the XVIth. All the antient Cathedrals are in this kind. See ARCHITECTURE.

The Inventors of the *Gothic* Architecture thought, no doubt, they had far surpass'd the *Greek* Architects: A *Greek* Building has not a single Ornament but what adds a Beauty to the Whole. The Parts necessary to sustain, or shelter it, as the Column, Corniches, *etc*. derive all their Beauty from their Proportions: Every thing is simple, measur'd, and restrain'd to the Use it is intended for. No daring, out of the way Strokes; nothing quaint to impose on the Eye: The Proportions are so just, that nothing appears very grand of itself, tho' the whole be grand. On the contrary, in the *Gothic* Architectrue, we see huge Vaults rais'd on slender Pillars, which one would expect every Minute to tumble down, tho' they will stand for many Ages. Every thing is crammed with Windows, Roses, Crosses, Figures, *etc*.

GOTHIC *Column*, is any round Pillar, in a *Gothic* Building, either too thick, or too small for its Height. See COLUMN.

There are some found even of 20 Diameters high; without either Diminution, or Swelling. See PILLAR, DIMINUTION, *etc*.

GRADATION, in Architecture, signifies an artful Disposition of several Parts, as it were, by Steps, or Degrees, after the manner of an Amphitheater; so that those placed before, do no Disservice, but rather Service to those behind.

The Painters also use the Word *Gradation* for an insensible Change of Colour, by the Diminution of the Teints, and Shades. See DEGRADATION.

GRASS-*Plots*, and *Walks*, make a considerable Article in Gardening, *etc*. See WALKS, *etc*.

Grass, or *Green Plots* are had either by sowing of Hay-seed, or laying of Turf: For the first, which is the cheapest way, the Seed of the finest Upland Pastures is to be chose, well sifted and cleansed.

For the second, the Turf should be cut on a Down, or Green, or Common, or Sheep-walk, where the *Grass* is short and fine; if there be any Knobs, or Roughnesses, the Place must be cleansed and roll'd after a Shower, before it be cut up. The Turf is cut in Squares, mark'd out with Lines, rais'd with a Knife, and roll'd up; about three Inches thick.

The Quarters, or Verges are to be prepared with a fine Coat of poor Earth to lay the Turf on; and after laying, the Turf must be well water'd, rolled, *etc*.

Small Pieces of *Grass*-work, as Knots, Shell-work, and Volutes of Parterres, Cut-work, Verges about Basons, *etc*. must always be laid with

Turf.—

To sow *Grass* for Plots, *etc.* the Ground must first be dug or broke with a Spade, then dress'd, laid even, raked fine, and cover'd an Inch thick with good Mould to facilitate the Growth of the Seed. Then the Seed to be sown pretty thick, that it may come up close and short; and lastly, to be raked up, and cover'd. The best Season is the latter End of *August*: When it is well come up, it must be mow'd, and this often repeated; since the oftener 'tis mow'd, the thicker and handsomer it grows. It must also be rolled from time to time.

GRAVEL *Walk*, in Gardening.—To lay, or form a Walk with *Gravel*, all the good Soil is to be pared away, below the Roots of any Grass, or Weeds; then the Place to be filled two or three Inches with coarse *Gravel* unfearced, laying it highest in the Middle: Then rolling it; a new Stratum of finer *Gravel* to be thrown on, two, or three Inches thick: And the Rolling to be repeated again and again.

Note, the Sides next the Beds should be laid a Foot and an half, or two with Turf from whence the Heat of the Sun cannot be reflected as from *Gravel*, to the Prejudice of the neighbouring Flowers.

GREEK *Orders* in Architecture, are the *Doric*, *Ionic*, and *Corinthian*. See ORDER; see also DORIC, IONIC, and CORINTHIAN.

GREEN-*Houses*, as now built, serve not only as *Conservatories*, but likewise as Ornaments of Gardens; being usually large and beautiful Structures, in Form of Galleries, wherein the Plants are handsomely ranged in Cases for the Purpose. See GARDEN

The *Green-House*, Mr. *Mortimer* directs to be open to the south, or very little declining therefrom: The Heighth and Breadth to be about 12 Foot, and Length according to the Number of Plants intended to be kept therein. It should be situate on the driest Ground, and so contrived, as that nothing may obstruct the Sun's Rays in Winter. In the building it, Care must be taken, not to plaister it on the Inside with Lime and Hair; Dampness being observed to continue longer on such Plaister, than on Bricks or Wainscote. To preserve it the more from Moisture and Colds, an artificial Heat is to be used: In order to this, some hang up Fires, and others place Pans of Coals in Holes in the Ground. The better way is, to have a Stove behind the *Green-House*, and to convey the Heat thereof thro' Trills made under the Floor for that Purpose. The best Method of all, according to *Bradley*, is the new Stove invented by M. *Gauger*, and published in *English* by Dr. *Desaguliers*.

Some have Sashes, Casements, and Doors of Glass; others prefer moveable Canvas Doors, to be taken off, and put on at pleasure.

The Pots and Cases are to be ranged in the *Green-House* so, as not to incommode one another, or hinder any from readily receiving both Sun, Air, *etc.* As to the Management of the Plants herein, *Mortimer* recommends the opening of the Mould about them, from time to time, and sprinkling a little fresh Mould in them, and a little warm Dung on that; as also to water them when the Leaves begin to wither and curl, and not oftner, which would make them fade and be fickly; and to take off such Leaves as wither and grow dry.

Bradley advises, that in the colder Parts of *England*, the Front of the *Green-House* be built in a Sweep, or Semicircle, to receive the Sun's Rays, one part or other of it, all Day: That all the Windows, *etc.* in the Front, have close, thick, wooden Shutters, to be shut every Night in Winter, for fear of Frosts, Nipping Winds, *etc.*—*Dutch* Tiles he thinks the best Lining for the Walls, as being dry, and reflecting a deal of Heat; and square Tiles for the Pavement, which readily imbibe the Wet, and never sweat, as Marble and the harder Stones do.—The Plants in the *Green-House* are not to fill above a ¼ part of the Space thereof; the rest being left vacant for the Air to circulate about the Plants. If a *House* be crowded with Plants, the Effluvia they are continually emitting by Perspiration, and the Vapours from the Mould, will condense the Air, and cause Dampness. The Plants to be disposed, as much as possible, about the middle Parts of the *Green-House*, where they will be safer from the Cold, which is usually greatest near the Glasses and Walls: The tenderest Plants to be disposed nearest to where the Heat comes in.

'Tis a general Rule among Gardeners, to set Exotic Plants into the *Green-House*, about the second Week in *September*, and to take them out again about the middle of *May*; tho' there should be some Distinction made herein; the tenderest Sorts from Places near the Line, as the Aloes, *etc.* being to be confined more to the *Green-House*, than Oranges, Myrtles, *etc.*

GROTESQUE, or GROTESK, or GROTTESQUE, a wild whimsical Figure of a Painter, or Engraver; having something ridiculous, extravagant, and even monstrous in it.

The Name arises hence, that Figures of this Kind were antiently much used to adorn the Grotto's wherein the Tombs of eminent Persons, or Families were inclosed. Such was that of *Ovid*, whose Grotto was discover'd near *Rome* about 50 Years ago. See GROTTO.

Calot a celebrated Engraver of *Lorrain*, had a wonderful Genius for designing *Grotesques*: the like is said of *Leonardo da Vinci*.

We also extend the Word to any Thing whimsical, or wildly pleasant, in a Person's Dress, Discourse, *etc.*—Masquerade Habits are the more valued, the more *Grotesque* they are: Our Theatres present us with Entertainments in *Grotesque* Characters, *i.e.* Persons quaintly dress'd; as Harlequins, Scaramouches, *etc.*

Planudes has given us a very *Grotesk* Picture of *Æsop*. *Ariosto* and the *Italian* Poets are full of *Grotesque* Descriptions.

GROTESQUE-*Work*, GROTESK-*Work*, or *Grottesco*, is a Work or Composition in Painting, and Sculpture, in the *Grotesque* Matter or

Taste; consisting either of Things which are merely imaginary, and have no Existence in Nature, or of Things turn'd and distorted out of the way of Nature, so as to raise Surprize and Ridicule.

Grotesque-Work is the same with what we sometimes call *Antique*. See ANTIQUE. See also MORISCO.

GROTESQUES, or GROTESKS, are particularly used for little fanciful Ornaments, of Animals, compounded with Foliages, Fruit, *etc.*

Such are those painted by *Raphael* in the Apartments of the Vatican; and those carved by *Michael Angelo* in the Ceilings of the Portico of the Capitol.

Vitruvius calls Compartments of this kind *Harpagenituli*.

GROTTO, or GROTTA, in natural History, a large deep Cavern or Den in a Moutain or Rock. See STONE.

The antient *Anchorites* retired into Dens, and *Grotto's*, to apply themselves the more attentively to Meditation, *etc.*

In *Grottos* are frequently found Crystals of the Rock, Stalactites, and other natural Conglaciations.

The Word is *Italian*, *Grotta*, form'd according to Menage, *etc.* from the Latin *Crypta. Du Cange* observes, that *Grotta* was used in the corrupt Latin.

M. *Homberg* conjectures, from several Circumstances, that the Marble Pillars in the *Grotto* of *Antiparos*, vegitate or grow.

That Author looks on this *Grotto* as a Garden, whereof the Pieces of Marble are the Plants; and endeavours to shew that they could only be produced by some vegetative Principles. *Memdel.* Acad. *Anno* 1702.

At *Foligno* in *Italy*, is another *Grotto*, consisting of Pillars and Orders of Architecture of Marble, with their Ornaments, *etc.* scarce inferior to those of Art; but they all grow downwards: So that if this too be a Garden, the Plants are turn'd upside down. *Memdel.* Acad. *Anno* 1711.

The *Zirchnitzer-Sea*, or Lake, in *Carniola*, famous for being full of Water, Fish, *etc.* the best part of the Year, and quite dry, and bearing Grass, Corn, *etc.* the rest; proceeds from some subterraneous *Grotto* or Lake; as is made highly probable by Mr. *Valvaser*; Philosoph. Trans. No. 191.

GROUND, in Painting, is the Surface, upon which the Figures, and other Objects, are raised or represented.

The *Ground* is properly understood of such Parts of the Piece as have nothing painted on them, but retain the original Colour upon which the other Colours are applied to make the Representations. See COLOURING.

A Drapery, Piece of Building, the like, are said to serve as a *Ground* to a Figure, when the Figure is painted on the Drapery or Building.

GROUND-*Sell*, or GROUND-*Plate*. See SELL.

GROUND-*Work*, in Building. See FOUNDATION.

GROUP, or GROUPPE, in Painting and Sculpture, an Assemblage or Knot of two or more Figures of Men, Beasts, Fruits, or the like, which have some apparent relation to each other.

In a good Painting, it is necessary that all the Figures be divided into 2 or 3 *Groups*, or separate Collections—Such and such a Thing make a *Group* with such and such others of different nature and Kind. The antique *Laomedon* is a fine *Group* of three beautiful Figures.

A *Group* has somewhat in it of the Nature of a Symphony or Concert of Voices: As, in the one the Voices must sustain each other, in order to fill the Ear with an agreeable Harmony from the whole; whence if any part were to cease, something would necessarily be miss'd: So, in the *Group*, if the Parts or Figures be not well balanced, something will be found disagreeable.

There are two Sorts of *Groups*, or two manners of considering *Groups*; with regard to the Design, and to the Clair obscure. The first is common both to Works of Painting, and those of Sculpture; the latter peculiar to Painting.

Groups, with regard to the Design, are Combinations of divers Figures, which have relation to each other, either on account of the Action, or of their Proximity, or of the Effect they have. These we conceive, in some measure, as representig so many different Subjects, or at least so many distinct Parts or Members of one greater Subject.

Groups, with regard to the Clair obscure, are Bodies of Figures wherein the Lights and Shadows are diffused in such manner that they strike the Eye together, and naturally lead it to consider them in one View. See CLAIR *obscure*.

The Word *Grouppe* is *French*, form'd of the *Italian Groppo*, a *Knot*.

In Architecture we say a *Group of Coloumns*, speaking of three or four Columns join'd together on the same Pedestal.

When there are but two together, we say a *couple*, not a *Group* of Columns.

GROVE, in Agriculture, *etc.* a little thick Wood.

The antient *Romans* had a sort of *Groves* near several of their Temples, which were consecrated to some God, and called *Luci*, by *Antiphrasis*, *a non lucendo*, as being shady and dark.

In very large and magnificent Gardens, a *Grove* is usually a Plot of Trees, inclosed with Palissades, consisting of tall Trees, as Elms, Horse-Chestnuts, *etc.* the Tops whereof make a tuft or Plume, and shade the Ground below.

At the Foot of the tall Trees, which generally run all along the Palissades at equal Distance, other lesser Trees are often planted, whose Tufts form a Resemblance of a sort of Copse within the former.

GUERITE, in Fortification, a small Tower of Wood or Stone, placed usually on the Point of a Bastion, or on the Angles of a Shoulder; to hold a Centinel, who is to take care of the Ditch, and watch out against a Surprize.

GUEULE, in Architecture. See GULE.

GUILD-HALL, or GILD-HALL, the great Court of Judicature for the City of *London*. See HALL.

In it are kept the Mayor's Court, the Sheriffs Court, the Court of Hustings, Court of Conscience, Court of Common-Council, Chamberlain's Court, *etc.*

Here also the Judges sit upon *Nisi prius*, etc.

GULE, GUEULE, or GOLA, in Architecture, a wavy Member, whose Contour resembles the Letter S; called by the *Greeks Cymatium*, *q.d.* a little Wave, and by our Workmen an *Ogee*. See CYMATIUM and OGEE.

This Member is of two Kinds, *recta* and *inversa*:

The *first*, and principal, has its Cavity above, and Convexity below. This always makes the Top of the Corona of the Cornice, jetting over the Drip of the Cornice like a Wave ready to fall.

It is called *Gula recta*, and by the *French Doucine*. Sometimes it is absolutely called the *Entablature*, as being the first or uppermost Member thereof. See DOUCINE, CORONA, CORNICHE, *etc.*

The *second* is just the reverse of the former, its Cavity being at the Bottom; so that it appears inverted, with regard to the former. This is used in the Architrave, and sometimes in the Cornice, along with the former, only separated by a Reglet.

Some derive the Word from the Resemblance these Members bear to the *Gula*, or Throat of a Man: Others from the Herald's Term *Gules*; as supposing the Moulding formed from the antient manner of wearing their Garments, which consisted of Slips or Swaths, alternately Fur and Stuff of various Colours; the Intervals between which, were called *Gules* or *Guales*.

GUTTÆ, in Architecture, are Ornaments in form of little Cones, used in the Plafond of the Doric Cornice, or on the Architrave, underneath the Triglyphs; representing a sort of Drops, or Bells; usually six in Number. See TRIGLYPH.

They are sometimes also called *Lachrymæ*, Tears; and *Campanæ* or *Campanulæ*, Bells. *Leon. Baptista Alberti* calls them *Nails*.

GUTTERS, in Building, are a kind of Vallies in the Roofs of Buildings, serving to drain and receive the Rain-Waters. See ROOF.

H

HABITATION, a Dwelling-Place, or House. See HOUSE, EDIFICE, BUILDING, *etc*.

HAIR, makes an Ingredient in the Composition of Plaister. See PLAISTER.

HALF-*Moon*, *Demi Lune*, in Fortification, is an Out-work, that hath only two Faces, forming together a saliant Angle, which is flanked by some Part of the Place, and of the other Bastions.

These *Half-Moons* are sometimes raised before the Curtain, when the Ditch is a little wider than it ought to be; in which Sense it is much the same with a Ravelin, only that the Gorge of an *Half-Moon* is made bending in like a Bow, or a Crescent, and is most Times used to cover the Point of a Bastion; whereas Ravelins are placed before the Curtain.—But they are defective, as being ill flanked.

HALL, was antiently used for a Mansion-House, or Habitation. See MANSION.

HALL, *Salle*, in Architecture, is a large Room, at the Entrance of a fine House, Palace, or the like. See PALACE, *etc*.

Vitruvius mentions three Sorts of *Halls*: The *Tetrastyle*, which has four Columns supporting the Plafond, or Ceiling: The *Corinthian*, which has Columns all around, let into the Wall, and is vaulted over: And the *Egyptian*, which had a Peristyle of insolated *Corinthian* Columns, bearing a second Order, wiht a Ceiling. These were call'd *Oeci*.

The *Hall* is properly the first, and finest Partition, or Member of an Apartment; and, in the Houses of Ministers of State, public Magistrates, *etc*. is that wherein they dispatch Business, and give Audience.

In very magnificent Buildings, where the *Hall* is larger and loftier than Ordinary, and placed in the middle of the House, it is call'd a *Saloon*.

A Royal Apartment is said to consist of a *Hall*, or Chamber of Guards, *Aula Prætoriana*; and Anti-chamber, *Procamera*; a Chamber, *Camera*; a Cabinet, *Conclave*; and a Gallery, *Porticus*. See APARTMENT.

HALL is also a public Building, erected for the Administration of the Policy and Justice of a City, or Corporation.—In this Sense we say the *Town-Hall*, a *Company's Hall*, etc.

Guild-HALL, is a stately building in the City of *London*, and the great Court of Judicature for that City. See GUILD-*Hall*.

Here, Meetings of the Citizens are held, for the Election of Officers, solemn Entertainments, *etc*.

HALL, is also particularly used for a Court of Justice; or an Edifice wherein there is one or more Tribunals.

In *Westminster Hall* are held the great Courts of this Kingdom, *viz.* the King's Bench, Chancery, Common-Pleas, and Exchequer.

Above Stairs, is likewise held the High Court of Parliament.

Westminster-Hall was the Royal Palace, or Place of Residence of our antient Kings; who ordinarily held their Parliaments, and Courts of Judicature, in their Dwelling-Houses, (as is still done by the Kings of *Spain*) and frequently sat the said Courts of Judicature, as they still do in Parliament.

A great Part of this Palace was burnt under *Henry* VIII. What remains is still reserved for the said Judicatories.—The great *Hall*, wherein the Courts of *Kings-Bench*, etc. are kept, is said to have been built by *William Rufus*; others say by *Richard* I. or II. 'Tis reckon'd superior, in Point of Dimensions, to any *Hall* in *Europe*, being 300 Foot long, and 100 broad.

HANGINGS, Linings for Rooms, made of Arras, Tapestry, *etc.* See TAPESTRY.

HANGS-*Over*, in Building.

HEAD, is used in Painting, Sculpture, *etc.* for a Picture, or Representation of that Part of the human Body; whether in Colours, Draught, Relievo, or Creux.

Those in Painting, taken from the Life, or supposed to bear a just Resemblance to the Person; are more properly call'd *Portraits*.

HEAD, in Architecture, *etc.* an Ornament of Sculpture, or carved Work, frequently serving as the Key of an Arch, Plat-band, and on other Occasions. See KEY and ARCH.

These Sort of *Heads* usually represent some of the Heathen Divinities, Virtues, Seasons, Ages, *etc.* with their Attributes.—As a Trident for *Neptune*, a Helmet for *Mars*, a Crown of Ears of Corn for *Ceres*, a Caduceus for *Mercury*, a Diadem for *Juno*.

The *Heads* of Beasts are also used in Places suitable thereto: As a Bullock's, or Sheep's *Head*, for a Shambles, or Market-house; a Dog's, for a Kennel; a Deer's, or Boar's, for a Park or Forest; a Horse's, for Equery, *etc.*

In the Metopes of the Freezes, and other Parts of certain *Antique*, *Doric* Temples, we meet with Representations of Bullocks, or Rams *Heads* slea'd; as a Symbol of the Sacrifices offer'd there.

HEALING, in Architecture, denotes the Covering the Roof of a Building. See ROOF and COVERING.

The *Healing* is various; as of *Lead*, *Tyles*, *Slate*, *Horsham Stone*, *Shingles*, *Reed* and *Straw*. See LEAD, TYLE, STONE, SHINGLES, *etc.*

HEARTH *Focus*. See CHIMNEY, *etc.*

HEARTH-*Stones*. See CHIMNEY.

HEDGE, in Agriculture, *etc.* a Fence, inclosing a Field, Garden, or the like; made of Branches of Trees interwoven. See FENCE.

The Word is form'd of the German *hag*, or *haag*, or the Anglo-Saxon *hegge*, or *hoeg*; which signify simply Inclosure, Circumference.

Quick-set Hedge, is that made of quick, or live Trees, which have taken Root; in Contradistinction to that made of Faggots, Hurdles, or dry Boughs, *etc.*

To plant a *Quick-Hedge* of Thorn, or the like, Mr. *Mortimer* directs, that the Sets be about as big as the Thumb, and cut within four of five Inches off the Ground: That if the *Hedge* have a Ditch, it be three Foot wide a-top, and one at Bottom, and two deep: That if the *Hedge* be without a Bank, or Ditch, the Sets be in two Rows, almost perpendicular, and at a Foot Distance: That the Turf be laid with the Grass Side downwards: That at every 30 Foot Distance, a young Oak, Elm, Crab, or the like be placed: That Stakes be driven into the loose Earth, at about two Foot and a half Distance, so low as to reach the firm Ground.

When the *Hedge* is eight or nine Years growth, it may be *splash'd* or laid down; by giving the Shoots, or Branches, a Cut with a Knife, or Bill, half thro'; and then weaving it about the Stakes, and trimming off the small superfluous Branches.

Instead of Building a Garden Wall facing the North East, Mr. *Lawrence* advises, that to save Charge, *etc.* a Crab-Tree *Hedge* of three Rows be planted; which will be a good Mound, and quickly grow up to be a better Fence than a Wall against the West and South-West Winds, which make the greatest Destruction in a Garden, and which blow two Parts in three of the whole Year: Beside the Stock of Fruit such a *Hedge*, grafted with Red streak, or Gennet Moyl, will yield.

HERSE, in Fortification, a Lattice, or Portcullice, in Form of an *Harrow*; beset with Iron Spikes.

It is usually hung by a Cord fasten'd to a Moulinet; to be cut, in Case of Surprize, or when the first Gate is broken with a Petard; that the *Herse* may fall, and stop up the Passage of the Gate, or other Entrance of a Fortress.

These *Herses* are also often laid in the Roads, to incommode the March as well of the Horse, as the Infantry.

The *Herse* is otherwise call'd a *Sarrasin*, or *Cataract*; and when it consists of strait Stakes, without any Cross-pieces, it is call'd *Orgues*.

The Word *Herse* is *French*, and literally signifies *Harrow*; being form'd of the Latin *Herpix*, or *Hirpex*; which denote the same Thing.

HEWING of *Timber*. See TIMBER.

HIPPODROME, HIPPODROMUS, in Antiquity, a List, or course, wherein their Horse Races were perform'd; as also their horses exercised.

HOLLOW, in Architecture, a Concave Moulding, about a Quadrant of a Circle; by some call'd *Casement*, by others an *Abacus*. See CASEMATE, *etc.*

HOLLOW *Tower*, in Fortification, is a Rounding made of the Remainder of two Brisures, to join the Curtain to the Orillon; where the small Shot are play'd, that they may not be so much expos'd to the View of the Enemy. See TOWER, *etc.*

HORN, in Architecture, is sometimes used for *Volute*. See VOLUTE.

HORN *Work*, in Fortification, a Sort of Out-work advancing toward the Field, to cover and defend a Curtain, Bastion, or other Place suspected to be weaker than the rest; as also to possess a Height, *etc.* See OUT-WORKS.

HORTICULTURE, the Art of Gardening. See GARDENING.
The Word is compounded of *Hortus*, Garden, and *colo*, I till, dress, *etc.*

HOSPITAL, popularly SPITTAL, a Place or Building erected, out of Charity, for the Reception and Support of the Poor, Aged, Infirm, and otherwise helpless.
The Word is form'd of the *Latin*, *Hospes*, Host, Stranger.
In the first Ages of the Church, the Bishop had the immediate Charge of all the Poor, both sound and diseas'd; as also of Widows, Orphans, Strangers, *etc.*—When the Churches came to have fix'd Revenues allotted them, it was decreed, that, at least, one fourth Part thereof should go to the Relief of the Poor; and to provide for them the more commodiously, divers Houses of Charity were built, which are since denominated *Hospitals*.
They were govern'd wholly by the Priests and Deacons, under the Inspection of the Bishop.
In Course of Time, separate Revenues were assign'd for the *Hospitals*; and particular Persons, out of Motives of Piety and Charity, gave Lands, and Money for erecting of *Hospitals*.
When the Church Discipline began to relax, the Priests, who till then had been the Administrators of *Hospitals*, converted them into a Sort of Benefices, which they held at Pleasure, without giving account thereof to any Body; reserving the greatest Part of the Income to their own Use; so that the Intentions of the Founders were frustrated.—To remove this Abuse, the Council of *Vienna* expressly prohibited the giving any *Hospital* to secular Priests in the Way of a Benefice; and directed the Administration thereof to be given to sufficient and responsible Laymen,

who should take an Oath, like that of Tutors, for the faithful Discharge thereof; and be accountable to the Ordinaries.—This Decree was executed and confirm'd by the Council of *Trent*.

In *England*, *Hospitals* founded for the mere Relief of the Indigent and Necessitous, are peculiarly call'd *Alms-houses*; the Name *Hospital* being reserv'd to those destin'd for the Sick, Aged, Young, *etc.*—The Principal of these are, The—

Royal HOSPITAL for disabled Soldiers, commonly call'd *Chelsea College*.

It was founded by King *Charles* II. carry'd on by King *James* II. and finish'd by King *William* and Queen *Mary*.

The Building is very spacious and magnificent: Its Figure is a π; the middle or front Part whereof consists of a Chapel and Hall; the other two Lines being four Stories high, divided into Wards or Galleries, two in each Story, containing each twenty-six distinct Apartments for the Foot Soldiers. At each of the four Corners of the main Building, there is a Pavilion, in one whereof is the Governor's Lodging and the Council Chamber; the other being Lodgings for several of the Officers of the House. Beside the main Building, there are four Wings or Out-Buildings; one for the *Infirmary*, another for several Officers of the House, another for old maim'd Officers of Horse and Foot; and the fourth for the Baker, Laundress, *etc.*

The Number of ordinary Pensioners is 476; beside the Officers and Servants of the House: The out, or extraordinary Pensioners, are also very numerous; and these, upon Occasion, do Duty in the several Garrisons, from whence Draughts are made for the Army, *etc.*

The Pensioners are all provided with Cloaths, Diet, Washing, Lodging, Firing, and one Day's Pay in every Week for spending Money.

The Qualifications requir'd to be admitted of this Body, are, that each Person bring a Certificate from his superior Officer, that he has been maim'd or disabled in the Sevice of the Crown; or that he has serv'd the Crown 20 Years, which must be make appear by Muster-Rolls.

To defray the Charges of this *Hospital*, there is a considerable Sum paid yearly out of the Poundage of the Army; beside one Day's Pay of each Officer, and each common Soldier, every Year; which, in Time of War, amonts to 13 or 14000*l*.

For the Administration of this *Hospital*, there is a Governor, Lieutenant-Governor, Major, Treasurer, *etc.*

Greenwich HOSPITAL, is a Retreat for Seamen, who, by Age, Wounds, or other Accidents, are disabled from Service; and for the Widows and Children of such as are slain in the Service.

This, in Point of Magnificence and Spaciousness, excels even *Chelsea Hospital*. A good Part of it was built in K. *Charles* II's Time, at the Expence of 36000*l*. It was much promoted by King *William*; and finish'd under Queen *Anne* and King *George*.

The Number of Pensioners entertain'd in this *Hospital* is 300. To each hundred Men are allow'd six Nurses, being the Widows of Seamen.

The Victualling is according to the Allowance of *Chelsea Hospital*, *viz.* four Men to a Mess; each Mess to contain four Pound of Flesh, a Gallon of Beer, *etc.*

It is administer'd by a Governor, Lieutenant-Governor, Captain, Lieutenant, Chaplain, Steward, Physician, *etc.*

Christ's HOSPITAL, popularly call'd the *Blue Coat* HOSPITAL, was anciently a Monastery of Grey Friars, founded by *Raihere*, the first Prior thereof, in the Time of *Henry* I. dissolv'd by *Henry* VIII. and converted by *Edward* VI. into an *Hospital* for poor Children, who are supply'd with all Necessaries and Conveniencies, cloath'd, dieted, and taught.

Since its first Endowment it has receiv'd abundance of new Donations. A great Part of it was burnt down by the great Fire; but is again rebuilt by the Care of the Governors, tho' not without incurring a great Debt, and anticipating the Revenues of the *Hospital*.

Formerly, a thosand poor Children, most of them Orphans, were maintain'd on this Foundation; and six or seven Score yearly put out Apprentices, and the Maids to Service; but the Number, thro' the Deficiency of the Funds, is now much less.

Here were two Mathematical Schools; the first founded by King *Charles* II; but they are now united. Youths are there taught several Parts of practical Mathematicks, particularly Navigation, to fit them for Apprentices to Masters of ships: To say nothing of the Grammar School, (whence the most pregnant Boys are yearly sent to the University) Writing School, Drawing School, *etc.*

The Officers of this *Hospital* are a President, Treasurer, Governors, *etc.*

St. *Bartholomew's* HOSPITAL adjoins to *Christ's Hospital*, and formerly belong'd to the same Grey Friars.

At the Dissolution of Monasteries, *Henry* VIII. left 500 Marks a Year to it for the Relief of poor People; but it was more largely endow'd for the Use of sick and lame Persons only, by *Edward* VI.

It is govern'd by a President, Treasurer, *etc.* with other Officers. It is furnish'd with two Physicians, and three Master Surgeons, beside as many Assistant Surgeons.

In this *Hospital*, with two others depending on it, the one at *Kingsland*, the other call'd the *Lock* in *Southwark*, are about 300 Patients, provided with Lodging, Diet, Physick, Chirurgery, *etc.*

St. *Thomas's* HOSPITAL, in *Southwark*, is for the same Purposes as that of St. *Bartholomew*.

It was originally founded an *Hospital*, by *Richard*, Prior of *Bermondsey*, in 1213; surrender'd to King *Henry* VIII; and given by *Edward* VI. to the Citizens of *London*, for an *Hospital* for Infirm and lame People.

It consists of four quadrangular Courts: In the first are six Wards for Women; in the second, two Chapels, the lesser for the private Use of the

Hospital, and the larger Parochial: In the same Court are the Houses of the Treasurer, and other Officers: In the third Court are six or seven Wards for Men. The fourth has also three Wards, Baths hot and cold, a Chirurgery, Apothecary's Shop, *etc.*

There are about 3000 Persons taken in and discharg'd out of this *Hospital* yearly.

The Governors of this *Hospital* are the Lord Mayor and Court of Alermen, with about 260 other Citizens. Among which are a President, Treasurer, *etc.* two Physicians, and three Surgeons.

Guy's HOSPTIAL, or the HOSPITAL *of the Incurables*, is the Foundation of that wealthy Citizen and Bookseller, *Thomas Guy*, Esq;

It is chiefly intended for reputed incurable Persons; and is even to take in yearly a certain Number of Patients turn'd out of the other *Hospitals*, particularly *Bethlehem*, as incurable.

The Founder bestow'd 50000*l.* on the Building in his Life-time; and, by Will, endow'd it with 200000*l.* at his Death, in 1724: The greatest Benefaction, without Dispute, that ever was left by any one private Man.

Its Officers are a President and Governors, most of which are the same with those of St. *Thomas's Hospital*, which is in the Neighbourhood thereof, with a Treasurer, two Physicians, two Surgeons, *etc.*

HOT-BED, a Piece of Earth or Soil plentifully enrich'd with Manure, and defended from cold Winds, *etc.* to forward the Growth of Plants, and force Vegetation, when the Season or the Climate of itself is not warm enough.

By means of *hot Beds* skilfully manag'd, we can so nearly imitate the Temperature of other Climates, that Seeds of Plants brought from any Country between the Tropicks, may be made to flourish even under the Poles.

Heat and Humidity being the great Instruments of Vegetation; to promote the Growth of any Plants, these must be duly proportion'd, so as neither to exceed nor come short of the Bounds Nature has allotted for it.—Too much Heat we find rather scorches a Plant than makes it grow; and too much Moisture frequently chills it, unless quickly exhaled from the Roots.—With us, a moderate Heat is found the most eligible; such as is rais'd by the Ferment of wet Straw and Horse-litter, which, from the Earth lying thereon, will send forth, for some time, a gentle Steam, impregnated with vegetative Salts.

The usual Way of making *hot* Beds, is of Horse-litter and Grass mix'd together, and left on an Heap for eight or ten Days to putrefy; and then remov'd into a Bed, and cover'd up with Glasses or Frames.—Others chuse to take Horse Dung a Month or six Weeks old, and make a Seed-Bed of it about four Foot high, and cover it up with Straw a Foot thick, which is to be remov'd in three or four Days, and its Place supply'd either with Cows Dung, or the last Years Ridges.

The Process of ordering a good serviceable *hot* Bed, for the customary raising of Colliflowers, Cucumbers, Melons, Radishes, and other tender Plants, and Flowers in *January* or *February*, is directed by Mr. *Mortimer* in Manner following:

Provide a warm Place defended from all Winds, by being inclos'd with a Pale or Hedge, made of Reeds or Straw, about six or seven Foot high, of such Distance or Capacity as Occasion requires.—Within this Inclosure raise a Bed two or three Foot high, and three Foot over, of fresh Horse Dung, about six or eight Days old; then tread it down very hard on the Top, make it level, and if you think fit, edge it round with Boards or Bricks, laying fine rich Mould about three or four Inches thick on it: When the extream Heat of the Bed is over, which you may perceive by thrusting in your Finger; plant your Seeds at pleasure, and set your Forks four or five Inches above the Bed, to support a Frame made of Sticks, and cover'd with Straw or Bass-mat, in order to secure the Seeds and Plants from Cold and Wet; only the Covering may be open'd in a warm Day, for an Hour before Noon, and an Hour after.—But take Care to earth up your Plants as they shoot in Height; and when able to bear the Cold, they may be transplanted.

In *Holland* they make use of *hot* Beds made of Sand; which are not so apt to raise unwholsome Damps as those of Horse Dung.—The *Dutch* likewise make *hot* Beds of Tanner's Bark, which, when once rightly prepar'd, will maintain an equable Heat for six Months.

Bradley, with very good Reason, proposes a Thermometer to be us'd to regulate the Heat of *hot* Beds.—For Plants that are either to be brought up in a colder Seson, or a colder Climate than what they naturally require; you are to take the Height at which the Thermometer stands in their proper Season or Clime, as a Standard; and by applying a Thermometer to the *hot* Bed, judge whether the Heat is to be intended or remitted.—Thus a *hot* Bed for Cucumbers must be kept, to raise the spirit in the Glass, to the same Height as the natural Temperature of the Weather will raise it to about the latter End of *May* and *June*, when Cucumbers will grow abroad without any artificial Heat or Shelter.

HOUSE, a Habitation, or Place built with Conveniences to live in; or a Building wherein to shelter a Man's Person and Goods from the Inclemencies of the Weather, and the Injuries of ill-dispos'd Persons. See BUILDING.

We say a Brick *House*, a Stone *House*, a *House* of two Stories, of three Stories, a Manor *House*, a Farm *House*, etc. See BRICK, STONE.

Ancient *Rome* consisted of 48000 insulated *Houses*. See INSULATED.

A Pleasure *House* or Country *House*, is that built for a Person to enjoy and divert himself occasionally in. This is the *Villa* of the ancient *Romans*; and what in *Spain* and *Portugal* they call *Quinta*; in *Provence*, *Cassine*; in some other Parts of *France*, *Closerie*; in *Italy*, *Vigna*.

In this Sense the Word *Vigna* is sometimes us'd in *English* to denote the Country Seats of the Noble *Romans*; as the *Vigna Farnese*, *Vigna Borghese*, etc.—The Citizens of *Paris* have also their *Maisons de Bouteilles*, Bottle-*houses*, to retire to, and entertain their Friends; which, in *Latin*, might be call'd *Micæ*, the Emperor *Domitian* having a House built for the like Purpose, mention'd under this Name by *Martial*, lib.II. *Epig*. 59.

'Tis a Thing principally to be aim'd at, in the Site or Situation of a Country *House* or Seat, that it have Wood and Water near it: If it cannot be conveniently built among Trees, yet there are few Places where Trees may not be speedily rais'd about it.

It is far better to have a *House* defended by Trees than Hills; for Trees yield a cooling, refreshing, sweet, and healthy Air, and Shade, during the Heat of the Summer, and very much break the cold Winds and Tempests from every Coast in the Winter.—The Hills, according as they are situated, defend only from some certain Winds; and if they are on the North Side of the *House*, as they defend from the cold Air in the Winter, so they also deprive you of the cool refreshing Breezes, which are commonly blown from thence in the Summer.—And if the Hills are situated on the South Side, it then proves also very inconvenient.

A *House* should not be too low seated, since this precludes the Convenience of Cellars.—If you cannot avoid building on low Grounds, set the first Floor above the Ground the higher, to supply what you want to sink in your Cellar in the Ground; for in such low and moist Grounds, in conduces much to the Dryness and Healthiness of the Air, to have Cellars under the *House*, so that the Floors be good, and ceiled underneath.

Houses built too high, in Places obvious to the Winds, and not well defended by Hills or Trees, require more Materials to build them, and more also of Reparations to maintain them; and are not so commodious to the Inhabitants as the lower built *Houses*, which may be at a much easier Rate, and also as compleat and beautiful as the other.

In Buildings or *Houses* not above two Stories with the Ground Room, and not exceeding twenty Foot to the Raison-Place, and upon a good Foundation; the Length of two Bricks, or eighteen Inches for the heading Course, will be sufficient for the Ground Work of any common Structure, and six or seven Courses above the Earth to a Water-Table, where the Thickness of the Walls are abated, or taken in, on either Side, the Thickness of a Brick, Namely two Inches and a Quarter.

For large and high *Houses* or Buildings of three, four, or five Stories with the Garrets, the Walls of such Edifices ought to be from the Foundation to the first Water-Table, three heading Courses of Brick, or 28 Inches at least; and at every Story a Water-Table, or taken in on the Inside for the Summers, Girders, and Joints to rest upon, laid into the Middle, or one Quarter of the Wall at least, for the better Bond.—But as for the innermost or partition Wall, a half Brick will be sufficiently thick; and for the upper Stories, nine Inches, or a Brick length, will suffice.

The Parts, Proportions, *etc.* of the *Houses* in *London*, are regulated by a Statute made for Rebuilding the City after Fire.—By this it is enacted, That the *Houses* of the *first and least Sort of Building*, fronting By-streets or Lanes, shall be two Stories high, beside Cellars and Garrets; the Cellars six Foot and an half high, if the Springs of Water hinder not; and the first Story nine Foot from the Floor to the Ceiling, and the second Story as much: That all the Walls in Front and Rear be, as high as the first Story, full the Thickness of the Length of two Bricks; and thence upwards to the Garrets, of the Thickness of one Brick and an half; and that the Thickness of the Garret Walls on the back Part, be left to the Discretion of the Builder, so that the same be not less than one Brick Length; and that the Thickness of the Party Wall in the Garret, be of the Thickness of the Length of one Brick at least.

That the *Houses* of the *second Sort of Building*, fronting Streets and Lanes of Note, and the River of *Thames*, consist of three Stories high, beside Cellars and Garrets; that the Cellars thereof be six Foot and an half high, if the Springs hinder not; that the first Story contain full ten Feet in Height from the Floor to the Ceiling; the second ten Foot, the third nine Foot; that all the said Walls in the Front and Rear, as high as the first Story, be two Bricks and an half thick, and from thence upward to the Garret Floor of one Brick and an half thick; and the Thickness of the Garret Walls on the back Part, be left to the Discretion of the Builder, so that the same may not be less than one Brick thick: And also that the Thickness of the Party Walls between every *House* of this second and larger Sort of Building, be two Bricks thick, as high as the first Story, and thence upward to the Garrets of the Thickness of one Brick and an half.

Also that the *Houses of the third Sort of Buildings*, fronting the high and principal Streets, shall consist of four Stories high, besides Cellars and Garrets; that the first Story contain full ten Foot in Height from the Floor to the Ceiling, the second ten Foot and an half, and the third nine Foot, the fourth eight Foot and an half; that all the said Walls in the Front and Rear, as high as the first Story, be two Bricks and an half in Thickness, and from thence upwards to the Garret Floor, of the Thickness of one Brick and an half; and that the Thickness of the Garret Walls on the back Part, be not less than one Brick; and also that the Party Walls between every *House* of this third and larger Sort of Building, be two Bricks thick as high as the first Floor, and thence upwards to the Garret Floor one Brick and an half.

Also, that in all *Houses of the fourth Sort of Buildings*, being Mansion *Houses*, and of the greatest Bigness, not fronting any Streets or Lanes as aforesaid, the Number of Stories, and the Height thereof, shall be left to the Discretion of the Builder, so as he exceed not five Stories.

The same Act also enjoins, that no Timber be laid within twelve Inches of the fore Side of the Chimney Jambs; and that all Joysts on the Back of any Chimney, be laid, with a Trimmer, at six Inches distant from the Back; as also, that no Timber be laid within the Tunnel of any

Chimney, upon Penalty to the Workman for every Day's Default, of 10*s.* and 10*s.* every Week it continues unreform'd.

Add, That as the Buildings of *London* join one upon another, and almost every several *House* hath a distinct *Proprietor*, the Parliament hath decreed, that the Wall dividing the Proprietors Ground, shall be built at the equal Charge of both the Owners: Whence it will not be impertinent to shew how these Party Walls are valu'd.

Now, all Brick-works, whether one, two, three, four, or any other Number of Brick Lengths in Thickness, are all to be reduc'd to the Thickness of a Brick and a half.

It hath been observ'd, that about 4500 of Bricks, of about 16*s. per* 1000; a hundred and a Quarter of Lime, at 10*s. per* Hundred; two Loads and a half of Sand, at 3*s. per* Load, will compleatly raise one Rod of Brick-work of a Brick and a half Thickness. See BRICK, *etc.*

And thus much will a Rod of Party Wall, the Materials only, reduc'd to a Brick and a half thick, amount to, at the former, suppos'd Rates; to which may be added for Workmanship 1*l.* 8*s.*

So that for every Rod of Party Wall they allow 3*l.* a-piece; whence, if a Party Wall be measur'd, and the Measure when reduc'd to a Brick and a half, be found to contain sixteen Rods; that sixteen multiply'd by 3*l.* giveth 48*l.* and so much is the one Proprietor to allow the other.

Work-HOUSE. See WORK-*House.*

Town-HOUSE, or *Hall*, is a Place where the Officers and Magistrates of a Town or City hold their Meetings, for the due Administration of their Laws and Policy. See HALL, GILD, *etc.*

HOUSE is also us'd for a Convent or Monastery. See CONVENT, *etc.*

The Chief of such an Order has so many *Houses* dependent on it.—There have been Reforms made of several religious *Houses.*

Regular Priests give the Name *Houses* to the Places they reside in, and not that of Convents or Monasteries, which properly belong to simple Friars.—Thus we say, the *Jesiuts House*, the *Barnabites* or *Theatins House.*

The Jesuits have both profess'd *Houses* and Colleges for Novices, which they call *Houses* of Probation.

They have also *Houses* of Retreat for Spiritual Exercises, where they receive secular Persons and Ecclesiasticks, dispos'd to practise the same with them, for eight or ten Days.

HOUSE is also us'd for one of the Estates of the Kingdom assembled in Parliament.

Thus we say, the *House* of Lords, the *House* of Commons, *etc.*

HURDLES, in Fortification, Twigs of Willows or Osiers, interwoven close together, and usually laden with Earth. Their use is to render Batteries firm, or to consolidate the Passages over muddy Ditches, or to cover Traverses and Lodgments for the Defence of the Workmen, against the artificial Fires or Stones that may be cast upon them.

HYPOCAUSTUM, among the *Greeks* and *Romans*, was a subterraneous Place, wherein was a Furnace that serv'd to heat the Baths.—*Vitruvius* calls it *Caldarium*.

The Ancients had properly two Sorts of *Hypocausta*; the one call'd by *Cicero Vaporarium*, and by others *Laconicum* or *Sudatio*; which was a large sweating Bath, in which were three brazen Vessels, call'd *Caldarium*, *Tepidarium*, and *Frigidarium*, according to the Water contain'd therein.

The other Sort of *Hypocaustum*, was a sort of *Fornax* or Kiln to heat their Winter Parlours, *Cœnatiunculæ Hybernæ*.

The latter *Hypocaustum* was call'd *Alveus* and *Fornax*; and the Man that tended the Fire *Fornacator*.

The *Hypocaustum* among Moderns, is that Part or Place where the Fire is kept that warms a Stove or Hot-house. See STOVE.

The Word is *Greek*, form'd of the Preposition *sub*, under, and the Verb *incendo*, I burn.

HYPODROME. See HIPPODROME.

4 *Cyclopaedia* Volume II

Title Page

The Preface: Part Two

CYCLOPÆDIA:

OR, AN

UNIVERSAL DICTIONARY

OF

ARTS and SCIENCES;

CONTAINING

The DEFINITIONS of the TERMS,

and ACCOUNTS of

The THINGS signify'd thereby,

In the several ARTS,

Both *LIBERAL* and *MECHANICAL*,

And the several SCIENCES,

HUMAN and *DIVINE*:

The Figures, Kinds, Properties, Productions, Preparations, and Uses, of Things NATURAL and ARTIFICIAL;

The Rise, Progress, and State of Things ECCLESIASTICAL, CIVIL, MILITARY, AND COMMERCIAL:

With the several Systems, Sects, Opinions, *etc.* among Philosophers, Divines, Mathematicians, Physicians, Antiquaries, Criticks, *etc.*

The Whole intended as a Course of Antient and Modern LEARNING

Compiled from the best authors, Dictionaries, Journals, Memoirs, Transactions, Ephemerides, *etc.* in several Languages.

In TWO VOLUMES.

By *E. CHAMBERS* Gent.

Floriferus ut apes in saltibus omnia libant,
Omnia nos——————————LUCRET.

VOLUME the SECOND.

LONDON:

Printed for *James* and *John Knapton, John Darby, Daniel Midwinter, Arthur Bettesworth, John Senex, Robert Gosling, John Pemberton, William* and *John Innys, John Osborn and Tho. Longman, Charles Rivington, John Hooke, Ranew Robinson, Francis Clay, Aaron Ward, Edward Symon, Daniel Browne, Andrew Johnston,* and *Thomas Osborn*, M.DCC.XXVIII.

THE

PREFACE.

I MIGHT here have ended my Preface; and perhaps the Reader would be willing enough to be thus dismiss'd. But something has been already started which will require a further Disquisition. —The Distribution we have made of Knowledge is founded on this, That the several Branches thereof commence either *Art* or *Science* according to the Agency or Non-agency of the human Mind in respect thereof: It remains to take the Matter up a little higher; and explain the Reason and Manner of this Operation. To consider Knowledge in its Principles, antecedent to such Intervention of ours; and even pursue it up to its Cause, and shew how it exists there, before it be Knowledge: And to trace the Progress of the Mind thro' the Whole, and the Order of the Modifications induced by it. This is a Desideratum, hitherto scarce attempted; but which we could not here decline entering upon, on account of its immediate Relation to the present Design. 'Tis the Basis of all Learning in general; the great, but obscure Hinge, on which the whole Encyclopædia turns.

TO be a little more explicit—*Words* are the next Matter of Knowledge; I mean, of Knowledge consider'd as it now stands, communicable, or capable of being transmitted from one to another. We should have known many Things without Language; but it would only have been such Things as we had seen or perceived our selves. The Observations of others could no way have been added to our own; but every Individual must have gone thro' a Course for himself, exclusive of all Advantages to, or from, Contemporaries, Predecessors, or Posterity. —'Tis evident that, in this Case, nothing like an *Art*, or *Science*, could ever have arose; not even in the Mind of the most sagacious Observer: The little System of Things which come immediately in one Man's way, wou'd but have afforded a slender Stock of Knowledge; especially to a Being whose Views were all to terminate in himself. Add, that, as the chief Occasions of his Observation would be of the same kind with those of other Animals; 'tis probable his Knowledge would not have been very different, whether we consider its Quantity or Quality. 'Tis confess'd that all our Knowledge, in its Origin, is no other than Sense; whence it should follow, that one Being has no natural Advantage over another in its Disposition for Knowledge, other than what it has in the superior Number, Extent, or Acuteness of its Senses.

'TIS, in effect, to Language that we are chiefly indebted for what we call *Science*. By means hereof our Ideas and Notices, tho things in their own nature merely personal, and adapted only to private use; are extended to others, to improve their stock: and thus, by a kind of second Sense, we get Perceptions of the Objects that are perceived by all Mankind; and are present, as it were by proxy, to things at all Distances from us: We hear Sounds made a thousand Years ago, and see Things that pass a thousand Miles off. If the Eagle really sees, the Raven smells, and the Hare hears, further and better than Man; their Sense, at best, is but narrow, in comparison of ours, which is extended, by the Artifice of Language, over the whole Globe. They see with their own Eyes only; we with those of the whole Species.—In effect, by Language, we are upon much the same footing, in respect of Knowledge, as if each individual had the natural Sense of a thousand: an Accession which alone must have set us far above any other Animals. But at the same time, this very accession of a multitude of Ideas more than naturally belong'd to us, must have been in great measure useless; without certain other Faculties of ordering and arranging 'em; of abstracting, or making one a Representative of a Number; of comparing 'em together, in order to learn their Relations; and of compounding, combining 'em, *etc.* to make 'em act jointly. The Effect hereof is what we call *Discoursing* and *Philosophizing*: And hence arise *Doctrines*, *Theories*, etc.

EVERY Word is supposed to stand for some Part, or Point of Knowledge; such as do not, have no business in the Language, and ought of Consequence to be thrown out of doors. It follows, that the Vocabulary of any Language, is representative of the several Notices of the People among whom it obtains: I mean of the primary or absolute Notices; for by the Construction of these Words with one another, a new Set of secundary or relative Notices are express'd.—To enter better into this, it is to be observ'd, that the several Objects of our Senses, with that other Set of Things analogous hereto, the proper Objects of the Imagination, are represented by fixed Names; denoting, some of 'em, Individuals; others, Kinds, *etc.* Now, these, which make the first or fundamental Part of a Language, 'tis obvious, are no other than a Representation of the Works of Nature, as they exist in a kind of still Life, or in a State of Independency one upon another. But in regard we do not consider the Creation as thus quiescent, but observe a great number of Mutations arise in the Things we are conversant among; we are hence put under a necessity of framing another Set of Words, to express these Variations, and the Actions to which they are owing, with the several Circumstances and Modifications thereof. By this means, Nature is remov'd out of her dormant Constitution, and shewn in Action;

and thus may occasional Descriptions be framed, accommodate to the present State of Things.

HENCE arise two Kinds of Knowledge; the one *absolute*, including the standing Phænomena: the other *relative*, or *occasional*, including what is done, or passes, with regard to them. The former is in some Sense permanent; the latter merely transient, or historical. The first is held forth, as already observed, in the Vocabulary: the second vague, and uncircumscrib'd by any Bounds; being what fills all the other Books extant. In effect, this last, being in some measure casual, may be said to be infinite: for that every new Case, *i.e.* every new Application and Combination of the former, furnishes a new Accession.

IN the wide Field of Knowledge, appear some Parts which have been more cultivated than the rest; either on account of the Goodness of the Soil, and its easy Tillage, or by reason they have fallen under the Hands of industrious and able Husbandmen. These Spots, being regularly laid out and planted, and conveniently circumscrib'd or fenced round, make what we call the *Arts*, and *Sciences*: And to these have the Labours, and Endeavours of the Men of Curiosity and Learning in all Ages, been chiefly confin'd. Their Bounds have been enlarg'd from time to time, and new Acquisitions made from the adjoining Waste; but still the Space of Ground they possess is but narrow; and there is room either to extend 'em vastly, or to lay out new ones. They shew like the *Cyclades* at a distance: *Apparent rari nantes in Gurgite vasto.*

THEY were divided, or canton'd out by their first Discoverers, into a number of Provinces, under distinct Names; and have thus remain'd for time immemorial, with little Alteration. And yet this Distribution of the Land of Science, like that of the Face of the Earth or Heavens, is wholly arbitrary and occasional; and might easily be broke thro', and alter'd, and perhaps not without advantage. Had not *Alexander* and *Cæsar* liv'd, the Division of the Globe had doubtless been very different from what we now find it; and the Case would have been the same with the World of Learning, had no such Person been born as *Aristotle*. The first Divisions of Knowledge were as scanty and ill concerted as those of the first Geographers; and for the like Reason: And tho future *Columbus's* and *Bacon's*, by opening new Tracts, have carried our Knowledge a great way further; yet the Regard we bear to the antient Adventurers, and the established Division, has made us take up with it, under all its Inconveniencies, and strain and stretch things, to make our Discoveries quadrate thereto. I do not know whether it might not be more for the general Interest of Learning, to have all the Inclosures and Partitions thrown down, and the whole laid in common again, under one undistinguish'd Name. Our Inquiries, in such case, would not be confin'd to so narrow a Channel; but we should be led to explore, and pursue many a rich Mine and Vein, now doom'd to lie neglected, because out of the way.

ART and *Science* are indeed two Words of familiar Use, and great importance; but, I doubt little understood. The Philosophers have long labour'd to explain and ascertain their Notion and Difference; but all their Explanation amounts to little more than the substituting one obscure Notion for another. Their Attempts usually terminate in some barren Definition, which rather casts Obscurity than Light on the Subject. Nor is the Reason far to seek, however it may have escap'd Notice; but evidently lies in an Abuse of Language, whereby those different Words come to be applied to Things of the same Nature; and each of 'em in their turn to Things wholly different. Whence, any Definition that can hold of 'em universally, must needs be very abstracted, and general; and may hold of almost any thing else; and of consequence can express very little of the Essence, and obvious Phænomena thereof: To come at which, we must be at the Pains of a new Investigation.

TO *SCIENCE*, then, belong such Things as Men may discover by the use of Reasoning, and Sense: Whatever the Mind descries in virtue of that Faculty whereby we perceive Things, and their Relations, is matter of Science: Such are the Laws of Nature, the Affections of Bodies, the Rules and Canons of Right and Wrong, Truth and Error, the Properties of Lines and Numbers, *etc.* *Science*, in effect, is the Result of mere Reason and Sense in their general or natural State, as imparted to all Men; and not modified, or circumstantiated by any thing peculiar in the Make of a Man's Mind, the Objects he has been conversant among, or the Ideas he has present to him. Consequently, Science is no other than a Series of Deductions, or Conclusions which every Person, endued with those Faculties, must, with a proper degree of Attention, see, nay and draw: And A SCIENCE, *i.e.* a form'd Science, is no more than a System of such Conclusions, relating to some one Subject, orderly and artfully laid down in Words, to save Others the Labour and Expence of making 'em at first hand. Thus a Person who has all the Ideas express'd in Euclid's Definitions, and sees the immediate Connexion of those in his Axioms; which no Man acquainted with the Language can be supposed without; has it in his Power, with Attention and Industry, to form all the Theorems and Problems that follow: He has nothing to do but to range those Ideas orderly in his Mind, compare 'em together, one by one, in all their Changes, and put down the immediate Relations observ'd in the Comparison, *i.e.* their parity, imparity, *etc.* And after the Relations of each to each are thus got; which make a kind of primary Propositions; to proceed to combine 'em, and take down the Relations resulting from a Comparison of several Combinations. By such means, without any other Helps than Penetration and Perseverance, might he make out an infinite Number of Propositions: more by half than *Euclid* has done; there being a new Relation, *i.e.* a new Proposition, resulting from every new Combination.

TO ART, on the other hand, belong such Things as mere general Reason would not have attained to: Things which lie out of the direct Path of Deduction, and which required a peculiar Cast, or Byass of Mind to see or arrive at. A Man might call these the Results of particular, or personal Reason, in opposition to the former; but that such a Denomination would be thought unphilosophical. It may, perhaps, be more just to consider the Reason, here, as modified, or tinctured with something in the Complexion, Humour, or Manner of thinking of the Person*; or as restrain'd or diverted, out of its proper course, by some Views, or Notices peculiar to him.—The Difference between Art and Science, amounts to much the same as between *Wit* and *Humour*; the former whereof is a general Faculty of exciting agreeable and surprizing Pictures in the Imagination†, by the associating of Ideas, which at the same time have both notable Diversity and a Congruity; and the latter, a particular one: The former is pure and absolute in its kind; the latter tinged with something foreign and complexional.

'TIS essential to *Art*, therefore, as to Humour, to partake of the Person from whom it proceeded; and consequently there are as many Arts, as Inventors of Methods of performing, or doing things. Hence, there is no coming at an Art, otherwise than by learning it. A person left to his own Thought; will scarce ever hit on the same thing, unless either we suppose a marvellous Agreement between the Characters and Circumstances of the Persons; or that the Art is in great measure scientifical, and partakes but little of the Genius and Humour of the Inventor.—There is no such thing, properly speaking, as studying an Art, or learning a Science: The first, every Man beside the Inventor must be taught; the latter, every man must find. In effect, to attain to an Art, there is some previous Knowledge requir'd, which a Man's own Reason would never have suggested; whereas a Science requires no more than clear Ideas, and close Attention. With these Helps a Man may of himself go the whole length of a Science, so far as it is properly a Science. Indeed if the Improvers, or rather Writers thereof, have gone a jot out of the common way, in compliance with their own personal Views; they have so far adulterated the Science, and put it on the footing of an Art. And to this very Cause are owing a great part of the Difficulties we meet withal in attaining the Sciences: The rest arise from want of Sense, *i.e.* of Clearness and Precision in our Perceptions, and want of Perseverance and Attention to 'em. These render Geometry it-self, little other than an Art: We want Preliminaries to it as to other Arts. And thus every Science is an Art to some People, and only to be attained, as we do mechanic Arts, by Habit, and Remembrance; instead of Contemplation and Deduction.—Reason, clogg'd and embarass'd by Genius and Complexion, can no more rise to the heights of Science, than when pure and refin'd, it can descend to the depths of Art.

AN Art and a Science, therefore, only differ as less and more pure: A *Science* is a System of Deductions made by Reason alone, undetermin'd

**Vid.* Bossu, Traité du Poëme Epique, l. i. c. I.

†Locke, *on Human Understand.* Lib. ii. c. II.

by any thing foreign or extrinsic to it-self: An *Art*, on the contrary, requires a number of Data and Postulata to be furnish'd from without, and never goes any length, but at every turn it needs new ones. 'Tis the Knowledge or Perception of these Data that in one Sense constitutes the Art; the rest, that is, the doctrinal Part, is of the nature of Science; which attentive Reason alone will descry.

AN Art, in this light, appears to be a Portion of Science, or general Knowledge, consider'd, not in it-self, as Science, but with relation to its Circumstances, or Appendages. In a Science, the Mind looks directly backwards and forwards, to the Premises and Conclusions: in an Art we also look laterally, to the concomitant Circumstances. A Science, in effect, is that to an Art, which a Stream running in a direct Channel, without regard to any thing but it-self and its own progress; is to the same Stream turn'd out of its proper Course, and running in a different one dispos'd into Cascades, Jets, Cisterns, Ponds, *etc.* and serving to water Gardens, turn Mills, and other particular Purposes. In which case, the Progress of the Stream is not considered with regard to it-self, but only as it concerns the Circumstances of the Works: every one of these Works, nay each part thereof, are so many Data, which modify the Course of the Stream, and vary it from its original Habitude. 'Tis easy to trace the Progress of the former, from its Rise to its Issue; in regard it flows consequentially: But a Man ever so well acquainted with this, will never be able, of himself, to discover that of the Second, for want of Acquaintance with the Circumstances, which his Reason can never find out; in regard they depend on the Genius, Humour and Caprice of the Engineer who laid the Design.

THESE are so many different Characters, or Conditions or Art and Science: But there is a Difference between 'em prior to any of these, and of which these are only Consequences. The Origin of 'em all lies higher, in the Principle of Action or Operation above specified; namely, as the Mind is either active or passive therein. With regard to this; those Things may be said to belong to *Science* which we only see, or perceive; which flow from the Nature and Constitution of Things, by the sole Agency of the Author thereof; subservient only to His general Purposes, exclusive of any immediate Agency or Intervention of Ours: And, on the contrary, those Things belong to *Art*, wherein such Science or Perception is further modified and circumstantiated in our Mind, and directed and applied by us, to particular Purposes and Occasions of our own.—From hence arise the several Differences above mention'd: For the Matters of Art are only *Personal*, as they are according to the Measure of the Artist's natural Faculties, in respect of Quantity and Degree; and to the Complexion and Cast of his moral Faculties, in respect of their Quality. The Perception, even of Matters of Art, is of the Nature of Science: so that thus far the two agree: And their Difference only commences from the superinducing a further Modification, in the Matter of such Perception; and the giving it a new Direction to some particular End. By means hereof, it becomes invested with a new Set of Conditions and Circumstances, wholly personal; as being all framed and adapted to the particular View and Aim of the Artist, (which is the mere Result of his particular Disposition, Humour, Manner of thinking, Situation, Occasion, *etc.*) and conducted according to his particular degree of Knowledge, and Address; which is the Effect of a particular Set of Objects, and a particular Organism of Body. In a word, in Art there is a moral View or Motive superadded to the natural Science, or Perception; which Motive is the proper Principle, or *primum Mobile* of Art: Perception is its Matter; and some Member of the Body its Organ or Instrument. And from such new Principle, *etc.* arise a new Set of secundary Perceptions, analogous to the natural and primary ones. The whole, therefore, ends in this, That Science arises from a natural Principle, Art from a moral one; or even, as moral Matters are also in one Sense natural, Science may be said to be of divine Original; Art, of human.

WE may define the Works or Productions of Art, therefore, to be all those Phænomena or Effects which would not have arose without the Agency or Intervention of Man. Now Man can only be said to act or intervene, so far as what he does is of his own Source or Principle, without being moved or directed by any established Law of Nature, *i.e.* so far as he is exempted from the Influences of any necessary Laws of Nature concurring, however remotely, to such Effect. So that if, as some Philosophers have maintained, Man were not really any truly a free Agent; there would be no such thing as Art, in the Sense here understood: but *Art* would only be a name given to that System or Series of Effects, to which Man is made by Nature, and in her hands, subservient, and might with equal reason be attributed to such Effects as any other natural Production, *e.g.* a Plant, or Mineral, is subservient.

HAD it not been for the inspired Writers, we should not have known but that the whole System of our World is a Production of Art; the Result of a new Application of Things made by some created Being, in virtue or consequence of some pre-establish'd Laws of the Almighty. Our general Laws of Nature, and Motion might only be particular Cases of some more universal one; special Instances, emerging out of some more general one, which it-self was not perhaps the first. Thus there might be an infinite Series or Subordination of Systems of Nature, each more universal, extensive, and, as we call it, more metaphysical, *i.e.* nearer the Source of Power and Action, than other.

WE see, then, how far Man is concern'd in the Productions of Art. Our Endeavours are contriv'd by Nature to be Means accessory to the Law's taking place, from whence the Effects are to arise. We are part of the Chain whereby the Effect is connected to the Cause. The Circumstances are in our Power on which such, and such Laws depend; and thus far we may be said to be *Active*, in the Case of Art: supposing that there is nothing higher, or further; and that the Chain ends with us; in a word, that our Agency is not subordinate, but collateral to that of the

Almighty. But if there be other superior Laws which respect those same Circumstances, and which are not in our Power, *i.e.* if the Circumstances necessary to the former Law, be themselves supposed necessary Laws, and the immediate Work of Nature; our Agency will dwindle into nothing. The utmost that can be said of us in such case, is, That we are Active in respect of the one, and Passive in that of the other; which to most People may appear a kind of Contradiction.—The Statue can't be form'd, unless our Desire or Inclination concur thereto; so far its Existence depends on us: But are our Desires and Inclinations with respect thereto of our own growth; or do they arise naturally, in consequence of an Apprehension of Good, and Advantage in the Subject? That is, does any thing appear good and advantageous to us absolutely and of it self; or only what the Creator represents to us as so? And do we desire or pursue this seeming Good, from any Principle or Tendency that is in us, other than what we owe to his Laws? The Difficulty seems to amount to this; whether between our Faculties of apprehending and willing, and their respective Objects, there be any Relation which he did not create or establish?—If any alledge, that 'tis such Relation constitutes the Faculty; and therefore that the Question ends in this, Whether our Faculties are from God or our selves: *i.e.* whether they can be the Causes of themselves? I should suspect some Sophism in the Case, which at present I have not leisure to detect.

THE *Means* and Measures of an Art, make a kind of preliminary Doctrine, necessary or conducive thereto, called the *Theory of the Art*; which, also, in one Sense, may be consider'd as another *Art*, distinct from the former: At least, to come at it is the Business of another Art.—If, for instance, a certain Position, or Set of Motions of the Body, be constituted by Nature the Occasion of a poetick Inspiration; and such and such Images and Ideas arising herefrom, be constituted, the occasions of such and such Passions in the Mind of a Reader, and such and such Views consequent thereon, *viz.* an Aversion to Enmity, and Contention: To form an Art productive of these Effects, we must first see and observe such or the like Effects, to arise from such or the like Causes; and argue or infer, that 'tis probable these Motions, or these Images, are the Occasions thereof: and consider and collect the Order, Manner, and Circumstances thereof, to form the *Art*, or *Method*.—So that we have here, as before, 1°, *Matter*, *viz.* Phænomena, first furnished by Sensation, and preserv'd in the Memory; 2°, *Form*, arising from the Moral View, which led us to frame an Art, and in order thereto, to consider and dwell on the Phænomena, compare 'em together, and infer something from 'em.—It appears, therefore, that we have two Arts of Poetry, very different from each other; coming from different Causes, tending to different Purposes, and rarely found, in any degree, in the same Person. The first Art *Homer* has in perfection, the second, *Aristotle*.

BUT for all their difference, the two are really of the same general Nature, and Kind; and only differ in point of Degree, and Subordination;

as they are nearer to, or further from, the Principle of all Knowledge and Art, Sensation.—*Homer*, we have shewn, was inspired: He derived his Art only from Nature acting on him in the ordinary Course of Things, and first presenting Objects to his Sense, then to his Imagination: And others are inspired from him, *i.e.* derive the Inspiration from Nature thro' his means: among whom is *Aristotle.* Nature, as she appears to the Senses, is *Homer's* Subject: as she shews her self in *Homer*, is *Aristotle's*; by which time the Inspiration is grown a degree cooler, and less forcible, and the Ideas thus excited at second hand moving the Mind less, it can attend more steadily to 'em, and perceive their Relations better. In the first it falls like Lightning, immediately from Heaven; the second may be compared to the Reflexion of the same in a Mirror. The reading of *Homer*, *i.e.* the exciting and calling up his Ideas and Images, does, as it were, impregnate *Aristotle's* Imagination; and transplant the Poet's whole Nursery into the Philosopher's Garden, to be further cultivated. Accordingly, *Aristotle* applying his Apprehension and Reason to 'em, and examining 'em closely on all Sides, perceives divers Relations and Analogies between 'em, which *Homer* was not aware of; and which the Warmth of his Imagination, and the quick Succession of new Ideas, would not give him room to attend to. These Analogies he calls *Rules*, or *Laws*; the Assemblage or System whereof, make what we call *Aristotle's Art of Poetry.*

THE like Process might be observ'd in the several other Arts. Those we have hitherto chiefly kept to, have been of the symbolical Kind: we shall here give an Instance in what we call the real Kind, *viz. Architecture.*—An *Athenian* Sculptor, then, observing an Acanthus shooting up under a Basket; is pleas'd with the Figure it presents; and taking the Hint, invents the Capital of a Column on the Model thereof: And by a number of like Steps, an entire Order gradually arose; and, in time, a whole Art.—Things thus advanced; and another Person seeing a Building framed after such manner; he attentively examines the several Members, their Forms, Proportions, *etc.* and puts 'em down in writing: And thus does another posterior Art arise. And between the two, there still remains the Subordination already observed between the Means, or Occasions of producing 'em; *i.e.* the Rules thus formed being couched in Words, or Language, supply the Office of the external Objects they were originally deriv'd from, and prove Occasions of raising Ideas or Images in the Imagination of future Artists, to be imitated in the proper Materials.

THE Arts, then, of Poetry and Architecture, come first in at *Homer's*, and *Callimachus's* Sense, in the simple Quality of natural Phænomena, or Objects; which meeting with other Ideas in the Memory, or Imagination, and coming to be compared and combined therewith, by the Agency of the Moral View or Principle which suggested the making of a Poem, *etc.* as advantageous and desirable; new Productions arise, *e.g.* a Poem, or a Building: which coming at length under the Cognizance and Consideration of Reason, certain Relations or Analogies are discovered therein, which tend to propagate, and produce the like at any time. —Reason returns Rules for Matter; which Rules, prove like the Philosopher's Stone, which tends to turn all Materials it is applied to, into Gold; and the Materials thus transmuted, like the pretended multiplicative Virtue of the same Gold, from every thing they are applied to, produce Rules again.

REASON, in effect, which is the last Faculty the Matter of Art arrives at, is the first from which the Form or Rules thereof, which are to propagate it, arise. In which view, Reason may be laid down as the Principle of this secondary Art, or Theory; as Imagination of the primary one, or of the Matter. We still see the Effect of the first Laws, even in the latter Art: External Objects strike the Sense and Imagination so strongly; that they reach to Reason; which, like an infinitely elastick Substance, reflects 'em back again; and thus they again grow into Objects of Sense: and so in a Circle.

THIS seems to make the two Arts differ very widely: And as Reason appears our highest Faculty, (inasmuch as 'tis this alone that tends to produce, and multiply) and accordingly, all our Knowledge appears proportionably higher and purer, as Reason is more concern'd therein: the Rules or Theory of an Art, appears of infinitely nearer Consequence than the Matter thereof. The Former is in some Sense active, and, like the Almighty Mind tends to produce new things, new Worlds, new Systems without end; the latter is mere Passion, and ends in bare Brute Perception.

HAVING thus discuss'd the Nature, and Characters of *Art* and *Science*; it remains to settle the Notion of *a TERM of Art*; a Diction as little understood as any thing in Language.—Art and Science, we have observed, are Denominations of Knowledge under this or that Habitude; and Words are Representatives of the several Parts thereof. The whole Compass of Words, in all their Cases, is suppos'd equivalent to the whole System of possible Science; tho 'tis only a small Part thereof that is actual, *i.e.* only a few of the possible Combinations are, or ever will be, made.

NOW, what we call *a Term*, consider'd as to its Nature and Origin, is no other than "a Word which denotes an Assemblage, or System of Ideas "relating to some one Point, which the Mind artfully complicates or "associates together, for the conveniency of its own Operations." Or, "a "Word which comprehends several Ideas under a certain Relation to each "other, whereby they represent some complex piece of Knowledge to the "Mind for the conveniency," *etc.* Or, "it is a Word, which holds several "different Ideas combin'd together in a Relation such as they appear'd "under when the Mind first consider'd 'em as a standing Phænomenon, "and took Measures to have 'em fix'd or retain'd in that Quality."

THUS much relates to what we may call *Terms of Knowledge*, which are one degree more simple than the Terms of an Art, or Science; and were, for that Reason, pitch'd upon to exhibit the common Nature, and Origin of both. These latter arise out of the former, by the Superaddition of some new Character, or Condition. They were before Members of the Commonwealth of Knowledge; but are now incorporated into some certain Province, or City thereof; where they become of farther Significance and Consideration than before: that is, some new Ideas and Circumstances are now taken into the Combination, which before did not belong to it.—A Term of Art, then, "is a Word that has a Meaning "beyond its general, or scientifical one; and this Meaning restrain'd to "some one Art." Or, it is "a Word used to denote a certain Combination "of Ideas, under some peculiar Relation; retained arbitrarily in some Art, "and either not used in any other Art, or for a different Combination, or "with other Relations and Circumstances."

THE Notion of a Term will receive some farther Light from that of a *DEFINITION*; which is, as it were, the Analysis thereof.—By *Definition* we undo, what was done in the *Term*; that is, we resolve the vague one. A *Definition*, then, may be defined, "an Enumeration of the several "simple Ideas couched under any Term, in the Relation wherein they "stand to one another."—We have already shewn, that Terms are Words which have peculiar and determinate Meanings, resulting from a certain Combination of Ideas; in which view, a Term may be said to be, "a Word "that is capable of Definition;" *i.e.* of having its Sense explain'd, and ascertained by an Enumeration of its Properties, and Relations: by which it is distinguish'd from other Words merely grammatical, whose Meanings are general and indeterminate, and may be used with equal propriety in a thousand Cases. We can explain a Term: A Word is inexplicable: all we can do towards this, amounts not to Definition, but only to *Substitution*.

'TIS the various Assemblage of simple Ideas denoted by common Words, that makes all the Variety of Terms; as 'tis of Simples in an Apothecary's Shop, that makes the Variety of his Medicines.—The Analogy goes farther; and it may be said that Terms, like Medicines, only differ from each other as their ingredients. Ideas, and the Relations thereof do differ.—If these be not all rehearsed in the Definition, the Term or Medicine is not specified, or distinguished from some other, which may have all except that one or two omitted. Consequently, such one or two are the Characteristicks of that Term; which may be explain'd in some sort by only enumerating those Characteristicks, and couching all the rest under that other Term. This amounts to little more than the *Substitution* abovemention'd; and yet to this is reducible all that the Schoolmen teach of *Genus*, *Species*, and *Difference*.

HAVING, therefore, dispatched the leading Words ART, SCIENCE, TERM, and DEFINITION; we proceed to consider the Nature of a *DICTIONARY*.—It were to be wished that the many Adventurers in Print, who publish their Thoughts under this or that Form and Denomination, would frame themselves a precise Notion of the Character and Laws thereof. There is something arbitrary, and artificial in all Writings: They are a kind of Draughts or Pictures, where the Aspect, Attitude, and Light, which the Objects are taken in, tho merely arbitrary, yet sway and direct the whole Representation. Books are, as it were, Plans or Prospects of Ideas, artfully arranged, and exhibited, not to the Eye, but to the Mind; and there is a kind of analogous Perspective which obtains in 'em, wherein we have something not much unlike Points of Sight, and of Distance. An Author, in effect, has some particular View or Design in drawing out his Ideas; either, nakedly to represent something, or distort and ridicule it, or amplify, or extenuate, or discover, or teach, or prove, *etc.* whence arise divers kinds of Pieces, under the Names of *Histories*, *Discourses*, *Treatises*, *Essays*, *Inquiries*, *Examinations*, *Paraphrases*, *Courses*, *Memoirs*, *Burlesques*, etc. In all which, tho the Matter or Subject may be the same, the Conduct or artificial Part is very different; as much as a Still-Life from a History, or a Grotesque, or a Nudity, or a Caricatour, or a Scene-work, or a Miniature, or a Profile, *etc.* Each of these Methods of Composition has its particular Characters, and Laws; and to form a Judgment of the Things represented, from the Pictures made of them, 'tis necessary we be able to unravel, or undo what is artificial in 'em, resolve 'em into their former State, and extricate what has been added to 'em in the Representation: That is, we should know that manner thereof; whether, *e.g.* they be mere Nature, thro' this or that Medium, in a fore, or side-View, withinside or without, to be seen from above or below; or Nature rais'd and improv'd, for the better, or the worse.—The Case amounts to the same as the viewing of Objects in a Mirror; where, unless the form of the Mirror be known, *viz.* whether it be plain, concave, convex, cylindrick, or conick, *etc.* we can make no Judgment of the Magnitude, Figure, *etc.* of the Objects.

'TIS beyond my purpose to enter into the Nature of the several Methods of Composition abovemention'd: I shall only note, by the way, that the first Writers in each, mark'd and chalked out the Measures for all that came after them. The several Manners of composing amount to so many *Arts*; which, we have already shewn, are things in great measure personal, and depend on the Genius or Humour of the Inventors.

WERE we to inquire who first led up the way of *Dictionaries*, of late so much frequented; some little Grammarian, would, probably, be found at the head thereof: And from his particular Views, Designs, *etc.* if known, one might probably deduce, not only the general Form, but even the particular Circumstances of the modern Productions under that name. The Relation, however, extends both ways; and if we can't deduce the Nature of a Dictionary from the Condition of the Author; we may the Conditions of the Author from the Nature of the Dictionary. Thus much, at least, we may say, that he was an Analyst; that his View was not to improve or advance Knowledge, but to teach, or convey it; and that he was hence led to unty the Complexions or Bundles of Ideas his Predecessors had made, and reduce 'em to their natural parity; which is all that is essential to a Dictionarist. Probably this was in the early Days of the *Phœnician* or *Egyptian* Sages, when Words were more complex and obscure than now; and mystic Symbols and Hieroglyphics obtain'd; so that an Explication of their Marks or Words, might amount to a Revelation of their whole inner Philosophy: In which Case, instead of a Grammarian, we must put perhaps a *Magus*, a *Mystes*, or *Brachman* at the head of Dictionaries. Indeed this seems the more probable; for that a grammatical Dictionary could only have place, where a Language was already become very copious, and many Synonyma's got into it; or where the People of one Language were desirous to learn that of another: which we have no reason to think could be very early, till much Commerce and Communication had made it necessary.

WHEN a Path is once made, Men are naturally disposed to follow it; even tho it be not the most convenient: Numbers will enlarge, and widen, or even make it straighter and easier; but 'tis odds they don't alter its Course. To deviate from it, is only for the Ignorant and Irregular; Persons who don't well know it, or are too licentious to keep it. And hence the Alterations and Improvements made in the several Arts, are chiefly owing to People of those Characters. These is scarce a more powerful Principle in Nature than that of Imitation, which not only leads us to do *what* we see others do, but *as* they do it. 'Tis true there are Exceptions from every Rule: there are Heteroclites, Persons in good measure exempted from the Influence of this Principle; and 'tis happy there are; witness such as *Paracelsus*, *Hobbes*, *Leibnitz*, etc. In effect, If an Art were first broached by an happy Genius, it is afterwards cultivated, on his Principles, to advantage; otherwise not: and it may wait long for the anomalous Hand of some Reformer, to set it to rights. Some of our Arts have met with such Hands, others still want 'em.

WERE we, now, to give an absolute and consistent Definition of a *Dictionary*; we should say, "It is a Collection of Definitions of the Words "of a Language."—Whence, according to the different kinds of Words and Definitions above laid down, *i.e.* according to the different Matter, and the different View wherein such Matter is considered, will arise different sorts of Dictionaries: *Grammatical*, as the common Dictionaries of Languages, which for one Word substitute another of equal import, but more obvious sense: *Philosophical*, which give the general Force or Effect of the Words, or what is common to 'em in all the Occasions where they occur: and *Technical*, which give the particular Sense attach'd to 'em in some one or more Arts.

BUT, in truth, this is a little chimerical; and is to forget what has been already said. Tho we have *Dictionaries* under all these Titles; it would perhaps be hard to find any conformable to this Partition; which is not so much taken from what really is, as what might, or should be. Dictionarists are far from considering their Subject so closely, or confining themselves to so narrow, tho direct, a Channel: They must have more room; and think themselves privileg'd by the general Quality of Lexicographers, to use all kinds of Definitions promiscuously. 'Tis no wonder they should not keep to Views which they had not, and which could only arise from Researches they never made. While the Notions of *Term* and *Art*, remain'd yet in the Rubbish they were left by the Schoolmen; those of *Definition* and *Dictionary* must needs be vague and arbitrary enough; and the Dictionarists and Expositors, profited by an Embarrass it was their Business to have remov'd. They have not only built on it, but improv'd it, by a continual varying and confounding of Views, imperfect Enumerations, *etc.*

THERE are two Manners of writing: In the one, which we may call *Scientifical*, we proceed from Ideas and Things, to Words; that is, first lay down the Thing, then the Name it is called by.—This is the way of Discovery, or Invention; for that the Thing ought to be first found before it be named. In this way, we come from Ignorance to Knowledge; from simple and common Ideas, to complex ones.

THE other, *Didactic*, just the Converse of the former; in which we go from Words, and Sounds, to Ideas, and Things; that is, begin with the Term, end with the Explanation. This is the historical Way, or the way of Teaching and Narration; of resolving the extraordinary Knowledge of one Person, into the ordinary of another; of distributing artificial Complications, into their simple Ideas: and thus razing and levelling again what Art had erected.

THE *Dictionary* comes under the latter Kind. It supposes the Advances and Discoveries made, and comes to explain or relate 'em. The Dictionarist, like an Historian, comes after the Affair, and gives a Description of what pass'd. The several *Terms*, are so many Subjects, supposed to be known to him; and which he imparts to others, by a Detail of the Particulars thereof.—Indeed, the Analogy between a *Dictionary* and a *History*, is closer than People at first sight may imagine: The Dictionarist relates what has pass'd with regard to each of our Ideas, in the Coalitions, or Combinations that have been made thereof: His Business is to deliver the Progresses made in the several Parts of Knowledge under his Consideration, by an orderly Retrospect and Deduction of the Terms, from their present complex, to their original simple State. The Dictionary of an Art, is the proper History of such Art: The Dictionary of a Language, the History of that Language. The one relates that such an Art, or such and such Parts thereof, stand so and so; are managed so and so; and the result so and so: The other, that such and such a Word is used as synonymous to such and such others. The Dictionarist is not supposed to have any hand in the Things he relates; he is no more concerned to make the Improvements, or establish the Significations, than the Historian to achieve the Transactions he relates.

THE difference between what we commonly call the *History* of an Art, and a *Dictionary* thereof, is only circumstantial; arising from the different Views of the two Authors: The one chiefly regards the Time and Order *when* each Step, each Advance, was first made, *i.e.* how it stood with respect to such and such Æras, or Periods of Time; and might more properly be called the *Chronology* of the Art: the other regarding chiefly the Object or Intention of the Art, relates its present Constitution, and *how* it proceeds to attain the End proposed. You may add, that the former primarily considers what is past, or already advanced; the other also what is present, or remains to be done: The one tells, *e.g.* how *Mercury* finding a dead Tortoise on the Shore, took its Shell, added Strings to it, and made it into a Lyre: The other, how a Lyre is, or may be made. And if you will likewise add this, that the History intermixes divers foreign, and accidental Circumstances with the Discovery; which the Dictionary abstracts and sets aside, and so reduces it nearer to Science: you will have the full and adequate Difference between 'em. Thus the making of the first Lyre is related with some Circumstances which have no place in the proper Structure of the Instrument, and are therefore to be omitted in the Dictionary, which only takes in what belongs to the Art, or Artists in general; not what belongs to some one of 'em.

THE whole, in effect, amounts to this, that the first time of doing a thing, is related by the Historian with the several Particulars which in any wise, tho occasionally only and remotely, affected it: Whereas the Dictionarist, coming afterward, keeps more closely and severely to the Point, and relates nothing but what is essential; that is, the first time, the thing is consider'd as now arising; a new Production or Phænomenon, from some analogous Principle; and therefore we attend to the foreign Causes that brought it forth: whereas afterwards, we consider it as arising from the pre-existing Theory, or Prescriptions of Artists, and thus resolve the Cause into the Art it self.

ANY other difference which there may seem to be between the two; is only as to more or less particular; which, indeed, is a thing that embarrasses and amuses us on many other occasions: Thus in mere civil Histories, if one relates the Series of a Campaign, another the Bombardment of a Town, and a third the Wounding and Death of a general Officer; tho the two latter Subjects be only Parts of the former, yet the first will be said to have composed a Piece of *History*, the second a Piece of *Fortification*, and the third a Piece of *Chirurgery*. And yet there is no other difference between them, than between the Geography of a Country, and the Topography of a Village, or a Hillock; the History of a Nation, and the Biography of a single Person.

TO say no more, the Dictionary of an Art stands in much the same Relation to the History thereof; that the History of a People, does to the Lives of all the considerable and active Persons therein. Their difference is only as to the Point of Sight; the Eye being supposed so near in the one Case, as to see the Parts distinctly, and in the other so far off, as to take in the Whole completely: whence the one gives you all the Incidents; the other only the greater. In effect, the one is all concerted to one point of view, most favourable to the Whole, and the great Parts; the other to many; the Eye being shifted for each Part, to furnish an adequate Representation thereof. In the one Case, it is supposed within the Work; so as only to see those Parts next it, which necessarily hide the rest; in the other, 'tis without, and can only take cognisance of those which lie outwards: So that the one chiefly discovers how things stand within; the other how they stand with regard to the adjacent ones.

TO explain a Term as a Term, we usually express the Circumstances wherewith it is attended in the Art to which it belongs, in their artful Names. This is agreeable to the manner of Artists, who writing of their respective Arts, use Terms as common Words, and suppose 'em to be known: and 'tis this that constitutes a technical Explanation; not the giving the general Effect or Force, in such Words as may equally agree to all other Arts.—And yet in some Cases we recede from this Rule, particularly in divers of the lower Class of Manual Arts, and the Structure of some Machines: Thus, *e.g.* in Turnery, we make no difficulty, for instance, instead of *Chuck*, to say a round piece of Wood, *etc.* The reason is, that where the several subordinate Terms of a Definition are themselves explain'd in their places, we may suppose 'em understood; but where the Term defined is it self so low, that we do not go lower to define the Parts couched under it; there we chuse, as more scientifical, to substitute some more obvious Name, or the general Meaning of the Word for the Term it self; and thus prefer the general or popular, to the technical Definition.

FOR it is to be observed, that the Dictionary has its Limits; it only carries Matters so low; to a certain pitch of Simplicity, where we suppose People may take 'em up, and carry 'em farther as they please. We bring 'em into their Sphere, and so leave 'em. So much Knowledge, *i.e.* such a number of complex Ideas, as we may presume 'em usually to have got in the common Occurrences of Life, we are willing to suppose, as a Footing: where these end, our Dictionary is to begin, which is to take in the rest.

I SHALL not here enter upon the Merits and Defects of the *English Tongue*, considered as a Language: A great deal has been said on that Head by others, for which the Reader may turn to the proper Article in the Dictionary it self. This Place we reserve, not for other Peoples Notions, but our own; and what we have to add, will be chiefly as it stands with regard to *Art*, and more particularly to a *Dictionary of Arts.*

I BELIEVE none will question but we met with Difficulties enough in the Course of this Work. The very Bulk and Dimensions of it confess as much, and the Variety and Uncertainty of its Matter still more. But these were in some sort natural Difficulties, and ought to be consider'd as necessarily appendent to the very Essence of the Design; and therefore did not afflict us so much as those that rose from it at second hand, or were superadded to it, as it were, by Accident. And such was the present wild State of our Language, which alone were sufficient to have baffled the best Scheme, and broke thro' the best Measures that could be form'd.

WE have already represented *Language* as something very important; and as having a near and necessary interest in Knowledge. *Names*, we here add, are solemn things, as they are Representatives of Ideas themselves, and used on most occasions in their stead: and *Terms*, or

Combinations of Ideas, are still more so; as much as complex Engines, are of farther and nicer Consideration than the simple mechanic Powers. But who would imagine this, to consider the wanton use we make of 'em; and with how little Fear, or Discretion, Words are treated among us? Every body think themselves privileg'd to alter, or set aside the old, and introduce new ones at pleasure. *England* is open to all Nations, at least in this respect; and our Traders in this Commodity, import their Wares from every Country in all security. The mercantile Humour seems to have possessed every Part of us, so that we are not only unwilling to be without the natural Produce of our Neighbours Countries, but we even envy 'em their Fashions, their Follies, and their Words. Scarce a petty Author that appears, but makes his Innovations: But when a Dictionary comes out, 'tis like an *East India* Fleet, and you are sure of a huge Cargo. The Effect is, that our Language is, and will continue in a perpetual flux; and no body knows whether he is master of it or no. The utmost he can say, is, that he had it for such a Day, exclusive of what has happen'd since.

A MAN never knows when he is at the end of the Terms, *e.g.* in Architecture. When he has got two or three Names, for some one Member, and thinks himself overstock'd, 'tis odds he has not half. 'Tis not enough he knows what it is named in the *English*; but he must likewise learn what the *French*, *Italians*, *Latins*, and *Greeks*, likewise call it, or frequently find himself at a stand. Thus it is in the Case of *Fillets*, *Lists*, *Listels*, *Reglets*, *Platbands*, *Bandeletts*, *Tænias*, and *Baguettes*; of *Chaplets*, *Astragals*, *Batoons*, and *Tores*; of *Gulas*, *Gueules*, *Doucines*, *Cimas*, *Cymatiums*, *Ogees*, and *Talons*; *Ovums*, *Ovolos*, *Echinus's*, *Quarter-rounds*, *Boultins*, etc. between which, there is no known, allowed differences; but they are either used indiscriminately, or distinguish'd arbitrarily; one Person making this distinction, and the next another, or perhaps none at all. So that if we come strictly to Dictionaries, we should have a different one for every Author.

BUT the Mischief does not end here: for as the antient Arts are in many respects different from the modern; the use of their Terms necessarily involves us in a new Confusion, and makes the same Work stand in an ancient Author for one thing, and in a modern for another. Thus it is in *Parastata*, *Orthostata*, *Anta*, etc. In effect, there is that Alteration continually making in the Language of Architecture, that there ought, in Propriety, to be a different Dictionary of it for every different Age.

THE Truth is, a fourth part of the Words in some of our popular Dictionaries, stand on no better Authority, than the single Practice of some one fanciful Author; who having an intemperate Desire to shew either his Learning or Breeding, has met with Dictionary-Writers fond enough to take his Fripperies off his hands, and expose 'em to the Publick for legitimate Goods. By such means, those Exotics have obtain'd a kind of Currency; so that a Dictionary would be thought defective without 'em. To omit even our Fopperies would be thought a Failing; and might even be esteemed by many as the most unpardonable of all.—On these accounts we have been oblig'd to temporise a little, how much soever against our Will; and thus perhaps have contributed to the still farther Establishment of a number of Words, which we had much rather have seen proscribed, or banish'd the Land.

UPON the whole, nothing could be more desirable than an *Index expurgatorius*, to clear the Language of our superfluous Words, and Equivocals; all the modern *French* and *Italian* Terms in the several Arts, where we have *Latin* and *Greek* ones; and even all the *Latin* and *Greek* ones, where we have *English* or *Saxon* ones, equal in Sound and Significancy. I think the learned Languages ought to have the preference to the modern, because every Person may be supposed to have read, but not to have travelled; and our Country Words I would prefer to any others, because there is the most analogy between 'em, and they usually retain more of the Origin and Etymology than those transplanted from other Languages.—Such a Reform would reduce our Dictionaries to more reasonable Dimensions; and disincumber the Arts from half the difficulty now to be surmounted in attaining 'em.

BUT, there is another Spring of Words no less prolifick than that hitherto spoke of, and which has produced a Swarm of spurious, mishapen Words, which no Nation but our own would ever have own'd: I mean the Itch or coining or making *English* Words, by a sort of analogy, from the *Latin* and *Greek* ones. This Fault the tribe of Lexicographers have carried to a strange excess. How must a Man stare, to see what detestable Stuff some late Writers of that Class have complimented us with: Words of their own manufacture, scarce fit to do any thing with, except cure Agues! Witness such as *Scopulosity*, *Siticulous*, *Scatebrosity*, *Siccific*, *Pugnacity*, *Segnity*, *Sputative*, *Mulierosity*, *Mugient*, *Gracility*, *Fastuousness*, *Exuccous*; and many thousand more, at the Reader's service, to be met withal in a Dictionary which few People are without. One would almost wish the Mold destroyed that such Grotesques were cast in, for fear of new Impressions. We are already overrun with this Author's Scarecrows: but what shall we be when, having thus anglicis'd all the *Greek* and *Latin* Words, he proceeds to do the same with the *Dutch*, *Irish*, *Welsh*, etc. Indeed, I am the less angry with him, for that he has carried the Abuse so far, as must not only save People from being seduc'd, but bring the Practice into Contempt. Such Monsters can't possibly live long: if they have escap'd the Midwife, who ought to have strangled 'em ere they came to light, yet if ever they stir abroad they must infallibly be knock'd o' the head.

HOW oddly will our Practice in this respect look, when confronted with that of our Neighbours? One of the most learned Men and greatest Critics of the last Age, M. *Menage*, incurr'd an infinite deal of Censure, for only endeavouring to introduce the single word *Prosateur*: and could not succeed in it, notwithstanding that a Word of that import was confessedly wanting in the *French*; and both the Sound and Analogy of the new Word were unexceptionable.

AND yet our Leisure and Curiosity have found means of making even these more cognizable than otherwise they are: we can, in some measure, alter the established Relation between our Faculties and their Objects; and make use of one Law of Nature, to undo or supersede another.—Thus we can magnify a little Sound or little Body, or a little Distance, *etc.* or we can diminish large ones; and thus make things in some measure adequate Objects, that naturally are not so.

BUT there is no great advantage in this: We only, by these means, come at a better apprehension of things which Nature seem'd to put out of our way for no other reason but because they did not concern us; lest we should be engaged to mistake, and run after things that had no relation to us, to the neglect of those which have.—Thus, Anatomy is really found of much less use than at first sight one would imagine; as being employ'd in taking things asunder and considering their Parts, which Nature chiefly intended to be considered and dealt together. There is I know not what secret Law, whereby the Effect of a thing is, as it were, attach'd to its integral State; so that in proportion as you either diminish it, by taking from it, or enlarge it by adding to it, its Effect is alter'd, in a manner beyond what we can well account for from the bare Consideration of Magnitude.

HENCE we discover why the Old is much more perfect in its kind than the New. The former has little to do but compare, order, methodize, *etc.* what is ready at hand; the latter has likewise to find; after which all the labour of the other still remains. The former takes Nature in all her Simplicity; the latter adds Art to her, and thus brings Nature into consideration in all her diversity: the former chiefly considers natural Bodies in their integral State; the latter divides, and analyses 'em: So that the former finds most of the principal Relations, the latter many more curious, and amusing ones. Hence, the former hastes to its Perfection, and can't long hold out; for that its Matter is limited: the latter can scarce ever arrive at it, since Experiments are endless. To say no more, to have Philosophy in its perfection, we should have the Order, Precision, and Distinctness of the Old; and the Matter, the Copia of the New.

THE modern is yet wild and unascertain'd. 'Tis not arrived at the Maturity of Method; the Mine is but just open'd, and the Adventurers are yet only sollicitous about the Matter to see what it affords. Circumstances do not yet come in course; and 'twill be long ere it arrive at a just extent to give room and leisure for reducing it to regularity. True, the Rules and Methods of the antient, are in some measure applicable to the new, and will go a good way towards the ordering and ascertaining of it; but the present Philosophers seem yet too warm and sanguine for such a Business; which must be left to the succeeding Age to think about. Add, that the farther they go on to dig Materials, still the more difficult will the ranging of 'em be; inasmuch as there is but one true and just order to lay them in; and the more of 'em, the more intricate that Order, and the harder to find. This a Man may be positive of, he never will see half the Experiments and Observations already made, laid up or used in a System of Physics.

BUT when that is done, a deal will still remain, ere we have the chief uses of it. For physical Knowledge, strictly consider'd, is only a Step, a Means of arriving at a higher and farther kind. Histories, Observations, and Experiments of the Kinds, Order, Strata, *etc.* for instance of Fossils, are very useful and laudable things, as they tend to lay in a Stock of sensible Phænomena, for the Mind to work upon, digest, and draw new Notices from, for the Improvement of our own Faculties, and the better Conduct of Life: But 'tis a Shortsightedness to forget this farther View, and look only to the Things themselves. The bare Acquisition of new Ideas is no real advantage, unless they be such as have some relation to our selves, and are in some sense adequate, and adapted to the Circumstances of our Wants, and Occasions, or capable of being made so. Knowledge, in the first State, is like Food in the Stomach, which may please and satisfy us, but is of no use to the Body till farther prepared. It must be brought nearer us, and made more our own, more homogeneous to our selves, ere it feeds us.—The modern Philosophy is not so properly a Philosophy, as the Adit or Opening of one. Its Matter has yet only undergone the first Concoction: we are yet only conversant about new physical Relations, learnt by Sensation; whereas to bring it to the Perfection requir'd, it must have undergone the farther Operations of Imagination, and Reason. Mere Physics, as such, do not make a Philosophy; those Physics must first be carried up to Metaphysics and Ethics, ere we can safely stop. So far as it is Physics, it is foreign to the Mind, and its Occasions; before it affect and influence our Reason and Judgment, it must be subtiliz'd vastly, and made more similar to the Metaphysical Nature of the Mind. While Physics, it remains under the Direction of the Author of Nature; and proceeds wholly by his Laws, and to execute his Purposes: ere it come under our Direction, and become subservient to our Will, it must have laid aside what was active, and necessary in it, and become passive to our Reason, *i.e.* it must have been transfer'd from the Dominion of the Almighty's Will, or Reason, and brought under ours; if that do not imply a Contradiction.

I WILL at least deal honestly with my Reader, and not be caught faulty in point of Morality, whatever I may be in any thing else.—What has been said hitherto, has been on the advantageous side of my Work; and I should not have acquitted my self, should I not likewise mention what may be alledg'd on the contrary Part.—The curious Reader, then, may expect, he will here meet with Omissions, and there with Redundancies: here the Method and Œcomomy are not kept to; there an Article is imperfectly treated: here, a Passage from some other Language is not sufficiently naturalized; there a Sentiment of some other Author is not sufficiently digested: There, in fine, the Author was asleep, and here the Printer.

ONE might palliate these Objections, by alledging, that "they are "things not peculiar to this Work, but extend to all of the Kind; that most "of 'em are things not foreign and accidental to it, but arise of necessity, "from the very Nature and form of a Dictionary; and that many of 'em, "are not peculiar even to a Dictionary, but agree to all extensive "Undertakings, and are appendant to the very best Part of the Design, its "Universality:" but instead of extenuating, I had rather be guilty of inflaming, and aggravating 'em.

FOR *Errors*, they cannot be very few, considering the Hands thro' which most Parts of our Knowledge have passed, and from whom we are obliged to take our Accounts. What one Author, upon the most particular Subject, will you produce, that has not his share of 'em? and what *Argus* could possibly see, and correct the Errors in all the Authors he had to do with? *Scaliger*, in his Exercitations against *Cardan*, has shewn some twenty thousand, in one small Work; and no body imagines he has pick'd it perfectly clean. Yet *Cardan* was no ill Author. *Bayle's* Dictionary has been called the Errata of *Morreri*; yet is not *Bayle* himself without his Errors.—The most we can say, is, that we hope there will be few found in the present Work, in comparison of others of the like kind. Many thousands we have corrected, both in the Dictionaries and other Writings we have collected from, by means of the Light which other Parts of Knowledge afforded: But after so large a Harvest, no doubt there remains a tolerable Gleaning. We flatter our selves, however, that what we have overlook'd, the Reader will frequently be enabled to correct, by the Means here afforded; and that there will be few Errors found in the Book, which the Book it self will not help to rectify.

AS to *Omissions*, there is scarce any avoiding 'em; and the more intelligent the Reader is, the more of this kind he will necessarily meet withal: they being only such in relation to his fulness. Indeed, I must own my self greatly a Debtor on this score; and tho at present insolvent, yet if the Reader will give me Credit, it shall be my endeavour to see all I owe discharged; if not in a Lump, yet by a Course of Payments.

FOR *Redundancies*, you know there cannot well be richness without 'em. After you have picked what you think fit of this kind, and laid it by; 'tis ten to one but the next Person that comes, will restore half of 'em to their places; and tax you Temerity, and want of Taste: and the next after him will go near to replace the other half.

AS to *Irregularities*, and breaches of Method, I will not claim Impunity on the Score of being the first that introduced any certain Rules, or Method into this way of writing at all: But there will be at least this Satisfaction attending my Case, that I cannot be indicted for the Breach of any Laws but my own.—Nor must it be forgot, that I pretend to have carried the *Dictionary-Way* to a pitch hitherto little thought of: So that if I have fallen short of the Mark on one side; it may be some Atonement that I have gone beyond it in another. I am sensible, however, there is no

Point I have been more delinquent in than this one of Method; and that I am at every turn forgetting my own View. The References, and necessary Connexions between the Parts, which should shew their Relation, and help the Imagination to put 'em together, are but too frequently dropt, and the Reader left without his Clue.

AS to *Jejuneness*, and Crudity; no doubt there must be a deal of that kind, considering the Time so great a Load of Fruit had to hang and ripen. Much of it was gathered ere it could possibly be matured; so that 'tis no wonder it now and then tastes of the Wood. But setting aside this; if a man may not be allow'd to say a good number of indifferent things, in the Compass of five hundred Sheets, I know not who would be an Author.

LASTLY, as to there being little in it *new*, and of my own growth; I must here change my Style; and from Confession, turn to Vindication. —The Work is, what it ought to be, a *Collection*; not the Produce of a single Brain, for that would go but a little way; but of a whole Commonwealth. If any Person will undertake to write a Dictionary, even of some one particular Art, from his own Fund, alone; a Man may safely undertake to prove it good for nothing. I do not pretend to entertain my Guests at this rate, with just what my own scanty Barns afford: The whole Country is ransack'd to make 'em the fuller Banquet. Call me what you will; a *Daw*, and I say I am stuck over with other Peoples Feathers: with all my Heart; but it would be altogether as just to compare me to the *Bee*, the Symbol of Industry, as that of Pride. For tho I pick up my Matters in a thousand Places; 'tis not to look gay my self, but to furnish you with Honey. I have rifled a thousand Flowers; prickly ones many of 'em, to load you Hive. No body that fell in my way, has been spared; Antient nor Modern, Foreign nor Domestick, Christian, nor Jew, nor Heathen: Philosophers, Divines, Mathematicians, Critics, Casuists, Grammarians, Physicians, Antiquaries, Mechanics, all are served alike. The Book is not mine, 'tis every body's; the mix'd Issue of a thousand Loins. The Prince of modern Authors, is pillaged to some purpose; and what Quarter then can any body else expect? If ever you wrote any thing your self; 'tis possible there is something in it of yours: so that you will at least allow something in it good.

NONE of our Predecessors can blame us for the use we have made of them; since it is their own Practice. It is a kind of Privilege attached to the Office of Lexicographer, if not by any formal Grant, yet by Connivance at least. We have already assumed the *Bee* for our Device; and who ever brought an Action of Trover or Trespass against that avowed Free-booter? If any body blames us, 'twill ten to one be some of those very Drones, who are sustained by our means.

'TIS idle to pretend any thing of Property in things of this nature. To offer a thing to the Publick, and yet pretend a Right reserved therein to one's self, if it be not absurd, yet it is sordid. The Words we speak: nay, the Breath we emit, are not more vague and common than our Thoughts, when divulged in print. You may as well prohibit People to use the Light that shines in their Eyes, because it comes from your Candle: E'en clap it in a dark Lanthorn, and let us not be amused, and dazzled by it; if we may not be the better for good things, let's not be the worse for the ill and indifferent ones mix'd with 'em.

WE see the same Thought, which was first started in one Author under a world of Crudity, borrow'd by another become farther improv'd and ripen'd; and at length transmitted to a third, yield Fruit in abundance. All Plants will not thrive in all Soils that will produce 'em; some languish in their Mother-Beds: whence the Gardner is under a frequent necessity of Replanting, Engrafting, *etc.*

TO do justice to a *Collection*, I mean a general and promiscuous one; it has its Advantages. Where numbers of things are thrown precariously together, we sometimes discover Relations among 'em, we should never have thought of looking for: As, the Painter's and Sculptor's Fancy, is frequently led on to the boldest and most masterly Designs, by something they spy in the fortuitous Sketches of Chance, or Nature: insomuch that a celebrated Author* makes no scruple to lay this down as the first Origin and Occasion of all these Arts. 'Tis certain most of our Knowledge is empirical, the Result of Accident, Occasion, and casual Experiment: 'Tis but very little we owe to Dogmatizing and Method; which, as already observ'd, are posterior Matters, and only come in play after the Game is started. 'Twas, in all probability, the hand of Chance that first threw Sulphur, Charcoal, and Salt-petre together; and what surprizing Effects have not arose from it; what Handle has it given to Art and Contrivance, to direct and apply this fortuitous Production?

WHAT Advantages may not Philosophy derive from such a Collection, or Farrago of Arts; when 'tis considered, that every Circumstance, every Article of an Art, ought to be look'd upon as a Datum, a Phænomenon, or Experiment in Philosophy; and that the least of 'em may possibly be the Foundation of a new System?—To consider only the *Tanning*, or *Currying* of Leather: what is the whole Process, but a Series of physical Effects, arising from new applications of Body to Body? And how many Lectures will the Philosopher have from Painting, Gardening, Agriculture, *etc.* touching *Planting*, *Engrafting*, *Pruning*, *Exposure*, *Expression*, *Walls*, etc. which might never have come in his way, but by such a chance? When a thing is once started, it may be applied infinite ways, and no body knows where it will stop.

THRO'OUT the Whole, we have had a particular regard, both in the Choice of the several Heads, and in dwelling or amplifying upon 'em; to the extending our Views, dilating our Knowledge, opening new Tracks, new Scents, new Vistas. We have endeavour'd not only to furnish the Mind; but to inlarge it, and make it in some measure co-extend with the Dimensions of all Minds, in all Ages and Places, and under all Situations and Circumstances: as Language, in some measure, makes our Senses

*Leon Battista Alberti, della Statua.

do. With which view, we have given the Sentiments, Notions, Manners, Customs, *etc.* of most People, that have any thing new, unusual, or handy in 'em.

SUCH a Variety of Views, Principles, and Manners of thinking, is a sure Remedy against being too violently attached to any one; and is the best way of preventing the making of Pedants, Bigots, *etc.* of any kind. It may be said, that every Art tends to give the Mind a particular Turn; and that the only way of maintaining it in its natural Rectitude, is by calling in other opposite ones, by way of Counter-ballance. Thus we find nothing more perverse and unsufferable than a mere Mathematician, mere Critic, Grammarian, Chymist, Poet, Herald, or the like; and the proper Disposition is only to be had from a just Temperament or Mixture of 'em all.

I OWN this is not the way to make a very great progress in any Art; but at the same time it is the only way to hinder our being spoil'd by any; and becoming Creatures rather of *Homer* or *Aristotle's* making, than God's: and receiving our Tastes, Views, Relishes, at second hand, rather than from Nature her self. This, however, is only to be understood with regard to personal Benefit. For no doubt the Publick is better provided for, by the mere Pursuers of particular Arts, than the general Appliers to all: since each is hereby brought to greater Perfection; and the Mixture and Temperament, wanting in the Individuals, is found in the Whole.

TO conclude, the ultimate View of a Work of this, or any other kind, should be, the forming a sound Mind, *i.e.* a System of Perceptions, and Notions agreeing to the System of Things, or in the Relation thereto, intended by its Author. The End of Learning and Study, is not the filling our Heads with other Mens Ideas; that is an Inrichment which may prove for the worse, if it carry any ill Quality with it: Richness is not the chief thing aim'd at; 'tis only a Circumstance, or Matter of a secondary Consideration: Soundness is the first. There are many Manures which the Husbandman dares not use, by reason they would corrupt the Land, at the same time they enriched it; and lay the Foundation of a Disease, which would in the End impoverish, and make it spend it self in unprofitable Weeds. A little pure Logic, or Theology, or Chymistry, in some Peoples Heads, what Mischief have they not produced?—But it must be owned, Mens Heads are not so soon fill'd: the Memory is not so tenacious as we imagine; Ideas are transient things, and seldom stay long enough with us to do us either much good, or harm: Ten to one but what we read to-day, is forgot again to-morrow. And what chiefly makes new Ideas of any significancy, is their extending and enlarging the Mind, and making it more capacious and susceptible.—But neither is this Enlargement the last Aim; but is chiefly of use, as it contributes to the increasing our Sensibility, to the making our Faculties more subtil, and adequate, and giving us a more exquisite Perception of things that occur; and thus enabling us to judge clearly, pronounce boldly, conclude readily, distinguish accurately, and to apprehend the manner and Reasons of our

Decisions. In which view, several things may be useful, that are not so much direct Matters of Knowledge, as subservient to the same End; for instance, much of the School Philosophy, which by exercising an exciting the Mind, has a kind of collateral tendency to sharpen its Faculties; and needs only be read, not retain'd, to produce its Effect.—But neither does the Matter end here: Even this does not amount to the full and adequate End of Knowledge: This is only improving the Organ; and there must be some farther End in such Improvement. No Man sharpens his Weapon on the sole Consideration of having it sharp, but to be the fitter for use. Briefly, then, our Faculties being only so many Inlets, whereby, and according to the measure whereof, we receive the Intimations of the Creator's Will, or rather, the Effects of his Power and Action; all the Improvements made in 'em, have a tendency to subject us more entirely to his Influence and Direction; and thus make us conspire, and move more in concert with the rest of his Works, to accomplish the great End of all things. In which our Happiness and Perfection consists; the Perfection of a single Nature, arising in proportion as it contributes to that of the ΤÒΠÃΝ.

5 *Cyclopaedia* Volume II

Architectural subjects portrayed in the letterpress articles

Our inventory of letterpress articles in the series *The Encyclopaedic-Dictionary in the Eighteenth Century: Architecture, Arts and Crafts* is composed of subjects derived from various branches of architecture and allied arts and crafts. The following short list of headwords is provided here to give the reader an outline impression of the range of material that is included in the letterpress articles:

Architectural terms, arch construction, building construction, building materials, cabinet work, carpentry, classical orders, decorative art, design principles including aesthetics, domes, drainage, fortification, furnishings, furniture, gardening, joinery, landscape terms, military art, mouldings, ornament, painting, perspective, sanitation, sculpture, surveying, terminology of the building trades and crafts, tools and equipment used in building construction, and vaults.

CYCLOPÆDIA:

OR, AN

UNIVERSAL DICTIONARY

OF

ARTS and SCIENCES.

VOLUME the SECOND

I

ICONOGRAPHIA, Description of Images, or of antient Statues of Marbles and Copper, of Busts and Semi-busts, of Penates, Paintings in Fresco, Mosaic Works, and antient Pieces of Mignature.

ICHONOGRAPHY in Perspective, is the View of any thing cut off by a Plane parallel to the Horizon, just at the Base or Bottom of it. In Architecture it is taken for the Geometrical Plan, or Platform of an Edifice, or the Ground-Plot of an House or Building delineated upon Paper, describing the Form of the several Apartments, Rooms, Windows, Chimneys, *etc.* and this is properly the Work of the Master Architect or Surveyor, being indeed the most abstruse and difficult of any. In Fortification it is, in like manner, the Plan or Representation of the Length and Breadth of a Fortress, the distinct Parts of which are marked out, either of the Ground itself, or upon Paper. The word is derived from the *Greek*, *Vestigium*, and *scribo*, as being a Description of the Footsteps or Traces of a Work. See *Plan*.

ILLUMINATION, the Action of a luminous Body, or Body that emits Light, or the Passion of an opake Body that receives it.

IMAGE, a natural, lively Representation of an Object, opposed to a smooth well-polished Surface. The *Latin* word *Imago*, comes originally from the *Greek imitari*. *Image* also signifies any artificial Representation performed by Man, as in Painting, Sculpture, and the like: In which sense, the word is now generally used in speaking of Things holy, or Things imagined to be so.

The Use, and Adoration of Images, are Things that have been a long time controverted in the World. The *Lutherans* condemn the *Calvinists*, for breaking the Images in the Churches of the Catholics, looking on it as a kind of Sacrilege; and yet they condemn the *Romanists*, (who are professed Image-worshippers) as Idolaters: nor can these last keep pace with the *Greeks*, who go far beyond them in this Point, which has occasioned abundance of Disputes among them. The *Jews* absolutely condemn all *Images*, and don't so much as suffer any Statues or Figures in their Houses, much less in their Synagogues, or Places of Worship. The *Mahometans* have a perfect Aversion to *Images*, which was what led them to destroy most of the beautiful Monuments of Antiquity, both Sacred and Profane, at *Constantinople*.

The noble *Romans* preserved the *Images* of their Ancestors with a great deal of Care and Concern, and had them carried in Procession at their Funerals and Triumphs. They were commonly made of Wax or Wood, tho' sometimes of Marble or Brass: They placed them in the Vestibles of their Houses, and they were to stay there, even if the Houses happened to be sold: it being accounted impious to displace them. *Appius Claudius* was the first who brought them into the Temples, in the Year of *Rome* 259, and he added Inscriptions to them, shewing the Origin of the Persons represented, and their brave and vertuous Atchievements. It was not, however, allowed for all, who had the Images of their Ancestors in their Houses, to have them carried at their Funerals; this was a thing only granted to such as had honourably discharged themselves of their Offices, for those who failed in this respect forfeited that Privilege, and in case they had been guilty of any great Crime, their *Images* were broken in pieces.

IMPOSTS, in Architecture, are what are sometimes called Chaptrels: being the Parts on which the Feet of Arches stand; or the Capitals of Pillars, that support Arches. These *Imposts* conform to their proper Orders. The Tuscan hath a Plinth only; the Dorick two Faces crown'd; the Ionick a Larmier, or Crown, over the two Faces, and its Mouldings may be carved; the Corinthian and Composite have a Larmier, Freeze, and other Mouldings. The Projectures of the *Imposts* must not exceed the Naked of the Pilaster. Sometimes the Entablature of the Order serves for the *Impost* of the Arch; and this looks very grand and stately. The *Impost* is a thing very essential to the Composition of the Ordonnances; insomuch that without it, in the place where the curve Line of the Arch meets with the perpendicular Line of the Pillar, there always seems a kind of Elbow.

INCRUSTATION, an Ornament in Architecture, consisting of hard polish'd Stones, or other brilliant Matter disposed in Compartiments in the Body of a Building. The *Incrustations* in the Castle at *Madrid* are of Potters Work: Those in the *Louvre* of Marble. We say 'tis an *Incrustation* of Stone or Marble, when a Wall is lined with them. The Antients call'd their Plaisters *Incrustations*.

INCRUSTED COLUMN, is a *Column* consisting of several Pieces or slender Branches of some precious Marble, masticated or cemented around a Mould or Brick, or other Matter; which is done, not only to save the precious Stone, whether it be Agat, Jasper, or the like, but also for the sake of shewing Pieces of it of uncommon Largeness, by the Neatness and Closeness of the *Incrustation*, which renders the Joints imperceptible, where the Mastic is of the same Colour.

INSTRUMENTS *of Sacrifice*, are Ornaments in the Antique Architecture; as Vases, Pateræ, Candlesticks, Knives, wherewith the Victims were kill'd, *etc*. Instances of which, we see in a Corinthian Freeze in the Remains of a Temple behind the Capital at *Rome*, etc.

INSULATED, by the *French* called *Isolee*, and the *Latins Insulatus*; is a Term apply'd to a Column that stands alone, or free from any contiguous Wall, *etc*. like an Island in the Sea; whence the Word is derived.

INTAGLIO'S, precious Stones, having the Heads of great Men, Inscriptions, and the like, engraven on 'em; such as we frequently see set in Rings, Seals, *etc*.

INTERCOLUMNS, or *Intercolumniation*, in Architecture, signify the Spaces between the Columns. The *Intercolumns* must always be proportion'd to the Height and Bulk of the Columns. *Vitruvius* calls it *Intercolumnium*: And according to that Author, the *Intercolumniation* is of five kinds; *viz*. Pychnostyle, Sistyle, Eustyle, Diastyle, and Aræostyle: which see explain'd in their places. For a Medium, some Authors have laid down the following Proportions for the *Intercolumns*. In the Tuscan Order, the *Intercolumn* must be four Diameters of the Body of the Column below; in the Doric three; in the Ionic two; in the Corinthian two, one quarter; in the Composite one and a half.

IONIC, in Architecture, is the Name of one of the five Orders. The first Idea of it was given by the People of *Ionia*; who, according to *Vitruvius*, form'd it on the Model of a young Woman, dress'd in her Hair, and of an easy elegant Shape: whereas the Doric had been form'd on the

Model of a strong, robust Man. The *Ionic* Column is the third in order; and is distinguish'd from the Composite, in that it has none of the Leaves of Acanthus in its Capital; and from the Tuscan, Doric and Corinthian, by the Volutes, or Rams-Horns, which adorn its Capital: and from the Tuscan and Doric, too, by the Channels or Flutings in its Shaft. These Flutings are in number 24. They are not always concave from the top of the Shaft to the bottom; but for that third of it next the Base, are fill'd up with a kind of Rods or Canes, by the *French* call'd *Batons*: and in the other two thirds, are left hollow, or striated, in imitation of the Folds or Plaits of a Garment. This Column is a Medium between the Massive and the Delicate Orders, the Simple and the Rich. Its Heights is 18 Modules, or 9 Diameters of the Column taken at the bottom. When it was first invented, its Height was but 16 Modules; but the Antients, to render it still more beautiful than the Doric, augmented its Height, by adding a Base to it, which was unknown in the Doric. M. *le Clerc* makes its Entablement 4 Modules and 10 Minutes, and its Pedestal 6 entire Modules: so that the whole Order makes 28 Modules 10 Minutes. 'Tis said, the Temple of *Diana* at *Ephesus*, the most celebrated Edifice of all Antiquity, was of this Order. At present it is properly used in Churches and Religious Houses, in Courts of Justice, and other Places of Tranquillity and Devotion. This Order has one advantage above any of the rest, and it consists in this, that the fore and hind-parts of its Capital are different from its Sides. But this is attended with an Inconvenience, when the Ordonnance it to turn from the Front of the Building to the Side: To obviate which, the Capital may be made Angular, as is done in the Temple of *Fortuna Virilis*. *Scamozzi*, and some other modern Architects, have introduced the upper part of the Composite Capital in lieu of the *Ionic*; imitating that of the *Temple of Concord*, whose four Sides are alike: to render it more beautiful, the Volute may be made a little oval and inclining. See ORDER and VOLUTE.

IRREGULAR COLUMN, in Architecture, is such a one as not only deviates from the Proportions of any of the five Orders, but whose Ornaments, whether in the Shaft or Capital, are absurd and ill-chosen.

ISLES, in Architecture, are Sides or Wings of a Building.

J

JAPANNING, the Art of varnishing and drawing Figures on Wood, *etc.* after the same manner as the Workmen do who are Natives of *Japan*, a famous Island not far from the Coast of *China*. The manner of it is this; they take a Pint of Spirit of Wine well dephlegmated, and four Ounces of Gum-Lacca (which last is first broke from the Sticks and Rubbish) and bruising it roughly in a Mortar, they put it to steep in Spring-water, tyed up in a Bag of coarse Linen, together with a little Castile-Soap, for the space of twelve Hours. This done, they rub out all the Tincture, and add to it a little Allum, and reserve it a-part; then add as much Mastick and white Amber, distilled in a Matrass, with the Spirit of Wine, by a two Days Digestion, frequently stirring it, that it don't stick to the Glass; then strain and press it out into another Vessel. This done, they take the Wood to be japanned, and cover it with a Layer of this Varnish, till it be sufficiently drenched with it; then taking some of the Colour, of which the Figures are to be, they incorporate it with seven times as much of the Varnish, and apply it with a Pencil, going over each part three several times, each a quarter of an Hour after the other: two Hours after this they polish it with Prestle, or *Dutch* Reeds. As to the Colours used in this Art, for a fair Red, they take *Spanish* Vermillion, with a fourth part of *Venice* Laque. Black, they make of Ivory calcined between two Crucibles; for Blue they use Ultramarine, and only twice as much Varnish as Colour. The rest are applied as above directed, except the Green, which is difficult to make fair and lively, and therefore seldom used. Night-*Japanning* is performed, by applying three or four Layers with the Colours first, then two of pure Varnish uncoloured, made according to the former Process. Before it be dry, they sift some Venturine, or Gold Wire, reduced to Powder, over it, and then cover it with as many Layers of pure Varnish, as render it like polished Glass; and, lastly, rub it over with Tripoli, Oil of Olive, or a Hatter's Felt.

JET D'EAU, a *French* Word signifying a Fountain, that casts up Water to any considerable Height in the Air. Mr. *Mariotte* saith, That a *Jet d'Eau* will never rise so high as its Reservatory, but always falls short of it by a Space, which is in a subduplicate Ratio of that Height; and this he proves by several Experiments. He saith also, That if a greater, branches out in many smaller ones, or is distributed thro' several *Jets*, the Square of the Diameter of the main Pipe, must be proportioned to the Sum of all the Expences of its Branches; and particularly, That if the Reservatory be 52 Foot high, and the Adjutage half an Inch in Diameter, the Pipe ought to be three Inches in Diameter. See FOUNTAIN.

JOINT, the Juncture, Articulation, or Assemblage of two or more things. *Joint*, in Architecture, is the Separation between the Stones, which is fill'd with Mortar, Plaister, or Cement. In Carpentery, *etc. Joint* signifies the several Manners of assembling or fitting Pieces of Wood together. Thus we say, *A Dove-tail Joint*, etc.

JOISTS, in Architecture, are those Pieces of Timber framed into the Girders and Summers, on which the Boards of the Floors are laid: They are from six to eight Inches square, and ought seldom to lie at a greater distance from each other than ten Inches, never than twelve; nor ought they ever to be longer than ten Foot, or to lie less into the Wall than eight Inches. Sometimes the Carpenters Furr their *Joists*, as they call it, that is, lay two Rows of *Joists* one over the other.

JONCTURE, or JOINTURE: The same with Joint.

JOYNERY, the Art of working in Wood, and of fitting or assembling various Parts or Members of it together: The *French* call it *Menuiserie*, from *Minutaria*, small Work, by which it is distinguish'd from Carpentry, which is conversant in the larger and less curious Works.

JUNCTURE, is any kind of Joint, or closing of two bodies.

K

KARKRONI, a Building where the Royal Manufactures of *Persia* are carried on. Here are made their Tapestries, Cloth of Gold, Silk, Wool and Brocades, Velvets, Taffeta's, Coats of Mail, Sabres, Bows, Arrows, and other Arms. There are also Painters in Miniature, Goldsmiths, Lapidaries, *etc.*

KEY, a little Iron Instrument for the opening of Locks. Its Use and Figures are too well known to need a particular Description. *Molineus* is of Opinion, that *Keys*, at first, only served for the untying certain Knots, wherewith they antiently secured their Doors; but the *Laconic Keys*, he maintains, were nearly a-kin in use to our own: they consisted of three single Teeth, and made the Figure of an E, of which Form there are some still to be seen in the Cabinets of the Curious.

KEY of an Arch or Vault is the last Stone placed a-top of a Vault, which being wider and fuller at the Top than Bottom, wedges, as it were, and binds in all the rest. This *Key* is different in the different Orders; in the *Tuscan* and *Doric* 'tis a Plain Stone, having a Projecture; in the *Ionic* 'tis cut and waved in Veins, after the manner of *Consoles*; in the *Corinthian* and *Composite*, 'tis a *Console* enriched with Sculpture, Foliages, *etc.*

L

LACUNAR, in Architecture, an arched Roof or Cieling, more especially the flanking or flooring above the Portico's and Piazza's.

LANDSKIP, or *Landschape*, the View or Prospect of a Country, extended as far as the Eye will carry. *Landskips* in Painting, are Pieces representing some Champaign or Rural Subject, as Hills, Vales, Rivers, Country Houses, *etc.* where human Figures are only introduced as Accidents or Circumstances. This is esteemed one of the lowest Branches of Painting.

LARMIER, in Architecture, a flat square massive Member of the Cornice, between the Cymatium and Ovolo; so called from its Use, which is to disperse the Water, and to cause it to fall at a distance from the Wall drop by drop, or as it were by Tears, *Larme* in *French* signifying a Tear. The *Larmier* is also call'd *Corona*: See ENTABLATURE and CORONA.

LATH, in Building, long, thin, narrow Slips of Wood used in Tyling and Walling. These are divided into three kinds, with regard to the different Woods they are made of, *viz.* Heart of Oak, Sap-Laths and Deal-Laths: the two last used only for Ceiling and Partitioning, and the first only for Tyling. Again, *Laths* are distinguish'd into three kinds more, with regard to their Length, *viz.* into 5 foot, 4 foot, and 3 foot *Laths*; tho the Statute allows but of two lengths, *viz.* those of 5 foot and of 3, each of which are to be a inch and half in breadth, and half an inch in thickness. These are what *Vitruvius* calls *Ambrices.*

LATHE, or *Leathe*, a very useful Engine for the Turning of Wood, Ivory, Metals, and other Materials. The Invention of the *Lathe* is very antient. *Diodorus Siculus* says, the first who used it was a Nephew of *Dædalus*, named *Talus*. *Pliny* ascribes it to *Theodore* of *Samos*, and mentions one *Thericles*, who render'd himself very famous by his Dexterity in managing the *Lathe.*

With this Instrument the Antients turn'd all kinds of Vases, many whereof they enrich'd with Figures and Ornaments in Basso Relievo. Thus *Virgil,*

Lenta quibus Torno facili superaddita Vitis.

The *Greek* and *Latin* Authors make frequent mention of the *Lathe*, and *Cicero* calls the Workmen who used it *Vascularii*. 'Twas a Proverb among the Antients, to say a thing was formed in the *Lathe*, to express its Delicacy and Justness.

The *Lathe* is composed of two wooden Cheeks or Sides, parallel to the Horizon, having a Groove or Opening between; perpendicular to these, are two other Pieces called Puppets, made to slide between the Cheeks, and to be fix'd down at any Point at pleasure.

LAYMAN, among Painters, is a Statue of Wood, whose Joints are so made that it may be put into any Posture. Its chief Use is for the casting and adjusting of Draperies for the clothing of Figures.

LAZARETTO, or LAZAR-HOUSE, a publick Building in form of an Hospital, for the Reception of poor Sick. In some Countries it is appointed for Persons coming from Places suspected of the Plague to quarantain in. This is usually a large Building, at a distance from any City, whose Apartments stand at a distance from each other, *etc.* where Ships are unladen, and their Equipage is laid up for forty Days, more or less, according to the Time and Place of Departure.

LEAD, a coarse, heavy, impure Metal, of all others the softest and most fusible, when purified; called by the Chymists *Saturn.* Those who have made an Analysis of it, find it contains a little Mercury, some Sulphur, and a great deal of bituminous Earth. *Lead* is found in various Countries, but abounds particularly in *England.* 'Tis found, too, in several kinds of Soils and Stones, some whereof, besides, contain Gold, some Silver, others Tin, *etc.* 'Tis melted on a Furnace provided for the purpose, with a strong Coal-Fire upon it; as it melts, it runs thro' a Canal on one Side of it, leaving the Earth, Stone, and Scoria, with the Ashes of the Coals. 'Tis purified by skimming it e'er cold, and throwing Suet and other fat Bodies into it.

Lead is much used in Building, particularly for Covering, Gutters, Pipes, and Glass. *Lead* is either cast into Sheets in a Mold, or milled; which last is found by much the least serviceable, not only on account of its Thinness, but also because 'tis so exceedingly stretched in milling; that when it comes to lie in the hot Sun, it shrinks and cracks, and consequently will not keep out the Water. The *Lead* used by Glaziers is first cast into slender Rods, twelve or fourteen Inches long, called *Canes*; which being afterwards drawn thro' their Vice, comes to have a Groove on either Side for the Panes of Glass; and this they call *Turned Lead.*

The Method of paling or soldering *Lead* for fitting on of Imboss'd Figures, *etc.* is by placing the Part whereon the Figure is to be paled, horizontal, and strewing on it some pulverized Rosin; under this place some Coals, till such time as the Rosin becomes reddish, and rises in Pimples; then apply the Figure, and rub some soft Solder in to the joining: when this is done, the Figure will be paled on, and as firm as if it had been cast on it.

LEVEL, a Mathematical Instrument, serving to draw a Line parallel to the Horizon, to lay off Floors, the Courses of Masonry, etc. horizontally, to measure the difference of Ascent or Descent between several Places, to convey Waters, drain Fens, etc. The Word comes from the *Latin Libella*, the Cross Beam that forms the Brachia of a Balance, which to be just, must stand horizontally. There are several Instruments of different Contrivance and Matter, that have been invented for the Perfection of Levelling; all of which, for the Practice, may be reduced to these that follow.

Water Level, which shews the horizontal Line by means of a Surface of Water, or other Liquid; founded on this Principle, that Water always naturally places itself *level*. The most simple is made of a long wooden Trough or Canal, whose Sides are parallel to its Base; so that being equally fill'd with Water, the Surface thereof shews the Line of *Level*: this is the *Chorobates* of the Antients, described by *Vitruvius*, *Lib*. 8. *Chap*. 6. This *Level* is also made with Cups fitted to the two Ends of a Pipe, 3 or 4 Foot long, about an Inch in Diameter; by means whereof the Water communicates from the one to the other Cup: and this Pipe being moveable on its Stand, by means of a Ball and Socket, when the two Cups become equally full of Water, their two Surfaces mark the Line of *Level*. This Instrument, instead of Cups, may be made with two short Cylinders of Glass 3 or 4 Inches long, fasten'd to each Extreme of the Pipe with Wax or Mastic. Into the Pipe is fill'd some common or colour'd Water, which shews itself through the Cylinders, by means whereof, the Line of *Level* is determined; the Height of the Water, with respect to the Center of the Earth, being always the same in both Cylinders. This *Level*, tho' very simple, is yet commodious for levelling of small Distances.

Air Level, that which shews the Line of *Level*, by means of a Bubble of Air inclosed with some Liquor in a Glass Tube of an indeterminate Length and Thickness, whose two Ends are sealed hermetically; that is, are closed with the Glass itself, by heating it with the Flame of a Lamp, till it become soft and tractable. When the Bubble of Air fixes itself at a certain Mark made exactly in the Middle of the Tube, the Plan of the Ruler wherein it is fixed is *level*. When it is not *level*, the Bubble will rise to one End. This Glass Tube may be set in another of Brass, having an Aperture in the Middle, whence the Bubble of Air may be observed. The Liquor wherewith the Tube is fill'd, is ordinarily either Oil of Tartar, or *Aqua secunda*; those not being liable to freezing as common Water, nor to Rarefaction and Condensation as Spirit of Wine. The Invention of this Instrument is ascribed to Mr. *Thevenors*.

Air Level with Sights, is an Improvement on that last described, which by the Addition of more Apparatus, becomes more commodious and exact: It consists of an *Air level* 1, (*Plat. Surveying*, *Fig*. 4.) about eight Inches long, and seven or eight Inches in Diameter, set in a Brass Tube 2, with an Aperture in the Middle. The Tubes are carried in a strong streight Ruler, a Foot long, at whose Ends are fixed two Sights exactly perpendicular to the Tubes, and of an equal Height, having a square Hole, form'd by two Fillets of Brass, crossing each other at right Angles, in the Middle whereof is drilled a very little Hole; through which little Hole, a Point on a *Level* with the Instrument is descried. The Brass Tube is fastened on the Ruler by means of two Screws, one whereof, mark'd 4, serves to raise or depress the Tube at pleasure, for bringing it towards a *Level*. The Top of the Ball and Socket is rivetted to a little Rule that springs, one End whereof is fasten'd with Screws to the great Ruler, and the other End has a Screw 5, serving to raise and depress the Instrument when nearly *level*. This Instrument is yet less commodious than the following one, because tho' the Holes be ever so small, yet they will still take in too great a Space to determine the Point of *Level* precisely.

Air Level with Telescope Sights. This *Level* (*Plate Surveying*, *Fig*. 5.) is like the last *Plate* Fig. 4. with this Difference, that instead of plain Sights, it carries a Telescope to determine exactly a Point of *Level* at a good distance. The Telescope is in a little Brass Tube, about 15 Inches long, fastened on the same Rule as the *Level*. At the End of the Tube of the Telescope, Mark'd 1, enters the little Tube 1, carrying the Eye Glass, and a Hair horizontally placed in the Focus of the Object-Glass 2; which little Tube may be drawn out, or push'd into the great one, for adjusting the Telescope to different Sights. At the other End of the Telescope is placed the Object-Glass: The Screw 3, is for raising or lowering the little Fork carrying the Hair, and making it agree with the Bubble of Air when the Instrument is *level*; and the Screw 4, is for making the Bubble of Air agree with the Telescope. The whole is fitted to a Ball and Socket. M. Huygens is said to have been the Inventor of this *Level*; which has this Advantage, that it may be invented, by turning the Ruler and Telescope half round. And if then the Hair cut the same Point that it did before the Turn, 'tis a Proof the Operation is just. It may be here observed, that one may add a Telescope to any kind of *Level*, by applying it upon, or parallel to the Base or Ruler, when there is occasion to take the *Level* of remote Objects.

Plumber Pendulum Level: That which shews the Horizontal Line by means of another Line perpendicular to that described by its Plummet or Pendulum. This Instrument (*Plate Surveying*, *Fig*. 6.) consists of two legs or Branches, join'd together at Right Angles; whereof that which carries the Thread and Plummet, is about a Foot and a half long. This Thread is hung towards the Top of the Branch, at the Point 2. The Middle of the Branch where the Thread passes is hollow, that so it may hang free every where but towards the Bottom of the Place 3, where there is a little Blade of Silver, whereon is drawn a Line perpendicular to the Telescope. The said Cavity is cover'd by two Pieces of Brass, making, as it were, a kind of Case, lest the Wind should agitate the Thread; for which reason the Silver Blade is covered with a Glass, to the end it may be seen when the Thread and Plummet play upon the Perpendicular. The Telescope 1, is fastened to the other Branch or Leg of the Instrument, and is about two Foot long, having a Hair placed horizontally across the Focus of the Object-Glass; which determines the Point of *Level*, when the String and Plummet hang against the Line of the Silver Blade. All the accuracy of this Instrument depends on the Telescopes being fitted at Right Angles to the Perpendicular. It has a Ball and Socket, by which it is fasten'd to its Foot, and is said to have been the Invention of M. *Picard*.

Reflecting Level: That made by means of a pretty long Surface of Water, representing the same Object inverted, which we see erect by the Eye; so that the Point where those two Objects appear to meet, is in a *Level* with the Place where the surface of Water is found. This is the Invention of M. *Marriotte*. There is also another *Reflecting Level*, consisting of a Mirror of Steel or the like, well polish'd, and placed a little before the Object-Glass of a Telescope suspended perpendicularly. The Mirror must make an Angle of 45 degrees with the Telescope; in which Case the Perpendicular Line of the said Telescope is converted into an Horizontal; which is the same with the Line of *Level*. The Invention of this is owing to Mr. *Cassini*.

We have another *Level* of Mr. *Huygen's* Invention, consisting of a Telescope *a*, (*Plate Surveying*, *Fig*. 7.) in form of a Cylinder, going through a Ferril, in which it is fasten'd by the middle. This Ferril has two flat Branches bb, one above, and the other below; at the ends whereof are fasten'd little moving Pieces, which carry two Rings, by one of which the Telescope is suspended to a Hook at the end of the Screw 3, and by the other a pretty heavy Weight is suspended, in order to keep the Telescope *in Equilibrio*. This Weight hangs in the Box 5, which is almost fill'd with Linseed Oil, Oil of Walnuts, or other Matter that will not easily coagulate, for more aptly settling the Balance of the Weight and Telescope. The Instrument carries two Telescopes close and very Parallel to each other, the Eye-Glass of the one being against the Object-Glass of the other, that one may see each way without turning the *Level*. In the Focus of the Object-Glass of each Telescope must a little Hair be strain'd horizontally, to be rais'd and lower'd as occasion requires, by a little Screw. If the Tube of the Telescope be not found level when suspended, a Ferril or Ring 4 is put on it, to be slid along, till it fixes to a *Level*. The Hook on which the Instrument is hung, is fix'd to a flat Wooden Cross, at the Ends of each Arm whereof, there is a Hook serving to keep the Telescope from too much Agitation in using or in carriage. To the said flat Cross is applied another hollow Cross, that serves as a Case for the Instrument; but the two Ends are left open, that the Telescope may be secured from the Weather, and always in a condition to use. The Foot of this Instrument is a round of Brass Plate, to which are fasten'd three Brass Ferrils moveable by means of Joints wherein are put Stoves; And on this foot is placed the Box.

Masons Level, is composed of three Rules, so joined as to form an Isosceles Rectangle, somewhat like a Roman A; at the Vertex whereof is fastened a Thread, from which hangs a Plummet; which passes over a fiducial Line marked in the Middle of the Base, when the thing to which the *Level* is applied, is horizontal; but declines from the Monk, when the Thing is lower on one side than the other.

Carpenters and Paviours Level, consists of a long Ruler, in the Middle whereof is fitted, at Right Angles, another somewhat bigger, at the Top whereof is fastened a Line with a Plummet; which, when it hangs over a fiducial Line at Right Angles with the Base, shews that the said Base is horizontal. These two last Levels, tho' very common, are esteemed the best for the Practice of Building, tho' the Operations made by 'em must needs be short.

LEVELLING, an Operation with a Level, for finding a Line parallel to the Horizon, at one or more Stations, in order to determine the Height of one Place with respect to another; for the laying off Grounds even, regulating of Descents, draining of Morasses, conducting of Waters, etc. One Place is said to be higher than another, or out of Level with it, when it is more remote from the Centre of the Earth; and a Line equally distant from the Centre of it in all its Points, is called the Line of true Level: whence, because the Earth is round, that line must be a Curve, and make a part of the Earth's Circumference, or an Arch concentrical with it, as the Line BCFG, *Plate Surveying*, *Fig*. 9. all the Points whereof are equally distant from the Centre of the Earth A. But the Line of Sight which the Operations of Levels give, is a Tangent or a Right Line perpendicular to the Semi-diameter: One Extreme of which Tangent, being the Point which determines it, will be above the Surface of the Earth, and of the true Level, as much as the Secant exceeds the Radius or Semi-diameter of the Earth.

The Operation of *Levelling* is as follows: Suppose the Height of the Point A (*Plate Surveying*, *Fig*. 10.) on the Top of a Mountain, above the Point B, and at the Foot thereof required: Place the Level about the Middle Distance, between the two Points, as in D, and Staffs in A and B, and let there be Persons instructed with Signals for raising and lowering on the said Staffs little Marks of Pasteboard, or other Matter. The *Level* being placed horizontally by the Bubble, etc. look towards the Staff A E, and cause the Mark to be raised or lowered, till the Middle, upper Edge, or other most conspicuous Part appear in the visual Ray. Then measuring exactly the perpendicular Height of the Point A above the Point E, which suppose 6 Feet 4 Inches, set down in your Book; then turn the Level horizontally about, that the Eye-Glass of the Telescope may be still next the Eye when you look the other way (if you have only plain Sights, the Instrument need not be turned) and cause the Person at the Staff B to raise or lower his Mark, till some conspicuous part of it fall in the visual Ray, as at C. Then measure the perpendicular Height of C above B, which suppose 16 Feet, 16 Inches; set this also down in the Book above the other Number of the first Observation; subtract the one from the other, the Remainder will be 10 Feet, 4 Inches, which is the Difference of Level between A and B, or the Height of the Point A above the Point B. Where Note, that if the Point D, where the Instrument is fixed, be in the middle between the two Points A and B, there will be no Necessity for reducing the apparent level to the true Level, the visual Ray in that case being raised equally above the true Level.

If it be further required to know whether there be a sufficient Descent for conveying Water from a Source A to the Point B, *Plate Surveying, Fig.* 11. Here in regard the Distance from A to B is considerable, 'tis required that several Options be made. Having then chosen a proper Place for the first Station, as at I, set up a Staff in the Point A near the Source, with a proper Mark to slide up and down the Staff as L, and measure the Distance from A to I, which suppose 2000 Yards. Then the Level being adjusted in the Point I, let the Mark L be raised and lowered till such time as you spy some conspicuous Part of it thro' the Telescope or Sights of the Level, and measure the Height AL, which suppose 13 Foot 5 Inches. But in regard the Distance AI is 2000 yards, you must have recourse to your Table for a Reduction, subtracting 11 Inches, which will leave the Height AL 12 Foot 6 Inches; and this note down in your Book. Now turn the Level horizontally about, so as the Eye-Glass of the Telescope may be towards the Staff at A; and fixing up another Staff at H, cause the mark G to be moved up and down, till you spy some conspicuous Part thro the Telescope or Sights. Measure the Height HG, which suppose 6 Yards, 4 Feet, 2 Inches. Measure likewise the Distance of the Points IH, which suppose 1300 Yards; for which Distance, according to The Table, 4 Inches, 8 Lines must be subtracted from the Height AG, which consequently will but leave 6 Yards, 3 Feet, 9 Inches, 4 Lines, to be taken down in your Book.

This done, remove the Level forwards to some other Eminence, as E, whence the Staff H may be view'd, as also another Staff at D, the Place whither the Water is to be convey'd. The Level being again adjusted in the Point E, look back to the Staff H; and managing the Mark as before, the visual Ray will give the Point F. Measure the Height HF, which suppose 11 Feet, 6 Inches. Measure likewise the Distance HE, which suppose 1000 Yards; for which Distance the Table gives 2 Inches, 3 Lines of Abatement; which being taken from the Height BD, there will remain 8 Feet, 9 Lines, which enter as before.

For the Manner of entering down Observations in your Book, observe that a proper Place or Station for the Level, between the two Points, has been pitched upon, you must write down the two Heights observed at that Station, in two different Columns, *viz.* under the first Column, those observed in looking thro' the Telescope when the Eye was from the Source, or towards the Point, which we may call *Back Sights*; and under the second Column those observed, when the Eye was next the Source, which we call *Fore-Sights*; in the manner following:

Back-Sights

	Foot		Inches		Lines
First Height Connected	12	:	6	:	0
Third Height	11	:	3	:	3
	23	:	9	:	3

Fore-Sights

	Foot		Inches		Lines
Second Height	21	:	9	:	4
Fourth Height	8	:	0	:	9
	29	:	10	:	1

Having summed up the Heights of each Column separately, subtract the lesser from the greater, the Remainder will be the Difference of Level between the Points A and B; as in this Example:

Feet		Inches		Lines	
29	:	10	:	1	
23	:	09	:	3	
6	:	00	:	10	*The Difference of Height or Level between the Points* A *and* B.

If the Distance of the two Points be required, add all the Distances measured, together, and dividing the Difference of Height by the Yards of the Distances, for each 200 Yards you will have a Descent of about 2 Inches, 9 Lines.

Levelling Staffs are Instruments used in *Levelling*, serving to carry the Marks to be observ'd, and at the same time to measure the Height of those Marks from the Ground. They usually consist each of two long square Wooden Rulers, made to slide over one another, and divided into Feet, Inches, etc.

For the Common Occasions of *Levelling* to be perform'd without much Apparatus of Instruments, Time, or Trouble, the following Method is recommended. Set a Pole upright in a Spring, Pond, River, or other Place whence Water is to be brought, and mark how many Feet and Inches of it are above Water. Then set up another Pole, of equal Length with the other, in the Place to which the Water is to come. Place the Center of a Quadrant on the top of this last Pole, the Plummet hanging free; spy thro' the Sights the top of the Pole that is in the Water, and if the Thread cuts any Degree of the Quadrant, the Water may be convey'd by a Pipe laid in the Earth. If you can't see from one Extreme to the other, the Operation may be repeated in the manner directed.

Dr. *Halley* suggests a new Method of *Levelling*, which has been put in practice by some of the *French* Academy: This is performed wholly by means of the Barometer, in which the Mercury is found to be suspended to so much the less Height, as the Place is further remote from the Centre of the Earth. Hence it follows, that the Different Heights of the Mercury in two Places give the Difference of Level. Mr. *Derham*, from some Observations he made at the top and bottom of the Monument, found that the Mercury fell $1/10$ of an Inch at every 82 Foot of perpendicular Ascent, when the Mercury is at 30 Inches. Dr. *Halley* allows $1/10$ of an Inch for every 30 Yards; which, considering how accurately the Barometers are now made, an inch in some of 'em being divided into an hundred or more Parts, all very sensible, he thinks this Method sufficiently exact to take the levels for the Conveyance of Water, and is less liable to Errors than the common Levels. The same Author found a Difference of 3 Inches 8 Tenths between the Height of the Mercury at the top and bottom of *Snowdon* Hill in *Wales*.

LIBRARY, an Apartment or Place destined for the placing of Books; or the Books themselves lodg'd in that Apartment. Some Authors refer the Origin of *Libraries* to the *Hebrews*, and observe that the Care they took for the Preservation of their Sacred Books, and the Memory of what concern'd the Actions of their Ancestors, became an Example to many other Nations, and particularly the *Egyptians*. *Osymandrias* King of *Egypt* is said to have taken the hint first, and had a *Library* built in his Palace. Nor were the *Ptolemys*, who reign'd in the same Country, less curious and magnificent in Books. The Scripture speaks of a *Library* of the Kings of *Persia*, *Esdras* V. 15. VI. 1. which some imagine to have consisted of the Historians of that Nation, and of Memoirs of the Affairs of the State; but, in effect, it appears rather to have been a Depository of Laws, Charters, and Ordinances of the Kings. The *Hebrew* Text calls it *the Book of Treasures*, and afterwards *the House of the Books of Treasures*. We may with more justice call that a *Library*, mention'd in the second of *Esdras* to have been built by *Nehemiah*, and in which were preserv'd the Books of the Prophets and of *David*, and the Letters of their Kings.

The first who erected a *Library* at *Athens*, was the Tyrant *Pisistratus*; and yet *Strabo* refers the Honour of it to *Aristotle*. That of *Pisistratus* was transported by *Xerxes* into *Persia*, and afterwards brought by *Seleucus Nicanor* to *Athens*. Long after, it was plundered by *Sylla*, and re-established by *Hadrian*. *Plutarch* says, that under *Eumenes* there was a *Library* at *Pergamus* containing 200,000 Books. *Tarannion*, a celebrated Grammarian, Contemporary with *Pompey*, had *Library* of 3,000 Volumes: That of *Ptolemy Philadelphus*, according to *Ammianus Marcellinus*, contain'd 7,000, all in Rolls, burnt by *Cæsar's* Soldiers. *Constantine* and his Successors erected a magnificent one at *Constantinople*, which in the eighth Century contain'd 300,000 Volumes, all burnt by order of *Leo Isauricus*; and among the rest, one wherein the *Iliad* and *Odyssee* were written in Letters of Gold on the Guts of a Serpent.

The most celebrated *Libraries* of antient *Rome* were the *Ulpian* and the *Palatin*. They also boast much of the *Libraries* of *Paulus Emilius*, who conquer'd *Persius*; of *Lucilius Lucullus*, of *Asinius Pollio*, *Atticus*, of *Julius Severus*, *Domitian*, *Seranus*, *Pamphylus*, *Martyr*, and the Emperors *Gordian* and *Trajan*.

Antiently every large Church had its *Library*; as appears by the Writings of St. *Jerom*, *Anastasius*, and others. Pope *Nicholas* laid the first Foundation of that of the *Vatican* in 1450. It was destroy'd by the Constable *Bourbon* in the sacking of *Rome*, and restored by Pope *Sixtus* V. and has been considerably enrich'd with the Ruins of that of *Heidelberg*, plunder'd by Count *Tilly* in 1622.

One of the most compleat *Libraries* in *Europe* is said to be that erected at *Florence* by *Cosmo de Medicis*; over the Gate whereof is wrote, *Labor absque Labore*. Tho it is now exceeded by that of the

French King; begun by *Francis* I. augmented by Cardinal *Richlieu*, and compleated by M. *Colbert*. The Emperor's *Library*, according to *Lambecius*, consists of 80,000 Volumes, and 15,940 Curious Medals.

The *Bodleian Library* at *Oxford*, built of the Foundation of that of Duke *Humphry*, exceeds that of any University in *Europe*, and even those of all the Sovereigns of *Europe*, except the Emperor's and *French* King's, which are each of 'em older by a Hundred Years. It was first open'd in 1602, and has since found a great number of Benefactors; particularly Sir *Rob. Cotton*, Sir *H. Savil*, Archbishop *Laud*, Sir *Kenelm Digby*, Mr. *Allen*, Dr. *Pocock*, Mr. *Selden*, and others. The *Vatican*, the *Medicean*, that of *Bessarioni* at *Venice*, and those just mention'd, exceed the *Bodleian* in *Greek* Manuscripts; which yet outdoes 'em all in Oriental Manuscripts. As to printed Books, the *Ambrosian* at *Milan*, and that of *Wolfenbuttel*, are two of the most famous, and yet both inferior to the *Bodleian*. The *Cotton* Library consists wholly of Manuscripts, particularly of such as relate to the History and Antiquities of *England*; which, as they are now bound, make about 1000 Volumes.

LICENCES, in Painting, are the Liberties which the Painter takes in dispensing with the Rules of Perspective, and the other Laws of his Art.

LIGHTS, in Architecture, are understood of the Openings of Gates and Windows, and other Places thro' which the Air and *Light* have a Passage. In the Pantheon all the *Light* comes from on high; it has no *Lights* but in the Dome.

LIGHTS, in Painting, are those Parts of a Piece that are illumined, or that lie open to the Luminary, by which the Piece is supposed to be enlightened; and that for this reason are painted in bright vivid Colours: in which sense *Light* is opposed to Shadow. *Light* is also used for the luminous Body that emits it. There are various kinds of *Lights*; general *Lights*, as the Air; particular *Lights*, as a Fire, a Candle, and even the Sun. Different *Lights* have very different Effects on the Piece, and occasion a Difference in the Management of every Part. A great deal therefore depends on the Painters chusing a Proper *Light* for his Piece to be illumined by, and a great deal more in the Conduct of the *Lights* and Shadows when the Luminary is pitched upon. The Strength and Relievo of a Figure, as well as its Gracefulness, depends entirely on the Management of the *Lights*, and the joining of those to the Shadows. The *Light* a Figure receives, is either direct of reflected, to each of which special Regard must be had. The Doctrine of *Lights* and Shadows makes that Part of Painting, called *Clair-obscure*, which see.

LIME, Calcined Stone, Marble, Free-stone, Chalk, or other Matter burnt by a large Fire in a Kiln or Furnace built express; to be afterwards used in the Composition of Mortar, for building; the Fire taking away all its Humidity, and opening its Pores, so that it becomes easily reducible to Powder. Quick *Lime*, or Unslack'd *Lime* is that which comes out of the Furnace; Slack'd *Lime* is that wash'd or steep'd in Water, and reserv'd for the making of Mortar. The best *Lime* is that made of the hardest and firmest Stones, and which is slack'd at its coming out of the Furnace. Sir *H. Wotton* looks on it as a great Error in the *English* to make *Lime* as they do, of Refuse, and Stuff without any Choice; whereas the *Italians*, at this day, and much more the Antients, burnt their firmest Stones, and even Fragments of Marble where it was plentiful; which in time became almost Marble again for its hardness, as appears in their standing Theatres, *etc*.

We have two kinds of *Lime* in common use in *England*; the one made of Stone, and the other of Chalk; whereof the former is much the strongest. That made of soft Stone or Chalk is fittest for plaistering of Ceilings and Walls within Doors; and that made of hard Stones, for Buildings and for plaistering without doors. Good *Lime* may also be made of Mill-stone, not coarse and sandy, but fine and greasy; as also of all kinds of Flints; tho 'tis hard to burn 'em, unless in a Reverberatory Furnace, as being apt to run to Glass.

Dieussant recommends *Lime* made of Sea-shells as the best, but *Goldman* finds fault with it, as being impatient of Moisture, and therefore easily peeling off from the Outsides of Walls: It is, however, the common *Lime* used in the *Indies*.

Before the Stones be thrown into the Kiln, they are to be broke in pieces; otherwise the Air contain'd in their Cavities, too much expanded by the Heat, makes 'em fly with so much violence as to damage the Kiln. According to *Alberti* and *Palladio*, *Lime* will not be sufficiently burnt in less than sixty Hours intense Heat.

The Marks of a well-burnt *Lime*, according to *Alberti*, are, that its Weight is to that of the Stone in a sesqui-alternate Proportion; that it is white, light, and sonorous; that when slaked it sticks to the Sides of the Vessel: To which *Boeckler* adds, that when slaked it sends forth a copious thick Smoak; and *Dieussant*, that it needs a great deal of Water to slake it.

To preserve *Lime* several Years, slake and work it up; dig a Pit under ground, into which let it pass thro a hole open at the bottom of the Vessel. As soon as the Pit is full, cover it up with Sand, to prevent its drying; thus keeping it moist till it be used. *Boeckler* gives another Method: Cover a Stratum of *Lime* two or three foot high with another of Sand of the like height; pour on Water enough to slake the *Lime*, but not to reduce it to dust after slaking. If the Sand cleave into Chinks, as the Smoak ascends, cover 'em up, so as no Vent may be given thereto. This *Lime*, he adds, kept ten or twelve Years, will be like Glue; and will, further, be of particular Use in painting Walls, as being no way prejudicial to the Colours. See MORTAR.

Lime is much used by Tanners, Skinners, *etc*. in the Preparation of their Leather.

Lime, or *Calx*, in Chymistry, is a kind of fine Powder, or Ashes remaining of Metals or Minerals, that have been a long time in a violent Fire.

Lime is of some medicinal Use, being applied externally in Desiccative and Consumptive Medicines. Its Water is said to be an excellent Medicine, taken internally. M. *Burlet* has written an ample Account of its Effects in the *French Memoirs*, in good measure from his own Experience. But he observes it succeeded much better in *Holland*, etc. than in *France*. It is a powerful Alterant, and like a pure Alkaline Water, fitted to blunt and destroy acid Ferments, which are the Principles of all Obstructions, and the Cause of most Chronic Diseases. Its principal Use is in a Cachexy, Green-sickness, Dropsy, Scurvy, Obstructions of the Liver, Spleen, *etc.* It is made by pouring six Pounds of hot Water on one of Quick-Lime, leaving them to soak and macerate for the space of 24 hours.

LIME-STONE, is a Stone of a whitish Colour, which being burnt in a Kiln, enters the Composition of Mortar, Plaister, *etc.* which see.

LIMNING, the Art of Painting in Water-Colours; in contradistinction to Painting properly so called, which is done in Oil-Colours.

Limning is much the more Antient Kind of Painting: Till a *Flemish* Painter, one *John van Eyck*, better known by the Name of *John* of *Bruges*, found out the Art of Painting in Oil, the Painters all painted in Water and in Fresco alone, both on their Walls, on Wooden Boards, and elsewhere. When they made use of Boards, they usually glued a fine Linnen Cloth over 'em, to prevent their opening; then laid on a Ground of White; lastly, they mix'd up their Colours with Water and Size, or with Water and Yolks of Eggs, well beaten with the Branches of a Fig-Tree, the Juice whereof thus mix'd with the Eggs; and with this Mixture they painted their Pieces.

In *Limning* all the Colours are proper enough, excepting the White made of Lime, which is only used in Fresco. But the Azure and Ultramarine must always be mix'd up with Size, or with Gum, in regard the Yolks of Eggs give yellow Colours a greenish Tincture. But there are always applied two Lays of hot Size e'er the Colours, mix'd even with Size, are laid on: the Composition made with Eggs, and the Juice of the Fig-tree, being only used for touching up and finishing, and to prevent the Necessity of having a Fire always at hand to keep the Size hot; yet 'tis certain, that the Size-Colours hold the best, and are accordingly always used in Cartoons, *etc.* This Size is made of Shreds of thin Leather, or of Parchment.

To *limn* on Linnen, they chuse that which is old, half worn, and close. This they stamp with White Lead, or with a fine Plaister beaten up with Size; which once dry, they go over it with a Lay of the same Size.

The Colours are all ground in Water, each by itself; and in proportion as they are required in working, are diluted with Size-Water. If the Yolks of Eggs are desired, they dilute 'em with Water made of equal Quantities of common Water and Vinegar, with the Yolk, White and Shell of an Egg, and the Ends of the little Branches of a Fig-tree cut small, all well beaten together in an Earthen Pan.

LINE, in Fortification, is sometimes taken for a Ditch bordered with its Parapet, and sometimes for a Row of Gabions, or Sacks of Earth, extended in Length on the Ground, to serve as a Shelter against the Enemies Fire. Thus they say, when the Trenches were carried on within thirty Paces of the Glacis, they drew two *Lines*, one on the Right, and the other on the Left, for a Place of Arms. *Line* fundamental is the first *Line* drawn for the Plan of a Place, and which shews its Area. *Line Capital* is that which is drawn from the Angle of the Gorge to the Angle of the Bastion. *Line Cogritel* is that which is drawn from the Angle of the Center to that of the Bastion. *Line of Defence*, is that which represents the Course of the Bullet of any sort of Fire-Arms, more especially of a Musquet-Ball, according to the Situation which it ought to have to defend the Face of the Bastion. *Line of Defence Fixed*, or *Fichant*, is that which is drawn from the Angle of the Curtain to the flank'd Angle of the opposite Bastions, but yet without touching the Face of the Bastion. This must never exceed 800 Feet, which they reckon the Distance at which a Musquet-Ball will do Execution. *Line of Defence Razant*, is that which being drawn from a certain Point of its Curtain, razeth the Face of the opposite Bastion; this is also called the *Line of Defence*, *Stringent*, or *Flanking*. *Line of Approach*, or *Attack*, signifies the Work which the Besiegers carry on under Covert, to gain the Moat, and the Body of the Place. *Line of Circumvallation*, is a *Line* or Trench cut by the Besiegers within Cannon-Shot of the Place, which rangeth round their Camp, and secures its Quarters against the Relief of the Besieged. *Line of Contravallation* is a Ditch bordered with a Parapet, which serves to cover the Besiegers on the Side of the Place, and to stop the Sallies of the Garison. *Lines of Communication* are those which run from one Work to another; but the *Line of Communication*, more especially so call'd, is a continued Trench, with which a Circumvallation, or Contravallation, is surrounded, and which maintains a Communication with all its Forts, Redoubts, and Tenailles. *Line of the Base* is a right *Line*, which joins the Points of the two nearest Bastions. To *line* a Work, is to strengthen a Rampart with a firm Wall, or to encompass a Parapet or Moat with good Turf, *etc.*

LINTEL, in Architecture, the Piece of Timber that lies horizontally over the Door-Posts and Window-Jaums, as well to bear the thickness of the Wall over it, as bind the Sides of the Walls together.

LIST, or LISTEL, in Architecture, called also a *Cincture*, *Fillet*, and *Reglet*; a little square Moulding, dispos'd in certain Parts of Columns, serving to crown or accompany larger Mouldings, and to separate the Flutings of Columns.

LOCK, a little Instrument used for the shutting of Doors, Chests, *etc.* The *Lock* is reckon'd the Masterpiece in Smithery; a great deal of Art and Delicacy being required in contriving and varying the Wards, Springs, Bolts, *etc.* and adjusting them to the Places where they are to be used, and the various Occasions of using them.

From the various Structure of *Locks*, accommodated to their different Intentions, they acquire various Names. Those placed on outer Doors are called *Stock-Locks*; those on Chamber-Doors, *Spring-Locks*; those on Trunks, *Trunk-Locks*, *Pad-Locks*, etc. Of these, the *Spring-Lock* is the most considerable, both for its Frequency and the Curiosity of its Structure. Its principal Parts are, the Main-Plate, the Cover Plate, and the Pin-hole: To the Main-Plate, belong the Key-Hole, Top-hook, Cross-wards, Bolt, Bolt-toe, or Bolt-nab, Drawback Spring, Tumbler, Pin of the Tumbler, and the Staples; to the Cover-Plate belong the Pin, Main-ward, Cross-ward, Step-ward or Dap-ward; to the Pin-hole belong the Hook-ward, Main Cross-ward, Shank, the Pot, or Bread, Bow-ward, and Bit.

LODGMENT, in Military Affairs, is sometimes an Incampment made by an Army; but oftener, a Retrenchment dug for a Cover or Shelter, when the Counterscarp, or some other Post is gain'd. It is also taken for the Place where the Soldiers quarter among the Burghers, in Huts, Barracks, or Tents. *Lodgment* of an Attack, is a Work cast up by the Besiegers, during their Approaches in a dangerous Post, where it is absolutely necessary to secure themselves against the Enemies Fire; as in a Covert-Way, in a Breach, in the Bottom of the Moat, *etc.* This *Lodgment* consists of all the Materials that are capable to make resistance, *viz.* Barrels, and Gabions of Earth, Pallisadoes, Wool-packs, Mantelets, Faggots, *etc.*

LUNETTES, in Fortification, are Enveloppes, Counterguards, or Mounts of Earth cast up before the Curtain, about five Fathom in breadth. They are usually made in Ditches full of Water, and Serve to the same purpose as False Braies. These *Lunettes* are composed of two Faces, which form a re-entering Angle; and their Platform being only twelve foot wide, is a little raised above the Level of the Water; and hath a Parapet three Fathom thick.

LUTHERN, or *Dormer*, a kind or Window over the Cornice, in the Roof of a Building; standing perpendicularly over the Naked of the Wall; and serving to illumine the Upper-Story. The *French* Architects distinguish these into various kinds, according to their various Forms; as

Square, Semicircular, Bulls Eyes, flat Arches, *Flemish Lutherns*, etc. The Word comes from the *Latin Lucerna*, Light or Lanthorn.

LYCEUM, the Name of a celebrated School at *Athens*, where *Aristotle* explain'd his Philosophy. The Place was compos'd of Portico's, and Trees planted in the Figure of an V. Hence the Philosophy of the *Lyceum* is us'd to signify the Philosophy of *Aristotle*, or the *Peripatetic* Philosophy.

Suidas observes, that the Place took its Name from its having been originally a Temple of *Apollo*, or rather a Portico or Gallery built by *Lycus*, Son of *Apollo*; but others mention it to have been built by *Pisistratus* or *Pericles*.

M

MAGAZINE, or Arsenal, is the Place in fortified Towns where all sorts of Stores are kept, and where Carpenters, Wheelwrights, Smiths, *etc.* are employed in making all things needful to furnish out the Train of Artillery.

MANAGE, or *Manege*, a Riding Academy, or Place for riding the Great Horse, and for breaking Horses to the proper Motions and Actions. In a *Manage* is a Place destined for vaulting round a Pillar, a Course or Carrier for running the Ring, and on the Side are Pillars, between which are placed the Horses destined for high Airs. *Manage* is also used for the Exercise itself, either of the Horse or the Rider. The Word is borrowed from the *French Menage*, and that from the *Italian Maneggio*, or, as some will have it, *à Manu agendo*.

MANEQUIN, in Painting, is understood of a little Statue, or Model, ordinarily made of Wax, and sometimes of Wood; the Junctures whereof are so contrived, as that it may be put into any Attitude one pleases, and its Draperies and their Folds be disposed at Discretion. The word *Manequin* is a Diminutive of Man, and properly signifies a little Man.

MANSE, is a Parsonage or Vicaridge-House, for the Incumbent to live in; this was originally, and is now, and essential Part of the Endowment of a Parish-Church, together with the Glebe and Tythes. It is sometimes called the *Presbyterium*.

MANSION, from *manendo*, a Dwelling-House, or Country Habitation; commonly used for the Lord's Chief Dwelling-House, within his Fee, otherwise called the *Capital Messuage*, or chief *Manor* Place. *Mansio*, or *Mansus*, were sometimes used in the same Sense with *Hide*;

that is, as much Land as one Plow could till in a Year.

MANNER, in Painting, *etc.* is used to express that particular Character observable in the Works of Painters, Poets, and other Men of Art, by which their Pencil, Hand, or Style, are distinguish'd. Thus the Curious in Paintings know the *Manners* of the Painters; and distinguish readily between the *Manners* of *Rubens*, *Titian*, or *Da Vinci*; between the antient and the new *Manner* of the same Painter; the *Flemish* and the *Italian Manner*. The *Manner* is used with respect both to the Invention, the Design, and the Colouring. The *Manner* of *Michael Angelo*, or *Raphael*, may also be known in their Scholars. Thus we say, such a Piece is of *Raphael's* School, *etc.*

MANTLE, or *Mantle-Tree*, in Architecture, is the lower-part of a Chimney, or that part laid a-cross the Jaumbs, and which sustains the Compartiment of the Chimney-piece.

MANTELETS, in War, are a kind of moveable Pent-houses, or Parapets made of Pieces of Timber, saw'd into Planks, about three Inches thick, and nailed one over another to the height of almost six Foot. They are generally cased with Tin, and set upon little Wheels; so that in a Siege, they may be driven before the Pioneers, and serve as Blinds to shelter them from the Enemies small Shot. There are also other sorts of *Mantelets*, covered on the top, whereof the Miners make use, to approach the Walls of a Town or Castle. It appears from *Vegetius*, that these were in use among the Antients, but they were built slighter, and yet larger than ours, being eight or nine Foot high, as many broad, and sixteen long: They were defended by a double Covering, the one of Boards, the other of Faggots, with the Ribs of Osiers, and cased without with Skins steep'd in Water, to prevent Fire.

MANUFACTURE, a Place where several Artists and Workmen are employ'd in the same kind of Work; or make a Commodity of the same kind. The Word is also popularly used for the Work itself; and by Extension for the like Work, carried on independently in different parts of a Country. In this sense, we say the Woollen *Manufacture*, Silk *Manufacture*, Velvet *Manufacture*, Tapestry *Manufacture*, Muslin *Manufacture*, etc. *Manufacture* of Hats, Stockings, *etc.* See *Wool*, *Silk*, *Velvet*, *Tapistry*, *Muslin*, etc. The Word comes from the *Latin*, *Manufactus*.

MARBLE, a Kind of Stone, extremely hard, firm and solid, dug out of Pits or Quarries: It takes a beautiful Polish, cuts very hardly, and is much used in Ornaments of fine Buildings, as Columns, Altars, Statues, *etc.* There are an infinite Number of different Kinds of *Marble*, usually denominated either from their Colour, their Country, or their Defects;

some are of one simple Colour, as white or black, others streak'd or variegated with Stains, Clouds, Waves, Veins, *etc.* All *Marbles* are opake, excepting the white, which, when cut into thin Slices, becomes transparent. They are also different in Weight and Hardness, and are to be consider'd with regard to their Colour, their Country, their Grain, and their Defects.

Under the Genus of *Marble* are comprehended *Porphyry*, which is the hardest, and which was antiently brought from *Numidia* in *Africa*; the most beautiful is that, whose Red is the most vivid, and the Stains the whitest and the smallest. See PORPHYRY. The *Serpentine*, which is a greenish brown, so call'd, because figur'd with little Stains. It is form'd of a great Number of Grains of Sand condens'd; it is of various Kinds, *viz.* *Egyptian*, *Italian*, *Violet*, and *Green*. See GRANATE, and SERPENTINE. *Jasper*, of which there are various Kinds, the Antient, the Florid, the Black, White, *etc.* See JASPER. *Alabaster*, of which there are various Kinds, both White and Variegated. They are all soft when taken out of the Quarry, but harden in the air. *Marbles* again may be consider'd either as Antient or Modern. By Antient we mean those, whose Quarries are lost or inaccessible to us, and whereof we have only some Pieces remaining. The Modern are those, whose Quarries are still open, and out of which Blocks continues to be dug.

Marbles denominated from their Countries.

African Marble is either of a reddish Brown, streak'd with Veins of White, or of a Carnation, with Veins of Green. *English White Marble* is vein'd with Red. *Marble* of *Auvergne* in *France*, is of a pale Red, mingled with Violet, Green, and Yellow. *Marble* of *Brabançon* in *Hainault*, is Black, vein'd with White. *Marble* of *Bresse*, in *Italy*, is Yellow with Spots of White. *Marble Brocatelle*, is mingled with little Shades of *Isabella*, Yellow, Pale and Gray. It comes from *Tortosa* in *Spain*, where it is dug out of an antient Quarry: There is also another Kind of antient *Brocatelle* dug near *Adrianople*. *Marble* of *Carrara*, on the Coasts of *Genoa*, is very white, and the fittest of all others for Works of Sculpture. *Marble* of *Champagne*, resembles the *Brocatelle*, being mix'd with blue in round Stains like Partridges Eyes. *Cipollino*, or *Cipollin Marble*, is of a Sea-green Colour, mix'd with large Waves or Clouds of white or pale green. *Scamozzi* takes this to be the same with that which the Antients call'd *Augustum et Tiberium Marmor*; because discover'd in *Egypt* in the Times of the Emperors *Augustus* and *Tiberius*. *Marble* of *Dinan*, near *Liege*, is of a pure black, very beautiful, and very common. *Marble* of *Guachenet*, near *Dinan* is of a reddish Brown, with white Spots and Veins. *Marble* of *Languedoc*, is of a vivid red, with large white Veins or Stains, and is very common; there is some, whose White borders pretty much on the Blue, but this is of less Values. *Lumachello Marble*, *Marble* so call'd, because mingled with Spots, gray, black, and white, wreath'd somewhat like Periwinkle-Shells. This is antient, and its Quarry is lost. *Marble* of *Margosse*, in the *Milaneze*, has a white Ground with brownish Veins, resembling the Colour of Iron-Rust. This is very common, and extremely hard. *Marble* of *Lavee*, in *Maine*, has a black Ground, with little narrow Veins of white; there is another Kind of it red, with Veins of a dirty white. *Marble* of *Namur* is black, like that of *Dinan*, but less beautiful, as inclining a little to the blue, and travers'd with little Streaks of grey. This is very common, and is frequently used in Paving. *Parian Marble* is Antique, and much celebrated in Authors; it is of a beautiful White: The greatest part of the *Grecian* Statues were made of it. *Varro* calls it *Lychnites*, because the Workmen dug it out of the Quarry by Lamp-Light. *Marble* of *Porta Santa*, at *Rome* call'd *Serna*, is mingled with large Clouds and Veins of red, yellow, and grey. *Marble Portor* has a black Ground, with Clouds and Veins of yellow. It is dug out of the foot of the *Alps* towards *Carrara*. *Marble* of *Ratrice*, in *Hainault*, is of a dirty red, mix'd with blue and white Clouds and Veins: this is pretty common, but is different in Beauty. *Marble* of *Savoy*, is a deep red mix'd with other Colours; each Piece whereof seems cemented on to the rest. *Marble* of *Sicily* is a brownish red, stain'd with oblong Squares of White and *Isabella*, like striped Taffeta's. The Antient has very vivid Colours, and the Modern comes pretty near it. *Marble* of *Signan*, in the *Pyreneans*, is ordinarily of a greenish brown, with red Stains; tho this is somewhat various in its Colours. *Marble* of *Theu*, near *Namur* in *Liege*, is a pure black, soft and easy to work, and receives a more beautiful Polish than those of *Namur* and *Dinan*.

Marble Bigio Nero, or black-grey, is antique. White-vein'd *Marble* has large Veins, with grey and blue Stains on a white Ground. It comes from *Carrara*. *White Marble*; that dug out of the *Pyreneans* on the side of *Bayonne*, is inferior to that of *Carrara*, its Grains being larger, and shining, like a kind of Salt. It is something like the antient white *Greek Marble*, whereof their Statues were made, but is not so hard or beautiful. *Antient Black and white Marble* is now very rare, its Quarries being entirely lost; it is divided between a pure white and a bright black in Plates. *Blue Turquin Marble*, is mix'd with a dirty kind of white, and comes from the Coast of *Genoa*. *Marble Fior di Persica*, comes from *Italy*, consists of red and white Stains, somewhat yellowish. *Yellow Marble*, is a kind of yellow *Isabella* without Veins; it is antique, and now very rare. *Black Attic Marble*, is of a pure black, without Stains, and softer than the modern black. There was some of it brought from *Greece*, call'd *Marmor Luculleum*; but this was not so much prized as that which the *Egyptians* brought from *Æthiopia*, approaching to an Iron Colour, and call'd *Basaltes*, or Touch-stone, because it served them for the Tryal of Metals. *White and black Marble* has a pure black Ground, with some very white Veins. *Marble Occhio di Pavone*, or Peacocks Eye, is mingled with red, white and blueish Clouds, somewhat resembling the Eyes at the end of a Peacock's Tail. *Green Marble antique*, is a Mixture of Grass-green and Black, in Clouds of unequal Forms and Bigness, and is very rare, the Quarries being lost. The *modern Green*, improperly called *Egyptian*, is brought from *Carrara*, on the Coast of *Genoa*; it is a deep Green spotted with Grey.

Marble denominated from its Defects.

Rigid Marble, that which, being too hard, works with difficulty, and is liable to splinter, as the Black of *Namur*. *Thready Marble*, is that full of Threads or Filaments. *Brittle Marble*, is that which crumbles under the Instrument, as the white *Greek Marble*, that of the *Pyreneans*, etc. *Terras Marble*, that with soft Places in it, which must be filled up with Cement, as that of *Languedoc*.

There are two Defects frequent in *Marbles*, which augment the Difficulty of cutting and polishing them. The one, what they sometimes call Nails, answering to the Knots in Wood; the other, call'd Emeril, is a Mixture of Copper or other Metals, making black Stains in the *Marble*. The Knots are common to all *Marbles*, the Emeril only in the white.

The Stuck whereof they make Statues, Busts, Basso-Relievos, and other Ornaments of Architecture, is only *Marble* pulveriz'd, mix'd in a certain Proportion with Plaister; the whole well sifted, work'd up with Water, and used like common Plaister. There is also a kind of artificial *Marble*, made of *Gypsum*, or a transparent Stone, resembling Marle; which becomes very hard, receives a tolerable Polish, and may deceive the Eye. There is also a kind of artificial *Marble* form'd by corrosive Tinctures, which penetrating into white *Marble*, to the depth of a Line, imitates the various Colours of other *Marbles*. Polish'd *Marble* is that which, being well rubb'd with Free-stone, and afterwards with Pumice-stone, is at last polish'd with Emery, if the *Marble* be of several Colours, and with Tin, if it be white. In *Italy* they polish with a Piece of Lead and Emery.

There are various Ways of polishing *Marble*. Some lay three or four Blocks in a Row, and with another, fix'd to a broad Beetle, and a Handle fix'd at oblique Angles, with Sand and Water between, work the upper Stone backwards and forwards on the lower ones, till the Strokes of the Ax are wore off; after which, they polish them with Emery and Putty. Father *Kircher* shews the Manner of applying Colours on *Marble*, so as to make them penetrate its whole Substance; insomuch that if the *Marble* be slit into several parallel Tables or Planks, the same Image will be found on each, that was painted on the first. Spots of Oil penetrate white *Marble*, so as they cannot be taken out. The word *Marble* comes from the *Latin Marmor*, and that from the *Greek*, *to shine*.

To *marble*, is to paint, or dispose Colours in such a manner, as that they may represent *Marble*. Thus we *marble* Paper, Wood, *etc*.

MARBLED, something resembling Marble.

MARQUETRY, *in-laid Work*; a Work composed of several Pieces of hard, fine Wood of different Colours, fasten'd, in thin Slices, on a Ground, and sometimes enrich'd with other Matters, as Tortoise-shell, Ivory, Tin, and Brass. There is another kind of *Marquetry* made, instead of Wood, of Glasses of various Colours; and a third, where nothing but precious Stones, and the richest Marbles, are used: but these are more properly called *Mosaic* Work. See MOSAIC.

The Art of Inlaying is very antient, and is supposed to have pass'd from the East to the West, as one of the Spoils brought by the *Romans* from *Asia*; indeed it was then but a simple thing: nor did it arrive at any tolerable Perfection, till the fifteenth Century, among the *Italians*; it seems however to have arrived at its height in the seventeenth Century among the *French*. Till *John* of *Verona*, a Cotemporary with *Raphael*, the finest Works of this kind were only black and white, which are what we now call *Moresco's*; but that Religious, who had a Genius for Painting, stain'd his Woods with Dyes or boiled Oils, which penetrated them. But he went no further, than the representing Buildings and Perspectives, which require no great Variety of Colours. Those who succeeded him, not only improved on the Invention of dying the Woods, by a Secret which they found of burning them without consuming, which served exceedingly well for the Shadows; but had also the Advantage of a number of fine new Woods of naturally bright Colours, by the Discovery of *America*. With these Assistances the Art is now capable of imitating any thing; whence some call it the *Art of Painting in Wood*.

The Ground whereon the Pieces are to be arranged and glued, is ordinarily of Oak or Fir well dried; and to prevent warping, is composed of several Pieces glued together. The Wood to be used being reduced into Leaves, of the Thickness of a Line, is either stain'd with some Colour, or made black for Shadow; which some effect, by putting it in Sand extremely heated over the fire, others by steeping it in Lime-Water and Sublimate, and others in Oil of Sulphur. Thus colour'd, the Contours of the Pieces are form'd, according to the Parts of the Design they are to represent. This last is the most difficult part of *Marquetry*, and that wherein most Patience and Attention are required. The two chief Instruments used herein, are the Saw and the Vice; the one, to hold the Matters to be form'd; the other, to take off from the Extremes, according to occasion. The Vice is of Wood, having one of its Chaps fix'd, the other moveable, and is open'd and shut by the Foot, by means of a Cord fasten'd to a Treadle. Its Structure is very ingenious, yet simple enough, and will be easily conceived from the Figure (*Tab. Miscellany*, *fig*. 1) The Leaves to be form'd (for there are frequently three or four of the same Kind form'd together) are put within the Chaps of the Vice, after being glued on the outermost part of the Design, whose Profile they are to follow; then the Workman pressing the Treadle, and thus holding fast the Piece, with his Saw runs over all the Out-lines of the Design. By thus joining and forming three or four Pieces together, they not only gain time, but the Matter is likewise the better enabled to sustain the Effort of the Saw; which, how delicate soever it may be, and how lightly soever the Workman may conduct it, without such a Precaution, would be apt to raise Splinters, to ruin the Beauty of the Work.

When the *Marquetry* is to consist of one single kind of Wood, or of Tortoise-shell, on a Copper or Tin Ground, or *vice versa*; they only form two Leaves on one another, *i.e.* a Leaf of Metal, and a Leaf of Wood or Shell: this they call sawing in Counter-parts; for by filling the Vacuities of one of the Leaves by the Pieces coming out of the other, the Metal may serve as a Ground to the Wood, and the Wood to the Metal.

All the Pieces thus formed with the Saw, and marked, to know 'em again, and the Shadow given in the manner already mention'd, they vaneer or fasten each in its Place on the common Ground; using for that purpose the best *English* Glue. The whole is put in a Press to dry, planed over, and polish'd with the Skin of the Sea-Dog, Wax, and Shave-Grass, as in simple VANEERING, which see. With this Difference, however, that in *Marquetry* the fine Branches, and several of the more delicate Parts of the Figures, are touch'd up and finish'd with a Graver. 'Tis the Cabinet-Makers, Joiners, and Toy-Men, among us, who work in *Marquetry*; 'tis the Enamellers and Stone-Cutters, who deal in *Mosaic* Work: the Instruments used in the former are mostly the same with those used by the Ebonists.

MASON, a Person employ'd under the Direction of an Architect, in the raising of a Stone-Building. The chief Business of a *Mason*, is to make the Mortar, raise the Walls from the Foundation to the Top, with the necessary Retreats and Perpendiculars, form the Vaults, and employ the Stones as deliver'd to him. When the Stones are large, the Business of hewing or cutting them belongs to the Stone-Cutters; tho these are frequently confounded with the *Masons*. The Ornaments of Sculpture are perform'd by Carvers in Stone, or Sculptors. For the Materials whereon *Masons* work, see STONE, MARBLE, *etc*. For the Manner of Working, see MASONRY, BUILDING, *etc*. For their Works themselves, see WALL, *etc*. The Tools or Implements principally used by them, are, the Square, Level, Plumb-line, Bevel, Compass, Hammer, Chissel, Mallet, Saw, Trowel, *etc*. Besides the common Instruments used in the Hand, they have likewise Machines for the raising of great Burdens, the conducting of large Stones, *etc*. The principal of these are the Lever, Wheel, Pully, *etc*. *Masonry* is sometimes used in a more general and unlimited Sense; so as, besides proper *Masonry*, to include the hewing or sawing of larger Stones, to fit them for Building, properly call'd *Stone-Cutting*; and the carving of the Ornaments, on the Members and Mouldings, properly call'd *Sculpture*: which see. *Isidore* derives the Word from *Machio*, a Machinist, as being obliged to use Machines in Building. Some derive it from *Machino*. *Du Cange* from *Maceria*, because the long Fence-Walls that inclose Vineyards, *etc*. in which

Masons are supposed to have been first employ'd, were call'd *Maceriæ. Mason est Maceriarum Constructor.* M. *Huet* derives it from *Mas*, an old Word, signifying House; hence *Mason* is a Person who makes *Mas's*, that is, Houses. In the corrupt *Latin*, a *Mason* was call'd *Magister Comacinus*, which *Lindenbroek* derives from *Comacina*, an Island in *Romania*, where, in the Time of the *Lombards*, the best Architects were found.

MASONRY, a Branch of Architecture, consisting, as some define it, in the Art of hewing or squaring Stones, and cutting them level and perpendicular, for the Uses of the Building: Tho in the more limited Sense of the Word, *Masonry* is the Art of assembling and joining Stones together with Mortar. Whence there arise as many different Kinds of *Masonry*, as there are different Forms and Manners of laying or joining the Stones. *Vitruvius* makes mention of seven Kinds of *Masonry* among the Antients; three of hew'd Stone, *viz.* that in form of a *Net*, that in *Binding*, and that call'd the *Greek Masonry*: And three of unhewed Stones, *viz.* that of an *equal* Course, that of an *unequal* Course, and that fill'd up in the middle. The seventh was a Composition of all the rest.

Net Masonry, call'd *Reticulation*, from its Resemblance of the Mashes of a *Net*, consists of Stones squared in their Courses, and so disposed, as that their Joints go obliquely, and the Diagonals are, the one perpendicular, and the other level. This is the most agreeable *Masonry* to the Eye, but it is apt to crack. See *Plate* ARCHITECTURE, *Fig*. 1.

Bound Masonry, is that wherein the Stones were placed one over another, like Tiles; the Joints of the Beds being level, and the Mounters perpendicular. So that the Joint that mounts and separates two Stones, falls directly over the middle of the Stone below. This is less beautiful than the Net-Work, but more solid and durable. See *Plate* ARCH. *Fig*. 2.

Greek Masonry, according to *Vitruvius*, is that where, after we have laid two Stones, each of which makes a Course, another is laid at the end, which makes two Courses; and the same Order observed throughout the Building. This may be call'd *Double Building*, in regard the Binding is not only of Stones of the same Course with one another, but likewise of one Course with another Course. See *Plate* ARCH. *Fig*. 3.

Masonry by Equal Courses, by the Antients call'd *Isodomum*, differs in nothing from the *bound Masonry*, but only in this, that its Stones are not hewn. See *Plate* ARCH. *Fig*. 5.

Masonry by Unequal Courses, call'd *Pseudisodomum*, is also made of unhew'd Stones, and laid in *bound Work*: But then they are not of the same Thickness, nor is there any Equality observed, excepting in the several Courses; the Courses themselves being unequal to each other. See *Plate* ARCH. *Fig*. 4.

Masonry fill'd up in the Middle, call'd by the Antients *Emplecton*, is likewise made of unhew'd Stone, and by Courses; but the Stones are only set in order as to the Courses, the Middle being fill'd up with Stones thrown in at random among the Mortar. See *Plate* ARCH. *Fig*. 5.

Compound Masonry, is of *Vitruvius's* proposing; so call'd, as being form'd of all the rest. In this, the Courses are of hew'd Stone, and the middle Place left void, fill'd up with Mortar and Pebbles thrown in together. After this the Stones of one Course are bound to those of another Course, with Cramp-Irons fasten'd with melted Lead. See *Plate* ARCH. *Fig*. 6.

All the Kinds of *Masonry* now in use may be reduced to these five, *viz. Bound Masonry*; that of *Brick-Work*, where the Bodies and Projectures of the Stones inclose square Spaces or Pannels, *etc*. set with Bricks: That *de Moilon*, or small Work, where the Courses are equal, well squared, and their Edges or Beds rusticated: That where the Courses are unequal; and that fill'd up in the middle with little Stones and Mortar.

MASQUE, in Architecture, is understood of certain Pieces of Sculpture, representing some hideous Form, Grotesque, or Satyrs Faces, *etc*. used to fill up and adorn some vacant Places, as in Freezes, the Pannels of Doors, Keys of Arches, *etc*. but particularly in Grotto's.

MASSES, in Painting, are the Large Parts of a Picture, containing the great Lights and Shadows; and thus, when it is almost dark, we see only the *Masses* of a Picture, *i.e*. the Places of the greatest Lights and Shadows.

MASSIVE, something gross and solid, in opposition to Tenderness and Delicacy. Thus we say, a Wall, or a Building, is too *massive*, that is, its Walls are too thick, or the Lights and Openings too little in proportion. A *massive* Column is that, which is too short for the Order, whose Capital it bears.

MASTER-PIECE, an exquisite, or extraordinary Work or Performance, in any Art or Science.

MASTER-PIECE, or *Chef d'Oeuvre*, is particularly used among the *French*, for a particular Work, which those who aspire to be admitted Master in any Art of Trade, are to perform in presence of the Masters or Jurands of that Company, by way of Specimen of their Capacity. The *Master-Piece* of a Mason is a Descent bias'd either in the Head or in Talut, corrected by a full Arch. That of a Carpenter, a Rampant Curve of a Stair-Case, the Spiral well adjusted with the Descent. That of a Joiner, a Flat-bottom Chest, or a Door-Case, or a Mantle-Tree. That of a Tiler, a Luthern well conducted in the *Fourchette*, with a Ridge. That of a Plumber, a little Cistern *a Cul de Lampe*. That of a Glazier, a Pair of Compartiments of Glasses of several Colours, hollow'd, incastrated, and join'd with Lead of *Chef d'Oeuvre*. That of Paviors, a Rose in a Free-Stone or Flint-Pavement. That of a Cordwainer, a Turn-up Shoe, *etc*.

MAUSOLEUM, a magnificent Tomb, or funeral Monument, consisting of Architecture, and Sculpture, with an Epitaph, erected in honour of some Prince, or other illustrious Person: as the *Mausoleum* of *Augustus* at *Rome*. The Word is also used to signify the Decoration of a Tomb, or *Catafalcha*, in a funeral Pomp. The Word comes from *Mausolus* King of *Caria*, to whom *Artemisia*, his Widow, erected a most stately Monument, that has since been number'd among the Wonders of the World, calling it from his Name, *Mausoleum*.

MERLON, in Fortification, that part of the Parapet which lies betwixt two Embrasures. See EMBRASURE.

The Word comes from *Merula*, or *Merla*, which in the corrupt *Latin* was used for a Battlement.

MESSUAGE, in Law, a Dwelling-House with some Land assigned for its use, *etc*.

By this Name may a Garden, Shop, Mill, Chamber, or Cellar be called.

In *Scotland*, *Messuage* is what we call the *Manor-House*, *viz*. Principal Dwelling-House, within any Barony.

METOCHE, in the Antient Architecture, a Term used by *Vitruvius* to signify the Space or Interval between the Dentils. See DENTICLES.

Baldus observes, that in an antient MS. Copy of that Author, the Word *Metatome* is found for *Metoche*. Hence *Daviler* takes occasion to suspect that the common Text of *Vitruvius* is corrupted; and concludes, that it should not be *Metoche*, but *Metatome*, *q.d.* Section.

METOPE, or METOPA, in Architecture, the Interval, or square Space between the Triglyphs, in the Frieze of the *Doric* Order. See TRIGLYPH and FRIEZE.

The Antients used to adorn these Parts with carved Works, or Paintings, representing the Heads of Oxen, Vessels, Basons, and other Utensils of the Heathen Sacrifices.

As there is found some difficulty in disposing the Triglylphs and *Metopes* in that just Symmetry which the *Doric* Order requires; some Architects make it a Rule, never to use this Order but in Temples.

Semi METOPE is a Space somewhat less than half a *Metope*, in the Corner of the *Doric* Frieze.

The Word *Metope*, in the original *Greek*, signifies the distance between one Aperture or Hole and another, or between one Triglyph and another; the Triglyphs being supposed to be Solives or Joists that fill the Apertures.

METROPOLIS, the Capital of a Country, or Province; or the principal City, and, as it were, the *Mother* of all the rest.

MEZZANINE, a Term used by some Architects to signify an Entresole.

The Word is borrowed from the *Italians*, who call *Mezzanini* those little Windows, less in height than breadth, which serve to illuminate an Attic, or Entresole.

MEZZO-TINTO, in Sculpture, a particular manner of Engraving Figures on Copper. See ENGRAVING.

Mezzo-Tinto is said to have been first invented by Prince *Rupert*; and Mr. *Evelyn*, in his History of *Chalcography*, gives us a Head perform'd by that Prince in this way.

'Tis pretty different from the common way of Engraving. To Perform it, they rake, hatch, or punch the Surface of the Plate all over with a Knife, or Instrument for the purpose; first one way, them a-cross, *etc.* till the face of the Plate be thus entirely furrow'd with Lines or Furrows close and as it were contiguous to each other; so that if an impression were then taken from it, it would be one uniform blot or smut.

This done, the Design is drawn, or marked on the same Face: after which, they proceed with Burnishers, Scrapers, *etc.* to expunge and take out the Dents or Furrows in all the parts where the Lights of the Piece are to be; and that more or less, as the Lights are to be stronger or fainter: leaving those parts black which are to represent the Shadows or Deepnings of the Draught.

MILITARY *Architecture*, is the Art of Fortification. See ARCHITECTURE and FORTIFICATION.

MILL, in propriety, is a Machine used for *Grinding*: But the Word in its general Signification is used for all Machines whose Action depends on a circular Motion.

Of these there are various kinds, which acquire various Names according to the various manners in which the moving Power is apply'd; but they may all be reduced to three heads, *viz. Wind-Mills*, *Water-Mills*, and *Hand-Mills*; under which last, are also comprehended those worked by Horses, *etc.*

Water-MILLS are those turned by the force or fall of a River, *etc.* Of which, again, there are two kinds; those where the force of the Water is apply'd above the Wheel, call'd *Over-Shot*; and those where it is apply'd below the Wheel, call'd *Under-Shot Mills.*

Wind-MILLS, are those turned by the force of Wind gathered in their Sails. Of these, some are call'd *Vertical*, others *Horizontal*, according to the position of the Sails; or rather according to the direction of their Motion with regard to the Horizon.

For the best Form of Horizontal Sails, as also for determining the Position of the Axis of Wind-Mills.

Portative, or *Hand*-MILLS, are those kept in Motion by the Hand; or whose Mill-Stones are turn'd, or Pistons driven by the force of Horses, or other Beasts.

The Use of *Mills* and *Mill-stones*, according to *Pausanias*, was first invented by *Miletus*, Son of *Meleges*, first King of *Sparta*. Tho *Pliny* attributes the Invention of every thing belonging to Bread and Baking, to *Ceres*. *Polydore Virgil* was not able to discover the Author of so useful a Machine. 'Tis doubted whether or no Water-Mills were known to the *Romans*; there being no mention made in the Digest but of *Mills* turn'd by Slaves and Asses.

Salmasius, however, and *Gothofrid*, will not allow *Water-Mills* to have been unknown to the antient *Romans*, though they were not in ordinary use.

Wind-Mills are of much more modern Invention: The first Model of these was brought from *Asia* into *Europe*, in the time of the Holy-Wars.

MILL is also used, in the general, for all Machines, which being moved by some external Force, serve to give a violent Impression on things apply'd thereto.

Mills in this sense, are Machines of vast use in the Manufactures, Arts, and Trades; for the making and preparing divers kinds of Merchandizes. The principal are those which follow.

Paper-MILL, a Water-Mill furnish'd with several Hammers, which beat, or pound the Rags or Cloth in a kind of Wooden-Trough: and thus by reducing them to little Pieces, turn them into a kind of Pulp, by means of Water convey'd into the Troughs by a Pipe for the purpose.

Fulling-MILL, is a Water-Mill which raises and beats down large wooden Pistons in proper Vessels, call'd Peels or Troughs; in order to full, scower, and cleanse Woollen Stuffs.

Linen-MILLS don't differ much from Fulling-Mills. Their Use is to scower Linens, after having been first cleans'd when taken out of the Lixivium, or Lye.

Some of these go by Water, the generality by Horses.

MILL in Coinage, is a Machine used to prepare the *Laminæ* or Plates of Metal, and to give them the proper thickness, hardness, and consistence before they be struck or stamped.

This Machine has not been long known among us; but is of some standing in *Germany*. It consists of several Wheels dented like those of Clocks, *etc.* which move two Cylinders of Steel, between which the Metal is passed to be brought to its proper thickness. It was first turned in order with Water, afterwards with Cattel, *etc.*

MILL, among the Gold Wire-Drawers, is a little Machine consisting of two Cylinders of Steel, serving to flatten the Gold, or Silver Wire, and reduce it into *Laminæ*, or Plates.

They have also *Mills* to wind the Gold Wire or Thread on the Silk; composed of several rows of Bobins all turned at the same time.

Sugar-MILLS, is a Machine that serves to bruise the Sugar-Canes, and express the Liquor or Juice contain'd therein.

The *Sugar-Mills* are very curious Contrivances. Of these there are four kinds; being turned either by Water, Wind, Men, or Horses.

Those turned by the hand were first in use; but are now laid aside, as being an intolerable Hardship on the poor Negroes, who were doom'd thereto; besides the slowness of their Progress.

Wind-Mills are the most Modern, but they are yet somewhat rare: excepting in *St. Christopher's* and *Barbadoes*, and among the *Portugueze*. These make good dispatch, but have this Inconvenience, that they are not easily stop'd; which proves frequently fatal to the Negroes who feed them.

Oil-MILLS, whether turn'd by Men, Water, Hand, or Horse, serve to bruise of break the Nuts, Olives, and other Fruits and Grains, whose Juice is to be taken by Expression to make Oil.

Tan, or *Bark*-MILLS, wrought by Water or Horses, serve to cut certain Barks into a coarse sort of Powder, proper for the Tanning of Hides, *etc*.

Sawing-MILL is a Water-Mill, serving to saw several Planks or Boards at the same time.

These are frequent in *France*, especially in *Dauphine*. They were lately prohibited in *England*, where they were begun to be introduced, from a view to the Ruin of the Sawyers, which must have ensued.

Forge-MILLS, turn'd by Water, serve to raise and let fall one or more huge Hammers, to beat and form the Iron into Bars, Anchors, or other massive Works.

MILLS *for Sword-Blades*, are likewise moved by Water. They are frequent at *Vienne* in *Dauphine*. By working heavy Hammers, they forge those excellent Sword-Blades, call'd *Blades of Vienne*.

Leather-MILLS, are used to full, and prepare with Oil, the Skins of Stags, Buffalo's, Elks, Bullocks, *etc*. to make what they call *Buff-Leather*, for the use of Military-Men.

This is effected by means of several large Pistons rising and falling on the Skins in large Wooden-Troughs, by means of a Wheel without-side, turn'd by the force of the Water.

Gun-Powder MILL, is that used to pound, and beat together the Ingredients whereof Gun-Powder is composed.

This is done in a kind of Iron or Brass Mortar by means of Iron Pestles wrought by a Wheel without the *Mill*, turn'd by the Water falling on it.

There are also *Silk*-MILLS, for spinning, throwing, and twisting Silks; which are large round Machines in form of Turrets, five or six Foot high, and six Yards in diameter; which being turn'd either by the force of Water, or that of Men, work at the same time an infinity of Bobins fastned thereto, whereon the Silk had been wound to be here spun, and twisted.

There are abundance of *Mills* of this kind in *France*, especially about *Lyons* and *Tours*, some of which are so disposed, as that three of them will go at the same time, and by the same Wheel wrought by Water, or by strength of Hand. That in the Hôpital *de la Charite* at *Lyons*, is wonderful, a single Man working no less than forty-eight of these *Mills*.

MINIATURE, or MIGNATURE, a delicate kind of Painting, consisting of little Points or Dots, instead of Lines; usually done on Velom, with very thin, simple Water-Colours.

Miniature is distinguished from other kinds of Painting by the smallness and delicacy of its Figures, the weakness of their Colours, and faintness of the Colouring, and in that it requires to be view'd very near.

Those Colours that have the least Body, are the best and most commodious for painting in *Miniature*; as *Carmine*, and *Ultramarine*, fine Lakes, and Green made of the Juices of several kinds of Herbs and Flowers.

Painting in *Miniature* is the nicest and most tedious of all others; being performed wholly with the Point of the Pencil.

There are some Painters who never use any white Colour in *Miniature*, but make the ground of the Velom serve to raise their Figures; in which Case, the Lights appear bright in proportion to the depth and strength of the Colours of the Figures. Others, before they go to work, give the Velom a light wash with white Lead, well prepared and purified.

When the Colours are laid on flat without dotting, tho' the Figures be small, and the ground either Velom or Paper, it is not call'd *Miniature*, but *Washing*.

The Colours for *Miniature* may be mix'd up with Water of Gum Arabic, or Gum Tragacanth.

The Word comes from the *Latin Minium*, Red Lead, that being a Colour much used in this kind of Painting. The *French* frequently call it Mignature, from *Mignon*, fine, pretty, on account of its smallness and delicacy.

MINUTE, in Architecture, is a 30th Part, or division of a Module. See MODULE.

MOAT, in Fortification, call'd also *Foss* and *Ditch*, a hollow Space or Ditch dug round a Town, or Fortress to be defended, on the outside the Wall or Rampart.

The Length and Breadth of the *Moat* often depends on the nature of the Soil; according as it is Marshy or Rocky.

A *Dry* MOAT is that which is destitute of Water, and ought to be deeper than one which is full of Water.

A *Lined* MOAT is that whose Scarp and Counterscarp are cased with a Wall of Mason's Work lying a-slope.

A *Flat-bottom'd* MOAT is that which hath no sloping, its Corners being somewhat rounded.

The Brink of the *Moat* next the Rampart in any Fortification is call'd the *Scarp*, and the opposite one the *Counterscarp*.

MODEL, an Original, or Pattern proposed for any one to copy or imitate.

St. *Paul's* Church is said to be built on the *Model* of St. *Peter's* at *Rome*.

MODEL is particularly used in Building for an Artificial Pattern, made of Wood, Stone, Plaister, or other Matter, with all its Parts and Proportions; in order for the better Conducting and Executing of some great Work, and to give an Idea of the Effect it will have in Large.

In all large Buildings, it is much the surest way to make a *Model* in *Relievo*; and not to trust a bare Design, or Draught.

There are also *Models* for the Building of Ships, *etc.* for extraordinary Stair-Cases, *etc.*

MODEL, in Painting and Sculpture, is any thing proposed to be imitated.

Hence in the Academies they give the Term *Model* to a naked Man, disposed in several Postures, to give an Opportunity for the Scholars to design him various Views and Attitudes.

The Sculptors have little *Models* of Clay or Wax to assist them in their Designs of others that are larger in Marble, *etc.* and to judge of the Attitude and Correctness of a Figure.

The Statuaries likewise give the Name *Model* to certain Figures of Clay or Wax, which are but just fashion'd, to serve by way of Guide for the making of larger, whether of Marble, or other Matter.

MODERN, something new, or of our Time; in opposition to any thing Antique, or *Antient*.

MODERN *Authors*, according to *Nauda*, are all those that have wrote since *Boetius*. The *Modern* Philosophy commences with *Galilæus*. The *Modern* Astronomy with *Copernicus*.

MODERN *Medals* are all those that have been struck within these 300 Years.

MODERN in *Architecture*, is improperly apply'd to the present, or *Italian* Manner of Building; as being according to the Rules of the *Antique*. See ANTIQUE. Nor is the Term less abused, when attributed to Architecture purely *Gothic*. See GOTHIC.

Modern Architecture, in strictness, is only applicable to that which partakes partly of the Antique, retaining somewhat of its Delicacy and Solidity; and partly of the Gothic, whence it borrows Members and Ornaments, without Proportion or Judgment. See ARCHITECTURE.

MODILLIONS, in Architecture, Ornaments in the Cornish of the Ionic, Corinthian, and Composite Columns.

The *Modillions* are little Consoles or Brackets under the Soffit of the Cornish, seeming to support the Larmier, tho in reality they are no more than Ornaments.

They ought always to be placed over the middle of the Column. They are particularly affected in the Corinthian order, where they are usually enrich'd with Sculpture

The *Modillion* is usually in form of an S inverted, and fitted to the Soffit of the Cornish.

The Proportions of *Modillions* must be so adjusted, as to produce a Regularity in the Parts of the Soffits.

The *Inter-Modillions, i.e.* the Distances between them, depend on the Inter-Columns, which oblige the *Modillions* to be made of a certain Length and Breadth, in order to make the Intervals perfect Squares, which are always found to have better effect than Parallelograms. To this it must be added, that in adjusting the *Modillions*, Care should be taken that they have such a Proportion, as that when the Orders are placed over one another, there be the same Number in the upper Order as in the lower, and that they fall perpendicularly over each other.

Modillions are also used under the Cornishes of Pediments; tho *Vitruvius* observes that they were not allow'd of in his time, in regard *Modillions* were intended to represent the Ends of Rafters, which could not be suppos'd to be used in a Pediment.

Some will have the *Modillions* of a Pediment to represent Purlins; and those at the Eaves to represent Rafters. *Daviler* rather takes them for a kind of inverted Consoles or Corbels.

The *Modillion* is sometimes also call'd *Mutule*, tho Use has introduc'd a little difference between the Idea of a *Modillion* and a Mutule; the Mutule being peculiar to the Doric Order; and the *Modillions* to the higher Orders. See MUTULE.

The Word comes from the *Italian Modiglione*, a little Measure.

MODULE, in Architecture, a certain Measure, or Bigness taken at pleasure, for regulating the Proportions of Columns, and the Symmetry or Distribution of the whole Building. See COLUMN.

Architects usually chuse the Semi-diameter of the bottom of the Column for their *Module*; and this they subdivide into Parts, or *Minutes*.

Vignola divides his *Module*, which is a Semi-diameter, into twelve Parts in the Tuscan and Doric, and into eighteen for the other Orders.

The *Module* of *Palladio*, *Scamozzi*, M. *Cambray*, *Desgodetz*, *Le Clerc*, etc. which is also equal to the Semi-diameter, is divided into thirty Parts or Minutes in all the Orders.

Some divide the whole height of the Column into 20 Parts for the Doric, 22½ for the Ionic, 25 for the Roman, *etc.* and one of these Parts they make a *Module*, to regulate the rest of the Building by.

There are two Ways of determining the Measures, or Proportions of Buildings; the first by a fix'd Standard Measure, which is usually the Diameter of the lower part of the Column, call'd a *Module*, subdivided into 60th Parts, call'd *Minutes*. In the second, there are no Minutes, nor any certain and stated Division of the *Module*; but it is divided occasionally into as many Parts as are judg'd necessary. Thus, the height of the Attic Base, which is half the *Module*, is divided either into three, to have the height of the Plinth; or into four, for that of the greater Torus; or into six, for that of the lesser. Both these Manners have been practised by the antient, as well as the modern Architects: But the second, which was that chiefly used among the Antients, is in my Opinion preferable. *Perrault.*

As *Vitruvius*, in the Doric Order, has lessen'd his *Module*, which in the other Orders is the Diameter of the lower part of the Column; and has reduced that great *Module* to a mean one, which is the Semidiameter: I here reduce the *Module* to a third part for the same reason, *viz.* to determine the several Measures without a Fraction. For in the Doric Order, beside that the height of the Base, as in the other Orders, is determin'd by one of these *mean Modules*; the same *Module* gives likewise the heights of the Capital, Architrave, Triglyphs, and Metopes. But our little *Module*, taken from the third of the Diameter of the lower part of the Column, has Uses much more extensive; for, by this the heights of Pedestals, of Columns, and Entablatures, in all the Orders, are determin'd without a Fraction.

As then the *great Module*, or Diameter of the Column, has 60 Minutes; and the *mean Module*, or half the Diameter, 30 Minutes; our *little* Module has 20. *Id.*

MOINEAU, in Fortification, is a flat Bastion raised in the middle of a Curtain when it is too long, and the Bastions of the Angles too remote to be able to defend one another.

MONASTERY, a *Convent*, or House built for the Reception of Religious, whether it be *Abbey*, *Priory*, *Nunnery*, or the like. See ABBEY.

Monastery is in a more immediate sense apply'd to the Houses of Mendicault Friars, and Nuns. The rest are more properly call'd *Convents*. See CONVENT.

MONUMENT in Architecture, a Building destin'd to preserve the Memory, *etc.* of the Person who rais'd it, or for whom it was rais'd.

Such is a Triumphal Arch, a Mausoleum, a Pyramid, *etc.* See MAUSOLEUM, *etc.*

The Word comes from the *Latin Monere*, to advise, and advertize.

The first *Monuments* which the Antients erected, were the Stones which they laid over their Tombs, whereon they wrote the Names and Actions of the deceas'd. See TOMB.

These Stones were distinguish'd by various Names according as their Figures were different. The *Greeks* gave the Name *Steles* to such as were Square in their Base, and preserv'd the same Depth throughout their whole Length; whence were deriv'd our Square Pilasters, or Attic Columns. See PILASTER.

They call'd those *Styles*, which being round in their Base, ended in a Point at top, which gave occasion to the Invention of diminish'd Columns.

The Name *Pyramids* they gave to those which were square at the Foot, and terminated in a point at top, in manner of a Funeral Pile. See PYRAMID.

And the Name *Obelisc*, to those whose Bases were more in Length than in Breadth, and which rose, still lessening, to a very great Height, resembling the Figure of the Spits or Instruments used by the Antients in roasting the Flesh of their Sacrifices, which they call'd *Obeles*. See OBELISK.

The MONUMENT, absolutely so call'd among us, denotes a Magnificent Pillar erected by Order of Parliament, in Memory of the Burning of the City, *Anno* 1666, in the very place where the Fire began.

It is of the Tuscan Order, 202 Foot high from the Ground, and 15 Foot in Diameter, all of solid *Portland* Stone, with a Stair-Case in the Middle of black Marble. The Pedestal is 21 Foot Square, and 40 high; the Front being enrich'd with curious Basso Relievo's.

MORISCO, or MORISK, a kind of Painting, Carving, *etc.* done after the Manner of the *Moors*; consisting of several Grotesque Pieces and Compartiments promiscuously intermingled, not containing any perfect Figure of a Man, or other Animal, but a wild Resemblance of Birds, Beasts, Trees, *etc.* See GROTESKE.

MORTAR, or MORTER, in Architecture, is a Preparation of Lime, Sand, *etc.* mixed up with Water; serving, as a Cement, to bind the Stones, *etc.* of a Building. See BUILDING, CEMENT, *etc.*

The Antients had a kind of *Mortar* so very hard and binding, that, after so long a Duration, 'tis next to impossible to separate the Parts of some of their Buildings; tho' there are some who ascribe that excessive Strength to Time, and the Influences of certain Properties in the Air, which is found to harden some Bodies very surprizingly.

De Lorme observes, that the best *Mortar*, is that made of *Puzzuoli*; adding, that it penetrates black Flints, and turns 'em white.

The Lime used in the antient *Mortar* is said to be burnt from the hardest Stones, and even Fragments of Marble. See LIME.

Mr. *Worledge* observes, that fine Dust makes weak *Mortar*; and the rounder the Sand, the stronger the *Mortar*. He therefore advises the Sand to be washed e'er mixed; and adds, that dirty Water weakens the *Mortar* considerably.

The Proportion of Lime and Sand in our common *Mortar* is extremely variable. *Vitruvius* prescribes three parts of Pit-Sand and two of River-Sand to one of Lime; but the Sand here seems to be over-dosed. About *London*, the Proportion of Sand to quick Lime is as 36 to 25. In some Parts they use equal Quantities of each.

Wolfius observes, that the Sand should be dry and sharp, so as to prick the Hands when rubbed; yet not earthy, so as to foul the Water it is wash'd in.

Vitruvius observes, that Fossil-Sands dry sooner than those taken out of Rivers. Whence, he adds, the latter is fittest for the Insides, the former for the Outsides of a Building. He subjoins, that Fossil-Sand lying long in the Air, becomes earthy. *Palladio* takes notice, that of all Sands white ones are the worst; the reason is owing to their want of Asperity.

Mixing and blending of MORTAR;

M. *Felibien* observes that the Antient Masons were so very scrupulous herein, that the *Greeks* kept ten Men constantly employ'd for a long space of time, to each Bason, which render'd it of such prodigious Hardness, that *Vitruvius* tells us the Pieces of Plaister falling off from old Walls served to make Tables. *Felibien* adds, it is a Maxim among old Masons to their Labourers, that they should dilute with the Sweat of their Brow, *i.e.* labour it a long time, instead of drowning it with Water, to have done the sooner.

Besides the common *Mortar* used in laying of Stones, Bricks, *etc.* there are several other kinds. As,

White MORTAR, used in plaistering the Walls and Ceilings; made of Ox-Hair mixed with Lime and Water, without any Sand.

The MORTAR, used in making of Water-Courses, Cisterns, *etc.* is very hard and durable, made of Lime and Hog's Grease, sometimes mix'd with the Juice of Figs, and sometimes with liquid Pitch: after Application it is washed over with Linseed-Oil.

MORTAR *for Furnaces*, etc. is made with red Clay wrought in Water, wherein Horse-Dung and Chimney-Soot has been steep'd.

MORTAR *for Sun-Dials* on Walls may be made of Lime and Sand temper'd with Linseed Oil; or, for want of that, with scumm'd Milk. This will grow to the Hardness of a Stone.

In Buildings, one part of wash'd Soap-Ashes mixed with another of Lime and Sand, make a very durable *Mortar*.

MORTOISE, or MORITSE, in Carpentry, *etc.* a kind of Joint, wherein a Hole or Incision, of a certain Depth, is made in the Thickness of a Piece of Wood, which is to receive another Piece, call'd a *Tenon*. See TENON.

The Word is originally *French*. *Borel* derives it from *mordere*, to bite.

MOSAIC, MOSAIC-*Work*, or, as some chuse to call it, MUSAIC, from the *Latin Opus Musivum*; an Assemblage of little Pieces of Glass, Marble, precious Stones, *etc.* of various Colours, cut square and cemented on a Ground of Stuc, *etc.* imitating the natural Colours and Degradations of Painting.

In this sense *Mosaic Work* includes *Marquetry*, or *Inlaid Work*,

Vaneering, etc. But in the restrain'd sense of the Word, it only takes in Works of Stone, Metals and Glass; those of Wood being distinguish'd by the Name of *Marquetry* or *Inlaying*.

Others distinguish otherwise between *Mosaic*, and *Marquetry*. In that properly call'd *Mosaic*, the several Stones are all of the same Colour; and the Changes, and Diminutions of Colours, and Shades, are made by applying different Stones one on another, but all of the same Colour. *Marquetry*, on the contrary, consists of Stones of different Colours; and by these the several Colours, Shades, Degradations, *etc*. are express'd.

The Critics are divided as to the Origin and Reason of the Name *Mosaic*. Some derive it from *Mosaicum*, a Corruption of *Musaicum*, as that is of *Musivum*, as it was call'd among the *Romans*. *Scaliger* derives it from the *Greek*, and imagines the Name was given to this sort of Works, as being very fine, and ingenioùs. *Nebricensis* is of opinion they were so call'd, because *ex illis Picturis ornabantur Musea*.

Mosaic seems to have taken its Origin from *Paving*: the fine Effect and Use of Pavements, composed of Pieces of Marble of different Colours so well join'd together, as that when dry'd they might be polish'd, and the whole make a very beautiful and solid Body, which continually trodden upon, and washed with Water, was not at all damaged; gave the Painter the hint; who soon carry'd the Art to a much greater Perfection: so as to represent Foliages, Masques, and other Grotesque Pieces of various Colours, on a Ground of black or white Marble. In fine, observing the good Effect which this kind of Work had in Pavements, and finding that it resisted Water, they proceeded to line Walls therewith, and to make various Figures for the Ornament or their Temples and public Buildings.

But Nature not producing Variety of Colours enough for 'em in Marbles, to paint all kinds of Objects, they be thought of counterfeiting 'em with Glass and Metal Colours; which succeeded so well with 'em, that having given all manner of Teints to an infinite number of little Pieces of these two Matters, to counterfeit Stones of various Colours, in order to get more Colours; the Workmen arranged 'em with so much Art, that their *Mosaic* seem'd almost to dispute with Paintings. This way of representing Objects having this advantage, that it resists the Injuries of the Air as well as Marble itself; and even grows more beautiful with Time, which effaces all other kind of Painting.

But the Moderns are gone further, and setting aside Glass and Metals, as too mean Materials, have introduced, along with the finest Marbles, the richest of precious Stones, as Lapis, Agat, Cornelians, Emeralds, Turquoises, *etc*.

Of these three Kinds of *Mosaic*-Work, that of colour'd Glass and Metals is now little in use, tho of a surprizing Lustre and Durableness: of the other two, that of Marbles alone is in common Use; the *Mosaic* of precious Stone being so very dear, that the few Workmen who apply themselves to it, make little else but petty Works, as Ornaments for Altar-

Pieces, Tables for rich Cabinets, *etc.* Tho out of these must be excepted that sumptuous Chapel of the Dukes of *Tuscany*, which has been so long in hand, and which, if ever it be finish'd, will be a noble Monument of the Magnificence and Piety of those Princes, as well as of the Patience and Address of the Workmen employ'd therein.

We shall however enter into some Detail of the Manner of working in those three Kinds of *Mosaic*; to which we shall add a fourth much newer, yet equally ingenious with any of the rest, made with a kind of *Gypsum* or Talc, found in Stone-Quarries about *Paris*.

MOSAIC *Work of Glass.*

This Kind of *Mosaic* they begin with little Pieces of Glass, which they provide of as many different Colours as possible. To this end, the Glassmen's Furnaces being disposed, and their Pots or Crucibles full of the Matter of which Glass is made, or rather of Glass already made, they put what Colour or Dye they think fit in each Crucible, always beginning with the weakest, and augmenting the Strength of the Colours from Crucible to Crucible, till they come to the deepest Die, as in mixing of Colours on a Palette to paint in Oil. When the Glass is well boil'd, and all the Colours in their perfection, they take out the Glass hot, as it is, and lay it on a smooth Marble, flatting it down with another like Marble, and then cutting it into Slices of equal bigness, and about the thickness of an Inch and half. They then with an Instrument, which the *Italians* call *Bocca di Cane*, make other Pieces square, and others of different Figures and Sizes, as occasion requires; these they dispose orderly in Cases; as in Painting *in Fresco*, 'tis usual to range all the different Teints in Shells, and according to their Colour.

If 'tis desir'd to have Gold, either in the Ground of the Painting, or in the Ornaments, or the Draperies, they take some of the Pieces of Glass, form'd and cut in the manner just mention'd. These they moisten on one side with Gum-Water, and afterwards lay 'em over with Gold-Leaf. They then put this Piece, or several Pieces at a time, on a Fire-shovel, which they place in the mouth of the Furnace, after having first cover'd them with another hollow Piece of Glass. Here they continue till such time as they become red-hot; after which the Shovel is drawn out, all at once, and the Gold becomes so firmly attach'd to the Glass, that it will never afterwards leave it.

Now, to apply these several Pieces, and out of 'em to form a Picture, they first make a Cartoon, or Design; this they transfer on the Ground or Plaister, by calquing, as in Painting in Fresco.

As this Plaister is to be laid thick on the Wall, it will continue fresh and soft a considerable time, so that there may be enough prepar'd at once to serve three of four days. It is compos'd of Lime made of hard Stone, with Brick-dust very fine, Gum-Tragacanth, and Whites of Eggs; when it is thus prepar'd, and laid on the Wall, and made the Design of what is to be represented, with Plyers they take out the little Pieces of

Glass, ranging 'em one after another, and still keeping strictly to the light Shadow, different Teints and Colours represented in the Design before; pressing or flatting 'em down with a Ruler, which serves both to sink 'em within the Ground, and to render the Surface even.

Thus, in a long time, and with an infinite deal of trouble, they finish the Work, which is still the more beautiful, as the Pieces of Glass are more uniform, and rang'd at more equal heights. Some of these are executed with so much Justness, that they appear as smooth as a Table of Marble, and as finish'd and masterly as a Painting in Fresco; with this advantage, that they have a fine Lustre, and will hold almost for ever.

The finest Works of this kind, that have descended to us, and those whereon the Moderns have retriev'd the Art, almost lost, are those of the Church of St. *Agnes*, formerly the Temple of *Bacchus* at *Rome*; at *Pisa*, *Florence*, and other Cities of *Italy*. The most esteem'd among the Works of the Moderns, are those of *Joseph Pine*, and the Chevalier *Lanfranc*, in the Church of St. *Peter* at *Rome*. There are very good ones likewise at *Venice*.

MOSAIC *Work of Marble, and precious Stones.*

These two Kinds of *Mosaics* bear so near a relation to each other, as to the Manner of Working, that to avoid Repetition, we shall give 'em both under one; observing, by the way, wherein the one differs from the other, either in the sawing or the ranging of the Stones.

Mosaic of Marble is used in large Works, as in Pavements of Churches, Basilics, and Palaces; and in the Incrustation and Vaneering of the Walls of the same Edifices. As to that of Stones, especially *precious Stones*, 'tis only used in small Works, as before observ'd.

The Ground of *Mosaic*-Works wholly Marble, is ordinarily a massive Marble, either white or black. On this Ground the Design is cut with a Chissel, having been first calqued. When 'tis dug of a sufficient Depth, *i.e.* an Inch or more, 'tis fill'd up with Marble of a proper Colour, first contourn'd, or fashion'd to the Design, and reduc'd to the Thickness of the Indentures, with various Instruments. To make the Pieces, thus inserted into the Indentures, hold, whose several Colours are to imitate those of the Design, they use a Stuc, compos'd of Lime and Marble Dust; or a Mastic, which each Workman prepares differently: after which, the Work is half polish'd with a soft kind of Stone.

The Figures thus mark'd out, the Painter, or Sculptor himself draws, with a Pencil, the Colours of the Figures not determin'd by the Ground, and in the same manner makes Strokes or Hatchings, in the Places where Shadows are to be; and when he has engrav'd with the Chissel all the Strokes thus drawn, he fills them up with a black Mastic, compos'd partly of *Burgundy-Pitch*, pour'd on hot; taking off, afterwards, what is superfluous, with a piece of soft Stone or Brick, which with Water and beaten Cement, takes away the Mastic, polishes the Marble, and renders the whole so even, one would imagine it only consisted of one Piece: 'Tis this kind of *Mosaic* we see in the pompous Church of the Invalids at *Paris*, and the fine Chapel at *Versailles*; and wherewith some entire Apartments of that Palace are incrustated.

For Mosaic Work of precious Stones; there are required other and more delicate Instruments than those used in Marble; as Wheels, Drills, Tin-Plates, *etc.* used by Lapidaries, and Engravers on Stone.

As none but the richest Marbles and Stones enter this Work, to make 'em go the further, they are sawn into the thinnest Leaves imaginable, scarce exceeding half a Line in Thickness; the Block to be saw'd, is fasten'd firmly with Cords, on the Bench, only rais'd a little on a piece of Wood, one or two Inches high. Two Iron Pins, which are on one side the Block, and which serve to fasten it, serve also to direct the Saw, which with Pieces thus saw'd, are put into a Vice contrived for the purpose; and with a kind of Saw or Bow made of fine Brass Wire, bent on a Piece of springy Wood, together with Emery steep'd in Water, the Leaf is gradually fashion'd, by following the Strokes of the Design made on Paper, and glued on the Piece.

When there are Pieces enough fasten'd to form an entire Flower, or some other Part of the Design, they are apply'd. The Ground that supports this Mosaic, is usually of Free Stone. The Matter wherewith the Stones are join'd together, is a Mastic, or kind of Stuc, laid very thin on the Leaves as they are fashion'd; and the Leaves in this State, apply'd with Plyers. If any Contour, or Side of a Leaf, be not either rounded enough, or squared enough, to fit the Place where it is to be used, when 'tis too large, 'tis brought down with a Brass File or Rasp, and when too small, is managed with a Drill, and other Lapidary-Instruments.

Manner of making MOSAIC *Work of Gypsum.*

The Gypsum is a kind of coarse Talc, or a shining transparent Stone, found in the Quarries of *Montmartre* near *Paris*, among the Stones thence dug to make the Plaister of *Paris*: 'tis different from the Plaister, but retains the Name the *Romans* gave the Plaister, *viz. Gypsum*. Of this Stone, calcined in a Kiln, and beaten in a Mortar, and pass'd thro a Sieve, they make a kind of artificial Marbles, imitating precious Stones, and of these compose a kind of *Mosaic* Work, which comes little short either of the Durableness or Vivacity of the natural Stones; and which has this advantage, that it admits of continu'd Pieces, or Paintings of entire Compartiments, without any Joining visible.

Some make the Ground of Plaister of *Paris*, others of Free-Stone: if the former, 'tis spread in a wooden Frame, of the length and breadth of the intended Work, and about an Inch and half thick. This Frame is so contriv'd, as, the Tenons being only join'd to the Mortoises by single Pins, they may be taken asunder, and the Frame be dismounted when the Plaister is dry. This Frame they cover on one side with a strong Linnen Cloth, nail'd all around, which being plac'd horizontally, with the Linnen at bottom is fill'd with Plaister, pass'd thro a wide Sieve. The Plaister being half dry, the Frame is set perpendicular, and left till it be quite dry; then taken out, by dismounting the Frame. In this *Mosaic*, the Ground is the most important Part.

Now to prepare the sifted Gypsum to be apply'd on this Ground, they dissolve and boil it in the best *English* Glue, and after mixing with it the Colour it is to bear, the whole is work'd up together into the ordinary Consistence of Plaister; and then taken and spread on the Ground, five or six Inches thick. It must be observ'd, that if the Work be such, as that Mouldings are requir'd, they are form'd with Gouges and other Instruments.

'Tis on this Plaister, thus colour'd like Marble or precious Stone, and which is to serve as a Ground to a Work either of Lapis, Agat, Alabaster, or the like, that the Design to be represented is drawn; having been first pounced or calqued. To hollow or impress the Design, they use the same Instruments with the Sculptors; the Ground whereon they are to work, not being much less hard than Marble itself. The Cavities thus made in the Ground, are fill'd up with the same Gypsum boil'd in Glue, only differently colour'd; and thus are the several Colours of the Original represented. To have the necessary Colours and Teints at hand, they temper Quantities of the Gypsum with the several Colours, in little Pots. When the Design is thus fill'd, and render'd visible, by half polishing it with Brick or soft Stone; they go over it again, cutting such Places as are either to be weaker, or more shadow'd, and filling 'em with Gypsum; which is repeated till all the Colours, added one after another, represent the Original to the Life. The Work being finish'd, is scower'd with soft Stone, Sand and Water; then with Pumice-stone, and lastly, polish'd with a wooden Mullet and Emery. Lastly, the Lustre is given it, by smeering it over with Oil, and rubbing it a long time with the Palm of the Hand; which gives it a Lustre nothing inferior to that of natural Marble.

If 'tis only requir'd to make a variegated Table, or other Work of several Colours; without *Mosaic* Figures, the Process is somewhat different. To this end, they only prepare separately, in large Bowls, as many different Colours as Nature shews in the Marble to be imitated, and after incorporating 'em with the Gypsum and Glue-Water, they take a Truel-full of each, and dispose 'em in a Trough, without any order; then without mingling 'em, and only by cutting or crossing the Gypsum of each Truel once or twice with each of the rest, they give 'em that beautiful Confusion, which makes the Value of natural Marbles: Of these they then make their Tables, or lay 'em in a Mold, according to the Work to be done.

As to *Marquetry or Inlaid-Work*, the Antients were well acquainted with it, and used it for the adorning of their Beds, Tables, and other Moveables; using for this purpose Ivory, and the richest Woods. But Friar *John* of *Verona* seems to have contributed the most to its Perfection, by discovering the Secret of dying Woods of all Colours and Degrees; by which means he was enabled to imitate Painting, and even to represent

Architecture in Perspective.

They begin with sawing their Woods into Leaves, of the Thickness of one or two Tenths of an Inch; then take Pieces of the Design they are to follow, and fasten 'em to those Leaves, and with a little Steel saw the Contour of the Design. All that is necessary being taken off with the Saw, they give the Shadow to those Places that require it, by placing the Piece in a hot Sand, or otherwise, with the Direction necessary to shadow it more or less. This done, they lay each Piece in its place, on a Ground of another Wood, as dry Oak, and there fasten them with strong Glue.

There are two other Branches of *Mosaic* Work; the one call'd *Damaskeening*, or *Damask-Work*, consisting in an Assemblage of Gold or Silver Threads, of which are sometimes form'd flat Works, and sometimes Basso-Relievos

The other is call'd *Shell-Work*, consisting of Shells, artificial Congelations, Petrifications, *etc.* used in Grottos.

MOSQUE, among the *Mahometans*, is a Temple set apart for the Exercises of Religion. See TEMPLE.

There are Royal *Mosques* founded by the Emperors, as *Solimania* and *Velidea* at *Constantinople*; and Private *Mosques* founded by Mufties, Viziers, Baffas, *etc.* they are built like large Halls, with Isles, Galleries, and Domes; and are adorned on the inside with Compartiments, and Pieces of *Arabesque* Work. On one side is always found a Pool with several Cocks.

The *Turks* have converted most of the Christian Churches into *Mosques*. At the top is always placed a Crescent.

The Word comes from the *Turkish Meschit*, which properly signifies a Temple built of Wood, such as the *Turks* first used. Hence the *Spaniards* derive their *Mescheta*, and the *Italians Moscheta*, and the *French* and *English Mosque. Borel* derives the Word from the *Greek Calf*, because of the frequent mention of a Cow in the *Alcoran*. Others, and with the greatest Appearance of Reason, derive it from the *Arabic Maschiad*, a Place of Worship.

MOULD, in the Mechanic Arts, *etc.* a Cavity artfully cut, with design to give its Form, or Impression to some softer Matter apply'd therein.

Moulds are Instruments of great Use in Sculpture, Foundery, *etc.*

The Workmen employ'd in melting the Mineral or Metallic Glebe dug out of Mines, have each their several *Mould*, to receive the melted Metal as it comes out of the Furnace; but different according to the diversity of Metals and Works. In Gold-Mines, they have *Moulds* for Ingotts. In Silver-Mines for Bars. In Copper and Lead-Mines for Pigs or Salmons: On Tin-Mines for Pigs and Ingotts: And in Iron-Mines for Sows, Chimney-Backs, Anvils, Caldrons, Pots, and other large Utensils and Merchandises of Iron, which are here cast, as it were, at first hand.

The *Founders of large Works*, as Statues, Bells, Guns, and other brazen Works, have *Moulds* of Wax, supported within side by what they call a *Core*, and cover'd without side with a Cap or Case. 'Tis in the Space which the Wax took up, which is now melted to leave it free, that the liquid Metal runs, and the Work is form'd; being carried thither thro' a great Number of little Canals which cover the whole *Mould*.

MOULDS *of Founders of small Works* are like the Frames of Coiners. 'Tis in these Frames, which are likewise fill'd with Sand, that their several Works are fashion'd; into which, when the two Frames whereof the *Mould* is composed, are rejoin'd, the melted Brass is run.

The Glasiers have two kinds of *Moulds*, both serving to cast their Lead. In the one they cast the Lead into long Rods or Canes fit to be drawn thro' the Vice, and the Grooves formed therein. This they sometimes call *Ingot-Mould*.

In the other they *Mould* those little Pieces of Lead a Line thick, and two Lines broad, fasten'd to the Iron-Bars. These may be also cast in the Vice.

The Goldsmiths use the Bones of the Cuttle-Fish to make *Moulds* for their small Works; which they do by pressing the Pattern between two Bones, and leaving a Jet or Hole to convey their Silver thro' after the Pattern has been taken out.

MOULD, among Masons, a Piece of hard Wood or Iron, hollowed within side, answerable to the Contours of the Mouldings or Cornishes, *etc.* to be form'd. This is otherwise call'd Calliber.

MOULDS among Plumbers, the Tables whereon they cast their Sheets of Lead. These they sometimes call simply *Tables*. Besides these, they have other real *Moulds* wherewith they cast Pipes without Soldering.

MOULDS among Glass-Grinders, are Wooden-Frames whereon they make the Tubes wherewith they fit their Perspectives, Telescopes, and other Optic Machines. These *Moulds* are Cylinders, of a Length and Diameter according to the Use they are to be apply'd to, but always thicker at one end than the other, to facilitate the sliding. The Tubes made on these *Moulds* are of two kinds; the one simply Paste-Board and Paper; the other of thin Leaves of Wood join'd to the Paste-Board. To make of these Tubes to draw out, only the last or innermost is form'd on the *Mould*; each Tube made afterwards serving as a *Mould* to that which is to go over it; but without taking out the *Mould* from the first.

MOULDS among Tallow-Chandlers are of two kinds: The first for the common dipp'd Candles, being the Vessel wherein the melted Tallow is disposed, and the Wick dipp'd. This is of Wood of a triangular Form, and supported on one of its Angles, so that it has an opening of near a Foot a-top. The other used in the Fabric of *Mould* Candles, is of Brass, Pewter, or Tin: Here each Candle has its several *Mould*.

MOULD among Gold-Beaters, a certain Number of Leaves of Velom, or pieces of Guts, cut square, of a certain size, and laid over one another, between which they put the Leaves of Gold and Silver which they beat on the Marble with the Hammer. They have four kinds of *Moulds*; two whereof are of Velom, and two of Gut. The Smallest of those of Velom consists of forty of fifty Leaves, the largest contains an hundred. For the others, each contains five hundred Leaves.

The *Moulds* have all their several Cases, consisting of two pieces of Parchment, serving to keep the Leaves of the *Mould* in their Place, and prevent their being disorder'd with beating.

MOULDINGS, or *Ornaments*, in Architecture, are Projectures beyond the Naked of a Wall, Column, Wainscot, *etc.* the Assemblage whereof forms *Cornishes*, *Door-Cases*, and other Pieces of Architecture.

Some *Mouldings* are crowned with a Fillet; others are without, as the Doucine, Talon, Ovolo, Torus, Scotia, Astragal, Gula, Corona, *etc.* See each under its proper Article.

Again, some are adorned with Sculpture, either hollow'd, or in Relievo.

Mouldings are in Architecture what Letters are in Writing. By the various Dispositions and Combinations of *Mouldings*, may be made an infinite Number of different Profiles for all sorts of Orders and Compositions, Regular and Irregular; and yet all the kinds of *Mouldings* may be reduced to threee, *viz. Square*; *Round*; and *Mixed*, *i.e.* composed of the other two.

For this Reason, those who invented the *Gothic* Architecture, resolving to recede from these perfect Figures, and affecting to use others less perfect, to distinguish their Architecture from the Antique, introduced a new Set of whimsical *Mouldings* and Ornaments.

Regular *Mouldings* are either large, as *Doucines*, *Ovolos*, *Gulas*, *Talons*, *Torus's*, *Scotias*, etc. or small, as *Fillets*, *Astragals*, *Conges*, etc.

MOUND, a Term used for a Bank, Rampart, or other Fence, particularly of Earth.

MOYENAU, in Fortification, is a small flat Bastion, commonly plac'd in the middle of an over-long Curtain; where the Bastions at the Extremities are not well defended from the small Shot, by reason of their Distance: This Work is proper for lodging a Body of Musqueteers, to fire upon the Enemy from all sides.

MUNIMENTS, or MINIMENTS, the Evidences or Writings, whereby a Man is enabled to defend the Title of his Estate. See MUNIMENT-*House*.

Wangford says, the word *Muniment* includes all manner of Evidence.

MUNIMENT-*House*, a little, strong Apartment in Cathedral and Collegiate Churches, Castles, Colleges, or the like, destined for keeping the Seal, Evidences, Charters, *etc.* of such Church, College, *etc.* call'd *Muniments*, or *Miniments*.

The Word is form'd of the *Latin Munio*, I defend.

MUNIONS, in Architecture, are the short upright Posts or Bars that divide the several Lights in a Window-Frame. See WINDOW.

MURAL, something belonging to a *Wall*; which the *Latins* call *Murus*. See WALL.

MURAL *Arch* is a Wall, or walled Arch placed exactly in the Plane of the Meridian, *i.e.* upon the Meridian-Line; for the fixing of a large Quadrant, Sextant, or other Instrument, to observe the Meridian Altitudes, *etc.* of the heavenly Bodies.

Tycho de Brabe was the first who used a *Mural-Arch* in his Observations; after him Mr. *Flamstead, de la Hire*, etc. used the same means.

MURING, the Walling, or raising the Walls of a Building. See WALL and WALLING.

MUTILATION, the retrenching, or cutting away of any Member of the Body.

The use of the Word is also extended to Statues and Buildings, where any part is wanting, or the Projecture of any Member is broke off.

MUTULE, MUTULUS, in Architecture, an Ornament, otherwise call'd *Modillion*. See MODILLIONS.

The only difference between *Mutule* and *Modillion* consists in this, that the former is used in speaking of the *Doric* Order, and the latter in the *Corinthian*. See DORIC, *etc.*

The *Mutules* in the *Doric* answer to the *Triglyphs*; whence some make *Guttæ*, or Drops to hang. See GUTTÆ.

N

NAILS *in Building*, etc. are little Metalline Members, serving to bind or fasten the Parts together, *etc.*

The several Kinds of *Nails* are very numerous. As *Back* and *Bottom Nails*; made with flat Shanks to hold fast, and not open the Wood: *Clamp Nails*, those proper to fasten the Clamps in Buildings, *etc. Clasp Nails*, whose Heads clasping and sticking into the Wood, render the Work smooth, so as to admit a Plane over it: They are of two Kinds, *viz. long*, proper for fine Buildings of Fir, *etc.* and *strong*, fit for Oak, and other hard Wood. *Clench Nails*, those used by Boat, Barge, *etc.* Builders; proper for boarded Buildings, that are to be taken down, because they will drive without splitting, and draw without breaking. *Clout Nails*, those ordinarily used for nailing on of Clouts to Axle-Trees. *Deck Nails*, those proper for fastning of Decks in Ships, doubling of Shipping, and Floors laid with Planks. *Dog Nails*, proper for fastening of Hinges to Doors, *etc*. *Flat Points* are of two kinds, *viz*. *Longs*, much used in Shipping, and proper where there is occasion to draw, and hold fast, yet no necessity of Clenching. *Jobent Nails*, those commonly used to nail thin Plates of Iron to Wood. *Lead Nails*, used to nail Lead, Leather and Canvas to hard Wood. *Port-Nails*, commonly used to nail Hinges to the Ports of Ships. *Pound Nails*, are four-squre in the Shank; much used in *Norfolk*, *Suffolk* and *Essex*, tho' scarce elsewhere, except for paling. *Ribbing Nails*, used to fasten the Ribbing, to keep the Ribs of Ships in their place in Building. *Rose Nails*, are drawn four-square in the Shank, and commonly in a round Tool. *Rother Nails*, chiefly used to fasten Rother-Irons to Ships. *Round-head Nails*, proper to fasten in Hinges, or other Uses, where a neat Head is required. *Scupper Nails*, much used to fasten Leather and Canvas to Wood. *Sharp Nails*, much used, especially in the *West-Indies*, made with sharp Points, and flat Shanks. *Sheething Nails*, used to fasten Sheething-Boards to Ships: the Rule for their Length, is to have them full three times as long as the Board is thick. *Square Nails*, of the same shape as sharp *Nails*, chiefly used for hard Woods. *Tacks*, the *smallest* of *Nails*, are to fasten Paper to Wood; *Middling*, for Wool-Cards and Oars; *Larger*, for Upholsterers, and Pumps.

In Lathing, 500 *Nails* are ordinarily allowed to a Bundel of five-foot Laths. In Flooring, 200, *i.e.* 240 are allowed for a Square of Flooring.

Nails are said to be toughened, when too brittle, by heating them in a Fire-Shovel, and putting some Tallow or Grease among them.

NAKED, in Architecture. The NAKED *of a Wall*, etc. is the Surface, or Plain, from whence the Projectures arise; or which serves as a Ground to the Projectures. See PROJECTURE.

Thus, we say, a Pilaster ought to exceed the *Naked* of the Wall by so many Inches; and that the Foliages of a Capital ought to answer to the *Naked* of the Column. See COLUMN, *etc.*

NAVE, in Architecture, *Navis Ecclesiæ*, the Body of a Church; or the Place where the People are disposed; reaching from the Rail or Balluster of the Choir to the chief Door. The Antient *Greeks* call'd it *Pronaos*.

The *Nave* of a Church belongs to the Parishoners; 'tis they are to repair it, *etc.*

Baldus derives the Word from the *Greek* Temple; which *Salmasius* brings from *Ship*; by reason the Vault or Roof of a Church bears resemblance to a Ship.

NERVES, in Architecture, are the Mouldings of the Projecting-Arches of Vaults; or those arising from the Branches of Ogives, and crossing each other diagonally in *Gothic* Vaults, serving to separate the Pendentives. See VAULT, *etc.*

NEWEL, in Architecture, the upright Post, which a Pair of Winding-Stairs turn about; or that part of the Stair-Case which sustains the Steps. See STAIR.

The *Newel* is a Cylinder of Stone, which bears on the Ground, and is form'd by the ends of the Steps of the Winding-Stairs.

There are also *Newels* of Wood, which are pieces of Wood placed perpendicularly, receiving the Tenants of the Steps of Wooden-Stairs into their Mortices, and wherein are fitted the Shafts and Rests of the Stair-Case, and the Flights of each Story.

NICHE, in Architecture, a Cavity, or Sinking, in the Thickness of a Wall; to place a Figure, or Statue in.

The larger *Niches* serve for Groups of Figures; the small ones for single Statues, sometimes only for Busts.

The Word comes from the *Italian Nicchio*, Sea-Shell; in regard the Statue is here inclosed as in a Shell; or, perhaps, by reason of the Shell wherewith the Top of some of them is adorn'd.

Great Care must be taken to proportion the *Niches* to the Figures; and that the Pedestals of the Figures be proportion'd to the *Niches*.

Round NICHE, is that whose Plan and Circumference are Circular.

Square NICHE, That where they are Square.

Angular NICHE, That form'd in a Corner of the Building.

Ground NICHE, That which, instead of bearing on a Massive, has its Rise from the Ground; as the *Niches* of the *Portico* of the Pantheon at *Rome*: These are otherwise call'd Tribunals.

Niches are sometimes made with Rustic-work, sometimes with Shell-work, and sometimes of Crail'd, or Arbor-work.

NUDITIES, in Painting, and Sculpture, those Parts of a human Figure not cover'd with any Drapery; or those Parts where the Carnations appear. See CARNATION.

NUEL, or rather *Newel* of a Stair-Case, See NEWEL and STAIRS.

NURSERY, in Gardening, is generally used in the same sense as *Seminary*, *viz.* for a Seed-plot for the raising of young Trees, or Plants.

Some Authors, however, make a difference beween the two; holding *Nursery*, properly, not to be a place wherein Plants are sown; but a place for the reception and rearing of young Plants, which are removed, or transplanted hither from the Seminary, *etc.*

Mr. *Lawrence* recommends the having several *Nurseries*, for the several Kinds of Trees: One for tall *Standards*; *viz.* Apples, Ashes, Elms, Limes, Oaks, Pears, Sycamores, *etc.* Another for *Dwarfs*, *viz.* such as are intended for Apricocks, Cherries, Peaches, Plumbs, *etc.* And a third for *Ever-Greens.*

The *Nursery for Standards* should be in a rich, light Soil; sown, with the proper Seeds, in *October*, or *November.* For *Apples* and *Pears*, Crab- and Wild-Pear Kernels, are to be preferred for Stocks: *Elms* and *Limes* are to be raised from planted Suckers: *Walnuts* to be sown with the green Shell upon 'em, to preserve them from Mice. This *Nursery*, if it be well managed and weeded for two Years, the Crabs and Pears will be fit for Grafting and Inoculating the third Year.

Firs and Pines are to be raised from those little Seeds, taken out of their large Apples.

The *Nursery for Dwarfs* does best by itself, that it may not be over-top'd by taller Trees. Stones of Apricocks and Peaches are not proper to raise those Trees; but in lieu thereof, sow the Stones of the Pear-Plumb, Mussel or *Bonum Magnum* Plumb; which prove better and more lasting than the former. For Stocks for all sorts of Cherries, black Cherry-Stones do best.

Mr. *Mortimer* directs all Stone-Fruit to be sown quickly after gathering; for that if they be kept, they will be two Years e'er they come up. Add, that if they have not all the Moisture of the Winter to rot the Shells, the Kernel will scarce come up at all.

To furnish the *Nursery of Ever-Greens*, the several sorts of Seeds or Berries, as Yew, Holly, Juniper, *etc.* are to be put in so many distinct Pots or Boxes, with fine Mould over them, and thus buried for a year; after which, they are to be taken out and sown.

If they were to be sown when gather'd like other Seeds, they would not come up the first year, nor grow so kindly.

O

OBELISK, OBELISCUS, a quadrangular Pyramid, very slender, and high; rais'd as an Ornament, in some public Place, to shew some Stone of enormous Size, and frequently charg'd with Inscriptions and Hieroglyphics.

The difference between *Obelisks* and Pyramids, according to some, consists in this, that the latter have large Bases, and the former very small ones.

Tho *Cardan* makes the difference to consist in this, that *Obelisks* are to be all of a piece, or consist of a single Stone, and Pyramids of several.

The Proportions of the Heighth and Thickness are nearly the same in all *Obelisks*; that is, their Height is nine, or nine and a half, sometimes ten times their Thickness; and their Thickness, or Diameter a-top is never less than half, nor greater than three fourths of that at bottom.

This kind of Monument appears very antient; and we are told was first made use of to transmit to Posterity the principal Precepts of Philosophy, which were engraven in hieroglyphical Characters hereon.

In After-times they were used to immortalize the Actions of Heroes, and the Memory of Persons beloved.

The first *Obelisk* we know of, was that raised by *Ramises*, King of *Egypt*, in the time of the *Trojan* War. It was 40 Cubits high, and, according to *Herodotus*, employ'd 20,000 Men in the Building. *Phius*, another King of *Egypt*, raised one of 45 Cubits; and *Ptolemy Philadelphus* another of 88 Cubits, in memory of *Arsince*.

Augustus erected one at *Rome* in the *Campus Martius*, which serv'd to mark the Hours on a horizontal Dial drawn on the Pavement.

F. *Kircher* reckons up 14 *Obelisks* celebrated above the rest, *viz.* that of *Alexandria*, that of the *Barberins*, those of *Constantinople*, of the *Mons Esquilinus*, of the *Campus Flaminius*, of *Florence*, of *Heliopolis*, of *Ludovisio*, of S. *Mabut*, of the *Medici*, of the *Vatican*, of M. *Cœlius*, and that of *Pamphila*.

The *Egyptian* Priests call'd *Obelisks* the *Sun's Fingers*, because serving as Styles, or Gnomons to mark the Hours on the Ground. The *Arabs* still call them *Pharaoh's Needles*: whence the *Italians* call them *Aguglia*; and the *French*, *Aiguilles*.

Borel derives the Word from the *Greek*, a *Broach*, Spindle, Point, or a kind of long Javelin.

Pliny says, the *Egyptians* cut their *Obelisks* in form of Sun-Beams; and that in the *Phœnician* Language, the word *Obelisk* signifies *Ray*.

OBSERVATORY, a Place destined for *observing* the Heavenly Bodies; or, a Building usually in form of a Tower, raised on some Eminence, for making of Astronomical Observations.

The more celebrated *Observatories*, are, 1. The *Greenwich Observatory*, built in 1676, by Order of King *Charles* II. at the solicitation of Sir *Jonas Moor* and Sir *Christopher Wren*; and furnished with the most accurate Instruments by the same; particularly a noble Sextant of 7 Feet Radius, with Telescope-Sights.

The Person to whom the Province of Observing was first committed, was Mr. *J. Flamstead*; a Man, who, as Dr. *Halley* expresses it, seem'd born for the Employment. For the space of fourteen Years, with unwearied Pains he watch'd the Motions of the Planets; chiefly those of the Moon, as was given him in charge; that a new Theory of that Planet, exhibiting all her Irregularities, being found, the Longitude might thence be determined.

In the Year 1690, having provided himself of a Mural-Arch of 7 feet Diameter, well fix'd in the Plane of the Meridian, he began to verify his Catalogue of the fixed Stars, which hitherto depended altogether on the Distances measured with the Sextant, after a new and very different Manner, *viz.* by taking the Meridian-Altitudes, and the Moments of Culmination, or the right Ascension and Declination.

This Instrument he was so pleas'd with, that he laid the use of the Sextant almost wholly aside. Thus was the Astronomer Royal employ'd for thirty Years; in the Course of which Time, nothing had appear'd in public, worthy so much Expence and Preparation: So that the Observer seem'd rather to have been employ'd for his own sake, and that of a few Friends, than for the Public; tho it was notorious, the Observations that had been made were very numerous, and the Papers swell'd to a great Bulk.

This occasion'd Prince *George* of *Denmark*, in the Year 1704, to appoint certain Members of the *Royal Society*, *viz.* the Honourable *Fr. Roberts*, Sir *C. Wren*, Sir *I. Newton*, Dr. *Gregory*, and Dr. *Arbuthnot*, to inspect *Flamstead's* Papers, and chuse out of them such as they should think fit for the Press; purposing to print them at his own Expence: But the Patron of the Work dying, e'er the Impression was half finished, it lay still for some time; till at length it was resum'd by order of Queen *Anne*; and the Care of the Press committed to Dr. *Arbuthnot*; and that of Correcting, and supplying the Copy, to Dr. *Halley*.

Such was the Rise and Progress of the *Historia Cœlestis*; the principal Part whereof is the Catalogue of fixed Stars, call'd also the *Greenwich Catalogue*.

The *Greenwich Observatory* is found by very accurate Observation to lie in 51°, 28', 30", North Latitude.

2. The *Paris Observatory* built by the late *Louis* XIV. in the *Fauxbourg St. Jaques*.

It is a very singular, but withall a very magnificent Building; the Design of M. *Perrault*. 'Tis 80 feet high, and a-top is a Terras. 'Tis here M. *de la Hire* has been employ'd.

The Difference in Longitude between this and the *Greenwich Observatory* is 2°, 20' West.

In it is a Cave, or Cellar, 170 feet descent, for Experiments that are to be made far from the Sun, *etc.* particularly such as relate to Congelations, Refrigerations, Indurations, Conservations, *etc.*

3. *Tycho Brahe's* Observatory was in the little Island *Ween*, or *Scarlet Island*; between the Coasts of *Schonen* and *Zeland*, in the *Baltic*.

It was erected and furnished with Instruments at his own Expence; and call'd by him *Uraniburg*.

Here he spent twenty Years in observing the Stars. The Result is his Catalogue.

Mr. *Gordon*, in *Phil. Trans.* observes, that this was none of the fittest Places for some kind of Observations, particularly the Risings and Settings; as lying too low, and being land-lock'd on all the Points of the Compass but three; and the Land-Horizon exceedingly rugged and uneven.

4. *Pekin Observatory.* Father *Le Compte* describes a very magnificent *Observatory* erected and furnished by the late Emperor of *China*, in his Capital, at the Intercession of some Jesuit Missionaries, chiefly Father

Venbiest, whom he made his chief Observer.

The Instruments are exceedingly large; but the Divisions less Accurate, and the Contrivance, in some respects, less Commodious than those of the *Europeans*. The chief are an Armillary, Zodiacal Sphere of 6 *Paris* feet Diameter, an Equinoctial Sphere 6 feet Diameter, an Azimuthal Horizon 6 feet Diameter, a large Quadrant 6 feet Radius, a Sextant 8 feet Radius, and a Celestial Globe 6 feet Diameter.

OCTAGON, in Fortification, is a Place that has eight Bastions.

OCTOSTYLE, in the antient Architecture, the Face of a Building or Ordonnance, containing 8 Columns.

The 8 Columns of the Octostyle may either be disposed in a right Line, as in the *Pseudodiptere* Temple of *Vitruvius*, and in the *Pantheon*; or in a Circle, as in the round *Monoptere* Temple of *Apollo Pythius* at *Delphi*, etc.

ODEUM, among the Antients, was a Place destined for the Rehearsal of the Music to be sung on the Theatre.

ODEUM was also used for other Buildings that had no relation to the Theatre: *Pericles* built an *Odeum* at *Athens*, where musical Prizes were contended for. *Pausanias* says, that *Herod* the *Athenian* built a magnificent *Odeum* for the Sepulchre of his Wife.

The *Latin* Writers also use the Word *Odeum* for the Choir of a Church. Grammarians are exceedingly in the dark about the signification of the Word among the Antients.

OFF-SETS, in Surveying, are Perpendiculars let fall, and measur'd from the Stationary-Lines, or the Lines between one Station and another, to the Hedge, Fence, or Extremity of the Inclosure.

OFFICES, with regard to Architecture, are all the Lodges and Apartments that serve for the necessary Services and Occasions of a Great House or Palace; particularly those which have a relation to Eating; as Kitchens, Parterres, Brew-Houses, Confectionaries, Fruiteries, Granaries, *etc.* as also Wood-houses, Equeries, *etc.* See HOUSE, BUILDING, *etc.*

The *Offices* are commonly in the Bassecours; sometimes they are sunk under-ground, and well vaulted, *etc.*

OGEE, or O G, as it is frequently wrote, or OGIVE, in Architecture, a Moulding, consisting of two Members, the one concave, and the other convex; the same with *Cymatium*. See CYMATIUM.

Vitruvius makes each Member a Quadrant of a Circle; *Scamozzi*, and some others, make them somewhat flatter, and strike them from two equilateral Triangles.

Its Form bears some resemblance to that of an S.

OGIVE is also used for an Arch, or Branch of a *Gothic* Vault, which, in lieu of being Circular, passes diagonally from one Angle to another, and forms a Cross between the other Arches which makes the side of the Square, whereof the Arches are Diagonals. See ARCH and VAULT.

The middle, where the *Ogives* cut or cross each other, is call'd the Key, which is sometimes cut in form of a Rose, or a *Cul de Lampe*. The Members or Mouldings of the *Ogives* are call'd Nerves, Branches, or Reins; and the Arches which separate the *Ogives*, double Arches. See VAULT.

OLITORY, a Kitchen-Garden, or a Garden of Herbs, Roots, *etc*. for Food. See GARDEN.

OPENING *Flank*, in Fortification, is that part of the Flank which is cover'd by the Orillon.

OPENING of *Trenches*, is the first Breaking of Ground by the Besiegers, in order to carry on their Approaches towards a Place.

ORANGERY, a Gallery in a Garden, or Parterre; expos'd to the South; but well closed with a Glass Window, to preserve the Oranges in, during the Winter Season.

Orangery is also used for the Parterre, where the *Oranges* are expos'd in kindly Weather. The *Orangery* of *Versailles* is the most magnificent that ever was built.

ORCHESTRA, in the Drama, the lower part of the antient Theatre; made in form of a Semi-Circle, and surrounded by the Seats.

It was so call'd because in the *Grecian* Theatres it was a Place where they kept their Balls; which signifies *to dance*.

The *Orchestra* among the *Greeks* made a part of the Scene; but on the *Roman* Theatres, none of the Actors went down to the *Orchestra*, which was taken up with Seats for the Senators; answering nearly to the *Pit* in our Theatre.

ORDER, in Architecture, a System of the several Members, Ornaments, and Proportions of Columns, and Pilasters: Or, a regular Arrangement of the projecting Parts of a Building; especially those of a Column: so as to form one beautiful Whole. See COLUMN.

Perrault defines an *Order* to be what is regulated by the Ordonnance, when it prescribes the Proportions of entire Columns, and determines the Figure of certain Parts proper to them, according to the different Characters their different Uses and Ends require.

M. *le Clerc* defines an *Order* to be a Column charg'd with an Entablature, and supported on a Pedestal.

The Definitions *Vitruvius*, *Barbaro*, *Scamozzi*, etc. give of *Order*, are so obscure, that it were vain to repeat them: without dwelling, therefore, on the Definition of a Word, which Custom has establish'd; 'tis sufficient to observe, that there are five *Orders* of Columns; three whereof are *Greek*, *viz.* the *Doric*, *Ionic*, and *Corinthian*; and two *Italic*, *viz.* the *Tuscan*, and *Composite*.

The three *Greek Orders* represent the three different Manners of Building, *viz.* the *Solid*, *Delicate*, and *Mean*; the two *Italic* ones are imperfect Productions thereof. The little regard the *Romans* had for these last, appears hence, that we don't meet with one Instance in the Antique, where they are intermix'd. That Abuse the Moderns have introduc'd, by the Mixture of the *Greek* and *Latin Orders*, *Daviler* observes, arises from their want of Reflection on the Use made thereof by the Antients.

The Origin of *Orders* is almost as antient as human Society: The Rigour of the Seasons first led Men to make little Cabins, to retire into; at first, half under ground, and the half above, cover'd with Stubble: At length growing more expert, they planted Trunks of Trees on-end, laying others a-cross, to sustain the Covering. See ARCHITECTURE.

Hence they took the Hint of a more regular Architecture; for the Trunks of Trees, upright, represent Columns: the Girts, or Bands, which serv'd to keep the Trunks from bursting, express'd Bases and Capitals; and the Summers laid a-cross, gave the Hint of Entablatures; as the Coverings, ending in Points, did of Pediments. This is *Vitruvius's* Hypothesis; which we find very well illustrated by M. *Blondel.* See CAPITAL, PEDESTAL, *etc.*

Others take it, that Columns took their Rise from Pyramids, which the Antients erected over their Tombs; and that the Urns, wherein they enclos'd their Ashes, represented the Capitals, whose Abacus was a Brick, laid thereon to cover the Urns: But *Vitruvius's* Account appears the more natural.

At length, the *Greeks* regulated the Height of their Columns on the foot of the Proportions of the human Body: The *Doric* represented a Man of a strong, robust Make; the *Ionic* that of a Woman; and the *Corinthian* that of a Girl: Their Bases and Capitals were their Head-Dress, their Shoes, *etc.* See BASE, *etc.*

These *Orders* took their Names from the People, among whom they were invented: *Scamozzi* uses significative Terms to express their Character; when he calls the *Tuscan*, the *Gigantic*; the *Doric*, the *Herculean*; the *Ionic*, the *Matronal*; the *Composite*, the *Heroic*; and the *Corinthian*, the *Virginal*. *See each Order under its proper Article*, TUSCAN, *etc.*

To give a general Idea of the *Orders*; it must be observ'd, that the whole is compos'd of two parts at the least, *viz.* the *Column*, and *Entablature*; and of four parts at the most; when there is a *Pedestal* under the Column, and one *Acroter*, or little Pedestal, a-top of the Entablature: That the Column has three parts, *viz.* the *Base*, the *Shaft*, and the *Capital*;

the Entablature has three likewise, *viz*. the *Architrave*, the *Frise*, and *Cornich*: Which Parts are all different in the several Orders. *See each Part under its proper Article.*

Tuscan ORDER is the first, most simple, and solid; its Column is seven Diameters high; and its Capital, Base, and Entablature, have but few Mouldings for Ornaments. See TUSCAN.

Doric ORDER, is the second, and the most agreeable to Nature. It has no Ornament on its Base, or on its Capital. Its height is eight Diameters. Its Frise is distinguish'd by Triglyphs and Metopes. See DORIC.

Ionic ORDER, is the third; and a kind of mean Proportional between the solid and delicate Manner. Its Capital is adorn'd with Volutes, and its Cornich with Denticles. See IONIC.

Mich. Angelo, contrary to all other Authors, gives the *Ionic* a single Row of Leaves at the bottom of the Capital.

Corinthian ORDER, invented by *Callimachus*, is the fourth, the richest, and most delicate. Its Capital is adorn'd with two Rows of Leaves, and eight Volutes, which sustain the Abacus. Its Column is ten Diameters high, and its Cornich has Modillions. See CORINTHIAN.

Composite ORDER, the fifth and last, (tho' *Scamozzi* and *le Clerc* make it the fourth) is so call'd, because its Capital is compos'd out of those of the other *Orders*; having the two Rows of Leaves of the *Corinthian*, and the Volutes of the *Ionic*. It is also call'd the *Roman*, because invented among that People. Its Column is ten Diameters high; and its Cornich has Denticles, or simple Modillions. See COMPOSITE.

Rustic ORDER, is that adorn'd with Rustic Quoins, Bossages, *etc*. See RUSTIC.

Attic ORDER, is a little *Order* of low Pilasters, with an architrav'd Cornich for its Entablature; as that of the Castle of *Versailles* over the *Ionic* on the side of the Garden. See ATTIC.

M. *Blondel* calls the little Pilasters of Attics and Mezzanines, *false Orders*.

Persian ORDER, is that which has Figures of *Persian* Slaves, instead of Columns, to support the Entablature. See PERSIAN.

Caryatic ORDER, is that whose Entablature is supported with Figures of Women, instead of Columns. See CARYATIDES.

Gothic ORDER, that which deviates from the Ornaments and Proportions of the Antique; and whose Columns are either too massive, in manner of Pillars; or too slender, like Poles: its Capitals out of all measure; and carv'd with Leaves of wild Acanthus, Thistles, Cabbage, or the like. See GOTHIC.

French ORDER, is a new contriv'd *Order*, wherein the Capital consists of Attributes agreeing to that People; as Cocks-Heads, Flowers-de-Lys, *etc*.

Its Proportions are *Corinthian*: Such is that of M. *le Brun*, in the Grand Gallery of *Versailles*; and that of M. *le Clerc*.

M. *le Clerc* gives a second *Tuscan Order*, and a *Spanish Order*, besides his *French Order*. The *Tuscan* he ranks between the first *Tuscan* and *Doric*. Its height he makes 23 Semi-diameters, 22 Minutes. The Column to have 15, the Pedestal 5, and the Entablature 3, and 22 Minutes: And he proposes its Frise to be adorn'd with Turtles, which are the Arms of *Tuscany*.

The *Spanish Order* he places between the *Corinthian* and *Composite*. The whole *Order* he makes 30 Semi-diameters, 28 Minutes; whereof the Column has 19, and 25 Min. the Pedestal 16, and 18 Min. and the Entablature 4, and 15 Min. The Horns of the Abacus he sustains with little Volutes; the middle, in lieu of a Rose, has a Lion's Snout: that Animal being the Symbol of *Spain*, and expressing the Strength, Gravity, and Prudence of that Nation.

ORDONNANCE, or ORDINANCE, in Painting, the Disposition of the Parts of a Painting, either with regard to the whole Piece, or to the several Parts; as the Groups, Masses, Contrasts, Aspects, *etc*.

The Doctrine of ORDONNANCE *is compriz'd in the following Rules.*

In the *Ordonnance* there are three things regarded, *viz*. the Place, or Scene, where; the Distribution, how; and the Contrast.

In the first, regard is to be had to the *Disposition* of things to serve as a Ground-Work; and to the *Plan*, or Position of Bodies: Under the former whereof, come, 1. The *Landskip*; whether an uninhabited Place, where there is a full liberty of representing all the Extravagancies of Nature; or inhabited, where the Marks of Cultivation, *etc*. must be exhibited. See LANDSKIP.

2. The *Building*, whether Rustic; wherein the Painter's Fancy is at liberty: Or *Regular*; wherein a nice Attention is required to the Orders. See ORDER.

3. The *Mixture* of both; wherein 'tis a Maxim to compose in great Pieces, and make the Ground-Plot big enough; to neglect some little Places, in order to bestow more on the whole Mass; and to shew the more considerable Places with the more advantage; and to make some Agitation in all the things that move.

As to the *Plan* of Bodies; they are either solid; which again, are either so by Nature, and which must be proportion'd to their Places; or Artificial, where regard must be had to the Rules of Geometry, Perspective, Architecture, *etc*. Or they move; which they do, either by a voluntary Motion, wherein care must be had to proportion 'em to their Situation, and to strengthen 'em by the regarding Equilibrium; or by some extraordinary Power, as Machines, *etc*. where the Causes of their Motions must appear. Or they are things at a distance. In all which, an even Plane must still be proposed to find their precise Situation, and settle their Place by sudden breaks and distances agreeable to their Perspective. See PERSPECTIVE, *etc*.

In placing the Figures, regard must be had, 1. To the *Group*, which connects the Subject, and stays the Sight. In this are to be consider'd the Knot, or *Nodus*, which binds the Group; and the Nearness of Figures, which we may call the Chain, as it holds 'em together; that the Group be sustain'd by something loose and distinct from it; and by the same join'd and continued to the other Groups: And that the Lights and Shadows be so disposed, as that we may at once see the Effects of all the Parts of the Composition.

2. To the *Actions*, wherein forced Attitudes are to be avoided; and simple Nature shewn in her most advantageous Postures. In weak and lean Figures, the Nudities not to be shewn; but occasions of covering 'em sought. In all human Figures, special Care to be taken, the Head be well placed between the Shoulders; the Trunk on the Haunches, and the whole on the Feet.

3. To the *Drapery*; which is to be adjusted so, as it may appear real Garments, and not Stuffs loosely thrown on. The Folds to be so disposed, as to leave the great parts wherein the Nudity may appear, free; ranging the little Folds about the Joints, and avoiding 'em on the Relievo of the Members. And, lastly, to dispose the Draperies, by raising the Stuff, and letting it fall lightly, that the Air sustaining the Folds, may let 'em fall soft.

Lastly, in the *Contrast*, are to be consider'd the Actions, which vary infinitely: The Aspects, which, in Actions of the same kind, may, by their difference, make a Contrast: The Situation, according as they meet above or under the Sight, far or near. And, lastly, the Custom; which extends, indeed, to all parts of Painting, but is particularly to be regarded in the *Ordonnance*; to be managed discretely, however, and stiffness and formality avoided.

ORDONNANCE, in Architecture, is nearly the same as in Painting; *viz.* the Disposition of the Parts of a Building, both with regard to the whole, and to one another. See BUILDING.

Vitruvius defines *Ordonnance* to be that which regulates the Size of all the Parts of a Building with respect to their Use.

This Definition is censur'd by M. *Perrault*, who takes the *Ordonnance* to consist in the Division of the Plan or Spot of Ground whereon the Building is to be rais'd; *i.e.* in the apportioning or laying it out, agreeable with the intended Dimensions of the whole Fabric; which Mr. *Evelyn* expresses in fewer Words by the determining of the Measures of what is assign'd to compose the several Apartments.

On this Foundation, *Ordonnance* is the judicious Contrivance of the Plan or Model; as when the Court, Hall, Lodgings, *etc.* are neither too large nor too little: But the Court, *v. gr.* affords convenient Light to the Apartments about it, and is large enough for usual Access. The Hall of fit Capacity to receive Company; the Bed-Chambers accordingly, *etc.* When these Divisions are either too great or too small with respect to the Place, as a large Court to a little House, or a little Chamber in a magnificent Palace; the fault is in the *Ordonnance*.

This the Antients call'd *Taxis*; and distinguished it from *Diathesis*, Disposition; which is, where all the Parts and Members of a Building are assign'd their just Place and Situation with regard to their Quality, Office, Rank, *etc.* without any regard to the Dimensions or Quantity: As, that the Vestible or Porch precede the Hall, the Hall the Parlor, that the Withdrawing Room, *etc.* The Bed-Chambers, again, to the Sun-rising; Libraries, Galleries of Paintings, and Cabinets of Curiosities, *etc.* to the North.

ORIGINAL, a first Draught, Design, or Autograph of any thing; serving as a Model, or Exemplar, to be imitated, or copied. See DESIGN, MODEL, *etc.*

ORILLON, in Fortification, a small rounding of Earth, lined with a Wall; raised on the Shoulder of those Bastions that have Casements; to cover the Cannon in the retired Flank, and prevent their being dismounted by the Enemy.

There are other sorts of *Orillons*, properly call'd *Shoulderings*, or *Epaulements*, almost of a square Figure.

ORLE, ORLET, or ORLO, in Architecture, a Fillet under the *Ovolo*, or Quarter-round of a Capital. See FILLET.

When 'tis at the Top or Bottom of the Shaft, 'tis call'd *Cincture*. See CINCTURE.

Palladio also uses the Word for the Plinth of the Bases of Columns and Pedestals. See PLINTH.

The Word is *French*; form'd from the *Latin*, *Orletum*, or *Orlum*; of *Ora*, a Border, or List.

ORNAMENTS, in Architecture, express all the Sculpture, or Carved Work, wherewith a Piece of Architecture is enrich'd. See SCULPTURE.

Vitruvius and *Vignola* also use the Word to signify the *Entablature*. See ENTABLATURE.

Ornaments, in *Relievo*, are those cut in the Contours of Mouldings; as *Leaves*, *Shells*, *Scrolls*, *Flowers*, etc. *Ornaments*, *in Creux*, those cut within the Mouldings; as *Eggs*, *Flutes*, etc. See MOULDINGS.

ORTEIL, in Fortification. See BERME.

ORTHOGRAPHY, in Architecture, is the Elevation of a Building. See ELEVATION.

The *Orthography* is either External, or Internal.

External Orthography is a Delineation of the external Face or Front of a Building; exhibiting the principal Wall, with its Apertures, Roof, Ornaments, and every thing visible to an Eye placed before the Building.

Internal Orthography, call'd also a *Section*, is a Delineation, or

Draught of a Building, such as it would appear, were the external Wall removed. See SECTION.

To lay down the ORTHOGRAPHY *of a Building.*

Draw a right Line for a Basis or ground Line, (Tab. PERSPECTIVE, *Fig.* 13.) A B, and at one end erect a Perpendicular A D. Upon A B set off the widths and distances of the Gates, Windows, *etc.* On the right Line A D, set off the Heights of the several Parts visible in the Face of the Building, *v. gr.* of the Doors, Windows, the Roof, Chimneys, *etc.* and apply a Ruler to each Point of Division. The common Intersections of the right Lines drawn from three Points parallel to the Lines A B and A D, determine the external *Orthography* of the Building: And after the same manner is the internal *Orthography* laid down. See PERSPECTIVE.

ORTHOGRAPHY, in Geometry, the Art of Drawing, or delineating the fore-right Plan of any Object, and of expressing the Heights or Elevations of each Part.

It is call'd *Orthography* from its determining things by perpendicular Lines falling on the Geometrical Plane.

ORTHOGRAPHY, in Perspective, a Delineation of the fore-right Plane of any Object. See PERSPECTIVE.

ORTHOGRAPHY, in Fortification, is the Profile, or Representation of a fortify'd Place; or a Draught so conducted, as that the length, breadth, height, and thickness of the several Parts are express'd; such as they would appear, if it were perpendicularly cut from Top to Bottom.

OUT-WORKS, in Fortification, those Works made without side the Ditch of a fortify'd Place, to cover and defend it. See FORTIFICATION.

The most ordinary of these are Ravelins, or Half-Moons, form'd between the two Bastions, on the flanquant Angle of the Counterscarpe, and before the Curtain, to cover the Gates and Bridges.

OVA, in Architecture, are Ornaments in form of Eggs, carved on the Contour of the *Ovolo*, or Quarter-round; and separated from each other by Anchors or Arrows Heads. See OVOLO.

The *English* usually call these Ornaments *Eggs and Anchors*.

Instead of Eggs, the Antients sometimes used *Hearts*; on which Foundation it was, that they introduced Arrows; to symbolize with Love.

OVICULUM, in the antient Architecture, a little *Ovum*, or Egg.

Baldus will have this to be the *Lesbian Astragal* of *Vitruvius*. See ASTRAGAL.

Some use the Word *Oviculum* for *Ovolo*. See OVOLO.

OVILIA, or SEPTA, in antient *Rome*, a Place in the *Campus Marius*, at first rail'd in, like a Sheep's Pen; whence its Name: Afterwards it was mounted with Marble, beautify'd with Walks and Galleries; as also with a Tribunal, or Seat of Justice: within which Precinct the People frequently assembled, to give their Suffrages for the Election of Magistrates.

The Ascent into the *Ovilia* was not by Stairs, but by Bridges made for the time; every Parish, Tribe, and Century, as the Assembly was Centurate or Tribute, *etc.* having its proper Bridge. Whence the Proverb, *de Ponte de jiciendus*, where a Person is to be barred from giving his Vote.

OVOLO, *Ovum*, in Architecture, a round Moulding, whose Profile, or Sweep, in the *Ionic* and *Composite* Capitals, is usually a Quadrant of a Circle; whence it is also popularly call'd the *Quarter-Round.*

It is usually enrich'd with Sculptures among the Antients, in form of Chesnut-Shells; whence *Vitruvius*, and others of the Antients, call it *Echinus*, *Chesnut-Shell*. See ECHINUS.

Among us, it is usually cut with the Representation of Eggs, and Anchors, or Arrows-Heads, placed alternately; whence its *Italian* Name *Ovolo*, Latin *Ovum*, and French *Oeuf*.

P

PAINTING, the Art of representing natural Bodies, and giving 'em a Kind of Life, by the Turn of Lines, and the Degrees of Colours.

Painting is said to have had its Rise among the *Egyptians*: And the *Greeks*, who learn'd it of 'em, carried it to its Perfection; if we may believe the Stories related of their *Apelles*, and *Zeuxis*.

The *Romans* were not without considerable Masters in this Art, in the later Times of the Republic, and under the first Emperors; but the Inundation of *Barbarians*, who ruin'd *Italy*, proved fatal to Painting, and almost reduced it to its first Elements. It was in *Italy*, however, that the Art returned to its ancient Honour, and in the Beginning of the XVth Century when *Cimabue*, betaking himself to the Pencil, translated the poor Remains of the Art, from a *Greek* Painter or two, into his own Country.

He was seconded by some *Florentines*: The first who got any Reputation was *Ghirlandaio*, *Michael Angelo's* Master; *Pietro Perugino*, *Raphael Urbin's* Master; and *Andrea Verocchio*, *Leonardo Da Vinci's* Master.

But the Scholars far surpassed the Masters; they not only effaced all that had been done before 'em, but carried Painting to a Pitch from which it has ever since been declining.

'Twas not by their own noble Works alone that they advanc'd

Painting; but by the Number of Pupils they bred up, and the Schools they form'd.

Angelo, in particular, founded the School of *Florence*; *Raphael*, the School of *Rome*; and *Leonardo*, the School of *Milan*; to which must be added, the *Lumbard* School, establish'd about the same time, and which became very considerable under *Georgian* and *Titian*. See SCHOOL.

Besides the *Italian* Masters, there were others on this Side the *Alps*, who had no Communication with those of *Italy*; such were *Albert Durer*, in *Germany*; *Holbens*, in *Switzerland*; *Lucas*, in *Holland*; and others in *France* and *Flanders*: but *Italy*, and particularly *Rome*, was the Place where the Art was practis'd with the greatest Success; and where, from Time to Time, the greatest Masters were produced.

To *Raphael's* School, succeeded that of the *Caraches*, which has lasted, in its Scholars, almost to the present Time; wherein the *French* Painters, by the Munificence of the late *Louis* XIV. seem almost in Condition to vie with those of *Greece* or *Italy*.

In *Paris* they have two considerable Bodies of Painters, the one, the *Royal Academy of Painting and Sculpture*, the other the *Community of Masters in Painting*, *Sculpture*, etc.

The Art of *Painting* is divided, by *Fresnoy*, into three principal Parts, *Invention*, *Design*, and *Colouring*; to which some add a fourth, *viz. Disposition*.

Felibien divides *Painting* into the *Composition*, the *Design*, and *Colouring*.

Mons. *Testling*, Painter to the late King, divides it, somewhat more accurately, into the *Design* or *Draught*, the *Proportion*, the *Expression*, the *Clair-obscure*, the *Ordonnance*, and the *Colouring*; to which his *English* Translator adds the *Perspecitve*.

Painting is of various Kinds, with regard to the Materials us'd; the Matter whereon they are applied; and the Manner of applying 'em.

Hence come *Painting* in Oil; *Painting* in Water-Colours, or *Limning*; *Painting* in *Fresco*; *Painting* on *Glass*; *Painting* in *Enamel*; and *Painting* in *Miniature*.

PAINTING *in Oil*. The Art of Painting in Oil was unknown to the Ancients; and it was a *Flemish* Painter, one *John van Eyck*, or *John de Bruges*, who first discover'd and put it in practice in the Beginning of the XIVth Century: 'Till him, all the Painters wrought in *Fresco*, or in Water-Colours.

This was an Invention of the utmost Advantage to the Art; since, by means hereof, the Colours of a Painting are preserv'd much longer and better, and receive a Lustre and Sweetness which the Ancients cou'd never attain to, what Varnish soever they made use of to cover their Pieces.

The whole Secret only consists in grinding the Colours with Nut-Oil or Linseed Oil: But it must be own'd, the Manner of Working is very different from that in *Fresco*, or in Water; by reason the Oil does not dry near so fast; which gives the Painter an Opportunity of touching and retouching all the Parts of his Figures, as often as he pleases: Which, in the other Kinds, is a Thing impracticable.

The Figures too are here capable of more Force and Boldness; in as much as the Black becomes blacker, when ground with Oil than with Water; besides that, all the Colours, mixing better together, make the Colouring sweeter, more delicate and agreeable, and give an Union and Tenderness to the whole Work, inimitable in any of the other Manners.

Painting in Oil is perform'd on *Walls*, on *Wood*, *Canvas*, *Stones* and all Sorts of *Metals*.

To PAINT *on a Wall*: when well dry, they give it two or three Washes of boiling Oil; till the Plaster remain quite greasy, and will imbibe no more. Over this they apply dessiccative or drying Colours, *viz.* white Chalk, red Oker, or other Chalks beaten pretty stiff. This Couch or Lay being well dry, they sketch out, and design their Subject; and at last paint it over, mixing a little Varnish with their Colours; to save the Varnishing afterwards.

Others, to fortify their Wall the better against Moisture, cover it with a Plaster of Lime, Marble-Dust, or a Cement made of beaten Tiles soak'd with Linseed Oil; and at last prepare a Composition of *Greek* Pitch, Mastic and thick Varnish boil'd together, which they apply hot over the former Plaster: When dry, they lay on their Colours as before.

Others, in fine, make their Plaster with Lime-Mortar, Tile-Cement, and Sand; and this dry, apply another of Lime, Cement, and Macheser or Iron Scum; which being well beaten and incorporated with Whites of Eggs and Linseed Oil, makes an excellent Couch or Plaster. When dry, the Colours are applied as before.

To PAINT *on Wood*; they usually give their Ground a Couch or Lay of White temper'd with Size; or they apply the Oil abovementioned: The rest, as in Painting on Walls.

To PAINT *on Linnen or Canvas*; the Canvas being stretch'd on a Frame, they give it a Couch or Lay of Size: When dry, they go over it with a Pumice-Stone, to smooth off the Knots. By means of the Size the little Threads and Hairs are all laid close on the Cloth, and the little Holes stopp'd up, so as no Colour can pass thro'.

When the Cloth is dry, they lay on Oker, which is a natural Earth, and bears a Body; sometimes, mixing with it a little white Lead to make it dry the sooner. When dry, they go over it with the Pumice-Stone to make it smooth.

After this, they sometimes add a second Lay composed of white Lead, and a little Charcoal Black, to render the Ground of an Ash-Colour, observing in each Manner to lay on as little Colour as possible, that the Cloth may not break, and that the Colours, when they come to be painted over, may preserve the better.

In some Paintings of *Titian* and *Paolo Veronese* we find they made their Ground with Water, and painted over it with Oil; which contributed much to the Vivacity and Freshness of their Works: For the Water Ground, by imbibing the Oil of the Colours, leaves them the more beautiful; the Oil itself taking away a deal of their Vivacity.

As little Oil therefore is to be used as possible, if 'tis desir'd to have the Colours keep fresh: For this Reason some mix 'em up with Oil of Aspic, which evaporates immediately, yet serves to make 'em manageable with the Pencil.

To PAINT *on Stones or Metals*, 'tis not necessary to apply Size, as on Cloth; it suffices to add a slight Couch of Colours, before you draw your Design; nor is even this done, on Stones where 'tis desired the Ground shou'd appear, as in certain Marbles of extraordinary Colours.

All the Colours used in *Fresco* are good in *Oil*, except White of Lime and Marble-Dust. See COLOUR.

Those chiefly used are *White Lead* or *Ceruss*, yellow and white *Massicot*, *Orpiment*, *Black Lead*, *Cinnaber* or Vermillion, *Lacca*, blue and green *Ashes*, *Indigo*, *Lamp-Black*, Burnt *Ivory*, and *Verdigrease*, etc.

As to Oils, the best are those of Walnuts, of Linseed, Aspic, and Turpentine. The dessiccative or drying Oils, are a Nut-Oil boil'd with Litharge and Sandarac, others with Spirit of Wine, Mastic and Gum-Lacca.

To have a Varnish that shall dry readily, they mix Spirit of Wine with Turpentine.

PAINTING *in Water-Colours*. See LIMNING.

PAINTING *in Fresco*. See FRESCO.

PAINTING *in Miniature*. See MINIATURE.

PAINTING *in Enamel*, See ENAMEL.

PAINTING *in Mosaic*. See MOSAIC.

PALACE, *Royal-House*, a Name generally given to the Dwelling-Houses of Kings, and Princes. See HOUSE.

In Course of Time the Name has also been applied to the Houses of other Persons; taking different Epithets, according to the Quality of the Inhabitants; as *Imperial Palace*, *Royal Palace*, *Pontifical*, *Cardinal*, *Episcopal*, *Ducal Palace*, etc.

Procopius derives the Origin of the Word *Palace* from a *Greek*, called *Pallas*, who gave his own Name to a magnificent House he had built: Adding, that *Augustus* after him, gave the Name *Palatium* to the House of the *Roman* Emperors on the Hill; which, for that Reason, was called the *Palatine* Mount. Others take it the contrary Way; and say, that *Romulus's* House, wherein *Augustus* lived, was properly called *Palatium*, because situate on the *Palatine* Mount.

Be this as it will, 'tis certain, *Palatium*, from a proper Name, in Time, became common to all Houses of Kings.

And as the Kings usually heard and determined Causes in their Houses, in what Part of the Realms soever situate; hence also *Palatium* became a Name for a Court of Justice; thus the *French* have their *Palais*, etc.

PALÆSTRA, among the antient *Greeks*, a publick Building where the Youth exercised themselves in Wrestling, Running, Quoits, *etc*.

Some say it consisted of a College, and an Academy; the one for Exercises of the Mind, the other of the Body. But most Authors rather take *Palæstra* to be a *Xystus* or mere Academy for Bodily Exercises, according to the Etymology of the Word, Wrestling, one of the chief Exercises amongst the Ancients. See XYSTUS.

The Length of the *Palæstra* was mark'd out into *Stadia*, each equal to 125 Geometrical Paces; and the Name *Stadium* was given to the *Arena* whereon they ran.

PALE, a little pointed Stake, or Piece of Wood, used in making Inclosures, Separations, *etc*. See PALISSADE.

PALES, or *Piles*, in Carpentry, are Rows or Files of Stakes drove deep into the Ground, to make wooden Bridges over Rivers. See PALLIFICATION.

They serve to support the Beams which are laid a-cross them, from one Row to another; and are strongly bound together with Cross-Pieces.

PALISSADE or PALISSADO, in *Fortification*, an Inclosure with Stakes or *Pales* driven into the Ground, eight or nine Inches thick; and nine Foot long, three whereof are hid under Ground.

It is used to fortify the Avenues of open Forts, Gorges, Half-moons, the Bottoms of Ditches, and the Parapets of Covert-ways; on the Talut of Ramparts; the Top of the Out-works, *etc*.

There are *Palissadoes* made perpendicular; others are made inclining to the Ground, that the Ropes cast over 'em, to tear 'em up, may slip.

PALISSADES *turning*, are an Invention of Mr. *Coehornes*, in order to preserve the *Palisades* of the Parapet from the Besiegers Shot.

He orders them so, that as many of them, as stand in the Length of a Rod, or in about ten Foot, turn up and down like Traps; so as not to be in Sight of the Enemy till they just bring on their Attack; and yet are always ready to do the proper Service of *Pallisades*.

PALLADIUM, in Antiquity, a Statue of the Goddess *Pallas*, preserv'd in *Troy*; whereon the Fate of the City depended.

The Tradition was, that in building a Cittadel, in Honour of *Pallas*, and a Temple in the most elevated Part thereof; the *Palladium* dropp'd from Heaven, and mark'd out the Place, which the Goddess was pleas'd to possess. After this, *Apollo* gave an Oracle, importing, that *Troy* should never be taken while the *Palladium* was found within its Walls: Which occasioned *Diomed* and *Ulysses* to undertake the stealing thereof.

'Tis said, there was anciently a Statue of *Pallas* preserv'd at *Rome*, in the Temple of *Vesta*: which some pretended to be the true *Palladium* of *Troy*, brought into *Italy* by *Æneas*: It was kept among the sacred Things of the Temple, only known to the Priests and Vestals.

This Statue was esteem'd the Destiny of *Rome*; and there were several others made perfectly like it, to secure it from being stollen.

There was also a *Palladium* in the Citadel of *Athens*, plac'd there by Nicias.

These *Palladiums*, in all Probability, were no other than a Kind of *Talismans*.

PALLIER, or *Paillier*, in Building, a Landing-Place in a Stair-Case; or a Step, which, being broader than the rest, serves to rest upon. See STAIR-CASE.

The Term is pure *French*, and not much used in *English.* In Perrons, or large Stair-cases, where there are sometimes several Palliers in the same Range, or Line, they ought to have at least the Width of two Steps.

Those in the Turns of Stair-cases ought to be as broad as long. *Vitruvius* calls the *Palliers* or Landing-Places of Theatres *Diazomata.*

PALLIFICATION, in Architecture, is the piling of the Ground-work; or strengthening it with Piles, or Timber driven into the Ground; which is practised, when they build upon a moist or marshy Soil. See FOUNDATION.

PANEL, in Joinery, *etc.* See PANNEL.

PANEL, in Masonry, one of the Faces of a hewn Stone. See STONE.

PANNEL, or PANEL, in Joinery, *etc.* a Tympanum or square Piece of thin Wood, sometimes carv'd, fram'd, or groov'd in a larger Piece between two Montants or upright Pieces, and two Traverses or Cross-Pieces.

Hence *Panels* or Panes of Glass, are Compartiments or Pieces of Glass of various Forms, Square, Hexagonal, *etc.*

PANNIER, in Architecture. See CORBEL.

PANTHEA, among the *Romans*, were single Statues, composed of the Figures or Symbols of several different Divinities. See STATUE.

Father *Joubert*, who calls them *Pantheons*, and who has observed them on several Medals, says their Heads are most commonly adorn'd with Symbols belonging to several Gods.

An Instance hereof we have in a Medal of *Antoninus Pius*; which at the same time represents *Serapis*, by the Bushel it bears; the *Sun*, by the Crown of Rays; *Jupiter Ammon*, by the Ram's Horns; *Pluto*, by the large

Beard; and *Esculapius*, by the Serpent twisted in his Hand. See STATUE.

M. *Baudelot*, in a Dissertation on the *Lares*, will have the *Pantheons* to have had their Rise from the Superstition of those, who, taking several Gods for the Protectors of their Houses, united them all in the same Statue, by adorning it with the several Symbols, proper to each of those Deities.

PANTHEON, in Architecture, is a Temple, or Church of a circular Form; dedicated to all the Gods, or all the Saints. See TEMPLE and CHURCH.

It is thus named from the *Greek all*, and *God*.

The *Pantheon* of antient *Rome*, is, of all others, the most celebrated, and that whence they all take their Name. It was built by *Agrippa*, Son-in-law of *Augustus*, in his third Consulate, 25 Years before *Christ*. It was dedicated by him to *Jupiter Ultor*, *Jupiter* the Revenger; and had the Name *Pantheon*, by reason of the great Number of Statues of the Gods ranged in Niches all around; and because built round, to represent Heaven, the Residence of all the Gods. It had but one Door, and one Window; receiving all its Light from the Top of its Dome.

The Pope, obtaining it of the Emperor *Phocas*, converted it into a Church, without any Alteration in the Building, and dedicated it to the Virgin and all the Martyrs. And it still subsists at *Rome* under the Title of *Notre Dame della Rotonda*. See ROTONDO.

The *Pantheon of Nismes*, was a Temple in that City, wherein were 12 Niches or Statues, supposed to have been destin'd for the 12 great Gods.

In the Escurial is a magnificent Chapel, call'd *Pantheon*, 35 Feet in Diameter, and 38 high, from the Pavement which is of Marble and Jasper inlay'd. The whole Inside of the Chapel is of black Marble, excepting the *Luthern* and some Ornaments of Jasper and red Marble.

In this Chapel are deposited the Bodies of the Kings and Queens: there are only Places made for 26; eight of which are already fill'd.

PARAPET, in Fortification, a Defence or Skreen, on the Outside of a Rampart, raised six or seven Foot high, to cover the Soldiers and the Cannon from the Enemy's Fire.

Parapets are raised an all Works, both within and without the Place; and even on the Approaches.

The *Parapet Royal*, or that of the Rampart, is to be of Earth, Cannon-proof, from 18 to 20 Foot thick. See RAMPART. Before the *Parapet* is a *Banquette*, or little Eminence, a Foot and half high, for the Soldiers to stand on.

The *Parapet* of the Wall is of Stone. The *Parapet* of the Trenches is either made of the Earth dug up, Gabions, Fascines, Sacks of Earth, or the like.

The *Parapet* of the Corridor is the *Glacis* or *Esplanade*.

PARAPET, is also a little Wall, breast-high, rais'd on the Brinks of Bridges, Keys, or high Buildings; to serve as a Stay; and prevent People's tumbling over.

Borel gives us, from *Jos. Maria Subresius*, a curious Collection of Names, which the Antients and Moderns have given to this Kind of *Parapets*: The *Latins* called them *Subarræ*, and *Bastiæ*, whence the Names *Bastion* and *Bastile*. They also called them *Pagineumata Loricæ*, and *Antemuralia*. The *Spaniards* called them *Barbacanes*; the *Italians*, *Parapetti*, because of their defending the Breast, *petto*; whence our *Parapet*.

PARASCENIUM, or *Postscenium*, among the *Romans*, was a Place behind the Theatre, whither the Actors withdrew to a dress, undress, *etc*. See THEATER.

PARGETING, in Building, is used for the Plaistering of Walls; sometimes for the Plaister itself. See PLASTER.

Pargeting is of various Kinds; as, 1. White Lime and Hair Mortar laid on bare Walls. 2. On bare Laths, as in partitioning and Plain Ceiling. 3. Rendring the Insides of Walls or Double Partition-Walls. 4. Rough-casting on Hearth-Laths. 5. Plaistering on Brick-work, in Imitation of Stone-work; and the like upon Hearth-Laths.

PARISH, the Precinct or Territory of a *Parish* Church. See CHURCH.

In the antient Church, there was one large Building in each City, for the People to meet in; and this they call'd *Parish*. But the Signification of the Word was afterwards enlarged, and by *Parish* was meant a Diocese, or the Jurisdiction of a Bishop, consisting of several Churches; unless we will suppose, as some do, that those Bishops were only Pastors of single Churches.

At least, the Word now retains its original Meaning.

Du-pin observes, that Country *Parishes* had not their Origin before the IVth Century; but those of Cities are more antient. The City of *Alexandria* is said to have been the first that was divided into *Parishes*. *Baronius* says, that in the Time of Pope *Cornelius* there were 46 *Parishes* in *Rome*.

The Division of *England* into *Parishes*, is attributed to *Honerius* Archbishop of *Canterbury*, in 636. *Cambden* reckons 9284 *Parishes* in *England*. *Chamberlayn* makes, at present, 9913.

PARK, an Inclosure stock'd with wild Beasts, *tam Sylvestres*, *quam Campestres*; say our old Lawyers.

Crompton observes, that a Subject may hold a *Park* by Prescription, or the King's Grant, which he can't do a Forest.

A *Park* differs from a *Chase* or *Warren*; for that a *Park* must be enclosed; if it lie open, it is a good Cause of seizing it into the King's Hand; as a free Chase may be, if it be enclosed. Nor can the Owner have any Action against such as hunt in his *Park*, if it lie open.

Du Cange refers the Invention of *Parks* to King *Henry* I. of *England*: But *Spelman* shews, 'tis much more antient; and was in use among the *Anglo Saxons*.

Zozimus assures us, the antient Kings of *Persia* had *Parks*.

The Word is originally *Celtic*, where it signifies an Inclosure, or Place shut up with Walls.

PARTERRE, in Gardening, that open Part of a Garden into which we enter, coming out of the House; usually, set with Flowers, or divided into Beds, incompassed with Plat-bands, *etc*. See GARDEN.

The *Parterre* is a level Division of Ground, which, for the most Part, faces the South and best Front of a House, and is generally furnished with Greens, Flowers, *etc*.

There are divers Kinds of *Parterres*, as *Bowling-Green* or *plain Parterres*; *Parterres of Embroidery*; *cut in Shell*, and *Scroll-work*, etc. with Sand-Allies between them.

An oblong, or long Square is accounted the most proper Figure for a *Parterre*; the Sides whereof, to be as two, or two and a half to one.

PASTIL, among Painters, *etc*. a Sort of Paste, made of several Colours, ground up with Gum-Water, either together or separately; in order to make Crayons to paint with on Paper or Parchment.

PASTIL, is also us'd for a dry Composition, yielding a fragrant Smell when burnt in a perfuming Pan, to clear and Scent the Air of a Chamber.

It is composed of odorous Resins, mixt with Aromatic Woods, or Drugs pulveriz'd, and incorporated with Mucilages of Gum Tragacanth. Some call 'em *Osselets of Cyprus*.

PAVEMENT, a Lay of Stone, or other Matter, serving to cover and strengthen the Ground of divers Places, for the more commodious walking on, or the Passage of Carriages.

The Word is form'd from the Latin *Parvimentum*, of *Parvire*, to beat down the Earth, in order to make it firm and strong.

In *England*, the *Pavements* of the Grand Streets, *etc*. are usually Flint, or Pebble; Courts, Equeries, Kitchins, Halls, Churches, *etc*. Tiles, Bricks, Flags, or Fire-Stone; sometimes a Kind of Free-Stone, and Rag-Stone. See STONE.

In some Cities, *E. gr. Venice*, the Streets, *etc*. are paved with Brick; Churches sometimes with Marble, and sometimes with Mosaic-Work, as the Churches of St. *Mark* at *Venice*.

In *France*, the public Roads, Streets, Courts, *etc*. are paved with *Gres*, a Kind of Free-Stone.

In *Amsterdam*, and the chief Cities of *Holland*, they call their Brick

Pavement the *Bourguer Masters Pavement*, to distinguish it from the Stone or Flint *Pavement*, which usually takes up the Middle of the Street, and which serves for Carriages; the Brick which borders it being destined for the Passage of People on Foot.

Pavements of Free-Stone, Flint, and Flags, in Streets, *etc.* are laid dry, *i.e.* are retained in a Bed of Sand; those of Courts, Equeries, Ground Rooms, *etc.* are laid in a Mortar of Lime and Sand; or in Lime and Ciment, especially if there be Vaults or Cellars underneath.

Some Masons, after laying a Floor dry, especially of Brick, spread a thin Mortar over it; sweeping it backwards and forwards to fill up the Joints.

Thirty two Statute Bricks laid flat pave a Yard square; 64 of Edge wise.

The Square Tiles us'd in Paving, call'd Paving-bricks, are of various Sizes, from 6 to 12 Inches Square. See BRICK.

Pavements of Churches, *etc.* frequently consist of Stones of several Colours; chiefly black and white, and in several forms, chiefly Square, and Lozange, artfully disposed.

Indeed, there needs no great variety of Colours to make a surprizing variety of Figures and Arrangements. *M. Truchet*, in the Memoirs of the *French* Academy, has shewn by the Rules of Combination, that two Square-Stones divided diagonally into two Colours, may be join'd together Checquer-wise 64 different Ways; which appears surprising enough; since two Letters or Figures are only combined two Ways.

The Reason is, that Letters only change their Situation with regard to first and second; the Top and Bottom remaining the same: But in the Arrangement of these Stones, each admits of four several Situations, in each whereof the other Square may be chang'd 16 Times, which gives 64 Combinations.

Indeed, from a further Examination of these 64 Combinations, he found there were only 32 different Figures; each Figure being repeated twice in the same Situation, tho' in a different Combination; so that the two only differ'd from each other by the Transposition of the darker or higher Squares.

Pavement of a Terrass, is that which serves for the covering of a Platform; whether it be over a Vault, or on a Wooden Floor. See TERRACE.

Those over Vaults are usually Stones squared, and bedded in Lead.

Those on Wood, call'd by the *Latins Pavimenta contignata*, are either Stones with Beds for Bridges; Tiles for Ceilings in Rooms, or Lays of Mortar, made of Ciment and Lime, with Flints or Bricks laid flat, as is still practis'd by the Eastern and Southern People a Top of their Houses. See PLATFORM.

All those *Pavements* which lye open, are call'd by the *Latins Pavimenta subdialia.*

PAVILLION, in Architecture, from the *Italian Padiglione*, Tent, of the *Latin Papilio*; signifies a Turret, or Building usually insulated, and under a single Roof; sometimes Square; and sometimes in form of a Dome.

Pavillions are sometimes also projecting Pieces, in the Facade of a Building, marking the Middle thereof; sometimes the *Pavillion* flanks a Corner, in which Case 'tis call'd *Angular Pavillion*. The *Louvre* is flank'd with four *Pavillions*. *Pavillions* are usually higher than the rest of the Building.

There are *Pavillions* built in Gardens; popularly call'd Summer-houses, Pleasure-houses, *etc*.

There are Castles or Forts which only consist in a single *Pavillion*.

PAVILLION, in War, a Tent, rais'd on Posts, to lodge under in the Summer-time.

PAVILLION, is also us'd in the General, for Flags, Colours, Ensigns, Standards, Banners, *etc*. all which, Authors usually confound with one another.

The Custom of bearing pointed *Pavillions*, as at present, first came from the *Mahometan Arabs*, at the Time they first seiz'd on *Spain*.

Till then, all Colours were stretch'd on cross Pieces like Thurch Banners; whence they were call'd in *Latin*, *Vexilla quasi Velilla*, a Diminutive of *Vela*, *Sails*.

PEDESTAL, in Architecture, the lowest Part of an order of Columns; being that which sustains the Column; and serves it as a Foot or Stand. See COLUMN.

The Word is form'd from the *Latin Pes Pedis*, Foot, and *Column*.

The *Pedestal*, call'd by the *Greeks*, *Stylobates* and *Stereo-bates*, consists of three principal Parts; *viz*. a square Trunk or *Dye*, which makes the Body; a *Corniche*, the Head; and a *Base* the Foot of the Pedestal.

The *Pedestal* is properly an Appendage to a Column; not an essential Part thereof; tho' M. *le Clerc* thinks 'tis Essential to a compleat Order. See ORDER.

The Proportions and Ornaments of the *Pedestal*, are different in the different Orders: *Vignola*, indeed, and most of the Moderns make the *Pedestal* and its Ornaments in all the Orders, one third of the Height of the Column, including the Base and Capital; but some deviate from this Rule.

M. *Perrault* makes the Proportions of the three constituent Parts of *Pedestals* the same in all the Orders, *viz*. the Base or Socle one fourth of the Pedestal; the Corniche an eighth Part; and the Socle or Plinth of the Base two thirds of the Base it self. The Height of the Dye is what remains of the whole Height of the *Pedestal*.

Tuscan PEDESTAL, is the simplest, and the lowest.

Palladio and *Scamozzi*, make it three Modules high; *Vignola* 5. See PROPORTION.

Its Members in *Vignola*, are only a *Plinth* for a Base, the *Dye*, and a

Talon crowned, for a Corniche.

The *Tuscan* Column has rarely any Base. See TUSCAN.

Doric PEDESTAL, *Palladio* makes four Modules, five Minutes high, *Vignola* [five] Modules four Minutes.

In the Antique, we not only don't meet with any *Pedestals*; but even, not with any Base in the *Doric* Order.

The Members in *Vignola's Doric Pedestal*, are the same with those in the *Tuscan*, with the Addition of a Mouchette in its Corniche. See DORIC.

Ionic PEDESTAL, in *Vignola* and *Serlio*, is 6 Modules high; in *Scamozzi* 5; in the Temple of *Fortuna Virilis* 'tis 7 Mod. 12 Min.

Its Members and Ornaments are mostly the same with those of the *Doric*, only a little richer.

The *Pedestal* now usually follow'd, is that of *Vitruvius*; tho' we don't find it in any Work of the Antique.

Some in lieu hereof Use the *Attic* Base, in imitation of the Ancients. See ATTIC.

Corinthian PEDESTAL is the richest and most delicate. In *Vignola* 'tis 7 Mod. high; in *Palladio* 5 Mod. 1 Min. in *Serlio* 6 Mod. 15 Min. in the *Coliseum* 4 Mod. 2 Min.

Its Members in *Vignola* are as follows: in the Base are a *Plinth* for a Socle, over that a *Tore* carved; then a *Reglet*, a *Gula* inverted and inrich'd, and an *Astragal.*

In the *Dye* are a *Reglet*, with the *Conge* over it, and near the Corniche a *Reglet* with a *Conge* underneath.

In the *Corniche* is an *Astragal* a *Frieze*, *Fillet*, *Astragal*, *Gorge*, *Talon*, and a *Fillet*. *See each under its proper Article.*

Composite PEDESTAL, in *Vignola*, is of the same Height with the *Corinthian*, *viz*. 7 Mod. in *Scamozzi* 6 Mod. 2 Min. in *Palladio* 6 Mod. 7 Min. in the *Goldsmiths* Arch 7 Mod. 8 Min.

Its Members in *Vignola*, are the same with those of the *Corinthian*; with this Difference, that whereas these are most of 'em enrich'd with Carvings in the *Corinthian*, they are all plain in the *Composite*.

Nor must it be omitted, that there is a Difference in the Profiles of the Base and Corniche in the Two Orders.

The Generality of Architects, *Daviler* observes, use Tables or Pannels, either in *Relievo* or *Creux*, in the Dyes of *Pedestals*; without any regard to the Character of the Order. Those in *Relievo*, he observes, only fit the *Tuscan* and *Doric*; the three others must be indented; which he adds, is a Thing the Ancients never practis'd, as being contrary to the Rules of Solidity.

Square PEDESTAL, is that whose Height and Width are equal; as that of the Arch of the Lyons at *Verona*, of the *Corinthian* Order; and such, some Followers of *Vitruvius*, as *Serlio*, *Philander*, etc. have given to their Tuscan Orders.

Double PEDESTAL, is that which supports two Columns, and is more in Width than Height.

Continued PEDESTAL, that which supports a Row of Columns without any Break or Interruption; such is that which sustains the fluted Ionic Columns of the Palace of the *Tuillieres* on the Side of the Garden.

PEDESTALS *of Statues*, are those serving to support Figures or Statues. See STATUE.

Vignola observes, there is no Part of Architecture more Arbitrary, and wherein more Liberty may be taken than in the *Pedestals* of Statues; there being no Laws prescribed by Antiquity; nor any even settled by the Moderns.

There is no settled Proportion for these *Pedestals*; but the Height depends on the Situation, and the Figure they sustain. Yet, when on the Ground, the *Pedestal* is usually two thirds, or two fifths of that of the Statue: But the more Massive the Statue, the stronger the *Pedestal.*

Their Form, Character, *etc.* are to be Extraordinary and Ingenious, far from the Regularity and Simplicity of the *Pedestals* of Columns. The same Author gives us a great Variety of Forms, Oval, Triangular, Multangular, *etc.*

PEDIMENT, in Architecture, a Kind of low Pinnacle; serving to crown an Ordinance, or finish a Frontispiece; and placed as an Ornament over Gates, Doors, Windows, Niches, Altars, *etc.*

The Pinnacles of the plainest Houses, *Vitruvius* observes, gave Architects the first Idea of this noble Part; which still retains the Appearance of its Original. See PINNACLE.

The Parts of the *Pediment* are, the *Tympanum*, and its *Corniche.*

The first is the Pannel naked, or Area of the *Pediment*, enclos'd between the Corniche, which crowns it, and the Entablature, which serves it as a Base or Socle. See TYMPAN, *etc.*

Architects have taken a deal of Liberty in the Form of this Member: Nor do they vary less as to the Proportion of the *Pediment.*

The most beautiful, according to *Daviler*, is that where its Height is about one fifth of the Length of its Base.

It is described thus: divide the Line *a b*, (Tab. Architecture Fig. II) which is the Length of the Base, into two equal Parts, in the Point *c*, by means of the Perpendicular *f d*; in this Perpendicular, take the Part *e d*, equal to *a c*; and from the Point *d*, as a Centre, describe the Arch *a e b*. The Point of the Perpendicular cut in *e*, will be the Top of the *Pediment* *a e b*; and the Corniche, and the Triangular Space included therein, the *Tympanum.*

Vitruvius calls the *Pediments*, *Fastigia*; a Word which signifies a Roof rais'd, or pointed in the Middle, which Form among the *Romans* was peculiar to Temples. All their Dwelling-houses are cover'd in the Plat-form manner; and *Salmasius* on *Solin* observes, that *Cæsar* was the first who obtain'd leave to Roof his House with a Ridge or Descent, after the manner of Temples. See PLATFORM.

Pliny tells us, that *Pediments* were First made to place Statues upon, whence they were call'd *Plastæ*.

The *Pediment* is usually Triangular, and sometimes an equilateral Triangle, call'd also a *pointed Pediment*; sometimes it is *Circular*; tho' *Felibien* observes, that we have no Instance of round *Pediments* in the Antique, beside those in the Chapels of the *Rotondo*.

Sometimes its upper Corniche is divided into three or four Sides, or right Lines. Sometimes the Corniche is cut, or open a Top; which is an Abuse introduced by the Moderns, particularly *Michael Angelo*; for the Design of this Part, at least over Doors, Windows, *etc.* being chiefly to shelter those underneath from the Rain; to leave it open in the Middle, is to frustrate its End.

Sometimes the *Pediment* is form'd of a couple of Rolls, or Wreathes, like two Consoles join'd together. See CONSOLE.

Sometimes the *Pediment* is without Base, or its lower Corniche is cut out, all but what is bestow'd on two Columns, or Pilasters, and on these an Arch or Sweep rais'd, in lieu of an Entablature; of which *Serlio* gives an Instance in the Antique, in a *Corinthian* Gate at *Foligny*, in *Umbria*; and *Daviler*, a more Modern one, in the Church of St. *Peter* at *Rome*.

Under this kind of *Pediments*, come those little arch'd Corniches, which form *Pediments* over Doors and Windows, supported by two Consoles, in lieu either of Entablature or Columns.

Sometimes the *Pediment* is made double, *i.e.* a less *Pediment* is made in the *Tympanum* of a larger, on account of some Projecture in the Middle; as in the Frontispiece of the Church of the Great *Jesus* at *Rome*; but this Repetition is an abuse in Architecture; tho' authoriz'd by very good Buildings; as the large Pavillion of the *Louvre*, where the *Caryatides* support three *Pediments* one in another.

Sometimes the *Tympanum* of the *Pediment* is cut out, or left open, to let in light; as we see under the Portico of the Capitol at Rome.

Lastly, this open *Pediment* is sometimes triangular, and enrich'd with *Sculpture*, as Roses, Leaves, etc. as we find it in most of the *Gothic* Churches.

M. le Clerc observes, that the Modillions in the Corniche of the *Pediment* shou'd always answer exactly over those of the Entablature. Indeed, *Vitruvius* says, that Ancients did not allow any Modillions at all in *Pediments*.

The same M. *le Clerc* observes, that the Corniche which serves the *Pediment* at the Base, shou'd have no *Cymatium*; by reason the *Cymatium* of the rest of the Entablature, when it meets the *Pediment*, passes over it.

This Change of Determination occasions a considerable Difficulty; the *Cymatium*, in this Case, appearing too broad in the Turn of the Angle. To remedy which, the Architects have recourse to several Expedients.

A Pointed Pediment may crown three Arches; but a *circular Pediment* can only crown one agreeably.

One wou'd never use above two *Tympana* over each other in the same

Frontispiece; and even where there are two, 'twou'd be proper to have the lower Circular, the upper pointed.

PENCIL, an Instrument used by Painters, for the Application of their Colours.

There are *Pencils* of various kinds, and made of various Matters: The most usual are of Badgers and Squirrels Hair, those of Swans-down, and those of Boars Bristles; which last are bound on to a Stick bigger, or less, according to the Uses they are destined for; and when large are call'd *Brushes*.

The others are inclos'd in the Barrel of a Quill.

The Word comes from the *Latin*, *Penicillum*, which signifies the same Thing. The Ancients, M. *Felibien* observes, had *Pencils* made of little Pieces of Spunge; whence doubtless, the Story of the Painter, who not able to express the Foam of a Dog, succeeded by throwing his *Spunge* at the Picture.

PENDENTIVE, in Architecture, the whole Body of a Vault, suspended out of the Perpendicular of the Walls, and bearing against the Arc-boutants. See VAULT.

Daviler describes it as a Portion of a Vault between the Arches of a Dome, usually enrich'd with Sculpture: And *Felibien*, as the Plain of the Vault, contain'd between the double Arches, the forming Arches and the Ogivcs. See OGEE.

The *Pendentives* are usually of Brick, or soft Stone; but Care must be taken that the Couches or Beds of Masonry be always laid level, and in right Lines proceeding from the Sweep whence the Rise is taken.

The Joints too must be made as small as possible, to save the Necessity of filling 'em up with Wood, or of using much Mortar.

PENTASTYLE, in Architecture, a Work wherein are five Rows of Columns.

Such was the *Portico* begun by the Emperor *Gallian*, and which was to have been continued from the *Flaminian-Gate* to the Bridge *Milvius*, *i.e.* from *Porto del Populo*, to the *Ponte-mole*.

PERAMBULATOR, in Surveying, an Instrument for the measuring of Distances, call'd also *Pedometer*, *Way-wiser*, and *Surveying Wheel*.

Its Advantages are its Handiness, and Expedition; its contrivance is such, as that it may be fitted to the Wheel of a Coach; in which it performs its Office, and measures the Road without any Trouble at all.

PERIPTERE, in the ancient Architecture, a Building encompass'd without-side by a Series of insulated Columns, forming a kind of Isle all around.

Such were the Basilic of *Antonine*, the Septizon of *Severus*, the

Portico of *Pompey*, *etc.* The *Peripteres* were properly Temples, which had Columns on all the four Sides; by which they were distinguish'd from the *Prostyles* and *Amphiprostyles*, the one of which had no Columns before, and the other none on the Sides. See PROSTYLE, *etc.*

M. *Perrault* observes, that *Periptere* in its general Sense, is the Name of a Genus, including all the Species of Temples which have Portico's of Columns all around; whether the Columns be Diptere, or Pseudo-diptere, or simply *Periptere*, which is a Species that bears the Name of the Genus, and which has its Columns distant from the Wall by the Breadth of an Intercolumniation. See TEMPLE.

For the Difference between Periptere and Peristyle. See PERISTYLE.

The Word is form'd of the *Greek circum*, about and *Ala*, Wing.

PERISTYLE, in the ancient Architecture, a Place, or Building, incompass'd with a Row of Columns on the Inside; by which it is distinguish'd from the *Periptere*, where the Columns are disposed without-side. See PERIPTERE.

Such was the Hypetre Temple of *Vitruvius*; and such are now some Basiliques in *Rome*, several Palaces in *Italy*, and most Cloisters of Religious.

Yet, the Word *Peristyle* is still used indifferently for a Range of Columns, either within or without a Building: As the Corinthian *Peristyle* of the Portail of the *Louvre*, etc.

The Word is form'd from the *Greek about*, and *Column*.

PERRON, in Architecture, a Stair-Case lying open, or without-side the Building; properly, the Steps in the Front of a Building, which lead into the first Story when rais'd a little above the Level of the Ground. See STAIRS.

Perrons are made of different Forms and Sizes, with regard to the Space and Height they are to lead to.

Sometimes the Steps are round, or Oval; more usually Square.

PERSIAN or PERSIC, in Architecture, a Name common to all Statues of Men, serving instead of Columns, to support Entablatures. See STATUE.

They only differ from Caryatides, in that those represent Statues of Women. See CARYATIDES.

The *Persian* is a Kind of Order of Columns, first practis'd among the *Athenians*; on occasion of a Victory their General *Pausanias* obtain'd over the *Persians*. As a Trophy of this Victory, the Figures of Men dress'd in the *Persian* Mode, with their Hands bound before them, and other Characters of Slavery, were charg'd with the Weight of Doric Entablatures; and made to do the Office of Doric Columns. See ORDER.

Persian Columns, M. *le Clerc* observes, are not always made with the

Marks of Slavery; but are frequently used as Symbols of Virtues, Vices, of Joy, Strength, Valour, *etc.* as when made in the Figure of *Hercules* to represent Strength, of *Mars*, *Mercury*, *Fauns*, *Satyres*, etc.

PERSPECTIVE, the Art of Delineating visible Objects on a plane Surface, such as they appear at a given Distance or Height, upon a transparent Plane, placed perpendicular to the Horizon, between the Eye and the Object.

This, we particularly call *linear Perspective*, as regarding the Position, Magnitude, Form, *etc.* of the several Lines, or Contours of Objects; and expressing their Diminution: In Opposition to the *Aerial Perspective*, which regards the Colour, Lustre, Strength, Boldness, *etc.* of distant Objects, consider'd as seen thro' a Column of Air; and expresses the Diminutions thereof.

The former is a Branch of Mathematicks: Some make it a Member of Opticks; others a Rivulet therefrom; and its Operations, are all Geometrical.

The latter is a Part of Painting, and consists wholly in the Conduct of the Colours, their different Teints, or Degrees, Force, Weakness, *etc.*

Some make a third Kind of *Perspective*, *viz. Specular Perspective*; which represents the Objects in Conical, Spherical, or other Mirrors, erect, and clear; whereas on Lawn and other Planes appear confused and irregular.

To conceive the Nature of *Perspective*; *i.e.* Linear *Perspective*: Suppose a Glass-plane *H I*, (Tab. *Perspective* Fig. 1.) rais'd perpendicularly on a horizontal Plane; and the Spectator *S*, directing his Eye *O*, to the Triangle *A B C*: If now we conceive the Rays *A O*, *O B*, *O C*, *etc.* in their Passage thro' the Plane, to leave their Traces or Vestigia, in *a*, *b*, *c*, *etc.* On the Plane; there will appear the Triangle *a b c*; which, as it strikes the Eye by the same Rays *a O*, *b O*, *c O*, by which the Species of the Triangle *A B C* is carried to the same; it will exhibit the true Appearance of the Triangle *A B C*, tho' the Object shou'd be remov'd; the same Distance and Height of the Eye being preserv'd.

The Business of *Perspective* then, is to shew by what certain Rules the Points *a*, *b*, *c*, *etc.* may be found Geometrically: And hence also, we have a mechanical Method of delineating any Object very accurately.

Perspective is either employ'd in representing the Ichnographies, or Ground-plats of Objects, as projected on *Perspective* Planes.

Or in Scenographies, or Representations of the Bodies themselves. See SCENOGRAPHY.

The general Laws of each, are subjoin'd; in order to which it is necessary to premise the following Lemmas in *Perspective*.

1. That the Appearance of a right Line is ever a right Line; whence, the two Extremes being given, the whole Line is given.

2. That if a Line be Perpendicular to any right Line drawn on a Plane, it will be Perpendicular to every other right Line drawn on the same

Plane.

3. And that the Height of a Point appearing on the Plain, is to the Height of the Eye, as the Distance of the objective Point from the Plane, to the Aggregate of that Distance and the Distance of the Eye.

Ichnographic PERSPECTIVE; *or the Laws of the Projection of Plane Figures.*

Perspective of a Point.

To exhibit the Appearance h *of an objective Point*, H. (Fig. 2.) From the given Point, draw a Perpendicular to the fundamental Line *D E*. From the fundamental *D E* cut off *I K* = *I H*; thro' the Point of Sight *F* draw a horizontal Line *E P*; and make *F P* equal to the Distance of the Eye *S L*: Lastly from the Point *I* to the Point of Sight *F* draw *F I*; and from *K* to the Point of Distance *P*, the Line *P K*. The Intersection *h* is the Appearance of the objective Point.

Hence, 1. Since, the Appearance of the extreme Points of a right Line being given, the Appearance of the whole Line is given; the Ichnographic Projection of any Rectilinear Figure may be had by this Method.

And, 2. Since any Number of Points of a Curve Line may by this means be projected on the *Perspective* Plane; the Projection of Curve Lines may likewise be effected after the same manner.

3. Therefore, this Method will likewise suffice for Mixtilinear Figures; and is consequently universal.

There are indeed other Methods deliver'd by other Authors, but this is the most usual. To conceive its force and effect, it will be proper to illustrate it with some Examples.

Perspective of a Triangle.

To find the Appearance of a Triangle, A B C, (Fig. 3.) whose Base *A B* is parallel to the fundamental Line *D E*. To the fundamental Line *D E* draw a Parallel at an Interval equal to the Altitude of the Eye. Assume a fundamental Point *V*, opposite to this either directly or obliquely, as the Case requires. Transfer the Distance of the Eye from *V* to *K*. From the several Angles of the Triangle *A C B*, let fall Perpendiculars *A* 1, *C* 2, *B* 3: set off these Perpendiculars upon the fundamental Line *D E* opposite to the Point of Distance *K*. From 1, 2, 3, draw right Lines to the fundamental or principal Point *V* 1, *V* 2, *V* 3. From the Points *A*, *B* and *C* of the fundamental Line *D E* draw other right Lines *A K*, *B K*, *C K*, to the Point of Distance *K*.

Since *a*, *b*, and are the Appearances of the Points *A*, *B* and *C c*; The right Lines *ca*, *ab* and *bc*, being drawn *a c b* will be the Appearance of the Triangle *A C B*.

After the same Manner is a Triangle projected on a Plane, where the Vertex *C* is opposed to the Eye: All here requir'd, is, that its Situation on the Geometrical Plane be changed, and the Vertex *C* turn'd towards the fundamental Line *D*.

Perspective of a Square.

To exhibit the Appearance of a Square, A B D C (Fig. 4.) *seen obliquely*, and having one of its Sides *A B* in the fundamental Line. The Square being view'd obliquely assume the principal Point *V* in the horizontal Line *H R*, in such manner as that a Perpendicular to the fundamental Line may fall without the Side of the Square *A B*, at least, may not bissect it; and make *V K* the Distance of the Eye. Transfer the Perpendiculars *A C* and *B D* to the fundamental Line *D E*; and draw the right Lines *K B*, *K D*, as also *A V*, *V C*. Then will *A* and *B* be their own Appearances; and *c* and *d* the Appearances of the Points *C* and *D*. Consequently *A c d B* is the Appearance of the Square *A B D C*.

If the Square *A C B D* shou'd be at a Distance from the fundamental Line *D E*; which yet rarely happens in Practice; the Distances of the Angles *A* and *B* must likewise be transferr'd to the fundamental Line: As is evident from the preceeding Problem. And since, even the oblique View is not very common; in what follows, we shall always suppose the Figure to be posited directly opposite to the Eye; unless, where the contrary is expressly mention'd.

4. *To exhibit the Appearance of a Square* A B C D (Fig. 5.) *whose Diagonal* A C *is Perpendicular to the fundamental Line.* Continue the Sides *D C* and *C B* till they meet the fundamental Line in 1 and 2. From the principal Point *V* set off the Distance of the Eye to *K* and *L*. From *K* to *A* and 1 draw right Lines *K A* and *K* 1; and from *L* to *A* and 2, the right Lines *L A*, *L* 2. The Intersections of these Lines will exhibit the Appearance of the Square *A B C D* view'd Angle-wise.

5. *To exhibit the Appearance of a Square* A B C D *Fig. 6. wherein another*, I M G H *is inscribed*; the Side of the greater, *A B*, being in the fundamental Line; and the Diagonal of the less, Perpendicular to the Fundamental. From the principal Point *V*, set off, each way, on the horizontal Line *H R*, the Distances *V L* and *V K*; draw *V A* and *V B*; and *K A* and *L B*; then will *A c d B* be the Appearance of the Square *A C D B*. Produce the Side of the inscrib'd Square *I H*, till it meet the fundamental Line in 1; and draw the right Lines *K* 1, and *K M*; then will *i h g M* be the Representation of the inscrib'd Square *I H G M*.

Hence is easily conceiv'd the Projection of any Figures inscrib'd in others.

Perspective of a Pavement.

5. *To project a Pavement consisting of square Stones, view'd directly.* Divide the Side *A B* transfer'd to the fundamental Line Fig. 7. into as many equal Parts as there are square Stones in one row. From the several points of Division, draw right Lines to the principal Point *V*; and from *A*

to the Point of Distance *K*, draw a right Line *A K*; and from *B* to the other Point of Distance *L*, draw another *L B*. Thro' the Points of the Intersections of the corresponding Lines, draw right Lines *V* on each Side, to be produced to the right Lines *A V*, and *B*. Then will *A f g B* be the Appearance of the *Pavement A F G B*.

Perspective of a Circle.

6. *To exhibit the Appearance of a Circle.* 1. If the Circle be small, circumscribe a Square about it. Draw Diagonals and Diameters *ha* and *de* (Fig. 8) intersecting each other at right Angles; and draw the right Lines *f g* and *b c* parallel to the Diameter *de* thro' *b* and *f*; as also thro' *c* and *g* draw right Lines meeting the fundamental Line *D E* in the Points 3 and 4. To the principal Point *V* draw right Lines *V* 1, *V* 3, *V* 4, *V* 2; and to the Points of Distance *L* and *K*, draw the right Lines *L* 2 and *K* 1. Lastly connect the Points of Intersection, *a*, *b*, *d*, *f*, *h*, *g*, *e*, *c* with the Arches *a b*, *b d*, *d f*, *etc.* Thus will *a b d f h g e c a* be the Appearance of the Circle.

If the Circle be large, on the Middle of the Fundamental *A B* (Fig. 9.) describe a Semicircle; and from the several Points of the Periphery, *C*, *F*, *G*, *H*, *I*, *etc.* to the fundamental Line, let fall perpendiculars *C* 1, *F* 2, *G* 3, *H* 4, *I* 5, *etc.* From the Points 1, 2, 3, 4, 5, *etc.* of *A B* draw right Lines to the principal Point *V*, as also a right Line from *B* to the Point of Distance *L*; and another from *A* to the Point of Distance *K*. Thro' the common Intersections, draw right Lines as in the preceeding Problem; thus shall we have the Points *c*, *f*, *g*, *h*, *i*, which are the Representations of these *A*. *C*, *F*, *G*, *H*, *I*, which being connected as before, give the Projection of the Circle.

Hence appears not only how any curvilinear Figure may be projected on a Plane; but also how any Pavement, consisting of any kind of Stones, may be delineated in *Perspective*.

Hence also, appears what Use the Square is of in *Perspective*, for even in the second Case we use a Square divided into certain Areolæ, and circumscribed about the Circle; tho' it be not delineated on the geometrical Plane in the Diagram.

Perspective of a regular Pentagon.

7. *To represent a regular Pentagon, having a broad Limb, terminated by Lines parallel thereto*: 1°. From the several Angles of the exterior Pentagon *A*, *B*, *C*, *D*, *E*, Fig. 10. to the fundamental Line *T S*, let fall Perpendiculars *A* 0, *B* 1, *C* 2, *D* 3, *E* 4; which, as in the former, transfer to the fundamental Line. Connect the Points 1, 2, 3, 4 to the principal Point *V*; and the Points 1, 2, 3, 4 to the Point of Distance *K*. Thus will the common Intersections represent the Appearance of the exterior Pentagon. 2. If now, from the inner Angles *G H L I*, the Perpendiculars *G* 0, *H* 5, *K* 6, *I* 7, *L* 8, be in the like manner let fall; and the rest be done as in the former; we shall have the Representation of the inner Pentagon. The

Pentagon *A B C D E*, therefore, with its Limb, is represented in *Perspective*.

This Problem is added for the sake of an Instance of the Projection of a Figure that has a broad Limb, or Edge.

It must be here observ'd, that if the Magnitudes of the several Parts of an Object, be given in Numbers, together with the Height and Distance of the Eye; its Figure is to be first constructed by a geometrical Scale; and the fundamental Point with the Point of Distance 1, to be determin'd by the same.

Nor is it always necessary, that the Object be delineated under the fundamental Line; in the Projection of Squares and Pavements 'tis best let alone. But where 'tis necessary, and Space is wanting; draw it a-part; find the Divisions in it, and transfer 'em to the fundamental Line in the Plane.

Threads being hung in the principal Point, and the Point of Distance, and stretch'd to the Points of the Divisions of the fundamental Line; the common Intersection of the Threads will give the Projection of the several Points without Confusion; a Thing much to be fear'd from the Multiplicity of Lines to be drawn.

Scenographic Perspective; or the Projection of Bodies on a Plane.

On the given Point C, (Fig. 11.) *to raise a* Perspective *Altitude, answerable to the given objective Altitude P Q.* On the fundamental Line, raise a Perpendicular *P Q*, equal to the given objective Altitude. From *P* and *Q* to any Point, as *T*, draw right Lines *P T* and *Q T*. From the given Point *C* draw a right Line *C K*, parallel to the fundamental Line *D E*; and meeting the right Line *Q T*, in *K*. In *K*, erect a Perpendicular to *K C*, *viz. I K*; this *I K* is the scenographic Altitude requir'd.

Perspective of a Solid.

To exhibit the Scenography of any Solid. Find the Projection of its Base in the Ichnographic *Perspective*; and in the several Points thereof erect the *Perspective* Altitude: Thus will the Scenography of the Solid be finish'd, except for what relates to the Shadow; which must be superadded from the Laws of Shadows, deliver'd under the Article *Shadow*. For an Example.

To exhibit the scenographic Perspective *of a Cube* view'd Angle-wise: Since the Base of a Cube view'd Angle-wise, and standing on a geometrical Plane is a Square view'd Angle-wise; draw a Square on the *perspective* Plane, after the manner laid down above, raise the Side of the Square *H I* (Fig. 12.) perpendicularly in some Point of the fundamental Line *D E*; and to any Point *V*, of the horizontal Line *H R*, draw right Lines *V I* and *V H*. From the Angles *d*, *b*, and *c*; draw *c* 1, *d* 2 parallel to the fundamental Line *D E*. From the Points 1 and 2 raise *L* 1 and *M* 2 Perpendicular to the same. Lastly, Since *H I* is the Altitude to be rais'd in *a*, *L* 1 in *c* and *h* and *M* 2 in *d*; in *a* raise *f a*, Perpendicular to *a E*; and in *b* and *c*, raise *b g* and *c e* Perpendicular to *b c* 1; and lastly raise *d h* Perpendicular to *d* 2; and let *a f* be equal to *H I*, *b g* and *e c* to *L* 1 and *h d* to *M* 2; If then the Points *g*, *h*, *e*, *f* be connected by right Lines, the Scenography will be finish'd.

This Method is general; but its Application is not equally obvious in every Case; see it further illustrated under the Article SCENOGRAPHY.

PERSPECTIVE *of Building*, etc.

In the Practice of the *Perspective* of Building, *etc.* Great regard is had to the Height of the horizontal Line; all above the horizontal, being seen in the upper Part, and all above it in the under Part, whence *Perspective* becomes divided into the *high* and *low* Sight; both which will be well illustrated by what follows.

To represent a Building (*v.g.* Palace, College, *etc.*) in *Perspective.* 1. Take the Ichnography, or Ground-plat of the Building; its Lengths, Breadths, and Depths; by actual measuring.

2. Make a Scale divided into two or three hundred equal Parts, either actually, or so as that each division signifie ten Parts: By this Scale lay down the *Ground-plot*, as in Figure 13.

This done, having a long Rule, and a Square, which by sliding on the Rule helps you to draw your Perpendiculars easilier, reduce it into *Perspective*, in its *Scenographick Appearance*.

Then having drawn a Line towards the Bottom of the Paper for the Front or Base Lineas; *F L* Fig. 14 divide it into as many equal Parts as you find the Building has in the Ichnography, or more if you please: This will serve for a Scale to determine the several Heights, *etc.* and to these Divisions, with a black Lead Pencil draw Lines from the Centre, when you have chosen it; which Choice requires Judgement on two accounts.

For, if the Centre be too nigh the Front-line, then the Depth of the whole Building will fore-shorten too much; if too far off, it will not fore-shorten enough. This may be illustrated thus; set an open Tankard, or the like on a Stand, so as that it be a little lower than your Eye; if you be a great Distance from it, you can see very little or nothing into it; if you come nigher to it by degrees, you will perceive the farther Edge seem to be rais'd a little higher than that next you, so that you may see a little Way into it; if you come very nigh it, you see too deep into it more than can well be exprest in Picture. We shall therefore find some one Place, which we must conclude the most convenient for the Draught, and which may be in general determin'd to be as far off the Front-line as the Front-line is long: This Rule, tho' it has just grounds, yet we sometimes dispence with it *pro re nata*; that we may express those Things with the better Appearance.

4. Consider how to place this Centre with such Advantage as that we may express those Things most, which we chiefly design to do; for as to the bottom and top Lines of the Sides of the Building that run from us in

or nigh the direct Line to the Centre, tho' you see the upper Part very well, yet the Sides that fall between the Ground line and Top, fall so very near one another, that 'twould be very difficult to express Particulars in them; so that the Centre must be well chosen in reference to this.

Those Buildings therefore, you would see most of, must be plac'd as far off as you think convenient from the direct Line that runs to the Centre; and the farther they are, the plainer they are.

Place then those Things you would see least of, nighest the direct Line; and see whether the others fall according to your Mind; but this must be done after you have drawn your Diagonal, which is the next Thing.

5. Having pitch'd on your Centre, and having from it drawn Lines to every Division of the Front-line, you are to determine your Diagonal, *A R*, thus: Having with a pair of Compasses measur'd the Length of the Front-line, take your Compasses, and putting one Foot in the Centre, see where the other will reach in the Horizon: (on both Sides if you please) where it rests, from that Point draw a thwart Line from it to the last Division of the Front; and this will be truely drawn, or pretty nigh to the Truth. That this is so, you may consider how it falls in respect of the two last Centre-lines: For if where the next Line from the last is intersected by the Diagonal, you draw a Parallel to the Front between them, as at *A* 10 you will have a *Rhombus*; if then all the Sides be pretty equal, you may be sure you are nigh the right; but if the Sides that run toward the Centre be too long, then Things will not fore-shorten enough; if the Sides be not long enough, then they will fore-shorten too much.

6. After the Front-line is thus divided, the Centre fixt, and the Diagonal placed, take the Breadth of the Chappel *A. B.* which in the *Ichnography* is shewn to be twenty Parts; because this Line is Perpendicular, it must run toward the Centre, therefore reckon twenty in the Diagonal, and the Rule laid parallel to the Front in that Point, will give you a Point in the Centre-line which will give the Breadth of the Chappel; Consequently a Line drawn from *A*. to *B*. puts it into the Ichnographick *Perspective*. The Length of the Chappel being seventy Divisions in the Front-line; reckon seventy from *B*. parallel to the Front-line, and there you will have a Point at *C*,—

The Depth of the Building from the Chappel Northward, being 115 from the Chappel, I reckon from *D*; (where it cuts the Diagonal at ten) onwards in the Diagonal; and at 115. in the Diagonal, with my Rule as before parallel in this Place in the Front, I have the Point *Z*. in the Central-line. Its Breadth being thirty, I reckon three Divisions, and there is the just Breadth there; and so on in every particular Part.

Having placed the Ichnography into *Perspective*, you may then give every Thing its proper Height thus:

7. The Height of the Chappel being 30, I reckon 30 on the Front-line, and with this Length by a Square clapt to the Front-line, I drop a Perpendicular to that Height, and so where the other Side of the Chappel

is plac'd, having reckon'd the Height upon a Supposed Parallel, there I draw another Line in that Height; then joining these several Heights by several Lines, you have the Profiles of each Building.

Now to diversifie these several Lines, that they confound you not, make the Ichnography, when you lay it into *Perspective*, in discontinued crooked Lines, the Heights in prick't Lines, and the Tops of each Building in continued Lines, as the Centre-lines are, in the Table. You will likewise find the Centre, tho' 'tis not here exprest, as likewise the Point of Distance, by continuing the Diagonal up to the supposed Horizon where it and the Eye is placed.

Having done thus, your Art must be employ'd for the particular Expressions of Things, by drawing and shadowing, which is the Life of this half-form'd Figure, and which we leave to the Painter.

It remains that we speak of the *Low-sight*: Here we suppose the Horizontal just the Height of the Eye, about 5 Foot from the Basis; tho' 'tis generally plac'd higher, even to a third Part of the Height of the Building, that the Side Buildings may be express'd more gracefully.

The Diagonal is best determined by dividing the last Division of the Basis-line into 5 Parts at *G*. Fig. 14 taking 4 of these, sometimes the whole 5, because we determin'd before, that the Length of the Front-line was the Distance of the Eye in the Horizon to the Point of Distance: But here I take 4, and then make this the Distance in the Horizon between the Eye and the Point of Distance. You may then either graduate the Plan at the several Intersections of the Diagonal with the Centre-lines, or else suppose it so; and then raise the Building, as you will find by *Perspectives* enough of this Sort every where to be met with.

PERSPECTIVE, is also used for a Kind of Picture or Painting, frequently seen in Gardens, and at the Ends of Galleries; design'd expressly to deceive the Sight by representing the Continuation of an Alley, a Building, Landskip, or the like.

PERSPECTIVE *Plane*, is a Glass, or other transparent Surface, suppos'd to be placed between the Eye and the Object, Perpendicular to the Horizon, unless the contrary be expressly mention'd.

Such is the Plane, *H I* (Fig. 15.) between the Eye *O* and the Object *A B C*, cutting the optic Rays in *a*, *b*, *c*.

This, some call the *Section*; some the *Table*, and others the *Glass*.

PHAROS, PHARE, or PHANAL, a *Light-house*; a Pile rais'd near a Port, where a Fire is kept burning in the Night to guide and direct Vessels near at Hand.

The *Pharos of Alexandria* was antiently very famous, in so much as to communicate its Name to all the rest; the *Collosus of Rhodes* serv'd as a *Pharos*.

Ozanam says, *Pharos* antiently signified a Streight, as the *Pharos* or *Pharo* of *Messina*.

PIAZZA, in Building popularly call'd *Piache*, an *Italian* Name for a Portico. See PORTICO.

The Word literally signifies a broad open Place, or Square; whence it became applied to the Walks, or Portico's around them.

PICKET, or PICQUET, or PIQUET, in Fortification, a Stake sharp at one End, and usually shod with Iron; used in laying out the Ground, to mark the several Measures and Angles thereof.

There are also larger *Pickets*, drove into the Earth to hold together Fascines, or Faggots, in any Work cast up in haste.

PICTURE, a Piece of Painting; or a Subject, represented in Colours, on Canvas, Wood, or the like, and inclosed in a Frame. See PAINTING.

PIEDOUCH, in Architecture, a little Stand, or Pedestal, either long or square, enrich'd with Mouldings; serving to support a Bust, or other little Figure. See PEDESTAL, BUST, *etc.*

The Word is *French*; form'd from the *Italian peduccio*, *little foot.*

PIEDROIT, in Architecture, a *Peer*; or a kind of square Pillar, part whereof is hid within a Wall. See PILLAR.

The only thing wherein it differs from a Pilaster, is, that the latter has a regular Base and Capital, which the other wants. See PILASTER.

PIEDROIT is also used for a Peer, or Jaumb of a Door, or Window; comprehending the Chambranle, Chamfering, Leaf, *etc.* See DOOR, WINDOW *etc.*

PIGMENTS, PIGMENTA, Preparations used by Painters, Dyers, *etc.* to impart Colours to Bodies, or to imitate particular Colours.

When Glass is stained, or coloured, as in Painting on Glass, or for the couterfeiting of Gems, or Precious Stones; the *Pigment* is usually of a Metalline, or a Mineral Nature.

PILA, *Peer*, in Building, a Massive of Masons Work in manner of a Pillar; usually hexagonal.

Such are those serving as Fulcra, to separate and support the Arches of a Stone Bridge, or the Beams of a wooden Bridge.

The *Peers* of a Stone Bridge are not to be less than one sixth Part of the Arch, nor more than a fourth. See BRIDGE.

The Word comes from the *Latin*, *Pila*, used in the same Sense by *Vitruvius*; and this, probably, from *cogo*, *coarcto.*

Evelyn observes, that *Pilæ* and their Quadras, as we see them in ancient Altars and Monuments, were used for Inscriptions; but the shorter and more massy serve for Arches and Buttresses to solid Work.— They were sometimes made half circular: But the Ancients preferr'd those pointed at right Angles, as best able to resist the Current.

PILASTER, in Architecture, a square Column, sometimes insulated, but more frequently let within a Wall, and only shewing a fourth or fifth Part of its Thickness. See COLUMN.

The *Pilaster* is different in different Orders; it borrows the Name of each Order, and has the same Proportions and the same Capitals, Members, and Ornaments, with the Columns themselves. See ORDER.

Pilasters are usually without either Swelling or Diminution, as broad at top as at the bottom; tho' some of the modern Architects, as M. *Mansard*, etc. diminish them at top, and even make them swell in the middle, like Columns; particularly when placed behind Columns.

Pilasters, M. *Perrault* observes, like Columns, become of different Kinds, according to the different Manner wherein they are applied to the Wall.—Some are wholly detach'd, call'd by *Vitruvius*, *Parastatæ*; others have three Faces clear out of the Wall; others two; and others only one; all call'd by *Vitruvius*, *Antæ*.

Insulated *Pilasters* are but rarely found in the Antique—The chief Use they made of *Pilasters* was at the Extremities of Portico's, to give the greater Strength to the Corners.

There are four principal Things to be regarded in *Pilasters*: *viz*. their Projecture out of the Wall, their Diminution, the Disposition of the Entablature when it happens to be common to them and to a Column, and their Flutings and Capitals.

1° Then, the Projecture of *Pilasters* which have only one Face out of the Wall, is to be one eighth of their Breadth; at most not above one sixth. When they receive Imposts against their Sides, the Projecture may be a Quarter of their Diameter.

2° *Pilasters* are but seldom diminish'd, when they have only one Face out of the Wall.—Indeed where they stand in the same Line with Columns, and the Entablature is continued over both, without any Break, the *Pilasters* are to have the same Diminution with the Columns; that is to say, on the Face respecting the Column; the Sides being left without any Diminution.

3° *Pilasters* are sometimes fluted, tho' the Columns they accompany be not; and, on the other hand, the Columns are sometimes fluted, when the *Pilasters* that accompany them are not.

The Flutings of *Pilasters* are always odd in Number, except in half *Pilasters*, meeting at inward angles; where four Flutings are made for three, *etc*. See FLUTES.

4° The Proportions of the Capitals of *Pilasters*, are the same as to Height with those of Columns, but differ in Width, the Leaves of the former being much wider, because *Pilasters*, tho' equal Extent, have only the same Number of Leaves for their Girt, *viz*. Eight.—Their usual Disposition is to have two in each Face, in the lower Row, and in the upper Row one in the middle, and two halves in the Angles, in the Turns whereof they meet.—Add to this, that the Rim of the Vase or Tambour is not strait as the lower Part is, but a little circular and prominent in the middle. See CAPITAL, *etc.*

In *Pilasters* that support Arches, the Proportions, *Palladio* shews, must be regulated by the Light they let in; and at Angles, by the Weight they are to sustain. For which Reason, says Sir *Henry Wotton*, a rustic Superficies best becomes them.

PILASTER-*Brick*. See BRICK.

PILE, in Antiquity, a Pyramid built of Wood, whereon were laid the Bodies of the deceased, to be burnt.

PILE is also used in Building, for great Stakes ramm'd into the Earth for a Foundation to build upon in marshy Ground. See FOUNDATION; see also PALLIFICATION.

Amsterdam, and some other Cities are wholly built upon *Piles*. The Stoppage of *Dagenham*-Breach is effected by Dove-tail *Piles*, i.e. *Piles* mortais'd into one another, by a Dove-tail Joint. See DOVE-TAIL.

PILE is also used, among Architects, for a Mass, or Body, of Building.

PILLAGE, among Builders, is sometimes used for a square Pillar, standing behind a Column to bear up the Arches; having a Base and Capital as a Pillar has. See PILLAR.

PILLAR, in Architecture, a kind of irregular Column, round and insulated, deviating from the Proportions of a just Column. See COLUMN.

Pillars are always either too massive or too slender for a regular Architecture. In effect, *Pillars* are not restrained to any Rules; their Parts and Proportions are arbitrary.

Such *e. gr.* are the *Pillars* which support *Gothic* Vaults and Buildings, *etc.*

A square *Pillar* is a massive Work, call'd also a Peer or Piedroit, serving to support Arches, *etc.*

Butting Pillar is a Buttress, or Body of Masonry, raised to prop or sustain the Shooting of a Vault, Arch, or other Work. See BUTTRESS.

PILLAR, in the Manage, signifies the Centre of the Volta, Ring, or Manage-Ground, round which a Horse turns; whether there be a wooden Pillar placed therein or not.

There are also other *Pillars* in Manages on the Circumference or Side; placed at certain Distances by two and two.—To distinguish these from that of the Center, they are called the *two Pillars.*

When these latter are spoken of, it is usual to say, *Work the Horse between the two Pillars.*—When the former, it is call'd, *working round the Pillar.*

The Use of the *Pillar* in the Center, is for regulating the Extent of Ground, that the Manage upon the Volts may be perform'd with Method and Justness; and that they may work in a Square by Rule and measure upon the four Lines of the Volts; and also to break unruly high mettled

Horses, without endangering the Rider.

The *two Pillars* are placed at the Distance of 2 or 3 Paces the one from the other.— The Horse is put between these, to teach him to raise before, and yerk out behind; and put himself upon rais'd Airs, *etc.* either by the Aids or Chastisement.

PINNACLE, in Architecture, the Top, or Roof, of a House, terminating in a Point. See ROOF.

The Word comes from the *Latin Pinna*, or *Pinnaculum.*—This kind of Roof among the Ancients was appropriated to Temples; their ordinary Roofs were all flat, or in the Platform Way. See PLATFORM.

'Twas from the *Pinnacle*, that the Pediment took its Rise. See PEDIMENT.

PINNING, in Building, the fastening of Tiles together, with Pins of Heart of Oak; for the Covering of a House, *etc.* See TYLE and COVERING.

PIPE, in Building, *etc.* a Canal or Conduit for the Conveyance of Water, and other Liquids.

Pipes for Water, Water-Engines, *etc.* are usually of Lead, Iron, Earth, or Wood.—Those of Timber are usually either Oak or Alder.

Those of Iron are cast in the Forges; their Length about two Foot and a half; several of which are piec'd together, by means of four Screws at each End, with Leather, or old Hat between them to stop the Water.

Those of Earth are made by the Potters.—These are fitted into one another, one End being always made wider than the other.—To join them the closer, and prevent their leaking, they are cover'd with Pitch and Tow.—Their Length is usually about the same with that of the Iron *Pipes*.

The Wooden *Pipes* are bored with large Iron Augers of different Sizes, succeeding one another from less to larger; the first pointed, the rest form'd like Spoons, increasing in Diameter from one Inch to six.—They are fitted into the Extremities of each other, and are sold by the Foot.

Leaden *Pipes* are of two kinds the one solder'd, the other not solder'd.

PITCH, in Building, is the *Angle*, a Gable End, and consequently the whole Roof of a Building, is set to. See GABLE.

If the Length of each Rafter be ¾ of the Breadth of the Building; the Roof is said to be *true pitch*.

If the Rafters are longer, 'tis said to be a *high* or *sharp pitch'd* Roof; if shorter, which seldom happens, it is said to be a *low* or *flat pitch'd* Roof. See ROOF, RAFTERS, *etc.*

PITT, *Cavea*, of a Theater, all that Space between the Amphitheatre, or Galleries, and Theater or Stage; call'd by the Ancients *Orchestra*, and by the *French Parterre*. See THEATER, *etc.*

This being the most commodious Part, it was here the *Roman* Senate was placed. See OKCHESTRA.

It has its Name *Pitt*, in *Latin Cavea*, from its being sunk below the Level of the Stage. See STAGE.

PLACARD, in Architecture, the Decoration of the Door of an Apartment; consisting of a Chambranle, crown'd with its Frieze or Gorge; and its Corniche sometimes supported by Consoles. See DOOR.

PLACE, in War, is a general Name for all kinds of Fortresses, where a Party may defend themselves; and may be defined to be a *Place* so disposed, as that the Parts which encompass, it defend and flank one another.

A *strong Place*, is a Place flank'd and cover'd with Bastions.

Regular Place, is that whose Angles, Sides, Bastions, and other Parts are equal; and is usually denominated from the Number of its Angles, as a *Pentagon*, a *Hexagon*, etc.

Palma nova, built by the *Venetians*, is a Dodecagon.

Irregular Place is that whose Sides and Angles are unequal.

PLACE *of Arms*, in Fortification, is a strong City, or Town, pitch'd upon for the chief Magazine of an Army.

PLACE *of Arms* in a City, is a large open Spot of Ground, where the Garrison holds its Rendezvous at Reviews, and in Cases of Alarm, to receive Orders from the Governor.

PLACE *of Arms* in a Siege is a spatious Place cover'd from the Enemy, where the Soldiers are kept ready to sustain those at work in the Trenches, and to be commanded to the Places where they are wanted.

PLACE *of Arms particular*, in a Garrison, is a Place near every Bastion, where the Soldiers, sent from the grand Place to the Quarters assigned them, relieve those that are either upon the Guard, or in sight.

PLACE *of Arms without*, is a Place allowed to the Covert Way, for the planting of Cannon, to oblige those who advance in their Approaches to retire.

PLACE *of Arms* in a Camp, is a large Space at the Head of the Camp, for the Army to be rang'd in and drawn up in Battalia.

There is also a *Place* for each particular Body to assemble in.

PLACE *of Arms* of a Troop, or Company, is the Spot of Ground on which the Troop, or Company, is drawn up.

PLAFOND, or PLATFOUND, in Architecture, the Ceiling of a Room, whether it be flat or arched; lin'd with Plaster, or Joiners-Work, and frequently enriched with Paintings, *etc.* See CEILING.

PLAFOND is also more particularly used for the Bottom of the Projecture of the Larmier of the Corniche; call'd also *Sofit*. See SOFITTO and LARMIER.

PLAISTER, in Building. See PLASTER, MORTAR, *etc.*

PLAN, a Representation of something drawn on a Plane.

Such are Maps, Charts, Ichnographies, *etc.*

PLAN, in Architecture, *etc.* is particularly used for a Draught of a Building, such as it appears, or is intended to appear, on the Ground; shewing the Extent, Division, and Distribution of its Area into Apartments, Rooms, Passages, *etc.* See BUILDING.

The Plan is the first Device or Sketch the Architect makes; it is also call'd the *Ground-Plot*, *Plat-Form*, and *Ichnography* of the Building.

The *Geometrical Plan* is that wherein the solid and vacant Parts are represented in their natural Proportion.

Rais'd Plan is that where the Elevation, or Upright, is shewn upon the *Geometrical Plan*, so as to hide the Distribution. See ELEVATION.

Perspective Plan is that conducted and exhibited by Degradations, or Diminutions, according to the Rules of Perspective.

To render the *Plans* intelligible, 'tis usual to distinguish the Massives with a black Wash. The Projectures on the Ground are drawn in full Lines, and those supposed over them in dotted Lines. The Augmentations or Alterations to be made, are distinguished by a Colour different from what is already built; and the Feints of each *Plan* made lighter as the Stories are rais'd.

In large Buildings 'tis usual to have so many several *Plans* for the first three Stories.

PLANCERE, in Architecture, the under Part of the Corona, or Drip; making the superior Part of the Cornice, between two Cymatiums. See CORONA, CORNICHE, *etc.*

PLANE, in Joinery, *etc.* an Edge-Instrument, used to pare or shave Woods smooth, even, *etc.*

It consists of a Piece of Wood, very smooth at bottom, serving as a Stock, or Shaft; in the middle whereof is an Aperture, thro' which passes a Steel Edge, or Chissel, obliquely placed, and very sharp, which takes off the Inequalities of the Wood it is slid along.

The Plane acquires various Names according to its various Forms, Sizes and Uses: as, 1° The *Fore-Plane*, which is very long, and is that commonly first used. The Edge of its Iron is not ground streight, but rises with a Convex-Arch in the middle, to bear being set the ranker; its Use being to take off the greater Irregularities of the Stuff, and to prepare it for the Smoothing *Plane*.

2° The *Smoothing Plane* is short and small, its Iron, fine; it takes off the greater Irregularities left by the Fore-Plane, and prepares the Wood for the Jointer.

3° The *Jointer* is the longest of all; its Edge very fine, not standing out above a Hair's Breadth; it comes after the Smoothing-Plane, and is chiefly intended to shoot the Edge of a Board perfectly streight for jointing smooth Tables, *etc.*

4° The *Strike-Block* is like the Jointer, but shorter; its Use, to shoot short Joints, *etc.*

5° *Rabbet-Plane*, used to cut the upper Edge of a Board, strait or Square, down into the Stuff; so as the Edge of another, cut after the same manner, may join in with it on the Square; it is also used to strike Fascia's in Mouldings. Its Iron is full as broad as its Stock, that the Angle may cut strait; and it delivers it Shavings at the Sides, not, like the others, at the Top.

6° The *Plow*, a narrow Rabbet-Plane, with the Addition of two Staves, whereon are Shoulders, and on the Shoulders a Fence.—Its Use is to plow a narrow square Groove on the Edge of a Board, *etc.*

7° *Moulding-Planes*; of these there are various Kinds, accommodated to the various Forms and Profiles of the Mouldings; (See MOULDINGS) as the *Round Plane*, the *Hollow*, the *OG*, the *Snipe's Bill*, etc. which are all of several Sizes, from half an Inch to an Inch and half. See MOULDINGS, *etc.*

To use the *Moulding Planes* on soft Wood, as Deal, Pear-Tree, *etc.* they set the Iron to an Angle of 45° with the Base or Sole of the *Plane.* On hard Wood, *v.gr.* Ebony, Box, *etc.* they set to an Angle of 80°; sometimes quite upright. To work on hard Wood, the Edge or Basil is ground to an Angle of 18 or 20 Deg: of soft Wood, to an Angle of about 12°. For the more acute the Basil, the smoother the Iron cuts; but the more obtuse, the stronger.

PLANTING, in Architecture, denotes the disposing of the first Courses of solid Stone on the Masonry of the Foundation, laid level according to the Measures, with all the Exactness possible. See FOUNDATION, BUILDING, HOUSE, *etc.*

PLASTER, or PLAISTER, in Building, a Composition of Lime, sometimes with Hair, sometimes with Sand, *etc.* to parget or cover the Nudities of a Building. See PARGETING.

PLASTER *of Paris*, is a Fossil-Stone, of the Nature of a Lime-stone; serving many Purposes in building; and used likewise in Sculpture, to mould and make Statues, Basso Relievo's, and other Decorations in Architecture.

It is dug out of Quarries, in several Parts of the Neighbourhood of *Paris*; whence its Name.—The finest is that of *Montmartze.*

This *Plaster* is of two kinds, *viz. Crude*, or in the Stone; and burnt, or beaten.

The *crude* is the native *Plaster* as it comes out of the Quarry; in which state 'tis used as Shards in the Foundations of Buildings.

The *burnt Plaster* is a Preparation of the former, by calcining it like Lime in a Kiln or Furnace, and then beating it into Powder, and diluting and working it. See LIME, *etc.* In this state it is used as Mortar or Cement in Buildings. See MORTAR.

This, when well sifted and reduced into an impalpable Powder; is used to make Figures and other Works of Sculpture; and is besides of some Use in taking out Spots of Grease, *etc.* in Stuffs and Silks.

In the *Plaster*-Quarries is also found a kind of false Talc, wherewith they conterfeit all kinds of Marble.

PLASTICE, or the PLASTICK *Art*, a Branch of Sculpture; being the Art of forming Figures of Men, Birds, Beasts, Fishes, Plants, *etc.* in Plaster, Clay, Stuc, *etc.* See SCULPTURE, PLASTER, *etc.*

The Workmen concerned herein are called *Plastæ.*

Plastice differs from Carving, in that here the Figures are made by Addition of what wants: But in Carving always by Subtraction of what is superfluous.

The *Plastic Art* is now chiefly used among us in Fret-Work-Cielings; but the *Italians* apply it to the Mantlings of Chimnies with great Figures.

PLAT-BAND, in Architecture, is any flat square Moulding, whose Height much exceeds its Projecture. See MOULDINGS.

Such are the Faces or Fasciæ of an Architrave, and the *Plat-band* of the Modillions of a Cornich. See ARCHITRAVE, CORNICHE, *etc.*

The *Plat-band* is signified in *Vitruvius*, and others, by the Words *Fascia*, *Tænia*, and *Corsa*. See FASCIA, TÆNIA, *etc.*

PLAT-BAND, of a Door or Window, is used for the Lintel, where that is made square, or not much arch'd. See LINTEL.

These *Plat-bands* are usually cross'd with Bars of Iron, when they have a great *Bearing*: But 'tis much better to ease them by Arches of Discharge built over them. See DOOR, WINDOW, *etc.*

PLAT-BANDS of *Flutings*, the Lists or Fillets between the Flutings of Columns. See FLUTES.

PLAT-BAND, in Gardening, a Border, or Bed of Flowers, along a Wall, or the side of a Parterre; frequently edg'd with Box, *etc.*

PLATFORM, in Architecture, is a Row of Beams, which support the Timber-Work of a Roof, and lie on the top of the Wall, where the Entablature ought to be raised.

The Word is also used for a kind of Terrass-Walk, or even Floor, on the top of a Building, from whence we may take a fair Prospect of the adjacent Country.

Hence an Edifice is said to be cover'd with a *Platform*, when it is flat a-top, or has no Ridge. See ROOF.

Most of the Oriental Buildings are thus cover'd; as were all those of the Ancients.—*Cæsar* was the first among the *Romans* who procured Leave to Build his House with a Ridge, or Pinnacle.

PLINTH, in Architecture, a flat square Member, in form of a Brick; whence its Name.

The Word comes from the *Greek*, *Brick*.

The *Plinth* is used as the Foot, or Foundation of Columns; being that flat square Table, under the Mouldings of Columns; being that flat square Table, under the Mouldings of the Base and Pedestal, at the Bottom of the whole Order; seeming to have been originally intended to keep the bottom of the primitive wooden Pillars from rotting. See BASE, PEDESTAL, COLUMN, *etc.*

The *Plinth* is also called *Orlo*. See ORLE.

Vitruvius calls the *Tuscan Abacus*, *Plinth*, from its resembling a square Brick. See ABACUS.

Plinth of a Statue, etc. is a Base, or Stand, either flat, round, or square; serving to support a Statue, *etc*. See STATUE, *etc.*

PLINTH *of a Wall*, is a Term used for two or three Rows of Bricks advancing out from the Wall; or, in the general, for any flat high Moulding, serving in a front Wall to mark the Floors; or to sustain the Eaves of a Wall, and the Larmier of a Chimney. See WALL, *etc.*

PLOTT, in Surveying, the Plan, or Draught, of any Parcel of Ground, *e. gr.* a Field, Farm, or Manor, survey'd with an Instrument, and laid down in the proper Figure and Dimensions. See PLOTTING.

PLOTTING, among Surveyors, the Art of describing, or laying down on Paper, *etc.* the several Angles and Lines of a Tract of Ground survey'd by a Theodolite, or the like Instrument, and a Chain.

In surveying with the Plain-Table, the *Plotting* is saved; the several Angles and Distances being laid down on the Spot as fast as they are taken.

But in working with the Theodolite, Semicircle, or Circumferentor, the Angles are taken in Degrees; and the Distances in Chains and Links.

So that there remains an After-Operation, to reduce those Numbers into Lines; and so to form a Draught, Plan, or Map.—This Operation is call'd *Plotting*.

Plotting, then is perform'd by means of two Instruments, the Protractor, and *Plotting*-Scale.—By the first, the several Angles observed in the Field with a Theodolite, or the like, and entred down in Degrees in the Field-Book, are protracted on Paper in their just Quantity.

By the latter, the several Distances measured with the Chain, and entered down, in like manner in the Field-Book, are laid down in their just Proportion.

Under those two Articles, are found, severally, the Use of those respective Instruments in the laying of Angles and Distances: We shall here give their Use conjointly, in the *Plotting* of a Field, survey'd either with the Circumferentor, or Theodolite.

PLUMBERY, the Art of casting, preparing, and working Lead; and of using it in Buildings, *etc.*

The Word is form'd of the *Latin Plumbum*, Lead.

The Lead used in *Plumbery* is furnish'd from the Lead-Works in large Ingots, or Blocks, call'd Pigs of Lead, ordinarily weighing about an hundred Pounds a-piece.

As this Lead melts very easily, 'tis easy to cast Figures hereof, of any kind, by running it into Moulds of Brass, Clay, Plaster, *etc.*

But the chief Article in *Plumbery*, is the Sheets and Pipes of Lead. —'Tis these make the Basis of the Plumbers Work in Building.

PORCH, a kind of Vestible supported by Columns; much used at the Entrance of the antient Churches, Temples, *etc.* See VESTIBLE.

In the antient Architecture, *Porch* was a Vestible, or a Disposition of insulated Columns, usually crown'd with a Pediment, forming a Covert Place before the principal Door of a Temple or Palace.

When it had four Columns in Front, it was call'd a *Tetrastyle*; when six, *Exastyle*; when eight, *Octostyle*; when ten, *Decastyle*, etc.

Vitruvius calls it Pronaos and Prodomos. When it has Iron Gates, it is call'd Propylæum.

PORTICO, in Architecture, a kind of Gallery built on the Ground; or a Piazza incompass'd with Arches, supported by Columns; where People walk under Covert. See PIAZZA.

The Roof is usually vaulted, sometimes flat. The Antients call'd it Lacunar. See LACUNAR.

Tho' the Word *Portico* be derived from *Porta*, Gate, Door; yet it is apply'd to any disposition of Columns which form a Gallery; without any immediate relation to Doors, or Gates.

The most celebrated *Portico's* of Antiquity were those of *Solomon's Temple*, which form'd the *Atrium*, and encompass'd the Sanctuary: That of *Athens*, built for the People to divert themselves in; and wherein the Philosophers held their Disputes and Conversations; which occasion'd the Disciples of *Zeno* to be call'd *Stoicks*, from the Greek, *Porticus*: And that of *Pompey* at *Rome*, raised merely for Magnificence; consisting of several Rows of Columns supporting a Plat-Form of vast Extent: a Design whereof, *Serlio* gives us in his Antique Buildings.

Among the modern *Portico's*, the most celebrated is the Piazza of St. *Peter* of the *Vatican*. That of *Covent-Garden*, London, the Work of *Inigo Jones*, is also much admired.

POSTERN, in Fortification, a *false Door*, usually made in the Angle of the Flank, or of the Curtain, or near the Orillon; descending into the Ditch, for the Convenience of private Sallies.

The Word is also used in the general for any private or back Door.

Potestas habere posternam in omni Curia penitus inhibeatur, sed unicus si Ingressus, etc. Fleta.

POSTIQUE, or POSTIC, in Architecture, *etc*. An Ornament of Sculpture is said to be *Postique*, when it is added after the Work itself is done.

A Table of Marble, or other Matter, is also said to be *Postique*, when it is incrusted in a Decoration of Architecture, *etc*.

The Word is form'd from the *Italian*, *Porticcio*, added.

POSTSCENIUM, in the antient Theatre. See PARASCENIUM.

POUDERINGS, in Building, a Term sometimes used for Devices serving to fill up vacant Spaces, in carv'd Works: As also, in Escutcheons, Writings, *etc*.

PROFILE, in Architecture, the Figure, or Draught of a Building, Fortification, or the like; wherein are express'd the several Heights, Widths, and Thicknesses, such as they would appear were the Building cut down, perpendicularly from the Roof to the Foundation; whence the *Profile* is also call'd the *Section*, and by *Vitruvius*, *Sciagraphy*. See SECTION, *etc*.

PROFILE, is sometimes also used for a Design, or Description; in opposition to a Plan, or *Ichnography*.

In which Sense, *Profiles* coincide with what we popularly call *Prospects*.

PROFILE is also used for the Contour, or Out-Line of a Building, or a Member of Architecture; as a Base, a Corniche, *etc*.

Hence *Profiling* is sometimes used for Designing, or describing the Member with Rule, Compass, *etc*.

PROFILE, in Sculpture, and Painting. A Head, a Portrait, *etc*. are said to be in *Profile* when they are represented side-ways, or in a side-view. As, when in a Portrait, there is but one side of the Face, one Eye, one Cheek, *etc*. shewn, and nothing of the other.

In almost all Medals the Faces are in *Profile*.

PROJECTION in *Building*, See PROJECTURE.

PROJECTURE, in Architecture, the Out-jetting, or Prominency, which the Mouldings and Members have, beyond the Naked of the Wall, Column, *etc*.

These the *Greeks* call *Ecphoræ*, the *Italians Sporti*, the *French Sailles*, our Workmen frequently *Sailings over*, and the *Latins Projecta*, from *projicio*, I cast forward; whence the *English*, *Projecture*.

Vitruvius gives it as a general Rule, that all the *projecting* Members in Buildings have their *Projectures* equal to their Heights: But this is not to be understood of the particular Members, or Mouldings, as Dentils, Corona's, the Fasciæ of Architraves, the Abacus of the *Tuscan* and *Doric* Capital, *etc.* but only of the *Projectures* of entire Cornices, *etc.* See CORNICHE, *etc.*

The great Point of Building, according to some modern Architects, consists in knowing how to vary the Proportions of *Projectures*, etc. agreeably to the Circumstances of the Building: Thus, say they, the nearness and remoteness making a difference in the View, requires different *Projectures*; but 'tis evident the Antients had no such Intention.

The *Projecture* of the Base and Cornice of Pedestals, M. *Perrault* observes, is greater in the antique than the modern Buildings by one third; which seems to follow, in good measure, from the Antients proportioning this *Projecture* to the height of the Pedestals; whereas the Moderns make the *Projecture* the same in all the Orders, tho' the height of the Pedestal be very different.

The reason of this Change which the Moderns have made of the antique, the same Author refers to a View to the Appearance of Solidity. See PEDESTAL.

PRONAOS, in the antient Architecture, a Porch to a Church, Palace, or other spacious Building. See PORCH.

PROPERTY, or PROPRIETY, in Law, strictly speaking, is the highest Right a Man can have to a thing; and such as no ways depends on any other Man's Courtesy.

In this Sense, none in our Kingdom have the *Property* of any Lands or Tenements, except the King, in Right of his Crown; all other Lands being of the nature of Fee and held of the King either mediately or immediately.

PROPORTION, in Architecture, the just Magnitude of the Members of each part of a Building, and the Relation of the several Parts to the Whole; *e. gr.* of the Dimensions of a Column, *etc.* with regard to the Ordonnance of the Building. See SYMMETRY. See also BUILDING.

One of the greatest Differences among Architects, M. *Perrault* observes, is in the *Proportions* of the Heights of Entablatures with respect to the Thickness of the Columns, to which they are always to be accommodated. See ENTABLATURE.

In effect, there is scarce any Work, either of the Antients of Moderns, wherein this *Proportion* is not different; some Entablatures are even near twice as high as others—Yet 'tis certain, this *Proportion* ought of all others to be most regulated; none being of greater Importance, as there is none wherein a Defect is sooner spied, nor any wherein it is more shocking. See COLUMN.

PROPORTION is likewise understood of the Magnitudes of the Members of Architecture, Statues, or the like, with regard to the Distance whence they are to be view'd.

The most celebrated Architects are much divided in their Opinions on this Subject: Some will have it, that the parts ought to be enlarg'd in *proportion* to their Elevation; and others, that they ought to remain in their natural Dimensions.

PROPYLÆUM, the Porch of a Temple, or Great Hall. See PORCH.

Hence the Word is used figuratively in Matters of Learning for an Introduction, Apparatus, or Prodromus to some greater Work.—In this sense we say, the *Propylæum* of the Jesuits at *Antwerp*, etc.

The Word is *Greek*, signifying the same thing.

PROSCENIUM, in the antient Theatre, the *Pulpitum*, or Eminence whereon the *Roman* Actors exhibited. See THEATER.

The *Proscenium* answer'd to our *Stage*.

It consisted of two Parts among the *Greeks*; one particularly so call'd, where the Actors perform'd. The other was the *Logeion*, where the Singers and the Mimicks acted their Parts.

Among the *Romans*, the *Proscenium* and *Pulpitum* were the same thing.

PROSPECT, PROSPECTIVE. See PERSPECTIVE.

PROSTYLE, in the antient *Greek* Architecture, a Range of Columns in the Front of a Temple. See TEMPLE.

The word is form'd from the *Greek*, *before*, and, *Column*.

PROTHYRUM, a Porch at the outward Door of a House, or Portal. See PORCH.

PROTHYRIS, PROTHYRIDE, is used by *Vignola* for a particular sort of a Key of an Arch, an Instance of which we have in his *Ionic* Order; consisting of a Roll of Water-Leaves, between two Reglets and two Fillets, crown'd with a *Doric Cymatium*; its Figure being much like that of a Modillion.

PROTHYRIS, in the antient Architecture, is sometimes also used for a Quoin, or Corner of a Wall; and sometimes for a cross Beam, or overthwart Rafter.

PSEUDO-DIPTERE, in the antient Architecture, a Temple with eight Columns in Front; and a single Row of Columns all around. See TEMPLE.

The Word signifies *false* or *imperfect Diptere*; and is used to distinguish this, from the *Diptere*; which had two Rows of Columns all around. See DIPTERE.

PULVINATA, PULVINATED, in the antient Architecture, a Term apply'd to a Freeze which swells, or bulges out, in manner of a Pillow; whence the Name. See FREEZE.

PUNCHIN, or PUNCHION, in Building. See PUNCHION.

PUNCHION, is used for several Iron Tools of various Sizes, and Figures used by the Engravers *in Creux*, on Metals—Seal-Gravers, particularly, use a great number; for the several pieces of Arms, *etc.* to be engraven—And many stamp the whole Seal from a single *Punchion*.

PUNCHION is also a common Name for all the Iron Instruments used by Stone-Cutters, Sculptors, Lock-Smiths, *etc.* for the cutting, inciding, or piercing their several Matters.

Those of Sculptors and Statuaries, serve for the repairing Statues, when taken out of the Moulds.

The Lock-Smiths use the greatest Variety of *Punchions*; some for piercing hot, others for piercing cold; some flat, some square, some round, others oval; each to pierce Holes of its respective Figure, in the several parts of Locks.

PUNCHION, or PUNCHIN, or NEEDLE, is Carpentry, is a piece of Timber placed upright between two Posts whose bearing is too great; serving, together with them, to sustain some large Weight.

The *Punchion* is usually lower and slighter than the Posts, and is join'd by a Brace, or the like, of Iron.

Punchion is also a piece of Timber raised upright under the Ridge of a Building, wherein the little Forces, *etc.* are jointed—*Vitruvius* calls the *Punchion*, *Columen*.

PURBECK *Stone*, see STONE.

PURLINS, in Building, those Pieces of Timber that lie a-cross the Rafters, on the inside; to keep them from sinking in the Middle of their Length. See RAFTERS.

By the Act of Parliament for Rebuilding *London*, it is provided, That all *Purlins* from fifteen Foot six Inches to eighteen Foot six Inches long, be, in their Square nine Inches and eight Inches—And all in length from eighteen Foot six Inches to twenty-one Foot six Inches, be in their Square twelve Inches and nine Inches.

PUTLOGS, in Building, short pieces of Timber, about seven Foot long, used in building Scaffolds—They lie at right Angles to the Wall with one of their ends resting upon the Ledgers or Poles, which lie parallel to the side of the Wall of the Building.

PYCNOSTYLE, or PYCHNOSTYLE, in the Antient Architecture, a Building where the Columns stand very close to one another; one

Diameter and a half of the Column being only allow'd for the Intercolumniation. See INTERCOLUMNS.

The *Pycnostyle* is the smallest of all the Intercolumniations mention'd by *Vitruvius*.

The Word is form'd from the *Greek*, *close*, and, *Column*.

Some make the *Pycnostyle* the same with the *Systyle*; others distinguish the latter, by its allowing half a Module more in the *Corinthian* Intercolumniation.

The *Pycnostyle*, Mr. *Evelyn* observes, chiefly belong'd to the *Composite* Order; and was used before the most magnificent Buildings, as at present in the *Peristyle* of St. *Peter's* at *Rome*, consisting of near 300 Columns; and such as yet remain of the Antients among the late discover'd Ruins of *Palmyra*.

PLYING *the Ground*, for Foundations. See FOUNDATION.

PYRAMID, in Architecture, is a solid massive Edifice; which from a square, triangular, or other Base, rises diminishing, to a Point, or Vertex.

Pyramids are sometimes used to preserve the Memory of singular Events; and sometimes to transmit to Posterity the Glory and Magnificence of Princes: but as they are the Symbol of Immortality, they are more commonly used as funeral Monuments. See MONUMENT.

Such is that of *Cestius* at *Rome*; and those other celebrated ones of *Egypt*, as famous for the Hugeness of their Size, as their Antiquity.

These last are all square in their Bases; and 'tis a thing has been frequently proposed, to establish a fixed Measure from them; to be thereby transmitted to Posterity.—See their Descriptions, Measures, *etc.* in *Thevenot*, *Pietro della Valle*, *etc.*

Among the *Egyptians*, the *Pyramid* is said to have been a Symbol of human Life; the Beginning whereof is represented by the Base, and the End by the Apex: On which account it was, they used to erect them on Sepulchres. HERODOTUS.

Some derive the Word from the *Greek*, *Wheat*, and *colligo*; pretending that the first Pyramids were built by the Patriarch *Joseph* for Granaries.—But *Villalpandus*, with much better reason, derives the Word from the *Greek*, *Fire*; because ending in a Point like Flame.

When they are very narrow at bottom, *i.e.* their Base very small; they are call'd *Obelisks* and *Needles*. See OBELISK.

Q

QUADRA, in Building, any square Border, or Frame, incompassing a Basso-Relievo, Pannel, Painter's or other Work.

The Word is also used abusively for a Frame or Border of another Form; as Round, Oval, or the like.

QUADRELS, in Building, a kind of Artificial Stones, perfectly square, whence their Name; made of a chalky, or whitish and pliable Earth, *etc.* dried in the Shade for two Years.

They were formerly in great request among the *Italian* Architects.

QUARRY, in our antient Writers call'd QUARERA, QUADRATA, QUARATIA, a Place under Ground, out of which are dug Marble, Free-Stone, Slate, Lime-Stone, or other Matters proper for Building.

For *Quarries of Free Stone*, they first open a Hole in manner of a Well, twelve or fourteen Foot in Diameter; and the Rubbish drawn out with a Windlass in large Osier Baskets, they heap up all around; placing their Wheel, which is to draw up the Stones, thereupon.

As the Hole advances, and their common Ladder becomes too short, they apply a particular Ladder for the purpose—When they have got through the Earth, and are arrived at the first Bank, or Stratum; they begin to apply their Wheel and Baskets to discharge the Stones as fast as they dig through 'em.

They usually find seven of these different Strata, or Beds of Stones, of different Heights, and serving for different Purposes; tho' the Number as well as Order wherein they follow is various.

As to the *drawing* of the Stone, *i.e.* the freeing it from the Bed; they find that common Stones, at least the softer kinds, as they lie, have two Grains; a cleaving Grain running parallel with the Horizon, and a breaking Grain perpendicular thereto—After *uncoping*, then, *i.e.* clearing the Earth from off it, they observe by the Grain where the Stone will cleave, and there drive in a good Number of Wedges till they have thus cleft it from the rest of the Rock.

This done, they proceed to break it: In order to which, applying the Ruler to it at both ends, (ten, *e. gr.* or twelve Inches a-part, according to the Uses the Stone is intended for) they strike a Line, and by this cut a little Channel with their Stone-Ax; and in the Channel set five or six Wedges (supposing the Stone three of four Foot) driving 'em in very carefully, with gentle blows, and still keeping 'em equally forward.

Having thus broke 'em in length, (which they are able to do to half an Inch of any size) applying a Square to the straight Side, they strike a Line, and proceed as before to break it in breadth.

This Method of drawing is found vastly preferable to that where the Stones are broken at random—One Load of the former is found to do the

Business of a Load and a half of the latter.

But it may be observ'd, that this cleaving Grain being generally wanting in the harder Stones; to break up these in the *Quarries*, they have great heavy Stone-Axes wherewith they work down a deep Channel into the Stone, and into this Channel, a-top, lay two Iron Bars, driving their Iron Wedges between these Bars.

QUARTERS, in Building, those slight upright Pieces of Timber, placed between the Punchions and Posts; used to lath upon.

They are of two kinds, *single* and *double*—The *single* Quarters are sawn to two Inches thick and four Inches broad; the *double* four Inches square.

QUEUE *d'aronde*, in Carpentry, a Method of Jointing. See DOVE-*Tail*.

QUINCUNX, q.d. *quinque unciæ*; a *Chequer*, or something that has five Inches, Ounces, or Parts.

The Term is chiefly used in Gardening for a Plantation of Trees, disposed originally in a Square, consisting of five Trees, one at each Corner, and a fifth in the Middle; which disposition repeated again and again, forms a regular Grove, Wood, or Wilderness, and when view'd by an Angle of the Square, or Parallelogram, presents equal and parallel Alleys.

Or the *Quincunx* is the Figure of a Plantation of Trees, disposed in several Rows, both length and breadthwise; in such manner as that the first Tree of the second Row commences in the Centre of the Square form'd by the two first Trees of the first Row, and the two first of the third; resembling the Figure of a five at Cards.—The finest manner of planting Trees to form a Grove, is in the *Quincunx*.

'Tis of this kind of *Quincunx* that *Cicero* speaks in his *Cato major*; and *Quintilian*, lib. 8. cap. 2.

The modern *Quincunxes*, *Daviler* observes, are made like those of the Antients, except for the fifth Tree, which is now generally disused; so that being, as it were, netted, and their Alleys view'd by the Side of the Rectangle, they form a perfect Chequer.

QUIRK, in Building, a Piece of Ground taken out of any regular Ground-Plat, or Floor—Thus, if the Ground-Plat were square, or Oblong; and a Piece be taken out of a Corner, to make a Court, or Yard, *etc.* the Piece is call'd a *Quirk*.

QUOINS, in Architecture, the Corners of Walls. See WALL.

The Word is particularly used for the Stones in the Corners of Brick-Buildings—when these stand out beyond the Brick-work (their Edges being chamfer'd off) they are call'd *Rustic Quoins*. See RUSTIC *Work*.

R

RABBETING, in Carpentry, the planing or cutting of Channels, or Grooves, in Boards. See PLANE.

In Ship-Carpentry it signifies the letting in of the Planks of the Ship into the Keel.

RAFTERS, in Building, are pieces of Timber, which stand by Pairs upon the Reason, meet in an Angle at the Top, and help to compose the Roof of a Building. See ROOF.

'Tis a Rule in Architecture, that no Rafters shou'd stand farther than twelve Inches from one another.

For the Sizes or Scantlings of *Rafters*, 'tis provided by Act of Parliament, that *principal Rafters* from 12 Foot 6 Inches, to 14 Foot 6 Inches long, be 5 Inches broad a-top, and 8 at the bottom, and 6 Inches thick.

Those from 14, 6 to 18, 6 long, to be 9 Inches broad at the Foot and 7 a-top, and 7 thick.

And those from 18, 6, to 21, 6, to be 10 Inches broad at the Foot, 8 a-top, and 8 Inches broad.

Single Rafters, 6 Foot 6 Inches long, to be 4 and 3 Inches in their Square. Those 8 Foot long must be 4 ½ and 3 ¾ Inches square.

RAIL, in Architecture, is applied variously; particularly, to those pieces of Timber, *etc.* which lie Horizontally between the Pannels of Wainscot; to those that lie over and under Ballusters in Balconies, Stair-Cases, *etc.* and to the pieces of Timber that lie horizontally from Post to Post, in Fences with Pales or without.

RAISER, in Building, a Board set on-edge under the fore-side of a Step, Stair, *etc.* See STAIRS, *etc.*

RAKING-*Table*, among Architects, a Member hollowed in the Square of a Pedestal, or elsewhere. See CAVETTO, SCOTIA.

RAMMER, an Instrument for driving down Stones or Piles into the Ground; or for the beating the Earth, and making it more solid for a Foundation. See FOUNDATION.

RAMPART, in Fortification, a Mount or Mass of Earth, rais'd about the Body of a Place, to cover it from the great Shot; and form'd into Bastions, Curtains, *etc.*

Upon the *Rampart* the Soldiers continually keep Guard, and Pieces of Artillery are planted for the Defence of the Place.

Hence, to Shelter the Guard from the Enemies Shot, the out-side of the

Rampart is built higher than the inside, *i.e.* a Parapet is rais'd upon it with a Platform.

Hence, also, Earth not being capable to be rais'd perpendicularly, like Stone; the *Rampart* is built with a *Talus* or Slope, both on the inner and outer-side.

The *Rampart* is sometime *lined*, *i.e.* fortified with a Stone Wall within-side; otherwise it has a *Berme*. See BERME.

'Tis incompassed with a Moat or Ditch, out of which the Earth that forms the *Rampart* is dug.

The Height of the *Rampart* must not exceed three Fathom; this being sufficient to cover the Houses from the Battery of the Cannon: Neither ought its Thickness to be above 10 or 12, unless more Earth be taken out of the Ditch, than can be otherways bestowed.

The *Ramparts* of Half-Moons are the better for being low; that the small Fire of the Defendants may the better reach the Bottom of the Ditch: but yet it must be so high, as not to be commanded by the Covert-Way.

The Word is form'd from the *Spanish Amparo*, Defence, Covering.

RAMPART is also used in Civil Architecture, for the Space left void between the Wall of a City and the next Houses.

This is what the *Romans* call'd *Pomœrium*, wherein it was forbid to build; and where they planted Rows of Trees, for the People to walk and amuse themselves under.

RANGING, in Building, the Side of a Work that runs straight, without breaking into Angles, is said to *range*, or *run range*.

RASANT, in Fortification. *Rasant Flank*, or *Line*, is that Part of the Curtain, or Flank, whence the Shot exploded raze or glance along the Face of the opposite Bastion.

RAVELIN, in Fortification, was antiently a flat Bastion, placed in the middle of a Curtain.

'Tis since made a detach'd Piece, compos'd only of two Faces, which make a Salient Angle, without any Flanks; and raised before the Curtains or Counterscarp, and commonly call'd an *Half-Moon*.

A *Ravelin* is a Triangular Work resembling the Point of a Bastion, with the Flanks cut off.

Its use before a Curtain, is to cover the opposite Flanks of the two next Bastions. 'Tis used also to cover a Bridge or a Gate; and is always placed without the Moat.

What the Engineers call a *Ravelin*, the Soldiers generally call a *Demi-Lune*, or *Half-Moon*.

There are also double *Ravelins*, which serve to defend each other. They are said to be *double* when they are join'd by a Curtain.

RECTORY, RECTORATE, a Parish-Church, Parsonage, or Spiritual Living, with all its Rights, Glebes, and Tithes.

REDENS, or REDANT, or REDAN, in Fortification, a Kind of Work indented in Form of the Teeth of a Saw, with Salient and Re-entering Angles, to the end that one Part may defend another.

It is also call'd Saw-work: The Faces Flank one another.

Redans are frequently used in the fortifying of Walls, where it is not necessary to be at the Expence of building Bastions; as when they stand on the Side of a River, a Marsh, the Sea, *etc.*

REDOUBT, REDUCT, in Fortification, a small square Fort, without any Defence but in Front; used in Trenches, Lines of Circumvallation, Contravallation, and Approach; as also for the lodging of *Corps de Garde*, and to defend Passages.

In marshy Grounds, *Redoubts* are often made of Mason's Work, for the Security of the Neighbourhood: Their Face consists of from ten to fifteen Fathom; the Ditch round them from eight to nine Foot broad and deep; and their Parapets have the same thickness.

The Word is form'd from the *Latin*, *reductus*.

REDUCT, in Building, a Quirk, or little Place, taken out of a larger, to make it more uniform and regular; or for some other Convenience, as for little Cabinets aside of Chimneys, Alcoves, *etc.*

REFECTORY, or REFECTUARY, a spacious Hall in Convents, and other Communities, where the Monks, Nuns, *etc.* take their Refections or Meals in Common.

The *Refectory* of the *Benedictines* of *St. George* at *Venice*, designed by *Palladio*, is one of the finest in the World. *Daviler.*

REGLET, or RIGLET, in Architecture, a little flat narrow Moulding, used chiefly in Compartiments, and Pannels, to separate the Parts or Members from one another, and to form Knots, Frets, and other Ornaments. See MOULDINGS, FRET, *etc.*

The Riglet, according to *Daviler*, differs from the Fillet and Listel, in that it projects equally, like a Ruler. See FILLET and LIST.

The Word is a diminutive of the *French*, *Regle*, Rule.

REGULA, in Architecture. See ORLE.

REJOYNTING, in Architecture, the filling up the Joints of the Stones in old Buildings, *etc.* when wore Hollow by Course of Time, or Water.

Rejoynting is perform'd with the best Mortar; as that of Lime, and

Ciment; sometimes, also with Plaister, as in the Joints of Vaults, *etc*. See MORTAR, *etc*.

RELIEVO, or RELIEF, *Imbossment*, in Sculpture, is applied to a Figure which projects, or stands out, prominent, from the Ground or Plain whereon it is form'd; whether that Figure be cut with the Chissel, moulded, or cast. There are three Kinds of *Relievo*; *viz. Alto*, *Basso*, and *Demi-Relievo*.

The *Alto Relievo*, *Haut-Relief*, or, *high Relievo*, is when the Figure is form'd after Nature, and projects as much as the Life.

Basso-Relievo, *Bass-Relief*, or *low Relievo*, is when the Work is but rais'd a little from its Ground; as we see in Medals, and in the Frontispieces of Buildings; particularly the Histories, Festoons, Foliages, and other Ornaments in Friezes.

Demi Relievo is when one half of the Figure rises from the Plan, *i.e.* when the Body of the Figure seems cut in two, and one half is clapp'd on a Ground.—When in a *Basso Relievo* there are some Parts that stand clear out, detach'd from the rest, the Work is call'd a *Demi-Bosse*.

RELIEVO, in Architecture, is the Projecture of any Ornament. See PROJECTURE.

This, *Daviler* observes, is always to be proportioned to the Magnitude of the Building it adorns; and the Distance at which it is to be viewed. If the Work be insolated and terminated on all Sides, it is call'd a *Figure in Relievo*, or a *round Imbossment*. Such are Statues, Acroters, *etc*.

RELIEVO, in Painting, is the Degree of Force or Boldness wherewith the Figures seem, at a due Distance, to stand out from the Ground of the Painting, as if really imboss'd.

The *Relievo* depends much on the Depth of the Shadow, and the Strength of the Light; or on the Heighth of the different Colours, bordering on one another; particularly on the difference of the Colour of the Figure from that of the Ground.

When the Light is well chosen, to make the nearest Parts or Figures advance; and well diffused on the Masses, still diminishing insensibly, and terminating in a large spacious Shadow, brought off insensibly, the *Relievo* is said to be *bold*, and the *Clair-obscure*, *well understood*.

They also say *Embroideries in Relievo*; in opposition to those that are flat.

REPOSE, in Painting, is used for certain Masses or large Systems, or Assemblages of Light and Shade; which being well conducted, prevent the Confusion of Objects and Figures; by engaging and taking up the Eye so as it can't attend to the other Parts of the Painting, for some time: And thus leading it to consider the several Groups gradually, and as it were to proceed from Stage to Stage.

REPOSITORY, REPOSITORIUM, a Store-house or Place where

things are laid up and kept.—In this Sense we say the *Repository* of the Royal Society, *etc.*

RESEARCHING, in Sculpture, the Repairing of a cast Figure, *etc.* with proper Tools; or the finishing it with Art and Exactness, so as the minutest Parts may be well defined.

RESSAUT, in Architecture, the effect of a Body which either Projects or Sinks; *i.e.* stands either more or out, or in, than another; so as to be out of the Line or Level therewith.

Such is a Socle, an Intablature, a Cornice, *etc.* upon an Avant Corps, arriere Corps, or the like.

The Term is *French*; and but little used in *English*; though the want of a Word of equal Import, pleads for its Naturalization.

RETIRADE, in Fortification, a kind of Retrenchment made in the Body of a Bastion, or other Work, which is to be disputed Inch by Inch, after the first Defences are dismantled.

It usually consists of two Faces, which make a re-entering Angle.

When a Breach is made in a Bastion, the Enemy may also make a *Retirade*, or a new Fortification behind it.

RETRENCHMENT, in Architecture, Carpentry, *etc.* is used not only for what is cut off from a Piece when too large, in order to a better proportioning it, or some other Convenience; but also for the Projectures taken out of Streets, publick Ways, *etc.* to render them more even, and in a Line.

RETRENCHMENT, in War, the Fortification of a Camp; or any Kind of Work cast up to strengthen or defend a Post against the Enemy.

Such are Ditches, with Parapets, Gabions, Fascines, *etc.* for a covering, *etc.*

The Enemy came with Design to oblige them to raise the Siege, but could not force the *Retrenchments*.

RETRENCHMENT is particularly used for a simple Retirade made on a Horn-work or Bastion; when 'tis intended to Dispute the Ground Inch by Inch.

'Tis usually a re-entering Angle, whose Faces Flank each other; and fortified with Ditches, Parapets, Gabions, *etc.*

REVERBERATION, in Physicks, the Act of a Body repelling or reflecting another, after its unpinging thereon.

Echo's are occasioned by the *Reverberation* of Sounds from arch'd Obstacles.

The Word is form'd from the *Latin*, *re* and *verbero*, I beat again.

Reverberation and *Resilition* refer to the same Action; only the one to the Agent, the other to the Patient.—A polished Body *reverberates* the Rays all around: the *Resilition* of the Rays does not arise from their striking against the solid Parts of Bodies.

REVERSED *Talon*, in Architecture. See TALON.

RIDEAU, in Fortification, a small Elevation of Earth, extending itself lengthways on a Plain; serving to cover a Camp, or give an Advantage to a Post.—It is also convenient for those who would besiege a Place at a Distance, and to secure the Workmen in their Approaches to the Foot of a Fortress.

RIDEAU, is sometimes also used for a Ditch, the Earth whereof is thrown upon its Side.

The Word in its original *French* signifies a Curtain or Cover; form'd from the *Latin*, *Ridelium.—Borel* derives it from *ridere*.

RIDGE, in Building, the highest Part of the Roof or Covering of a House. See ROOF.

RONDEL, in Fortification, a round Tower, sometimes erected at the Foot of a Bastion.

ROOD, a Quantity of Land, equal to the fourth Part of an Acre; and containing 40 square Perches or Poles.

ROOF, in Architecture, the uppermost Part of a Building. See BUILDING.

The*Roof* contains the Timber-Work, and its Furniture of Slate or Tile, wherewith a House is cover'd, or that which serves it as a Cover. —Though Carpenters usually restrain *Roof* to the Timber-Work only. See COVERING and RIDGE.

The Form of the *Roof* is various; sometimes 'tis *pointed*, in which Case the most beautiful Proportion is to have its Profile an equilateral Triangle.

Sometimes *Square*, that is, the Angle of the Ridge, is a Right Angle; which, therefore, is a mean Proportional between the *pointed* and

Flat Roof, which is in the same Proportion as a triangular Pediment. See PEDIMENT.—This is chiefly practiced in *Italy*, and the hot Countries, where little Snow falls.

Sometimes the *Roof* is in the *Pinnacle Form*. See PINNACLE.

Sometimes it has a *double Ridge*.—Sometimes 'tis *cut*, or *mutilated*, that is, consists of a *true* and a *false Roof*, which is laid over the former: This last is particularly call'd a *Mansard*, from its Inventor M. *Mansard*, a famous *French* Architect.

Sometimes 'tis in Form of a *Platform*; as in most of the Eastern Building. See PLATFORM.

Sometimes 'tis *truncated*; that is, instead of terminating in a Ridge or Angle, 'tis cut square off at a certain Heighth, and cover'd with a Terrass, and sometimes incompassed with Ballustrade.

Sometimes 'tis in Manner of a *Dome*, that is, its Plan Square, and the Contour Circular. See DOME, CUPOLA, *etc.*

Sometimes it is *round*, that is, the Plan is Round or Oval, and the Profile a direct Descent.—Sometimes the Base being very large, 'tis cut off to diminish its Heighth, and cover'd with a Terrass of Lead, rais'd a little in the middle, with Sky-Lights from Space to Space, to give Light to some Corridor, or other intermediate Pieces, which without such an Expedient would be too dark. See HOUSE, *etc.*

ROOF-*Trees*, or RUFF-*Trees*, are the Timbers in a Ship which go from the Half-Deck to the Fore Castle.

The Term is also used for the upper Timbers of any Building; whence in the Northern Counties, it is common to signify a whole Family, by saying, all under such a one's *Roof-Tree*.

ROOM, in Building.—See BUILDING, HOUSE, PARTITION, APARTMENT, CHAMBER, *etc.*

ROSTRALIS *Columna*. See COLUMN.

ROTONDO, ROTUNDO, in Architecture, a popular Term for any Building that is round both within and without-side; whether it be a Church, a Salon, a Vestible, or the like. See BUILDING, *etc.*

The most celebrated *Rotondo* of Antiquity is the Pantheon at *Rome*, dedicated to *Cybele*, and all the Gods, by *Agrippa*, Son-in-law of *Augustus*; but since consecrated by Pope *Boniface* IV. to the Virgin, and all the Saints, under the Title of *Sta. Maria della Rotondo*. See PANTHEON.

The Chapel of the *Escurial*, which is the Burying-Place of the Kings of *Spain*, is also a *Rotondo*; and in Imitation of that at *Rome*, is also call'd *Pantheon*.

ROUNDS, among Masons, *Rounds* are broken Pieces of Statues.

RUDENTURE, in Architecture, the Figure of a Rope or Staff, sometimes plain sometimes carv'd, wherewith a third Part of the Flutings of Columns are frequently filled up. See FLUTES.

It is thus call'd from the *Latin*, *rudens*, *Cable*; whence some call it a *Cabling*; and the Columns whose Flutings are thus fill'd, *rudented*, or *cabled Columns*. See COLUMN.

There are also *Rudentures* in *Relievo*, laid on the naked of Pilasters not fluted; an Instance of which we have in the Church of St. *Sapienza* at *Rome*.

RUDERATION, in Building, a Term used by *Vitruvius* for the laying of a Pavement with Pebbles or little Stones. See PAVEMENT.

To perform the *Ruderaton*, 'tis necessary the Ground be first well beaten, to make it firm, and prevent its cracking.—Then a Stratum of little Stones are laid, to be afterwards bound together with Mortar made of Lime and Sand, call'd by *Vitruvius, Statumen.*

If the Sand be new, its Proportion to the Lime may be as 3 to 1; if dug out of old Pavements of Walls, as 5 to 2. See MORTAR, *etc.*

RUDERATION, *Daviler* observes, is also used by *Vitruvius*, Lib. 7. Cap. 1. for the coarsest and most artless Kind of Masonry; where a Wall is, as it were, cobled up. See MASONRY.

RUINS, a Term particularly used for magnificent Buildings fallen to decay by length of Time; and whereof there only remains a confused Heap of Materials.

Such are the *Ruins* of the Tower of *Babel*, or Tower of *Belus*, two Days Journey from *Bagdat*, in *Syria*, on the Banks of the *Euphrates*; which are now no more than Heap of Bricks, cemented with Bitumen; and whereof we only perceive the Plan to have been Square.

Such, also, are the *Ruins* of a famous Temple or Palace near *Schiras* in *Persia*; which the Antiquaries will have to have been built by *Ahassuerus*; and which the *Persians* now call *Tchelminar*, or *Chelminar*, *q.d.* the forty Columns; because there are so many Columns remaining pretty entire, with the Traces of others; a great Quantity of *Basso Relievo's*, and unknown Characters, sufficient to shew the Magnificence of the Antique Architecture. See CHILMINAR.

RULE, REGULA, or RULER, a very simple Instrument, ordinarily of hard Wood, thin, narrow, and straight; serving to draw right Lines withal.

The *Rule* is of principal use in all the mechanical Arts.—To prove whether or no it be just; draw a Line by it on a Paper; then turn the *Rule* about, the Right end to the Left; and apply the same Edge this way to the Line: If the Edge now agree exactly with the Line, the *Ruler* is true.

Desmarets has a fine Poem on the Amours of the *Rule* and Compass.—The *Stone-Cutters-Rule* is usually four Foot long; and divided into Feet and Inches.

The *Mason's-Rule* is 12 or 15 Foot long, and is applied under the Level, to regulate the Courses, to make the Piedroits equal, *etc.*

RULE is also used for certain Instruments which have other considerable Uses beside that of drawing Lines.—Such are the Carpenters *Joint-Rule*, *Everard's*, and *Coggleshal's Sliding-Rules*, etc.

Carpenters-Joint-RULE, is an Instrument usually of Box, 24 Inches long, and one and a half broad; each Inch being subdivided into eight Parts.—On the same Side with these Divisions, is usually added *Gunter's* Line of Numbers.

On the other Side are the Lines of *Timber* and *Board-Measure*; the first beginning at 82, and continued to 36, near the other End: The latter is numbered from 7 to 36, four Inches from the other End.

Use of the Carpenter's Joint-RULE.

The Application of the Inches in measuring Lengths, Breadths, *etc.* is obvious.—The use of the other Side is all we need here meddle with.

1. *The Breadth of any Surface, as Board, Glass,* etc. *being given; to find how much in Length makes a Square Foot.*—Find the Number of Inches the Surface is broad, in the Line of Board Measure; and right against it is the Number of Inches required.

Thus if the Surface were 8 Inches broad, 18 Inches will be found to make a superficial Foot.

Or, more readily thus:- Apply the *Rule* to the Breadth of the Board or Glass; that End mark'd 36 being even with the Edge; the other Edge of the Surface will shew the Inches and Quarters of Inches which go to a square Foot.

2. *Use of the Table at the end of the Board Measure.*—If a Surface be one Inch broad, how many Inches long will make a superficial Foot? Look in the upper Row of Figures for 1 Inch, and under it in the second Row is 12 Inches, the answer to the Question.

3. *Use of the Line of Timber Measure.*—This resembles the former; for having learnt how much the Piece is Square, look for that Number on the Line of Timber-Measure: The Space thence to the End of the *Rule* is the Length, which at that Breadth makes a Foot of Timber.—Thus if the Piece be 9 Inches Square, the Length necessary to make a solid Foot of Timber, is 21⅔ Inches.—If the Timber be small, and under 9 Inches Square, seek the Square in the upper Rank of the Table; and immediately under it is the Feet and Inches that make a solid Foot.—Thus, if it be 7 Inches Square, 2 Foot 11 Inches will be found to make a solid Foot.

If the Piece be not exactly Square, but broader at one End than another; the Method is to add the two together, and take half the Sum for the Side of the Square.—For round Timber, the Method is to girt it round with a String, and to allow the fourth Part for the Side of the Square. —But this Method is erroneous; for hereby you lose above ⅕ of the true Solidity.

RUSTIC, in Architecture, a Manner of building in Imitation of Nature, rather than according to the Rules of Art. See BUILDING and ARCHITECTURE.

RUSTIC-*Work*, is where the Stones in the Face, *etc.* of a Building, instead of being smooth, are hatch'd or pick'd with the Point of a Hammer.

RUSTIC-*Quoins*, by *Vitruvius* call'd *Lapides Minentes.*

RUSTIC *Order*, is an Order with *Rustic* Quoins, *Rustic* Work, *etc.* See ORDER.—*Felibien* says, 'tis properly where the several Parts of the five Orders are not exactly observ'd; but this confounds *Rustic* with *Gothic*. See GOTHIC.

S

SAIGNER, in Fortification, a *French* Term, signifying to *bleed* or *drain*. Hence to *saigner* a Moat, is to empty or draw out the Water by Conveyances under Ground, that it may be passed over more easily, after laying Hurdles or Rushes on the Mud remaining.

SALLY, in Architecture, from the *French Sallie*, is what we more usually call *Projecture*. See PROJECTURE.

SALON, or *Saloon*, in Architecture, a very lofty, spacious Hall, vaulted at Top, and sometimes comprehending two Stories, or Ranges, of Windows: As that at *Blenheim* House. The *Salon* is a Grand Room in the Middle of a Building, or at the Head of a Gallery, *etc.* Its Faces, or Sides, are all to have a Symmetry with each other; and as it usually takes up the Height of two Stories, its Ceiling, *Daviler* observes, should be with a moderate Sweep. The *Salon* is a State Room. 'Tis much used in the Palaces in *Italy*; and from thence the Mode came to us. Embassadors, and other Great Visitors, are usually received in the *Salon*: 'Tis frequently built Square, sometimes Octogonal, as at *Marly*, and sometimes in other Forms.

SANCTUARY, among the *Jews*, the holiest and most retired Part of the Temple of *Jerusalem*; wherein was preserved the Ark of the Covenant; and into which No-Body was allowed to enter but the High Priest, and that only once a Year, to intercede for the People. The *Sanctuary*, called also *Sanctum Sanctorum*, or Holy of Holies, is supposed to be a Type of Figure of Heaven and of *Jesus Christ* the true High Priest, who is ascended thither to make Intercession for us. Some will have it, That the whole Temple was called the *Sanctuary*; and that the *Sanctum Sanctorum*, where the Ark was kept, was only a little Chapel or Oratory therein.

SAND, a fine, hard gravelly Earth, of great Use in Building, and many other Arts and Manufactures, as in the making of Glass, in Plumbery, Foundery, *etc.* There are Three Kinds of Sands, distinguished by the Places whence they are drawn: *viz. Pit-Sand*, *River-sand*, and *Sea-Sand*.

The Use of *Sand* in Building, is as an Ingredient in Mortar: See MORTAR. For this Use, *Pit-Sand* is of all others the Best; and of *Pit-Sand*, the whitest is always the worst. Of *River-Sand*, that found in the Falls of Waters is best, because most purged. *Sea-Sand* is the Worst. *Pit-*

Sand, as being fat and tough, is most used in Walls and Vaults. *River-Sand* serves for rough casting. All *Sand* is good in its Kind, if when squeezed and handled it crackles: and if being put on a white Cloth, it neither stains nor makes it foul. That *Sand* is naught, which, mixed with Water, makes it dirty and muddy, and which has been long in the Air; for such will retain much Earth and rotten Humour. Hence some Masons wash their *Sand* e'er they use it. The *Sand* of *Puzzuolo*, *De Lorme* observes, is the best in the World; especially for Maritime Building. Some distinguish a *Male Sand*, which is of a deeper Colour than another Sort in the same Bank or Bed, called *Female Sand*.

SAPHETA, in Architecture, is the Board over the Top of a Window, placed parallel and opposite to the Window-Board at the Bottom.

SARCOPHAGUS, or *Sarcophagum*, in Antiquity, a Tomb-Stone, wherein the Antients laid those they had not a mind to burn: The Word, derived from the *Greek*, literally signifies, one that eats Flesh; because at first, they used a Sort of Stones for the making of Tombs, which quickly consumed the Bodies: The Quarries from whence they dug them were in a City of *Troas*, named *Assum*: They had the Virtue to waste away a Body to nothing, save the Teeth, in Forty Days: This Stone resembled a Reddish Pumice Stone, and had a saltish Taste; they made Vessels of it to cure the Gout, into which they put their Feet, not suffereing them to continue there too long.

SAW, an Instrument serving to cleave, or divided into Pieces, divers solid Matters, as Wood, Stone, Marble, Ivory, *etc*. The *Saw* is one of the most useful Machines, in the Mechanic Arts, ever invented. The Fable, which is perhaps founded on some surer Tradition, attributes the Invention thereof to *Icarus*, who, vying with his Father *Dedalus*, enrich'd the rising Arts with several Discoveries. 'Tis added, he first contrived it on the Model of the Spine or Backbone of a flat Fish, such as the Soal. The *Saw* is made of Steel with Teeth, but those differently filed, and turned, according to the Use it is designed for. There are also Saws without Teeth, used in the sawing of Marbles and other Stones.

The best *Saws* are of Steel, ground bright and smooth; those of Iron are only Hammer-hardned: Hence, the first, besides their being stiffer, are likewise found smoother than the last. They are known to be well hammer'd by the stiff bending of the Blade; and well or evenly ground, by bending equally into a Bow. The Edge, wherein the Teeth are, is always thicker than the Back, in regard the Back is to follow the Edge. The Teeth are cut and sharpened by a Triangular File, first fixing the Blade of the *Saw* in a Whetting-block. When filed, the Teeth are to be *set*, that is, to be turned a-skew, or out of the right Line, to make the wider Kerf or Fissure, that the Back may follow the better: This is done by putting an Instrument, called a *Saw wrist*, between every other two

Teeth, and giving it a little Wrench, which turns one of the Teeth a little towards you, and the other a little from you. The Teeth are always set ranker for coarse cheap Stuff, than for hard and fine, in regard the ranker the Tooth is set, the more Stuff is lost in the Kerf; and if the Stuff be hard, the greater the Labour of sawing it.

The Workmen who make the greatest Use of the *Saw*, are, The Sawyers, Carpenters, Joiners, Ebonists, Stone-Cutters, Carvers, Sculptors, *etc.* The Lapidaries too have their *Saw*, as well as the Workers in *Mosaick*, but these bear little Resemblance to the common *Saws*.

But of all Mechanicks, there are none have so many as the Joiners, nor so many different Kinds. The chief are as follow. The *Pit-Saw*, a large two-handed *Saw*, used to saw Timber in Pits. 'Tis set rank for coarse Stuff, so as to make a Kerf or Fissure of almost a Quarter of an Inch; but for finer Stuff, finer. The *Whip-Saw*, which is likewise two handed, used to saw such large Pieces of Stuff as the *Hand-Saw* will not easily reach.—The *Hand-Saw* is made for a single Man's Use. Of these are various Kinds; as, The *Bow*, or *Frame-Saw*, furnished with Cheeks; by the twisted Cord and Tongue in the Middle thereof, the upper Ends are occasionally drawn close together, and the lower set the further apart. —The *Tenant-Saw*, which being very thin, has a Back to keep it from bending.—The *Compass-Saw*, which is very small, and its Teeth, usually, not set; its Use is to cut a round, or any other Compass-Kerf: Hence, the Edge is made broad, and the Back thin, that it may have a Compass to turn in.

SAWING, the Application of the *Saw*, in the dividing of Timber, *etc.* into Boards, *etc.* See SAW. There are Wind-mills, and Water-mills, which do the Office of *Sawing* Wood, with infinitely more Expedition and Ease than 'tis done by the Hand: They consist of several parallel Saws, which rise and fall perpendicularly, by means of one of the grand Principles of Motion. But a very few Hands are here needed, *viz.* only to push along the Timber, which are laid on Rollers, or suspended by Ropes, in Proportion as the *Sawing* advances. These are frequently found abroad, and were lately begun to be introduced in *England*; but the Parliament, in Consideration of this, that they would spoil the Sawyers Trade, and ruin great Numbers of Families, thought fit to prohibit them.

M. *Felibien*, in his Principles of Architecture, *etc.* mentions a Kind of long *Saws*, invented by one *Mission* Inspector of the Marble Quarries in the *Pyreneans*, by means whereof Stones are sawed even in the Rock itself whence they are taken. He adds, That some of them are twenty-three Foot long, but does not describe either their Form or Application: He only says, They are of Iron, without Teeth.

SCABELLUM, in the ancient Architecture, a kind of Pedestal, usually square, sometimes polygonous, very high and slender, commonly terminating in a kind of Sheath or Scabbard, or profiled in manner of a

Balluster. Its use is to bear *Busto's*, *Relivo's*, etc.

SCAFFOLD, a Timber-Work, raised in manner of an Amphitheatre to place Spectators upon for the commodious viewing of some Shew or Ceremony.—The Word is also used for a little Stage, or Theatre, raised in some publick Place, to execute Criminals upon, either by beheading or breaking upon the Wheel.—Also for an Assemblage of Planks and Boards sustained by Tressels, or by Pieces of Wood fixed in the Wall, whereon Masons, Sculptors, Painters, *etc.* stand to work in high Places, Wells, Ceilings, *etc. Menage* derives the Word from the *German Schawhauszch*, and *Guyet* derives it from the *Italian Catafalco*, which signifies the same thing. *Du Cange*, from *Eschafaudus*, Word in the corrupt *Latin*, signifying a Tribunal or Pulpit. He adds, That it might come from *Cata*, a Wooden Machine used to carry Earth to fill up Ditches, and carry over the Soldiers to the Attack; whence the *Italians* formed their *Catafalco*, the old *French* their *Chafaut*, and the *English* their *Scaffold.*

SCAMILLI IMPARES, in the Ancient Architecture, a Term much contended about among the Criticks; though, in effect, it signify no more than certain Benches, Zocco's or Blocks, serving to raise the rest of the Members of an Order, Column, Statue, or the like, and prevent their being lost to the Eye, which may chance to be placed below their Level; or below the Projecture of some of the Ornaments thereof. These *Scamilli* are well represented by the Pedestals of Statues.

SCANTLING, a Measure, Size, or Standard, whereby the Dimensions, *etc.* of Things are to be determined. The Word is form'd from the *French Eschantillon*, a Pattern or Specimen.

SCAPUS, in Architecture, the Fust or Shaft of a Column. See FUST.

SCARP, in Fortification, is the Foot of the Rampart-Wall, or the Sloping of the Wall from the Bottom of the Work to the Cordon on the Side of the Moat.

SCENE, in the ancient Drama, in its general Sense, was the Theatre whereon Dramatic Pieces, and other publick Shews, were represented. In its proper Sense, the ancient *Scene* was the Decoration of the Theatre. In the Extremity of the Theatre, were three large Gates or Openings, wherein appeared several perspective Decorations, *viz.* A Palace for Tragedies, Houses and Streets for Comedies, and Forests for Pastorals. These Decorations were all versatile, *i.e.* either they turn'd on Pivots, as described by *Vitruvius*, or trusile, *i.e.* slided along Grooves, as those in our Theatres; and as this or that Side, or Representation, was turned towards the Spectators, the *Scene* was called a *Tragic*, a *Comic*, or a

Pastoral Scene. Vitruvius observes, That the *Scene* was properly a large Face or Front of Building, adorned with Columns and Statues, among which were three large Apertures for perspective Representations in M. *Perrault's* Notes on *Vitruvius,* lib. 5. cap. 6. According to *Rosinus*, the *Scene*, in its proper and original Sense, is a Series of Trees disposed against each other, so as to form a continued Arch, and Shade, to defend those underneath it from the Injuries of the Weather: For in such Places it was, that, in ancient Times, e'er Theatres were built, they acted their Plays. Thus also *Cassiodorus* derives the Word *Scene*, from the close Shade of the Grove, where, in the Spring Time, the ancient Shepherds used to Sing and Play.

SCENOGRAPHY, in Perspective, a Representation of a Body on a perspective Plane; or, a Description thereof in all its Dimensions; such as it appears to the Eye. The *Ichnography* of a Building, *etc.* represents the Plan, or Ground work of the Building. The *Orthography*, the Front, or Upright, thereof. And the *Scenography*, the whole Building, Front, Sides, the Heighth and all. The Word is formed from the *Greek Scene*, and *Description.*

To exhibit the SCENOGRAPHY *of any Body.*

1°. Lay down the Basis, Ground-plot, or Plan of the Body, in the perspective *Ichnography*, according to the Method laid down under the Article PERSPECTIVE. 2°. Upon the several Points of the Plan raise the Perspective Heights: Thus will the *Scenography* of the Body be compleated; excepting that a proper Shade is to be added. The Method of raising the Heights is as follows.

On any Point given as C, (Tab. Perspect. Fig. 1.) *to raise a Perspective Altitude*, *answerable to an Objective Altitude* P Q. On the Terrestrial Line raise a Perpendicular P Q equal to the given objective Altitude. From P and Q, to any Point, as T, draw right Lines P T and Q T. From the given Point C draw a right Line C K, parallel to the Terrestrial Line D E, meeting the right Line Q T in K. In the Point K, upon the Line K C, erect a Perpendicular I K. This I K is the *Scenographic* Altitude required. The Application of this general Method of drawing the *Scenography* of a Body, is not so obvious, in every Case, but that it may be necessary to illustrate it a little by a few Examples.

To exhibit the SCENOGRAPHY *of a Cube*, *viewed by an Angle.*

1°. As the Basis of a Cube viewed by an Angle, standing on a Geometrical Plane, is a Square viewed by an Angle; Draw a Square, viewed Angular wise, on the Perspective Table, or Plane. 2°. Raise the Side H I (Fig. 2.) of the Square perpendicularly on each Point of the Terrestrial Line D E; and to any Point, as V, of the Horizontal Line H R, draw the right Line V I and V H. 3°. From the Angles d, b and c, draw c 1, d 2, *etc*. Parallel to the Terrestrial Line D E. 4°. From the Points 1 and 2, raise L 1, and M 2 perpendicular to the same. Lastly, Since H I is the Height to be raised in a, L I in c and b, and M 2 in d: In a raise the Line f a perpendicular to f a; in b and c, raise b g and c e perpendicular to b c 1; and Lastly, Raise d h, perpendicular to d 2: And make a f = H I, b g = e c = L 1, and h d= M 2: If then the Parts g h, e f be connected by right Lines, the *Scenography* will be compleat.

To exhibit the SCENOGRAPHY *of a Hollow* Quinquangular Prism.

1°. Since the Base of a hollow *Quinquangular Prism*, standing on a Geometrical Plane, is a Pentagon, with a Limb or Breadth of a certain Dimension; find the Appearance of this Pentagon on a Table, or Plane. See PERSPECTIVE. 2°. On any Point, as H, of the Terrestrial Line D E (Fig. 3.) raise a perpendicular H I, equal to the objective Altitude; and to any Point, as V, of the Horizontal Line H R, draw the Lines H V and I V. 3°. From the several Angles, a, b, d, e, c, of the Perspective *Ichnography*, both the Internal and External ones, draw right Lines, as b 2, d 3, *etc*. parallel to the Terrestrial Line; and from the Points 1, 2, 3, raise Perpendiculars to the same, as L 1, M 2 m 2, N 3 n 3. If these, then, be raised in the correspondent Points of the *Ichnography*, as in the preceding Article, the *Scenography* will be compleat.

To exhibit the SCENOGRAPHY *of a Cylinder.*

1°. Since the Base of a Cylinder, standing on a Geometrical Plane, is a Circle; seek the Appearance of a Circle. In the Points, a, b, d, f, h, g, e, c, (Fig. 4.) raise the apparent Altitudes, as in the preceding Articles. If now their upper Lines be connected by curve Lines, as in the Base, a, b, d, f, c, h, e, c, the *Scenography* of the Circle will be compleat.

'Tis evident that those Lines are to be omitted, both in the Plan and in the Elevation, which are not exposed to the Eye; though they are not to be disregarded, from the Beginning, as being necessary for the finding of other Lines. *E. gr.* In the *Scenography* of the Cube, viewed Angle wise, the Lines b d and d c, (Fig. 2.) in the Base, and the d h in the Elevation, are hid from the Eye, and are therefore omitted in the Description. But since the Point h is not to be found unless the Point d be had in the *Ichnography*; nor the Lines g h and d e be drawn without the Height d h; the Appearance of the Point d is as necessary to be determined in the Operation, as the Height h d.

To exhibit the SCENOGRAPHY *of a Pyramid, standing on its Base.*

Suppose, *e. gr.* it were required to delineate a Quadrangular Pyramid, viewed by an Angle: 1°. Since the Base of such Pyramid is a Square seen by an Angle, draw such a Square. 2°. To find the Vertex of the Pyramid, *i.e.* a Perpendicular let fall from the Vertex to the Base, draw Diagonals mutually intersecting each other in e (Fig. 5.) 3°. On any Point, as H, of the Terrestrial Line D E, raise the Altitude of the Pyramid H I; and drawing the right Lines H V, and I V to each Point of the Horizontal Line H R; produce the Diagonal a b, till it meet the Line V H in h. Lastly, From h draw h i parallel to H I. This being raised on the Point e, will give the Vertex of the Pyramid c d; consequently, the Lines d k, k a, and k b, will be determined at the same Time.—After the like Manner is the *Scenography* of a Cone delineated.

To exhibit the SCENOGRAPHY *of a Truncated Pyramid.*

Suppose the Truncated, Pyramid Quadrangular; 1°. then, If from the several Angles of the upper Base be conceiv'd Perpendiculars let fall to the lower Base, we shall have a Pentagon, with another inscribed therein, whose Sides are parallel to those of the former. This coincides with a Pentagon, furnished with a Rim or Breadth, *etc.* and may therefore be delineated in the same manner.—2°. Raising the Altitude of the Truncated Pyramid (I H, Fig. 6.) determine the *Scenographic* Altitudes, to be raised in the Points a, b, c, d. If now the Points f, g, h, i, k, be connected by right Lines, and the Lines, l k, f m, g n, h o, be drawn, the *Scenography* will be compleat.—By drawing two concentric Circles in a Geometrical Plan, and doing every Thing else, as in this Problem, the *Scenography* of a Truncated Pyramid will be drawn.

To exhibit the SCENOGRAPHY *of Walls, Columns, etc. or to raise them on the Pavement.*

1°. Suppose a Pavement A F H I (Fig. 7.) represented in a Plan, together with the Bases of the Columns, *etc.* if there be any . 2°. Upon the Terrestrial Line set off the Thickness of the Wall B A and 1, 3. 3°. Upon A and B, as also upon 3 and 1, raise Perpendiculars A D and B C, as also 3, 6, and 1, 7. 4°. Connect the Points D and 6 with the Principal Point V, by the right Lines D V and 6 V. 5°. Upon I and H raise Perpendiculars H G and E F. Thus will all the Walls be delineated. Now to raise the Pillars, *etc.* there needs nothing but, from their several Bases (whether Square, or Circular) projected on the Perspective Plan, to raise indefinite Perpendiculars; and on the Fundamental Line, where intersected by the Radius F a Passing through the Base, raise the true Altitude A D: for D V, being drawn as before, the *Scenographical* Altitudes will be determined.

To exhibit the SCENOGRAPHY *of a Door in a Building.*

Suppose a Door required to be delineated in a Wall D E F A (Fig. 7.) 1°. Upon the Fundamental Line set off its Distance A N from the Angle A, together with the Breadths of the Posts N I and L M, and the Breadth of the Gate itself L I. 2°. To the Point of Distance K, from the several Points N, I, L, M, draw right Lines K N, K I, K L, K M, which will determine the Breadth of the Door l i, and the Breadths of the Posts c n and m l. 3°. From A to O, set off the Height of the Gate A O, and from A to P, the Height of the Posts A P. 4°. Joyn O and P with the Principal Point, by the right Lines P V and O V. 5°. Then, from n, i, l, m, raise Perpendiculars, the Middle ones whereof are cut by the right Line O V in o, and the Extremes, by the right Line V P in p. Thus will the Door be delineated, with its Posts. If the Door were to have been exhibited in the Wall E F G H, the Method were nearly the same. For, 1°. Upon the Terrestrial Line, set off the Distance of the Door from the Angle, and thence also the Breadth of the Door R T. 2°. From R and T, draw right Lines to the Principal Point V, to have the Breadth r t in the perspective Plan. 3°. From r and t, raise indefinite Perpendiculars to F H. 4°. From A to O, set off the true Height A O. Lastly, From O, to the principal Point V, draw the right Line O V, intersecting E F in Z, and make r r and t t equal to T Z. Thus is the Door r r, t t, drawn, and the Posts are easily added, as before.

To exhibit the SCENOGRAPHY *of Windows in a Wall.*

When you know how to represent Doors, you'll find no Difficulty in adding Windows; all that is here further required, being to set off the Height of the Window from the Bottom of the Ground. The Whole Operation is as follows. 1°. From 1 to 2, set of the Thickness of the Wall at the Window; from 3 to 4, its Distance from the Angle 3; and from 4 to 5 its Breadth. 2°. From 4 and 5, to the Point of Distance L, draw the right Lines E 5 and L 4, which will give the Perspective Breadth 10, 9 of the Window. 3°. From 10 and 9 raise Lines perpendicular to the Pavement; *i.e.* draw the parallel Lines 6, 3. 4°. From 3 to 11 set off the Distance of the Window from the Pavement 3, 11; and, from 11 to 12, its Height 11, 12. Lastly, From 11 and 12, to the principal Point V, draw Lines V 11 and V 12; which intersecting the Perpendiculars 10, 13 and 9, 14, in 13 and 14, as also in 15 and 16, will exhibit the Appearance of the Window.

From these Examples, which are all no more than Applications of the first grand general Rule, it will be easily perceived what Method to take to delineate any other Thing, and at any Height from the Pavement.

A Mechanical Method of exhibiting the Scenography *of any Object*; See under the Article DESIGN.

SCHOOL, in Painting, is a Term used to distinguish the different Manners of Places, and Persons: As, the *Roman School*, the *Venetian School*, the *Flemish School*, etc. *Raphael's School*, *Titian's School*, *da Vinci's School*, etc. meaning their Disciples, Pupils, *etc.*

SCHOOL of *Athens*, in Painting, is a celebrated Piece of *Raphael*, now in the *Vatican*. It contains a great Number of Figures, representing Philosophers, Mathematicians, and other Persons, engaged in the Arts and the Sciences. Several Authors have wrote of this Painting, and given different Explications thereof. *Vasari* will have it to be, The Agreement of Philosophy, and Astronomy, with Theology. The Engravers, by the Inscription they add at the Bottom of the Print thereof, shew, That they take it for a Painting of St. *Paul* preaching at *Athens*. An *Augustin* of *Venice*, takes the Philosopher who is writing, for a St. *Mark*; and he at his Knees, for the Angel *Gabriel*. M. *de Piles* rejects all these Explications of the *School* of *Athens*, and especially the last: His Opinion is, That 'tis nothing more than the mere Image or Representation of Philosophy, which *Raphael* here designs under all the Philosophers he has painted. In Behalf of the *Venetian* Engravers it may be said, that they do not pretend to explain the Painting, but have only copied such of the Figures as they thought proper to represent St. *Mark*, St. *Gabriel*, *etc*.

SCIOGRAPHY, the Profile, or Section of a Building, to shew the Inside thereof. See SECTION and PROFILE.

SCOTIA, in Architecture, a Semicircular Cavity or Channel between the Tores, in the Bases of Columns. See BASE. The *Scotia* has an Effect just opposite to that of the Quarter-round. Our Workmen frequently call it the *Casement*. The *Scotia* is a concave, dark Moulding; whence its Name, *viz*. from *Obscurity*, *Darkness*. 'Tis also called *Trochilus*, partly from its Form. In the *Corinthian* Base, there are Two *Scotias*, the upper whereof is the smaller. According to *Felibien*, the *Cavetto* is a fourth Part of the *Scotia*. See CAVETTO.

SCRATCH-WORK, in *Italian*, *Sgrafitti*, was a Way of Painting in Fresco, by preparing a black Ground, on which was laid a white Plaister; which White being taken off with an Iron Bodkin, the Black appear'd through the Holes, and serv'd for Shadows: This kind of Work is lasting; but being very rough, is unpleasant to the Sight.

SCRIBING, in Joinery, *etc*. A Term used, when one side of a Piece of Stuff being to be fitted to the Side of some other Piece, which last is not regular; to make the Two, then, join close together all the Way, they *Scribe* it, thus: That is, they lay the Piece of Stuff to be *Scribed* close to the other Piece they intend to *Scribe* to, and open their Compasses to the greatest Distance the two Pieces any where stand from each other; then, bearing one of the Legs against the Side to be *Scribed* to, with the other Point they draw a Line on Stuff to be *Scribed*: Thus have they a Line on the irregular Piece parallel to the Edge of the Regular One; and if the Stuff be wrought away exactly to the Line, when the two Pieces are put together, they will seem a Joint.

SCROWLS, in Architecture. See VOLUTE.

SCULPTURE, the Art of Cutting, or Carving Wood, Stone, or other Matter, to form various Figures for Representations; as also of fashioning

Wax, Earth, Plaister, *etc.* to serve as Models, or Moulds, for the Casting of Metal Figures. *Sculpture*, in its Latitude, includes both the Art of working in Creux, properly called *Engraving*, and of working in Relievo, which is what we strictly call *Sculpture*.

The Antiquity of this Art is past Doubt; as the Sacred Writings, the most ancient and authentic Monuments we have of the earliest Ages, mention it in several Places: Witness *Laban's* Idols stollen away by *Rachel*, and the Golden Calf which the *Israelites* set up in the Desart, *etc.* But 'tis very difficult to fix the Original of the Art, and the first Artists, from prophane Authors; what we read thereof, being intermixed with Fables, after the Manner and Taste of those Ages: Some make a Porter of *Sicyon*, named *Dibutades*, the first Sculptor. Others say, The Art had its Origin in the Isle of *Samos*, where one *Ideocus* and *Theodorus*, perform'd Works of this Kind, long before *Dibutades's* Time. 'Tis added, That *Demuratus*, Father of *Tarquin the Elder*, first brought it into *Italy* upon his retiring thither; and that by Means of *Enciropus* and *Entygramma*, Two excellent Workmen herein, who communicated it chiefly to the *Tuscans*; among whom it was afterwards cultivated with great Success. They add, That *Tarquin* sent for *Tauranus*; one of the most eminent among them, to *Rome*, to make a Statue of *Jupiter*, *etc.* of baked Earth; for the Frontispiece of the Temple of that Deity.

About this Time, there were many *Sculptors*, both in *Greece* and *Italy*, who wrought altogether in Earth. Some of the most noted, are, *Calisthenes*, an *Athenian*, who made himself and his House famous, by the great Number of Earthen Figures he adorned it withal. *Demophilus* and *Gorsanus*, two Painters, who inrich'd the Temple of *Ceres* with great Variety of Painting and Earthen Images. In effect, all the first Statues of the Heathen Deities, were either of Earth or Wood; and 'twas not so much any Frailty of the Matter, or Unfitness for the purpose, as the Riches and Luxury of the People, that first induced them to make them of Marble, and other more precious Stones. Indeed, how rich soever the Matter were, whereon they wrought, yet they still used Earth, to form Models thereof. And to this Day, whether they be for cutting Marble Statues with the Chissel, or for casting them in Metal, they never undertake the one or the other, without first making a perfect Model thereof in Earth. Whence, doubtless, arose the Observation of *Praxiteles*, That the Art of Moulding Earthen Figures, was the natural Mother of that of making Marble and Metal Figures; which last never appeared in Perfection, till about 300 Years after the Building of *Rome*; though the First was at its Height long before.

Phidias of *Athens*, who came next, surpass'd all his Predecessors, both in Marble, in Ivory and Metals: And about the same Time, appeared several others, who carried *Sculpture* to the highest Perfection it ever arrived at; particularly *Policletes* at *Sicyon*; then *Myron*; *Lysippus*, who alone was allowed the Honour of casting *Alexander's* Image in Brass. *Praxiteles* and *Scopas*, who made those excellent Figures, now before the

Pope's Palace at *Montecavallo*: *Briaxis*, *Timotheus* and *Leochares*, who, with *Scopas*, wrought at the famous Tomb of *Mausolus*, King of *Caria*; *Cesisodorus*, *Canachus*, *Dedalus*, *Buthieus*, *Nyceratus*, *Euphanor*, *Theodorus*, *Xenocrates*, *Phiromachus*, *Stratonicus*, *Antigonus*, who wrote on the Subject of his Art; the famous Authors of the *Laocoon*, viz. *Agesander*, *Polydore*, and *Athenodorus*, and infinite Others, the Names of some whereof have passed to Posterity; those of others have perished with their Works: For though the Number of Statues in *Asia*, *Greece*, and *Italy*, were so immense, that in *Rome* alone, as we are informed, there were more than there were living Persons, yet we have but very few now left, at least very few of the finest. When *Marcus Scaurus* was Ædile, his Office obliging him to provide what was requisite towards the Publick Rejoycings, he adorned the stately Theatre, which he erected, with 3000 Brazen Statues; and though *L. Mummius*, and *Lucullus*, brought away a great Number out of *Asia* and *Greece*, yet, there were still above 3000 remaining in *Rhodes*, as many at *Athens*, and more at *Delphi*: but what is most extraordinary, is the Bigness of the Figures, which those ancient Artists had the Courage to undertake. Amongst those *Lucullus* brought to *Rome*, there was one of *Apollo*, 30 Cubits high; the *Colossus* of *Rhodes*, made by *Cares* of *Lyndos*, the Disciple of *Lycippus*, far exceeded it; *Nero's* Statue, made by *Xenodorus*, after that of *Mercury*, was also of an extraordinary Size, being 110 Feet high: *Sculpture*, however, did not continue above 150 Years, after *Phidias's* Time, till it began insensibly to decline; not, but that there were still some fine Pieces of Workmanship both in *Greece* and *Italy*, though not performed with so good a Fancy, and such exquisite Beauty.

Besides that the *Greek* Statues are most esteemed for the Work; there is a special Difference between them, and those of the *Romans*: in that the greatest Part of the first are naked, like those who wrestle or perform some other bodily Exercise, wherein the Youth of those Times placed all their Glory; whereas, the others are clad or armed, and particularly have the *Toga* on, which was the greatest Mark of Honour amongst the *Romans*.

To perform any thing in the Way of *Sculpture*, they begin with making a Model of Earth, or Wax. For Earthen Models, they use but few Instruments: Their Hands and Fingers do almost the Whole. For Waxen Models, to a Pound of Wax they put half a Pound of Colophony; some add Turpentine, melting the Whole with Oil of Olive: Some add a little Vermillion, or other Matter, to give it a Colour. 'Tis wrought and molded with the Fingers, like Earthen Models.

SCULPTURE *in Wood.*

For *Sculpture* in Wood, which we properly call *Carving*, the first Thing required, is, to chuse a Wood proper for the particular Kind of Work. If it be any thing large, and require a deal of Strength and Solidity, the hardest and most durable Wood is to be chosen, as Oak or Chesnut.

For smaller Works they use Pear-tree, and Service-tree; but as these Woods are very hard, for little delicate Works they use softer Woods, only close, and of a fine Grain: Such is the Linden-tree, which the Chissel is found to cut more easily, and cleanly, than any other Wood.

As to Statues, we find the Ancients have made them of almost all Kinds of Wood. At *Sicyon* was an Image of *Apollo* in Box; at *Ephesus*, that of *Diana* was in Cedar. As these two Kinds of Wood are very hard, and even held incorruptible, especially Cedar; *Pliny* observes, they were judged particularly expedient for Representations of the Deities. In a Temple, on Mount *Cyllenus*, dedicated to *Mercury*, was an Image of that God form'd of the Wood of the Lemon-Tree: Others there were of the Palm-Tree, Olive-Tree, Ebony, and even of the Vine.

For large Works, if it be only single Figures, 'tis better they consist of several Pieces, than of a single one, by reason of the Liableness of the Latter to warp; for every large Piece may probably not be dried to the Heart, however it may appear without Side. Observe, that the Wood will not be fit for working, till after it have been cut at least ten Years.

SCULPTURE *in Marble, and other Stones.*

The first Thing they do, is out of a great Block of Marble to saw another of the Size required, which is performed with a smooth Steel Saw without Teeth, casting Water and Sand thereon, from Time to Time: Then they fashion it, by taking off what is superfluous with a stubbed Point, and a heavy Mallet; after this, bringing it near the Measures required, they bring it still nearer with another finer Point. They now use a flat cutting Instrument, having two Notches in its Edge, or three Teeth; then a Chissel to take off the Scratches the former has left. This last Instrument they use with a deal of Delicacy, giving thereby a Softness, and Tenderness to their Figure; till, at length, taking Rasps of different Degrees of Fineness, by degrees they bring their Work into a Condition for Polishing. To Polish, or make the Parts smooth and sleek, they use Pumice-Stone, and Smalt; then Tripoli, and when a still greater Lustre is required, a Skin and burnt Straw.

When any considerable Work is undertaken, as a Statue, Basso Relievo, or the like, they always make a Model, before-hand, of Clay; but as this shrinks in drying, and easily cracks and breaks, they only use it to make a Mould of Plaister, or Stucco, wherein they make a Figure of the same Matter, which serves them thence-forth for a Model, and by which they adjust all their Measures and Proportions. To proceed the more regularly, on the Head of the Model they place an immoveable Circle, divided into Degrees, with a moveable Ruler, or Index, fastened in the Centre of the Circle, and divided likewise into equal Parts. From the End of the Ruler hangs a Thread with a Plummet, which serves to take all the Points to be transferr'd thence to the Block of Marble, from whose Top hangs another Plummet like that of the Model. Indeed there are some excellent *Sculptors* who disapprove of this Method, urging, That the smallest Motion of the Model changes their Measures, for which Reason they rather choose to take all their Measures with the Compasses.

SEALING, in Architecture, the fixing a Piece of Wood or Iron in a Wall, with Plaister, Mortar, Cement, Lead, or other solid Bindings. For Staples, Hinges and Joints, Plaister is very proper.

SEASONING of *Timber*. See TIMBER.

SECTION *of a Building*, is its *Profile*; or a Delineation of its Heights and Depths raised on the Plan; as if the Fabrick was cut asunder, to discover the Inside. See PROFILE.

Horizontal SECTION. See ICHNOGRAPHY.

SELL, in Building, is of Two kinds, *viz. Ground-Sell*, which is the lowest Piece of Timber in a Timber Building, and that whereon the whole Superstructure is raised; and *Window-Sell*, called also *Window-Soil*, the bottom Piece in a Window-Frame. See WINDOW.

SEPTIZON, in the Ancient Architecture, a Term almost appropriated to a famous Mausoleum, of the Family of the *Anthonins*; which *Aur. Victor* tells us, was built in the Tenth Region of the City of *Rome*: being a very large insulated Building, with Seven Stages or Stories of Columns. The Plan was square; and the Upper Stories of Columns, falling back much, render'd the Pile of a Pyramidal Form; terminated a-top, with the Statue of the Emperor *Septimus Severus*, who built it. It had its Name *Septizon*, *Septizonium*, from *Septem* and *Zona*, *q.d.* Seven Zones or Girdles, by reason of its being girt with Seven Rows of Columns. Historians make mention of another *Septizon*, more Ancient than that of *Severus*, built near the *Thermæ* of *Anthoninus*.

SEPULCHER, a Tomb, or Funeral Monument; or a Place destined for the Interment of the Deceas'd. The Term is chiefly used with regard to the Burying-Places of the Antients: Those of the Moderns we usually call *Tombs*. Besides, the usual *Sepulchers* for the Interment, either of the whole Body, or of the Ashes of the Body burnt; the Antients had a peculiar Kind, call'd *Cenotaphium*; being an empty *Sepulcher*, made in Honour of some Person, who, perhaps, had no Burial at all; from a Superstitious Opinion, That the Souls of those who wanted Burial, wandered an Hundred Years, e'er they were admitted to pass into the *Elysian* Fields. The Pyramids are built to serve as *Sepulchers* for the Kings of *Egypt*. And the *Obelisks* had generally the same Intention. See OBELISK. *Sepulchers* were held Sacred and Inviolable; and the Care taken thereof, deem'd a Religious Duty, grounded on the Fear of God, and the Belief of the Soul's Immortality. Those who search'd or violated them, were odious to all Nations, and severely punished. The *Egyptians* call'd their *Sepulchers*, *Eternal Houses*, in Contradistinction to their Houses and Palaces, which they call'd *Inns*; by reason of the short Sojourn we have in the one, in Comparison of the long Stay we have in the other. The *Eastern* Pilgrimages are all made with Design to visit the *Holy Sepulcher*, that is, the Tomb of *Jesus Christ*. No Body enters here but bare-footed, and with a World of Ceremonies. The *Turks* exact 24 Crowns of each Pilgrim, whom Devotion carries to the *Holy Sepulcher*.

SEPULCHRAL, something belonging to *Sepulchers* or Tombs. *Sepulchral Inscriptions* are the surest Monuments we have of Antiquity. *Sepulchral Lamps*, are those said to have been found burning in the Tombs of several Martyrs and others. *Sepulchral Column*, is a Column erected on a Tomb, with an Inscription on its Shaft.

SERPENTINE MARBLE, or *Stone*, a Kind of Marble called by the Ancients *Ophites*, from the *Greek Serpent*, as being speckled like a Serpent's Skin. The Ground is blackish, but covered with green and yellowish Stains, Streaks, *etc*. exceeding hard, precious and antique. The Scarceness of the *Serpentine* makes, that 'tis only used by Way of Incrustation. The largest Pieces we know of, are some Tables in the Compartiments of the Attic of the Pantheon; and Two Columns in the Church St. *Laurence* in *Lucina* at *Rome*. There is also a soft Kind of *Serpentine* brought from *Germany*, used to make Vessels of, but not in Building.

SEWERS, in Building, are Shores or Conduits, or Conveyances for the Suillage and Filth of a House. Sir *Henry Wotton* advises, that Art imitate Nature in these Ignoble Conveyances, and separate them from Sight, (where there wants running Water) into the most remote, lowest and thickest part of the Foundation, with secret Vents passing up through the Walls, to the wide Air, like Tunnels; which all the *Italian* Architects commend for the Discharge of noisom Vapours, though elsewhere little practised.

SGRAFIT, in Painting, a Term derived from the *Italian Sgrafficiata*, Scratch-work, signifying a Method of Painting in Black and White only, not in Fresco, yet such as will bear the Weather. *Sgrafit* is both the Design, and the Painting, all in one: 'Tis chiefly used, to embellish the Fronts of Palaces, and other magnificent Buildings.

SHAFT, in Building: the *Shaft of a Column*, is the Body thereof; thus call'd from its Streightness: but by Architects more frequently the *Fust*. See FUST. *Shaft*, is also used for the Spire of a Church Steeple. See SPIRE. And for the Tunnel of a Chimney. See CHIMNEY.

SHINGLES, in Building, small Pieces of Wood, or quarter'd oaken

Boards, saw'd to a certain Scantling; or more usually cleft to about an Inch thick at one End, and made like Wedges, Four of Five Inches broad, and Eight or Nine Inches long. They are used in Covering, especially Churches and Steeples, instead of Tiles, or Slates. This Covering is dear; yet where Tiles, *etc.* are very scarce, and a light Cover required, is preferable to Thatch. If made of good Oak, and cleft, not sawed, and well season'd, they make a sure, light, and durable Covering. The Building is first to be cover'd all over with Boards, and the *Shingles* nail'd thereon.

SHOULDERING, in Fortification, is a Retrenchment, opposed to the Enemies; or a Work cast up for a Defence on one Side: whether made of Heaps of Earth cast up, or of Gabions, and Fascines. A *Shouldring* also is a Square Orillon, sometimes made in the Bastions, on the Flank near the *Shoulder*, to cover the Cannon of a Casement. It is also taken for a Demi-Bastion, or Work, consisting of one Face, and one Flank, which ends in a Point, at the Head of an Horn-work, or Crown-work: Neither is it to be understood only of a small Flank added to the Sides of the Horn-work to defend them, when they are too long; but also of the Redents which are raised on a streight Line.

SHOULDERING-PIECE, in Building.

SILLON, in Fortification, is an Elevation of Earth made in the Middle of the Moat, to fortify it, when too broad. It is otherwise called *Envelope*, which is the more common Name.

SIMA, in Architecture, a Term used by *Wolfius*, and some other Writers, for what we otherwise call *Cymatium* or *Simatium*. See CYMATIUM.

SIMATIUM or SIMAISE, in Architecture: See CYMATIUM. *Simatium* and *Cymatium*, are generally confounded together; yet, in effect, they ought to be distinguished: the latter being the Genus, and the former the Species. *Simatium* of *Sima*, camous, according to *Felibien*, is the last and uppermost Member of grand Corniches, called particularly the great *Doucine* or *Gula recta*; and by the *Greeks*, *Epithides*. In the Antique Buildings, the *Simatium*, a-top of the Doric Corniche, is generally in Form of a Cavetto, or Semi-Scotia; as we see particularly in the Theatre of *Marcellus*: This, some modern Architects have imitated; but in the *Ionic* Order, the *Simatium* is always a *Doucine*. The *Simatium*, or *Doucine*, then, is distinguished from the other Kinds of *Cymatia*, by its being camous.

SLATE, a blue fossile Stone, very soft when dug out of the Quarry, and on that Account easily cut or saw'd into thin long Squares or Escallops, to serve in Lieu of Tiles for the Covering of Houses. 'Tis sometimes also used to make Tables of; to Pave withal, *etc.* The Ancients were unacquainted with the Use of *Slate*, and instead thereof cover'd their Houses with Shingle, as we read in *Pliny*. Besides the *Blue Slate*, we have in *England* a *Greyish Slate*, call'd also *Horsham Stone*, from a Town in *Sussex* of that Name, where the Greatest Quantities of it are found.

The *Blue Slate,* is a very light, lasting, and beautiful Covering but chargeable withal, in regard the Roof must be first Boarded over, the *Slates* hung on Tacks, and laid with finer Mortar than Tiles. The *Grey Slate* is chiefly used in the Covering of Churches, Chappels, Chancels, *etc.*

'Tis dearer than Tiles, but far more durable. The timber of the Roof needs to be very strong for these *Grey Slates*, it being almost double the Weight of Tiles.

To judge of the Goodness of *Slate*, Mr. *Colepress*, in the *Philosophical Transactions*, orders it to be knock'd against any hard Body, to make it yield a Sound: If the Sound be good and clear, the Stone is firm and good; otherwise, 'tis crazy.

SOCLE or ZOCLE, in Architecture, a flat, square Member under the Bases of Pedestals, of Statues, Vases, *etc.* which it serves as a Foot, or Stand.

The Word is *French*, form'd from the *Italian Zoccola*, or the *Latin Soccus*, the Shoe of the ancient Actors. *Vitruvius* calls it *Quadra.*

A *continued* SOCLE, is a kind of continued Stand, or Pedestal, without either Base, or Cornice ranging round the whole Building; called by *Vitruvius*, *Stereobata*, and the *French*, *Soubaissement.*

SOFFITO or SOFIT, in Architecture, any Plafond or Ceiling, form'd of cross Beams, or flying Cornices, the square Compartiments or Pannels whereof are inrich'd with Sculptures, Painting, or Gilding. Such are those we see in the Basilics and Palaces of *Italy*, in the Apartments at *Luxembourg*, at *Paris*, *etc.* See CEILING.

The word is *Italian*, and signifies the same with the *Latin*, *Lacunar* and *Laquear*; with this Difference, that *Lacunar* is used for any Ceiling, with square, hollow Pannels, called *Lacus*: and *Laquear* for Compartiments interlaced with Plat-Bands, after the manner of Knots or *Laquei*.

SOFFITO or *Sofit*, is particularly used for the under Side or Face of an Architrave; and for that of the Corona or Larmier, which the Ancients called *Lacunar*, and we usually, *Plafond*.

It is enrich'd with Compartiments of Roses; and in the *Doric* Order has 18 Drops, disposed in three Ranks, Six in each, placed to the Right of the Guttæ, at the Bottom of the Triglyphs. See GUTTÆ, DORIC, ARCHITRAVE, *etc.*

SPHINX, in Sculpture, *etc.* a Figure or Representation of a Monster of that Name, famed among the Ancients; now mostly used as an Ornament in Gardens, Terrasses, *etc.*

'Tis represented with the Head and Breasts of a Woman, the Wings of a Bird, the Claws of a Lion, and the rest of the Body like a Dog.

It is supposed to have been engender'd by *Typhon*, and sent by *Juno*, to be revenged on the *Thebans*. Its Office was to propose dark, enigmatical Questions to all Passers-by, and if they did not give the Explication thereof, to devour them.

It made horrible Ravages on Mount *Sphingion*, and could not by any means be destroyed, till after *Oedipus* had solved the following Riddle; What Animal is it, that in the Morning walks on four Legs; at Noon on Two, and at Night on Three?

Among the *Egyptians*, the *Sphinx* was the Symbol of Religion, by reason of the Obscurity of its Mysteries. And on the same account, the *Romans* placed a *Sphinx* in the *Pronaos* or Porch of their Temples.

SPIRAL, in Architecture, Sculpture, *etc.* is a Curve that ascends, winding about a Cone or *Spire*; so as all the Points thereof continually approach the Axis: By this it is distinguished from the *Helix*, which winds after the same Manner, around a Cylinder.

Ignorant Architects confound the Two: But the more knowing distinguish them carefully.

SPIRAL *Stairs*, in Building. See STAIRS.

SPIRE, SPIRA, in the ancient Architecture, is sometimes used for the Base of a Column; and sometimes for the *Astragal* or *Tore*.

From the *Latin*, *Spiræ*, the Folds of a Serpent, which when laid, bears some Resemblance thereto; or from the *Greek*, the Coils on a Rope.

SPITTLE-*House*, a Corruption of the Word *Hospital*. See HOSPITAL.

STADIUM, an ancient *Greek* long Measure, containing 125 Geometrical Paces, or 625 Feet; corresponding to our Furlong.

Eight of them make a Geometrical or *Italian* Mile; and 20, according to M. *Dacier*, a *French* League; according to others 24 make a League.

Guilletiere computes that the *Stadium* was only 600 *Athenian* Feet, which amount to 566 *French* Royal Feet, or 604 *English* Feet; so that the *Stadium* should only have been 113 Geometrical Paces.

It must be observed however, that the *Stadium* was different in different Times and Places.

The Word is form'd from the *Greek*, *Station*; and, 'tis said, on this Occasion, that *Hercules*, after running so far at one Breath, stood still.

STADIUM was also the Course or Career, wherein the *Greeks* run their Races.

Vitruvius describes it as an open Space 125 Paces long, terminated at the two Extremes with two Posts, called *Carcer* and *Meta.*

Along it was an Amphitheatre, where the Spectators were placed to see the *Athletæ* Exercise, Running, Wrestling, *etc.*

There were *Stada* likewise, covered over, and encompassed with Colonnades and Porticoes, serving for the same Exercises in ill Weather. Captive Children used to run the *Stadium.* Ablanc.

A more natural Derivation of the Word *Stadium*, from *Statio*, than that popular one mentioned in the last Article, is drawn from the *Athletæ* stopping and resting, when at the End of this Course: Whence the Name might be applied to the same Distance, measured in any other Place.

STAFF, in Surveying, a kind of Stand, whereon to mount a Theodolite, Circumferentor, plain Table, or the like, for Use. It consists of Three Legs of Wood, joyned together at one End, whereon the Instrument is placed; and made peeked at the other, to enter the Ground. Its upper End is usually filled with a Ball and Socket.

STAGE, in the Modern Drama, the Place of Action and Representation; included between the Pit and the Scenes.

The *Stage* answers to the *Proscenium* or *Pulpitum* of the ancient Theatre.

STAIRS, in Building, the Steps whereby we ascend and descend, from one Story of a House to another.

The Dimensions of *Stairs* are differently assigned by different Authors: In this however they agree, that they must not be more than Six, nor less than Four Inches high; not more than 18, nor less than 12 Inches Broad; not more than 16, nor less than 6 Foot long, each Stair. But these Measures have only regard to fine Buildings; ordinary Houses are excepted: Yet even in these, the *Stairs* are not to exceed Eight Inches in Height, nor less than Nine Inches in Breadth; nor Three in Length.

To reduce the Dimensions of *Stairs* to some natural, or at least geometrical Standard, *Vitruvius* borrows the Proportions of the Sides of a rectangle Triangle, which the ancient School express'd by the Numbers 3, 4 and 5. The first for the perpendicular Height; the second for the horizontal Breadth; and the third for the whole Slope or Inclination, from the Edge of onè *Stair* to that of another.

But this Rule is set aside, and with good Reason, by the modern Builders. For, on this Principle, the lower the Stairs, the narrower they must be; and *Stairs*, for Instance, Four Inches high (such as we find mentioned in ancient Architects) must be but 5⅓ Inches broad.

One Rule to be regarded in the making of *Stairs*, is, That they be laid according to the *Italian* Phrase, *Con un tantino da Scarpa*, i.e. *somewhat sloping*, or a little highest behind, that the Foot may, as it were, both ascend and descend at once; which, though observed by few, is found a secret and delicate Deception of the Pains in mounting.

STAIR-*Case*, an Ascent inclosed between Walls, or a Ballustrade, consisting of *Stairs* or Steps, with Landing-places and Rails; serving to make a Communication between the several Stories of a House. See STAIRS.

The Construction of a compleat *Stair-Case*, is one of the most curious Works in Architecture. The common Rules to be observed therein, are as follow;

1° That it have a full, free Light, to prevent Accidents of slipping, falling, *etc*. 2° That the Space over Head be large and airy, which the *Italians* call *un bel Sfogolo*, good Ventilation, in regard a Man spends much Breath in mounting. 3° That the Half-paces, or Landing-places, be conveniently distributed for reposing in the Way. 4° That to prevent Rencounters, *etc*. the *Stair-Case* ben't too narrow; but this last is to be regulated by the Quality of the Building. 5° That Care be taken in placing the *Stair-Case*, so as the *Stairs* may be distributed, without Prejudice to the rest of the Building.

The Kinds of *Stair-Cases* are various: In some, the *Stairs* are straight, others winding; in others both Ways.

Again, of *straight Stairs*, called also *Flyers*, some fly directly forwards, others are Square, others Triangular, others call'd *French Flights*. Of *winding Stairs*, call'd also *spiral* or *cockle Stairs*, some are Square, some Circular, or some Elliptical.

And these again, are various; some winding round a solid and others an open Newel.

Lastly, of *mix'd Stairs*, some are call'd *Dog-leg'd*: Others both wind about a solid Newel, and fly about a square, open Newel.

Stair-Cases are of that Importance in Building, that we can't excuse the not giving a more particular Account of each kind.

Straight STAIRS, then, are such as always *fly*, that is, proceed in a Right Line, and never wind, whence their Denomination. See FLYERS.

Of these there are several Kinds; as—*Direct Flyers* or *Plain Flyers*, which proceed directly from one Floor to another, without turning either to the Right or Left; seldom used, except for Garret or Cellar *Stairs*. —*Square Flyers*, which fly round the Sides of a Square Newel, either solid or open; having at every Corner of the Newel, a square, half Step, taking up ¼ of a Circle. So that they fly from one half Step to another, and the Length of the *Stairs* is perpendicular to the Side of the Newel. —*Triangular Flyers*, which fly round by the Sides of a triangular Newel, either solid or open, having at each Corner of the Newel a Trapezial Half-step, taking up ⅔ of a Circle: So they fly from one Half-step to another; and their Length is Perpendicular to the Side of the Newel.—*French Flyers*, which fly, first directly forwards, till they come within the Length of a *Stair* of the Wall; and then have a square half Pace, from which you immediately ascend to another half Pace, from which the *Stairs* fly directly back again, parallel to their first Flight.

Winding STAIRS, are such as always *wind*, and never fly: Of these there is great Variety; as,—*Circular winding Stairs*, whereof there are four Kinds, *viz.* such as wind about a solid Newel, the fore Edge of each being in a Right Line, pointing to the Centre of the Newel; commonly used in Church Steeples and great old Houses. Such as wind round an open Newel, the Fore-side of each being in a right Line, pointing to the Centre of the Newel, as those in the *Monument, London.* Such as wind round a solid Newel, only the Fore-side of each, an Arch of a Circle, either concave or convex, pointing near to the Circumference of the Newel. And such as resemble the last, in all other respects, save that they have an open Newel. Any of these *winding Stairs* take up less Room than any other Kind. In *Stairs* that wind round a solid Newel either $\frac{1}{6}$ or $\frac{1}{4}$, $\frac{1}{3}$ or $\frac{3}{7}$ of that of the *Stair-case*, according as that is in Bigness. If very small, the Newel is but $\frac{1}{6}$, and if large, $\frac{3}{7}$, *etc.*

In *Stairs* that wind round an open Newel, *Palladio* orders the Newel to be $\frac{1}{2}$ the Diameter of the *Stair-case*; though there does not appear any Reason why the Newel here should not be proportion'd to the *Stair-case*, as in the former. As to the Number of *Stairs* in each Revolution; *Palladio* orders, that in a *Stair-case* or 6 or 7 Foot Diameter, the *Stairs* in each Revolution be 12; if the Diameter be 8, the *Stairs* to be 16; if 9 or 10, the *Stairs* to be 20, and if 18, to be 24.—*Elliptical winding Stairs*, whereof there are two Kinds; the one winding round a Solid, the other round an open Newel. They are much of the same Nature as *circular Stairs*, excepting that in the one, the Newel is a Circle, and in the other an Ellipsis.—*Square winding Stairs*, are such as wind round a square Newel, either Solid or open; the Fore-side of each *Stair* being a right Line, pointing to the Centre of the Newel.—*Triangular winding Stairs*, are such as wind round a triangular Newel; the Fore-side of each being a right Line, pointing to the Centre of the Newel.—*Columnated window Stairs*: *Palladio* mentions a Stair-case in *Pompey's* Portico at *Rome*, set on Columns, so, as that the Light they received from above, might distribute itself to all Parts alike.—*Double winding Stairs*, *Scamozzi* mentions a *Stair-case* of this Form, made by *Piedro del Bergo*, and *Jean Cossin*, at *Sciamberg* in *France*, in the King's Palace.

They are so contrived, as that two Persons, the one ascending and the other descending, shall never Meet. Dr. *Grew* describes a Model of this kind of *Stair-case*, kept in the *Musæum* of the *Royal Society*. The Foot of one of the *Stair-cases*, he says, is opposite to that of the other, and both make a parallel Ascent, and within the same Cylinder. The Newel in the Middle is hollow, and built with long Apertures, to convey Light from Candles placed at the Bottom and on the Sides of the Newel, into both Cases.—*Quadruple winding Stairs*, *Palladio* mentions a *Stair-case* of this Form, in the Castle of *Chambor* near *Bloys*. It consists of four *Stair-cases*, carried up together, having each its several Entrance, and going up one over another, in such Manner, as that being in the middle of the

Building, the Four serve for four Apartments; so that the People of the one need not go up and down the *Stairs* of the other; yet being open in the Middle, they all see each other pass.

Mixt STAIRS, are such as partly fly, and partly wind; whence some call them *Flyers* and *Winders*. Of these there are several Kinds, as, —*Dog-leg'd Stairs*, which first fly directly forwards, then wind a Semi-circle, and then fly directly backwards, parallel to that.—*Square flyers and Winders*, have a square Newel, either Solid or open, and fly by the Sides of the Newel, winding a Quadrant of a Circle at each Corner. —*Solid and open Newel'd Flyers and Winders*, are of two Kinds; the one winds a Quadrant of a Circle about a Solid Newel; then flies by the Side of a square open Newel; then winds again, by the Side of a solid Newel then flies again, and so alternately. The other flies first, then winds, and then flies again, alternately.

STANCHEONS, in Building. See PUNCHIN.

STAR-*Fort*, in Fortification, is a Work with several Faces, generally composed of from 5 to 8 Points, with Saliant and Re-entring Angles flanking one another; every one of its Sides containing from 12 to 25 Fathoms.

STATUARY, a Branch of Sculpture, employ'd in the making of *Statues*. See STATUE.

Statuary is one of those Arts wherein the Ancients surpass'd the Moderns: Indeed 'twas much more popular, and more cultivated among the former than the latter. See SCULPTURE.

'Tis disputed between *Statuary* and *Painting*, which of the two is the most difficult and the most artful.

The Invention of *Statuary* was at first very coarse. *Leon Battist Alberti*, who has an express Treatise on *Statues*, imagines that it took its Rise from something casually observed in the Productions of Nature, that, with a little Help, might seem disposed to represent the Figure of some Animal. The common Story is, That a Maid, full of the Idea of her Lover, made the first Essay, by the Assistance of her Father's Implements, who was a Potter. This at least, is pretty certain, that Earth is the first Matter *Statuary* was practised upon.

STATUARY is also used for the Artificer, who makes *Statues*. In this Sense we say, *Phidias* was the greatest *Statuary* among the Ancients, and *Michael Angelo* among the Moderns.

STATUE, a Piece of Sculpture in full Relievo, representing a human Figure.

Daviler, more Scientifically defines *Statue* a Representation in high Relievo and Insulate, of some Person distinguish'd by his Birth, Merit or great Actions; placed as an Ornament in a fine Building, or exposed in a publick Place to preserve the Memory thereof.

Statues are form'd with the Chissel, of several Matters, as Stone, Marble, Plaister, *etc*.

They are also cast of various Kinds of Metals, particularly Gold, Silver, Brass and Lead.

Dedalus, the Son of *Eupalamus*, who lived not only before the Siege of Troy, but even before the Expedition of the *Argonauts*, among many other notable Contrivances ascribed to him, is said to have been the Inventor of *Statues*. And yet 'tis certain, there were *Statuaries* before him; only he 'twas first found how to give them Action and Motion, and to make them appear as if alive. Before him, they made them with the Feet joined together, never intending to express any Action. He first loosened the Feet of his, and gave them the Attitudes of People walking and acting.

The *Phœnicians* are said to have been the first who erected *Statues* to the Gods.

The *Greeks* succeeded in their *Statues* beyond the *Romans*; both the Workmanship and the Fancy of the *Roman Statues*, were inferior to the *Grecian*. Indeed we have very few remaining, that have escaped the Injuries of Time.

In Strictness, the Term *Statue* is only applied to Figures on Foot; the Word being form'd from the *Latin*, *Statura*, the Size of the Body; or from the *Stare*, to stand.

Statues are usually distinguish'd into four Kinds; the First, are those less than the Life; of which Kind we have several *Statues* of Men, of Kings, and Gods themselves.

The Second, of those equal with the Life; in which Manner it was, that the Ancients, at the publick Expence, used to make *Statues* of Persons eminent for Virtue, Learning, or the Services they had done.

The Third, of those that exceed the Life, among which, those which surpass'd the Life once and a half, were for Kings and Emperors; and those double the Life, for Heroes.

The Fourth Kind, were of those that exceeded the Life twice or thrice, or even more, and were called *Colossus's*.

Every *Statue* resembling the Person it represents, is called *Statua Iconica*.

Greek STATUE, is used for a Figure that is Naked and Antique; it being in this Manner the *Greeks* represented their Deities, *Athletæ* of the Olympic Games, and Heroes. The Reason of this Nudity, whereby the *Greek Statues* are distinguished, is, that those who exercised Wresting, wherein the *Greek* Youth placed their chief Glory, always perform'd naked.

The *Statues* of Heroes were particularly called *Achillean Statues*, by reason of the great Number of Figures of that Prince, in most of the Cities of *Greece*.

Roman STATUES. This Appellation was given to such as were

cloath'd, and which receive various Names from their various Dresses. Those of Emperors with long Gowns over their Armour, were called *Statuæ Paludatæ*; those of Captains and Cavaliers, with Coats of Arms, *Thoracatæ*: Those of Soldiers, with Cuirasses, *Loricatæ*; those of Senators and Augurs, *Trabeatæ*; those of Magistrates with long Robes, *Togatæ*; those of the People with a plain Tunica, *Tunicatæ*; and lastly, those of Women with long Trains, *Stolatæ*.

The *Romans* had another division of *Statues*, into *Divine*, which were those consecrated to the Gods, as *Jupiter*, *Mars*, *Apollo*, etc. Heroes, which were those of the Demi-Gods, as *Hercules*, *etc*. and *August*, which were those of the Emperors; as those two of *Cæsar* and *Augustus*, under the Portico of the Capitol.

Pedestrian STATUE, a *Statue* standing on foot; as that of King *Charles* II. in the *Royal Exchange*, and that of King *James* II. in the *Privy Gardens*.

Equestrian STATUE, that representing some illustrious Person on Horseback; as that famous one of *Marcus Aurelius* at *Rome*, and those of King *Charles* I. and II. at *Charing-Cross* and in *Stocks-Market, London*.

Curule STATUE. Those are thus called, which are represented in Chariots drawn by *Bigæ* or *Quadrigæ*, that is, by two or four Horses; of which Kind there were several in the *Circus's Hippodromes*, *etc*. or in Cars, as we see on triumphal Arches in antique Medals.

Allegorical STATUE, that which, under a human Figure, or other Symbol, represents something of another Kind, as a Part of the Earth, a Season, Age, Element, Temperament, Hour, *etc*.

Hydraulic STATUE, any Figure placed as an Ornament to a Fountain or Grotto; or that does the Office of a *Jet d'Eau*, a Cock, Pipe or the like, by any of its Parts, or by an Attribute it holds. The same is to be understood of any Animal serving for the same Use.

STEEPLE, an Appendage, raised on the western End of a Church, to hold the Bells.

Steeples are denominated from their Form, either *Spires* or *Towers*.

The First, are such as ascend continually diminishing either Conically or Pyramidically. The latter are mere Parallelipipeds, and are cover'd atop, Plat-form like. In each Kind, there is usually a sort of Windows or Apertures to let out the Sounds, and so contrived, at the same time, to drive it down.

Masius, in his Treatise of Bells, treats likewise of *Steeples*. The most remarkable in the World, is that at *Pisa*, which leans all on one Side, and appears every Moment ready to fall; yet without any Danger. This odd Disposition, he observes, is not owing to a Shock of an Earthquake, as is generally imagined; but it was contrived so at first by the Architect; as is evident from the Ceilings, Windows, Doors, *etc*. which are all in the Level.

STEP. See STAIRS.

STEREOBATA, in the ancient Architecture, the Basis or Foundation whereon a Wall or other Building is rais'd.

This answers pretty well to the *continued Socle* or Basement of the Moderns; Some confound it with the ancient *Stylobata*, or Pedestal; but in effect, the *Stereobata* is that to the *Stylobata*, which the *Stylobata* is to the *Spira* or Base.

The Word is form'd from the *Greek*, solid Prop.

STILES, in Carpentry, *etc.* are the upright Pieces, which go from the bottom to the top in any Wainscot.

STONE, in Natural History, a hard, Solid, mineral Body, neither fusible nor malleable, form'd, in Succession of Time, in the Body of the Earth.

Building STONE: Of this there are several kinds; the principal among us, are *Marble*, *Fire-Stone*, *Purbec-stone*, *Rag-stone*, *Alabaster*, *Free-stone*, and *common stone.*

For *Free-stone*, that dug in the Peninsula of *Portland*, and thence call'd *Portland-Stone*, is much used being softer and whiter than *Purbec-stone*, and is commonly raised out of the quarries in bigger blocks than that. Some also call *Rigate* or *Fire-stone*, *Free-stone*.

Mr. *Boyle* observes, That a competent knowledge of the Manner of the Sap or Juice found in *Stones* used in Building, is of the last Importance; the same *Stone* dug out of the same Quarry at one Season, being found to moulder away in a few Winters, which dug at another Season, will brave the Weather for many Ages; and that there are others, which though dug at the proper Season, yet make but ruinous Buildings, if used at an improper Season.

The same Author adds, That as there are some sort of *Stones* which will decay in a few Years, there are others will not have attain'd their full Hardness in 30 or 40 Years, or even much more.

For the drawing of STONE *out of the Quarry.* See QUARRY.

STONE-*Henge*, in Antiquity, a fam'd Pile or Monument, on *Salisbury* Plain, six Miles distant from that City.

It consists of three Ranks of huge rough *Stones*, ranged one within another, some of them Twenty-eight Feet high, and Seven broad; sustaining others, laid across their Heads, and fasten'd by Mortaises; so that the whole must have anciently hung together.

Antiquaries are strangely at a loss, as to the Origin, Use, Structure, *etc.* of this wonderful Fabrick. Most of them allow the *Stones* to be artificial, and to have been made on the Spot; which is the more probable, as we are pretty well assured the Ancients had the Art of making *Stones* with Sand and a strong Lime or Cement; and as the *Stones* are too big for any Land

Carriage; and yet are in a Plain, which for some Miles round, scarce affords any *Stones* at all.

The Legends give us various other Accounts; some will have them brought miraculously by St. *Patrick* from *Ireland*, others, *etc.*

As to its Use, some Antiquaries take it to have been an ancient Temple of the *Druids*, others of the *Romans*, dedicated to *Cœlus*; in which they are confirm'd by its having been open atop. Others, reading it *Stone Hengist*, maintain it to have been a Monument erected in Memory of *Hengist*, the first General of the *Saxons* in *England*: And others, to name no more, will have it a Funeral Monument, raised to that brave *Romano-Briton, Aurelius Ambrosius*; to which Opinion, some Circumstances of his Actions, and still remaining *Latin* Name of the Place (*Mons Ambrosii*) and that very ancient *Welsh* Proverb, *Mal givaish Emrys*, Like the Work of *Ambrosius*, give some Countenance.

STOVE, in Building, a Hot-house or Room. See CHIMNEY, *etc.*

Palladio observes, that the Ancients used to warm their Rooms with certain secret Pipes, which came through the Walls, conveying Heat to several Parts of the House, from one common Furnace. Whether this were a common Custom, says Sir *Henry Wotton*, or a Curiosity, we cannot determine; but it was certainly, both for Profit and Use, far beyond the *German Stoves*.

STRIAE, in the ancient Architecture, the Lists, Fillets or Rays which separate the Striges or Flutings of Columns. See LIST and FLUTES.

STRIGES, in the ancient Architecture, are what in the Modern we call *Flutings*. See FLUTES.

They are thus call'd, as supposed to have been originally intended to imitate the Folds of Plaits in Womens Robes; which the *Latins* call *Striges*.

The Fillets or Spaces between them, we call *Striae*.

STRUCTURE, in Architecture. See BUILDING.

SUBSTRUCTION, in Building. See FOUNDATION.

SUMMER, in Architecture, is a large Stone, the first that is laid over Columns and Pilasters, in beginning to make a cross Vault; or 'tis the Stone which being laid over a Piedroit or Column, is hollowed, to receive the first Haunce of a Plat-band.

The Word is form'd from the *French*, *Sommier*, which signifies the same Thing.

SUMMER, in Carpentry, is a large Piece of Timber, which being supported on two Stone Peers, or Posts, serves as a Lintel to a Door, Window, *etc.*

SURVEYING, the Art or Act of measuring Lands; *i.e.* of taking the Dimensions of any Track of Ground, laying down the same in a Map or Draught; and finding the Content or Area thereof.

Surveying, call'd also *Geodesia*, is a very ancient Art; 'tis even held to have been the First, or primitive Part of Geometry, and that which gave Occasion to, and laid the Foundation of, all the rest.

Surveying consists of Three Parts or Members; the First, is the taking of the necessary Measures, and making the necessary Observations on the Ground itself: The Second, is the laying down of these Measures and Observations on Paper; and the Third, the finding the Area or Quantity of the Ground thus laid down.

The First is what we properly call *Surveying*. The Second we call *Plotting* or *Protracting*, or *Mapping*: And the Third, *Casting up*.

The First, again, consists of Two Parts, *viz.* the making of Observations for the Angles, and the taking of Measures for the Distances.

The former of these is performed by some one or other of the following Instruments, *viz.* the *Theodolite*, *Circumferentor*, *Semi-circle*, *Plane Table* or *Compass*.

The latter is performed, by means either of the Chain or the Perambulator.

The second Branch of *Surveying*, is perform'd by means of the *Protractor* and *Plotting Scale.*

The Third is perform'd, by reducing the several Divisions, Inclosures, *etc.* into *Triangles*, *Squares*, *Trapeziums*, *Parallelograms*, *etc.* but especially *Triangles.*

SURVEYING *Cross*, is an Instrument little known, and less used in *England*; though in *France*, *etc.* it serves in lieu of a Theodolite or the like Instrument: It consists of a brass Circle, or rather a circular Limb, graduated, and again divided into Four equal Parts, by two Right Lines cutting each other at right Angles in the Centre. At each of the Four Extremities of the Lines, and in the Centre are fix'd Sights. The whole is mounted on a Staff.

SURVEYOR, one that hath the Over-sight and Care of considerable Works, Lands, *etc.* As, the *Surveyor General of the King's Manors*; *Surveyor of the King's Exchange*; *Surveyor General of the Works*; *Surveyors of the Highways*, etc.

SUSPIRAL, a Spring of Water, passing under Ground towards a Conduit or Cistern; also a Breathing-Hole, or Ventiduct.

SWALLOW-tail, in Fortification, is a single Tenaille, narrower towards the fortify'd Place, than towards the Country.

SYMBOLICAL Column, see COLUMN.

SYMMETRY, the Relation of Equality, in the Height, Length and Breadth of the Parts necessary to compose a beautiful Whole.

Symmetry, according to *Vitruvius*, consists in the Union and Conformity of Relation, of the Members of a Work, to their Whole, and of each of the separate Parts to the Beauty of the entire Work; Regard being had to some certain Measure: so that the Body is framed with *Symmetry*, by the Relation the Arm, Elbow, Hand, Fingers, *etc*. have to each other, and to their Whole.

Symmetry, arises from that Proportion, which the *Greeks* call Analogy, which is the Relation of Conformity of all the Parts of a Building, and of their Whole, to some certain Measure; whereon depends the Nature of *Symmetry*.

In Architecture, we call *Uniform Symmetry*, that where the Ordonnance reigns in the same Manner, throughout the whole Pourtour; and *Respective Symmetry*, that where only the opposite Sides are equal to each other.

The Word is form'd from the *Greek*, *with*, and *Measure*.

T

TABERNACLE, among the *Jews*, the Place wherein the Ark of the Covenant was lodg'd; both while they were in Tents, during their Journey from *Ægypt*; and when fix'd in *Jerusalem*, and the Ark kept in the Temple.

Philo describes the *Jewish Tabernacle* thus: It was a Building compos'd of 48 Cedar Boards, lined with massive Gold; under each whereof was a Silver Stand, or Foot, and at the Top a Capital of Gold: It was encompass'd with ten Pieces of rich Tapistry, of different Colours, Purple, Scarlet, Hyacinth, *etc*. The Length of the *Tabernacle* was 30 Cubits; its Breadth 10, *Josephus* says 12.

The Ark was placed in a secret Oratory, in the Middle of this *Tabernacle*; and was gilt both within-side and without: the Top whereof, being a Kind of Lid, was call'd *Propitiatory*, because it appeas'd the Wrath of God.

It was encompass'd with several Veils, stretch'd over it, with Hooks, and Buckles of Gold.

The Word is form'd from the *Latin Tabernaculum*, a Tent.

Feast of TABERNACLES.

TABERNACLE has also been used of late for a sort of temporary Church, or Chapel, contriv'd to serve the immediate Purposes of the Parishoners, *etc*. while their proper Church is Repairing, Rebuilding, or the like.

TABLE, *Tabula*, a Moveable, usually made or Wood, or Stone, supported on Pillars, or the like; for the commodious Reception of Things placed thereon.

Moses made a *Table* in the Tabernacle for the layng of the Shew Bread upon; describ'd by *Philo Judæus* as two Cubits long, one broad, and one and half high. Among Christians, the *Table*, or *Lord's Table*, signifies the Sacrament of the Eucharist.

TABLE, *in Architecture*, a smooth simple Member, or Ornament, of various Forms; but most usually in that of a long Square.

A *Projecting Table*, is such a one as stands out from the Naked of the Wall, Pedestal, or other Matter it adorns.—*Raked Table*, is that which is hollow'd in the Die of a Pedestal, or elsewhere, and which is usually encompass'd with a Moulding.—*Razed Table*, an Embossment in a Frontispiece, for the putting an Inscription or other Ornament in Sculpture. This is what M. *Perrault* understands by *Abacus* in *Vitruvius*.—*Crown'd Table*, that cover'd with a Cornice, and wherein is cut a Basso Relievo, or a Piece of black Marble incrustated for an Inscription.—*Rusticated Table*, that which is pick'd, and whose Surface seems rough as in Grotto's, *etc.*

Plain TABLE, a Surveying Instrument.

TABLE, in Perspective, is a plain Surface, suppos'd to be transparent, and perpendicular to the Horizon.

'Tis always imagin'd to be plac'd at a certain Distance between the Eye and Objects, for the Objects to be represented thereon by means of visual Rays passing from every Point thereof, thro' the Table to the Eye.

Whence it is also call'd *Perspective Plane*.

TÆNIA, or TENIA, in Architecture (from the *Greek*, a Bandelet, or little Band, call'd by the *Latins Vitta*, *Fascia*, *Zona*, and *Corona*,) a Member of the Doric Architrave, resembling a square Fillet, or Ruler, and serving in lieu of a Cymatium. See DORIC and CYMATIUM.

Barbaro calls it *Listel*; but *Palladio* used the old Name *Tænia*. See LIST.

Philander says, there are two Kinds, *viz.* the above-mention'd, which he calls the *lower*; and another, which serves as a Capital to the Triglylphs. *Leon Baptist. Alberti* calls the *Tæniæ*, *Regulæ* and *Fasciolæ*; and *Daviler*, *Bandelettes*.

TAIL *of the Trenches*, in the military Art, is the first Work which the Besiegers make at the Opening of the Trenches, as the Head of the Attack is carry'd on toward the Place.

TAILLOIR, in Architecture, a Term some of our Writers, after the *French*, use for *Abacus*. See ABACUS.

TALON, in Architecture, a kind of Astragal; or Moulding consisting of a square Fillet, crowning a Cymatium; frequently found to terminate Ornaments of Joyners-Work, as those of Doors, *etc.* See CYMATIUM.

The *Talon* is a Moulding concave at the Bottom, and convex a-top; having an Effect just opposite to the Doucine. See DOUCINE.

When the concave Part is a-top, 'tis call'd an *Inverted Talon.* The *Talon* is usually call'd by our *English* Workmen *Ogee*, or *O.G*; and by Authors an *upright* or *inverted Cymatium.* See OGEE.

The Word is *French*, and literally signifies *Heel.*

TALUS, or TALUT, in Architecture, the sensible Inclination or Slope of a Work; as of the Outside of a Wall, when its Thickness is diminish'd by Degrees, as it rises in Height, to make it the firmer. See WALL.

TALUS, in Fortification: The *Talus of a Bastion*, or *Rampart*, is the Slope allow'd to such a Work, whether it be of Earth or Stone, the better to support its Weight. See RAMPART.

The *Talus exterior* of a Work, is its Steepness on the Side of the Field, which is always made as little as possible, to prevent the Enemies Scalado; unless the Earth be bad, and then it is absolutely necessary to allow a considerable *Talus* for its Parapet.

The *Talus interior* of a Work, is its Steepness on the Inside towards the Place.

TAMBOUR, in Architecture, a Term apply'd to the *Corinthian* and Composite Capitals; as bearing some Resemblance to a Drum; which the *French* call *Tambour.* See CAPITAL.

Some chuse to call it the *Vase*, and others the *Campana*, Bell. See VASE, *etc.*

TAMBOUR is also us'd for a little Box of Timber-Work, cover'd with a Ceiling, within-side the Porch of certain Churches; both to prevent the View of Persons passing by, and to keep off the Wind, *etc.* by means of folding Doors, *etc.*

TAMBOUR is also a round Stone, or Course of Stones, several whereof form a Section of the Shaft of a Column, not so high as Diameter.

TAPER, *Tapering*, is understood of a Piece of Timber, or the like, when broad beneath, and sharp towards the Top; as is the Case in Pyramids, Cones, *etc.*

TAPESTRY, or TAPISTRY, a curious Kind of Manufacture, serving to adorn a Chamber, or other Apartment, by hanging, or covering the Walls thereof. See HANGINGS.

Some use *Tapestry* as a general Name for all Kinds of Hangings, whether wove, or wrought with the Needle, and whether Silken, Woollen, Linnen, Leather, or Paper; (in which they are countenanc'd by the Etymology of the Word, form'd from the *French Tapisser*, to line, of the

Latin Tapes, a Cover of a Bed, *etc.*) But in the common Use of our Language, the Term is now appropriated to a Kind of woven Hangings, of Wooll and Silk, frequently rais'd, and enrich'd with Gold and Silver, representing Figures of Men, Animals, Landskips, *etc.*

The Invention of *Tapestry* seems to have come from the *Levant*; and what makes this the more probable is, that formerly the Workmen concern'd herein, were call'd, at least in *France*, *Sarrazins*, or *Sarrasinois*.

'Tis suppos'd that the *English* and *Flemish*, who were the first that excell'd therein, might bring the Art with them from some of the Croisades, or Expeditions against the *Sarazens*.

Be this as it will, 'tis certain those two Nations, particularly the *English*, were the first who set on Foot this noble and rich Manufacture in *Europe*; now one of the finest Ornaments of Palaces, Basilicks, Churches, *etc.*

Hence, if they ben't allow'd the Inventors, they have, at least, the Glory of being the Restorers of so curious and admirable an Art, as gives a kind of Life to Woolls and Silks, in no Respect inferior to the Paintings of the best Masters.

'Twas late e'er the *French* apply'd themselves to *Tapistry*: The first Establishment of that Kind, was under *Henry* IV. in the Year 1607, in the *Fauxbourg*, *S. Michael*. But this fell with the Death of that Prince. Under *Lewis* XIV. the Manufacture was retriev'd by the Care and Address of the great M. *Colbert*, to whom the Establishment of the Gobelins is owing, a Royal *Tapistry* Manufacture, which has produc'd Works in this Kind, scarce inferior to the finest *English* or *Flemish Tapistry*, either with regard to the Designs, the Colours, or the Firmness.

The *Tapistry-men* distinguish two Kinds of Work, *viz. Tapistry* of the *high* and *low Warp*; tho' the Difference is rather in the Manner of working, than in the Work itself, which is, in effect, the same in both; only the Looms, and consequently the Warps, are differently situate: Those of the *low Warp* being placed flat, and parallel to the Horizon; and those, on the contrary, of the *high Warp*, erected perpendicularly.

The *English* anciently excell'd all the World in their *Tapistry of the high Warp*; and they still retain their former Reputation, tho' with some little Change: Their *low Warps* are still admir'd; but as for the *high ones*, they are quite laid aside in our Nation.

The *French* have three considerable *Tapistry* Manufactures besides that of the Gobelins; the first at *Aubusson* in *Auvergne*, the second at *Felletin* in the *Upper Marche*, and the third at *Beauvois*: They were all equally establish'd for the high and the low Warp, but have all laid aside the former, excepting the Gobelins.

There are admirable low Warps in *Flanders*, generally exceeding those of *France*; the chief, and almost only *Flemish* Manufactories, are at *Brussels*, *Antwerp*, *Oudenard*, *Lisle*, *Tournay*, *Bruges*, and *Valenciennes*.

At *Brussels* and *Antwerp* they succeed both in human Figures, in Animals, and Landskips; and that both with regard to the Designs and the

Workmanship. At *Oudenard* their Landskips and Animals are good, but their human Figures naught. *Lisle*, and the other Cities nam'd, come behind *Oudenard*. The *French* Manufacture of *Felletin* does tolerably well in Landskips, *Aubusson* in Figures, and *Beauvois* in both.

The usual Widths of *Tapistries*, are from two Ells to three Ells and half, *Paris* Measure.

TARRASS, or TERRASS, a Sort of Plaister, or strong Mortar, chiefly used to line Basons, Cisterns, Wells and other Reservoirs of Water. See MORTAR.

TASSELS, in a Building, are those Pieces of Boards that lie under the Ends of the Mantle-Trees. See MANTLE.

TAXIS, in the ancient Architecture, the same with *Ordonance* in the new, is describ'd by *Vitruvius* to be that which gives every Part of a Building its just Dimensions, with regard to its Uses. See ORDONNANCE.

TECHNICAL, something that relates to an Art. See ART.

In this Sense we see *Technical Words*, *Technical Verses*, etc. and in this Sense Dr. *Harris* intitles his Dictionary of Arts and Sciences, *Lexicon Technicum*.

TEINTS, and *Semi*-Teints, in Painting, the several Colours us'd in a Picture, consider'd as more or less high, or bright, or deep, or thin, or weaken'd, or diminish'd, *etc*. to give the proper Relievo, or Softness, or Distance, *etc*. of the several Objects. See COLOURING.

The Word is pure French, where it signifies the same Thing.

TEMPLE, a publick Building, erected in Honour of some Deity, either true, or false; and wherein the People meet to pay religious Worship to the same.

Clemens Alexandrinus and *Eusebius*, refer the Origin of *Temples* to the Sepulchres built for the Dead. *Herodotus* and *Strabo* will have the *Ægyptians* to be the first who built *Temples* to the Gods. The first, built in *Greece*, is ascrib'd to *Deucalion*, by *Apollonius*, *Argonaut*. Lib. III.

In Antiquity we meet with many People who would not build any *Temples* to their Gods, for fear of confining them to too narrow Bounds. They perform'd their Sacrifices in all Places indifferently, from a Persuasion, that the whole World is a *Temple* of God, and that he requires no other. This was the Doctrine of the *Magi*, follow'd by the *Persians*, the *Scythians*, the *Numidians*, and many other Nations mention'd by *Herodotus*, Lib. I. *Strabo* Lib: XV. and *Cicero* in his 2d Oration against *Verres*.

The *Persians*, who worshipp'd the Sun, believ'd it would wrong his

Power, to inclose Him in the Walls of a *Temple*, who had the whole World for his Habitation; And hence, when *Xerxes* ravag'd *Greece*, the *Magi* exhorted him to destroy all the *Temples* he met with.

The *Sicyonians* would built no *Temple* to their Goddess *Coronis*: Nor the *Athenians*, for the like Reason, erect any Statue to Clemency, who, they said, was to live in the Hearts of Men, not within Walls.

The *Bithynians* had no *Temples*, but the Mountains to worship on; nor the ancient *Germans* any other but the Woods.

Even some Philosophers have blamed the Use and Building of *Temples*, particularly *Diogenes*, *Zeno*, and his Followers the Stoicks: But, it may be said, That if God have no Need of *Temples*, Men have no Need of Places to meet in for the publick Offices of Religion: Accordingly, *Temples* may be traced back even into the remotest Antiquity. See *Hospinianus de Origne Templorum.*

The *Romans* had several Kinds of *Temples*; whereof those built by the Kings, *etc.* consecrated by the Augurs, and wherein the Exercise of Religion was regularly perform'd, were call'd, by Way of Eminence, *Templa*, *Temples.*

Those that were not consecrated, were call'd *Ædes.*

The little *Temples*, that were cover'd or roof'd, they call'd *Ædiculæ.*

Those open, *Sacella.*

Some other Edifices consecrated to particular Mysteries of Religion, they call'd *Fana* and *Delubra.*

All which Kinds of *Temples*, *Vitruvius* tells us, had other particular Denomiations, according to the Form and Manner of their Construction; as hereafter related.

Indeed, the *Romans* out-did all Nations in the Point of *Temples*; they not only built *Temples* to their Gods, to their Virtues, to their Diseases, *etc.* but also to their Emperors, and that in their Life-time; Instances whereof we meet withal in Medals, Inscriptions, and other Monuments. *Horace* compliments *Augustus* hereupon, and sets him above *Hercules*, and all the Hero's of Fable; in that those were only admitted into *Temples* after their Death, whereas *Augustus* had his *Temples* and Altars while living.

Viventi tibi maturos largimur honores;
Jurandasq; *tuum per nomen ponimus Aras.* Epist. ad August.

TEMPLE, in Architecture. The ancient *Temples* were distinguish'd, with regard to their Construction, into various Kinds; as,

TEMPLES of *Antæ*, or Simply *Antæ*: These, according to *Vitruvius*, were the most simple of all Temples; having only angular Pilasters, call'd *Antæ*, or *Parastatæ*, at the Corners, and two Tuscan Columns on each Side the Doors.

Tetrastyle-TEMPLE, or simply *Tetrastyle*, was a *Temple* that had four Columns in Front and as many behind; as that of *Fortuna Virilis* at *Rome.*

See TETRASTYE.

Prostyle-TEMPLE, that which had only Columns in its Front, or Foreside; as that of *Ceres* in *Eleusis*, in *Greece*. See PROSTYLE.

Amphiprostyle, or *double Prostyle*, that which had Columns both before and behind, and which was also Tetrastyle. See AMPHIPROSTYLE.

Periptere-TEMPLE, that which had four Rows of insolated Columns around, and was Hexastyle, *i.e.* had six Columns in Front; as the *Temple* of Honour at *Rome*. See PERIPTERE.

Diptere-TEMPLE, that which had eight Rows of Columns around, and was also Octostyle: Or had eight Columns in Front; as that of *Diana* at *Ephesus*. See DIPTERE.

The Word *Temple* is form'd from the *Latin Templum*, which some derive from the *Greek* signifying the same Thing; and others from *abscindo*, I cut off, I separate, in Regard a *Temple* is a Place separated from common Uses; others derive it from the old *Latin* Word *templare*, to contemplate.

The ancient Augurs gave the Name *Temples* to those Parts of the Heavens, which they mark'd out for the observing of the Flight of Birds. Their Formula was this: *Templa tesqua sunto*.

TENAILLE, in Fortification, a kind of Out-Work, consisting of Two Parallel Sides, with a Front, wherein is a re-entering Angle.

In Strictness, that Angle, and the Faces that compose it, are the *Tenaille*.

The Word is *French*, and signifies, literally, Pincers or Nippers.

The *Tenaille* is of two Kinds: *Simple* and *Double*.

Simple or *single Tenaille*, is a large Out-work, as D A B C E (Tab. Fortification, Fig. 8.) consisting of two Faces or Sides A B and C B, including a re-entering Angle B.

The great Defects of the *Tenaille* are, that it takes up too much Room, and on that Account is advantageous to the Enemy; that its Angle B is undefended; the Height of the Parapet hindering the seeing down into it, so that the Enemy can lodge there under Covert: and that the Sides A C and C D are not sufficiently defended.

For these Reasons, the *Tenaille* is now excluded out of Fortification by the best Engineers; and never made, but where there wants Time to form a Horn-work.

Double or *Flank'd* TENAILLE, is a large Out-work, consisting of two simple *Tenailles*, or two re-entring Angles F G H and H I K (Fig. 9). This is now, also, excluded out of Fortification for the same Reasons as the former.

TENAILLE *of the Place*, is the Face of the Place, comprehended between the Points of Two neighbouring Bastions; including the Curtain, the two Flanks raised on the Curtain, and the two Sides of the Bastions which face one another.

So that the *Tenaille* is the same with what is otherwise call'd, the *Face of a Fortress.*

TENAILLE *of the Ditch*, is a low Work raised before the Curtain, in the Middle of the Foss or Ditch. It is of three Sorts: The First is composed of a Curtain, two Flanks and two Faces. The Rampart of the Curtain, including the Parapet and Talus, is but five Fathom thick, but the Rampart of the Flanks and Faces Seven.

The Second, which *Vauban* saith, he found to be of very good Defence, is composed only of two Faces, made on the Lines of Defence, whose Rampart and Faces are parallel.

The third Sort differs from the Second, only in this, that its Rampart is parallel to the Curtain of the Place.

All three Sorts are good Defences for the Ditch, and lie so low, that they cannot be hurt by the Besiegers Cannon, till they are Masters of the Covert-way, and have planted their Cannon there.

TENANCY, a Habitation or House to live in, or a Tenement, or Possession held of another.

TENEMENT, TENANCY, in Law, a House or Lands, depending of a Manor or Lordship; or a Fee or Farm, held of a superior Lord, and which he may re-call, when the Term or Condition is expired.

Frank Tenement, is any Lands, Houses, Offices or the like, wherein a Man has Estate for Life, or in Fee.—*Base Tenement*, is where a Man holds Lands, *etc.* at the Will of the Lord.

Kitchin, *Briton*, *etc.* make *Frank Tenement* and *Base Tenement*, Opposites; so that *Frank Tenement* should be where the Tenant is at liberty to quit it when he pleases.

TENON, in Building, *etc.* the End of a Piece of Wood, or Metal, diminish'd by one Third of its Thickness, to be receiv'd into a Hole in another Piece, call'd the *Mortise*, for the jointing or fastening the two together. See MORTOISE.

Among Joiners, *etc.* the *Tenon* is made in various Forms, square, Dove-tail'd, for double Mortises, *etc.*

Vitruvius calls the *Tenons*, *Cardines*; Dove-tail'd *Tenons* he calls *Subscudes*, or *Securiclæ*. See DOVE-*Tail.*

TERM, *in the Arts*, or TERM *of Art*, is a Word, which, besides the literal and popular Meaning which it has, or may have, in common Language, bears a further and peculiar Meaning in some Art or Science.

TERRACE, or TERRAS, a Bank of Earth, raised in a Garden, Court, *etc.* above the Level of the Ground.

The *Terras* is an Earth-work usually lined and breasted with a strong Wall, in Compliance with the natural Inequality of the Ground.

Sometimes it is made in Talus, or a-slope, and cover'd with Turf. A *Counter-terrace*, is a *Terrace* raised over another, for the joining of the Ground, or the raising a Parterre.

TERRACE is also applied to the Roofs of Houses that are flat, and whereon one may walk; as also to Balconies that project. See ROOF.

The *Terrace* is properly the Covering of a Building, which is in Plat-form; as that of the Peristyle of the *Louvre*; or that of the Observatory, paved with Flint and Mortar. All the Buildings of the Oriental Nations, are cover'd with *Terrasses*, to take the fresh Air on, and even to lie on. See PLATFORM.

TERRE-*Plain*, in Fortification, the Top, Plat-form, or Horizontal Surface of the Rampart, whereon the Cannon are placed, and the Defenders perform their Office.

It is thus call'd, as lying level; having only a little Slope outwardly, to bear the Recoil of the Cannon.

It is terminated by the Parapet on that Side towards the Campaign; and by the inner Talus, on the Side towards the Place. Its Breadth is from 24 to 30 Feet.

TESSELATED *Pavement*, *Pavementum Tesselatum*, a rich Pavement of *Mosaick* Work, made of curious small square Marbles, Bricks or Tyles, call'd *Tesselæ*, from the Form of Dies. See MOSAIC *Work*.

These *Pavements* were much used in the Tents of the *Roman* Generals.

TETRASTYLE, in the ancient Architecture, a Building, and particularly a Temple, with four Columns both before and behind, *i.e.* in Front and Rear. See TEMPLE.

The Word is form'd from the *Greek four*, and *Columns*.

THEATER or THEATRE, THEATRUM, among the Ancients, a Public Edifice, for the exhibiting of Spectacles or Shews to the People.

Under the Word *Theater* was comprehended, not only the Eminence whereon the Actors appear'd, and the Action pass'd; but also the whole Area, or Extent of the Place, common to the Actors and Spectators.

In this Sense, the *Theater* was a Building encompass'd with Portico's, and furnished with Seats of Stone, disposed in a Semi-circle, and ascending by degrees over one another; which encompass'd a Space, call'd the *Orchestra*, in the Front whereof, was the *Proscenium*, or *Pulpitum*, whereon the Actors perform'd, and which is what we properly call the *Theater* or *Stage*.

On the *Proscenium* stood the *Scena*, a large Front, adorn'd with Orders of Architecture, behind which was the *Postscenium*, or Place where the Actors made themselves ready, retired, *etc*. So that the Scena, in its full Extent, comprehended all the Part belonging to the Actors.

In the *Greek Theaters*, the Orchestra made a Part of the Scena; but in the *Roman Theaters*, none of the Actors ever descended into the Orchestra; which was taken up by the Seats of the Senators.

The most celebrated *Theaters* remaining of Antiquity, are the *Theater* of *Marcellus*, and that of *Pompey*; which are also call'd *Amphitheaters*. See AMPHITHEATER.

At *Athens*, are still seen, the Remains of the Temple of *Bacchus*, which was the first *Theater* in the World, and a Master-piece in Architecture. All *Theaters* were consecrated to *Venus* and *Bacchus*.

The Word is form'd from the *Greek*, Spectacle, Shew, of *specto*, *video*, *I see*.

THEATER, among the Moderns, is the Stage, or Place whereon the Drama, or Play is exhibited; answering to the *Scena* of the Ancients.

In its full Latitude, however, the *Theater* includes the whole Play-House; in which Sense, it is a spacious Room, or Hall, part whereof is taken up by the *Scena*, which comprehends the Stage, the Decorations and the Machines; and the rest distributed into a Space, call'd the *Pit*, or *Parterre*, which is cover'd with Seats, Boxes, *etc*. and terminated with an Elevation of one, or two Galleries, disposed into Benches ascending over one another.

THEATER is also used in Architecture, chiefly among the *Italians*, for an Assemblage of several Buildings, which, by a happy Disposition and Elevation, represents an agreeable Scene to the Eye.

Such as are most of the Vineyards at *Rome*; but particularly that of *Monte Dragone*, at *Frescati*; and in *France*, the new Castle of St. *Germain en Laye*.

Anatomical THEATER, in a School of Medicine and Chirurgery, is a Hall, with several Rows of Seats, disposed in the Sweep of an Amphitheatre; having a Table, bearing on a Pivot, in the middle; for the Dissection of Bodies.

Such is the *Anatomical Theater* of the Royal Garden of Plants at *Paris*, *etc*.

The *Theater* at *Oxford* is a beautiful Building erected by Archbishop *Sheldon*, for the Use of Scholastic Exercises.

THEOLOGIUM, in the ancient Theater, was a Place, or little Stage, above that whereon the ordinary Actors appear'd.

The *Theologium* was the Place where the Gods appear'd; it also included the Machines whereon they descended, and from which they spoke.

There was a *Theologium* requir'd for the Representation of the *Ajax* of *Sophocles*, the *Hippolitus* of *Euripedes*, etc. Scal. Poet. lib. I. cap. 1.

THERMÆ, ancient Buildings, destined for Bathing in.

Among the noblest Monuments of ancient *Rome* are reckon'd the *Thermæ* or Baths of *Dioclesian*.

Among us, *Thermæ* is restrain'd to such Baths as are hot.

The Word is form'd from the *Greek*, hot.

Thermæ, 'tis commonly argued, owe their Heat to a Colluctation, or Effervescence of the Minerals in them. Though Dr. *Woodward* ascribes it to the subterraneous Heat, or Fire which communicates with them by some Spiracle or Canal whereby a greater Quantity of Heat is derived thither, than to ordinary Springs.

THRONE, a magnificent Seat or Chair; is a Royal Seat, inrich'd with Ornaments of Architecture and Sculpture, made of some precious Matter, rais'd on several Steps, and cover'd with a kind of Canopy.

Such as are in the Rooms of Audience of Kings and other Sovereigns.

TIGE, in Architecture, a *French* Term for the Shaft or Fust of a Column; comprehended between the Astragal and the Capital. See FUST and COLUMN.

TILE. See TYLE.

TIMBER, includes all Kinds of fell'd and season'd Woods, us'd in the several Parts of Building; as Carpentry, Joinery, Turnery, *etc.* See WOOD and BUILDING.

The Kinds of *Timber* are numerous: We shall only mention some of the most usual; from *Evelyn's Sylva*, etc. as,

1°. *Oak*; the Uses whereof heed no enumerating: To endure all Seasons and Weathers, there is no Wood like it: Hence its use in Pales, Shingles, Posts, Rails, Boards, *etc.* For Water-works it is second to none; and where it lies expos'd both to Air and Water, there is none equal to it.

2°. *Elm*: This fell'd between *November* and *February*, is all Spine or Heart, and no Sap; and is of singular Use in Places where it is always wet or dry: Its Toughness likewise makes it of Use to Wheel-wrights, Mill-wrights, *etc.* nor must it be omitted, that its not being liable to break and fly in Chips, makes it fit for Dressers and Planks to chop on.

3°. *Beech*: Its chief Use is in Turnery, Joinery, Upholstery, and the like, as being of a clean, white, fine Grain, not apt to bend or slit: Yet is it sometimes, especially of late, us'd for building *Timber*; and if it lie constantly wet, is judg'd to out-last Oak.

4°. *Ash*: Its Use is almost universal: 'Tis good for Building, or other Occasions where it may lie dry: It serves the Carpenter, Cooper, Turner, Plough-wright, Wheel-wright, Gardener; as also at Sea for Oars, Hand-Spikes, *etc.*

5°. *Fir*; commonly known by the Name of *Deal*, is of late much us'd in Building, especially within Doors, for Stairs, Floors, Wainscot, and most Works of Ornament.

6°. *Walnut-Tree*, is of universal Use; excepting for the Out-sides of Buildings: None better for the Joiners Use; it being of a more curious brown Colour than Beech, and less subject to Worms.

7°. *Chesnut-Tree*, next to Oak, is the *Timber* most sought for by Joiners and Carpenters: 'Tis very lasting.

8°. *Service-Tree*, us'd in Joinery, as being of a delicate Grain, and fit for Curiosities: it also yields Beams of considerable Bigness for Building.

9°. *Poplar*, *Abel*, and *Aspen*, differing very little from one another, are much us'd of late instead of Fir: They look as well, and are tougher and harder.

10°. *Alder*; much us'd for Sewers or Pipes to convey Water. When always wet it grows hard like a Stone; but where sometimes wet, and sometimes dry, it rots immediately.

TOILET, a fine Cloth, of Linnen, Silk, or Tapistry, spread over the Table in a Bed-chamber or Dressing Room, to undress and dress upon.

The Dressing-Box, wherein are kept the Paints, Pomatums, Essences, Patches, *etc.* the Pin-cushion, Powder-Box, Brushes, *etc.* are esteem'd Parts of the Equipage of the Ladies *Toilet.*

That of the Men, consists of a Comb-Case, Brushes, *etc.* To make a Visit to one at his *Toilet*, is to come to entertain him while he is dressing or undressing.

Satin, Velvet, Brocade Point de France, *etc.* are now ordinarily us'd for *Toilets*; anciently they were much plainer; whence the Name, which is form'd from the *French*, *Toilette*, a Diminutive of *Toile*, any thin Stuff.

TOL-BOOTH, or TOLL-BOOTH, a Place in a City where Goods are weigh'd, to ascertain the Duties or Imparts thereon.

TOMB, includes both the *Sepulchre*, wherein the Defunct is interr'd; and the *Monument* erected to preserve his Memory. See SEPULCHRE and MONUMENT.

Among the *Romans*, none but the Emperors, Vestals, and Persons signaliz'd by great Actions, were allow'd to have *Tombs* in the Cities; the rest were all in the Country, near the High-Roads; whence those common Words, *Siste*, *et abi*, *Viator*; which are still retain'd in the Inscriptions of our Monuments in Churches, *etc.* tho' the Occasion be no more.

At *Anchiales* is still seen the *Tomb* of *Sardanapalus*, with this Inscription in Verse: Sardanapalus *built* Anchiales *and* Tarsus *in one Day*: *Go, Passenger, eat, drink, and be merry*; *the rest is nothing.*

The Word *Tomb* is form'd from the *Greek*, *Tumulus*, Sepulchre; or, according to *Menage*, from the *Latin*, *Tombe*, which signified the same Thing.

The *Romans* had a kind of empty *Tomb*, call'd *Cenotaphium*, or an *honorary Tomb*, wherein no Deceas'd was laid. It was us'd to be built in Favour of Persons perishing at Sea, in remote Countries, *etc.* where no Sepulture could be had.

The Occasion hereof was a superstitious Notion, that the Souls of those whose Bodies were not bury'd, wander'd an hundred Years by the

Banks of the Rivers of Hell, e'er they were admitted to pass over.

The *Cenotaphia* were inscrib'd with the Words *ob Honorem*, or *Memoriæ Sacrum*; whereas other *Tombs*, wherein the Ashes were really deposited, were inscrib'd D.M.S. *q.d.* Sacred to the Manes.

When the Words *Tacito Omine* were added, it declar'd the Ashes inclos'd therein infamous for some Crime.

TONDIN, in Architecture. See TORE.

TORE, TORUS, in Architecture, a thick round Moulding us'd in the Bases of Columns. See BASE.

'Tis the Bigness that distinguishes the *Torus* from the *Astragal*. See ASTRAGAL.

The Bases of *Tuscan* and *Doric* Columns have but one *Tore*, which is between the Plinth and the Listel. In the *Attic* Base there are two; the *upper*, which is the smaller; and an *under*.

The Word is form'd from the *Greek*, a Cable, whereof it bears some Resemblance; or from the *Latin*, *Torus*, a Bed, as being suppos'd to represent the Edge of such Matters as swell out with the Weight of the incumbent Column.

It is also call'd a *thick Baton*; sometimes its Contour resembles half a Heart; in which Case it is call'd *Baguette*. See BAGUETTE.

TOREUMATOGRAPHY, a Term signifying the Knowledge, or rather, the Description of ancient Sculpture, and Basso-Relievo's.

The *Italian* Engravers have given a great deal of Light to this Science. See SCULPTURE and ENGRAVING.

TOREUTICE, that Part of Sculpture, call'd *Turning*. See SCULPTURE and TURNING.

TOWER, a tall Building, consisting of several Stories, and usually of a round Form.

Before the Invention of Guns, they us'd to fortify Places with *Towers*; and to attack them with moveable *Towers* of Wood, mounted on Wheels, to set the Besiegers on a Level with the Walls, and drive the Besieg'd from under the same.

These *Towers* were sometimes 20 Stories, and 30 Fathom high: they were cover'd with raw Skins, and an hundred Men, employ'd to move them.

Towers are also built to enable one, by their Elevaton, to view to a great Distance.

These are of all Figures; as, square, round, pentagonal, *etc.* See PHAROS.

In *China* is a famous *Tower* of *Porcelain*, whereof the *Dutch* relate Wonders.

Towers are also built for Fortresses, Prisons, *etc.* as the *Tower* of *London*, the *Towers* of the Bastille, *etc.*

The Tower of London is not only a Cittadel, to defend and command the City, River, *etc.* but also a Royal Palace, where our Kings with their Courts have sometimes lodg'd; a Royal Arsenal, wherein are Arms and Ammunition for 60000 Soldiers; a Treasury for the Jewels and Ornaments of the Crown; a Mint for coining of Money; the great Archive, wherein are preserv'd all the ancient Records of the Courts of *Westminster*, etc. and the chief Prison for State Criminals.

In the midst of it is the great square, white *Tower*, built by *William* the Conqueror;—Within the *Tower* is a Parochial Church, exempt from all Jurisdiction of the Archbishop, and a Royal Chapel, now disus'd.

The chief Officer of the *Tower* is the *Constable*, under whom is the *Lieutenant*, who acts by his Direction, and in his Absence. He has by Grant of several of our Kings, *Unam Lagenam*, two Gallons and a Pint of Wine, out of every Vessel; and a certain Quantity out of every Boat laden with Lobsters, Oysters, and other shell Fish, and double the Quantity out of every Alien's Boat, passing by the *Tower*: To him also belongs a Fee of 200*l.* for every Duke committed Prisoner; 100*l.* for every Peer not a Duke, and 50*l.* for every Commoner.

Hollow TOWER, in Fortification, is a Rounding made of the Remainder of a Brisure, to join the Courtin to the Orillon.

TRABEATION, in the ancient Architecture, the same with *Entablature*. See ENTABLATURE.

It consists of three principal Members or Divisions, *viz.* the Architrave, Freeze, and Cornice; each of which consists of divers lesser Members and Ornaments. See ARCHITRAVE, FREEZE, *etc.*

The Proportion, Members, *etc.* are different in the several Orders. See ORDER; see also each Order apart, TUSCAN, DORIC, *etc.*

TRAJAN *Column*, a famous historical Column, erected in *Rome*, in Honour of the Emperor *Trajan*. See COLUMN.

'Tis of the *Tuscan* Order, though somewhat Irregular: Its Height is Eight Diameters, and its Pedestal *Corinthian*; it was built in a large Square there, call'd the *Forum Romanum*.

Its Base consists of 12 Stones of an enormous Size, and it is raised on a Socle or Foot of Eight Steps. Within-side, is a Stair-case, illumined with 44 Windows. 'Tis 140 Foot high, which is 35 Foot short of the *Antonine Column*; but the Workmanship of the former is much more valued.

'Tis adorn'd from top to bottom with Bas-relievo's, representing the great Actions of that Emperor against the *Dacæ*.

Several learned Men have explain'd the Bas-reliefs of the *Trajan Column*, and among others *Ciaconius* and *Fabretti*.

The late King of *France*, *Louis* XIV. had Models of all the Bas-reliefs taken off in Plaister of *Paris*.

TRANSOM, among Builders, the Piece that is framed across a double light Window. See WINDOW.

TRENCH, a Ditch cut or dug in the Ground, to drain off the Waters in a Meadow, a Morass, or the like; or to divert the Course of a River.

Many of the Bogs in *Ireland* have been drain'd, and made good Ground, by only digging *Trenches* around them.

TRENCHES, in *Fortification*, are Moats or Ditches, which the Besiegers cut to approach more securely to the Place attack'd; whence they are also call'd Lines of Approach.

These *Trenches* are of several Sorts, according to the Nature of the Soil; for if the adjacent Territory be rocky, the *Trench* is only an Elevation of Bavins, Gabions, Woollpacks, or Shouldrings of Earth, cast round about the Place: But where the Ground may be easily open'd, the *Trench* is dug therein, and border'd with a Parapet on the Side of the Besieg'd.

The Breadth of the *Trench* is from eight to ten Foot, and the Depth from six to seven; and 'tis cut in Talus, or aslope.

The *Trenches* are to be carry'd on with winding Lines, in some Manner parallel to the Works of the Fortress, so as not to be in View of the Enemy, nor to expose its Length to their Shot, which they call *Enfilading*; for then it will be in Danger of being Enfiladed, or scoured by the Enemies Cannon: This carrying of the *Trenches* obliquely, they call carrying them by *Coudees* or *Traverses*.

They call it *Opening the Trenches*, when the Besiegers begin to work upon the Line of Approaches, which is usually done in the Night, and sometimes within Musket-shot, and sometimes within half, or whole Cannon-shot of the Place, if there be no rising Ground about it, the Garrison strong, and their Cannon well serv'd.

The Workmen that open the *Trenches*, are always supported by Bodies of Men against the Sallies of the Besieg'd; and sometimes those Bodies lie between them and the Place, as also on their Right and Left.

TRENCHING *Plough*, is an Instrument for cutting out the Sides of *Trenches* and Drains, or the Sides of Turf, *etc*.

TRIGLYPHS, in Architecture, a Sort of Ornaments repeating at equal Intervals in the *Doric Freeze*. See FREEZE, *etc*.

They consist of two entire Gutters or Channels, cut to a right Angle, call'd *Glyphes*, and separated by two Interstices, call'd, by *Vitruvius*, *Shanks*, from two Half Channels at the Sides. See GLYPHE.

The ordinary Proportion of *Triglyphs*, is to be a Module broad, and one and a Half high.—But this Proportion, M. *le Clerc* observes, sometimes occasions ill-proportion'd Intercolumniations in Portico's; for which Reason he chuses to accommodate the Proportion of his *Triglyphs* to that of the Intercolumns. See INTERCOLUMNS.

The Intervals between the *Triglyphs*, are called *Metopes*. See METOPE.

Under the Channels, or Glyphes, are placed Guttæ or Drops. See GUTTÆ.

The *Triglyphs* make the most distinguishing Character of the *Doric* Order.—Some imagine them originally intended to convey the Guttæ that are underneath them: Others fancy they bear some Resemblance to a Lyre, and thence conjecture the Order to have been originally invented for some Temple sacred to *Apollo*. See DORIC.

The Word is form'd from the *Greek*, three Engravings, from *Sculpo*.

TRIMMERS, in Architecture, Pieces of Timber, fram'd at right Angles to the Joints, against the Ways; for Chimneys and Well-holes for Stairs.

TRINGLE, in Architecture, a Name common to several little square Members, or Ornaments; as Reglets, Listels and Plat-bands. See REGLET, LIST, *etc*.

The Word is *French*, whre it signifies the same thing.

TRINGLE is particularly used for a little Member fix'd exactly over every Triglyph, under the Plat-band of the Architrave; from whence hang down the Guttæ or Pendant Drops. See TRIGLYPH and GUTTÆ.

TUILERIE, or TYLERY, a Tyle-work; being properly a large Building with a Drying-place, cover'd a-top, but furnished with Apertures on all Sides, through which the Wind having Admittance, dries the Tyles, Bricks, *etc.* in the Shade, which the Sun would crack, before they be in the Furnace. See BRICK and TYLE.

The Word is pure *French*, form'd from *Tuile*.

The Garden of the *Louvre* is call'd the *Tuileries*, as being a Place where Tyles were anciently made, *etc.*—But the Term *Tuileries* does not only include the Garden; but also a magnificent Palace, whose Face takes up the whole Length of the Garden: And hence it is, that they say, The King lodges in the *Tuileries*; the King has quitted the *Tuileries* for a few Days, to reside in the *Louvre*.

The Palace of the *Tuileries* is joined to the *Louvre* by a large Gallery, which runs all along the same, and has its Prospects thereon.

The *Tuileries* was begun in 1564, by *Catherine de Medicis*, Wife of *Henry* II. in the Time of her Regency; finished by *Henry* IV. and magnificently adorned by *Louis* XIV.—The Garden of the *Tuileries* was much improved by *Louis* XIII.

TURNING, a Branch of Sculpture; being the Art of fashioning hard Bodies, as Brass, Ivory, Wood, *etc.* into a round or oval Form, in a Lathe.

See LATHE.

Turning is performed, by putting the Substance to be *turned*, upon two Points, as an Axis; and moving it about on that Axis; while an edge Tool set steady to the Outside of the Substance Concentric to the Axis.

The Invention of *Turning* appears to be very ancient.—Some, indeed, to do Honour to the Age, will have it brought to Perfection by the Moderns; but if what *Pliny*, and some other ancient Authors relate, be true, that the Ancients *turned* those precious Vases, enrich'd with Figures and Ornaments in Relievo, which we still see in the Cabinets of the Curious; it must be owned, that all that has been added in these Ages, makes but a poor Amends for what we lost of the manner of *Turning* of the Ancients. See SCULPTURE.

The principal Instruments used in *Turning*, beside the Lathe, are Chissels and Mandrels of various Forms.

TUSCAN, in Architecture, the first, simplest, and most Massive of the five Orders. See ORDER.

The *Tuscan Order* takes its Name from an ancient People of *Lydia*, who coming out of *Asia* to people *Tuscany*, first executed it in some Temples, which they built in their new Plantations.

Vitruvius calls the *Tuscan*, the *Rustic Order*; with whom agrees M. *le Chambray*, who in his Parallel, says, it ought never to be used but in Country-houses and Palaces. M. *le Clerc* adds, that in the manner *Vitruvius*, *Palladio* and some others, have ordered it, it does not deserve to be used at all. But in *Vignola's* Manner of Composition, he allows it a Beauty, even in its Simplicity; and such as makes it proper not only for private Houses, but even for public Buildings, as in the Piazzas of Squares and Markets; in the Magazines and Granaries of Cities; and even in the Offices and lower Apartments in Palaces.

The *Tuscan* has its Character and Proportions, as well as the other Orders; but we have no ancient Monuments, to give us any regular *Tuscan* Order for a Standard.

M. *Perrault* observes, that the Characters of the *Tuscan* are nearly the same with those of the *Doric*; and adds, that the *Tuscan* is, in effect, no other than the *Doric*, made somewhat stronger, by the Shortning of the Shaft of the Column, and simpler, by the small Number, and Largeness of the Mouldings. See DORIC.

Vitruvius makes the whole Height of the Order 14 Modules, wherein he is followed by *Vignola*, M. *le Clerc*, *etc.*—*Serlio* only makes it 12. —*Palladio* gives us one *Tuscan* Profile, much the same as that of *Vitruvius*; and another too rich: On which Side *Scamozzi* is likewise faulty. Hence it is, that that of *Vignola*, who has made the Order very regular, is most follow'd by the Modern Architects. See COLUMN.

Of all the Orders, the *Tuscan* is the most easily executed; as having neither Triglyphs nor Dentils, nor Modillions to confine its Intercolumns.—On this account the Columns of this Order may be ranged in any of the five Manners of *Vitruvius*, *viz.* the *Picnostyle*, *Sistyle*, *Eustyle*, *Diastyle* and *Areostyle*. See INTERCOLUMNS.

For the Parts and Members of the Tuscan *Order, their Proportions, etc.* see CAPITAL, BASE, PEDESTAL, FREEZE, *etc.*

TWIST, is used in Architecture, for a Piece of Timber, otherwise called a *Girder*. See GIRDERS.

TYLE or TILE, in Building, a sort of thin factitous Stone, used in the Roofs, *etc.* of Houses; or more properly, it is a kind of fat, clayey Earth, knodden and moulded of a just Thickness, dried and burnt in a Kiln, like a Brick, and used in the Covering of Houses. See BRICK and COVERING.

It is thus called from the *French*, *Tuile*, of the *Latin*, *Tegula*.

Tyles are made, says M. *Leybourn*, of better Earth than Brick-earth, and something nearer akin to the Potter's Earth.—According to Stat. 17 *Edw.* IV. the Earth for *Tyles* should be cast up before the first of *November*, shired and turned before the first of *February*, and not made into *Tyles* before the first of *March*; And should likewise be tried and sever'd from Stones, Marle and Chalk.

For the Method of Burning them, see BRICK.

As to the applying of *Tyles*; some lay them dry, as they come from the Kiln, without Mortar or any thing else: Others lay them in a kind of Mortar made of Loam and Horse-dung.—In some Parts, as in *Kent*, they lay them in Moss.

There are various Kinds of *Tyles*, for the various Occasions of Building; as *Plain*, *Thack*, *Ridge*, *Roof*, *Crease*, *Gutter*, *Crooked*, *Flemish*, *Corner*, *Hip*, *Dorman*, *Dormar Scallop*, *Astragal*, *Traverse*, *Paving* and *Dutch Tyles*.

Plain or *thack Tyles*, are those in ordinary Use for the Covering of Houses. They are squeez'd flat, while yet soft, in a Mould.—They are of an oblong Figure, and by Stat. 17° *Edw.* IV. are to be 10½ Inches long; 6¼ broad, and Half an Inch and Half a Quarter thick. But these Dimensions are not over-strictly kept to.

Ridge, *Roof* or *Crease Tyles*, are those used to cover the Ridges of Houses; being made circular breadth-wise, like a half Cylinder.—These are what *Pliny* calls *Laterculi*; and are, by Statute, to be 13 Inches long, and of the same Thickness with the *Plain Tyles*.

Hip or *Corner Tyles*, are those which lie on the Hips or Corners of Roofs.—As to Form; they are first made flat, like *Plain Tyles*, but of a Quadrangular Figure, whose two Sides are right Lines, and two Ends, Arches of Circles, one End being a little Concave and the other Convex; the convex End to be about Seven times as broad as the concave End; so that they would be Triangular, but that one Corner is taken off: Then, before they are burnt, they are bent on a Mould, breadthwise, like *Ridge Tyles*.—They have a Hole at their narrow End, to nail them on by; and are laid with their narrow End upwards. By Statute, they are to be 10½ Inches long, and of a convenient Breadth and Thickness.

Gutter Tyles, are those which lie in Gutters or Valleys in Cross-buildings.—They are made like *Corner Tyles*, only the Corners of the broad End are Turned back again with two Wings.—They have no Holes in them, but are laid broad End upwards, without any nailing.—They are made in the same Mould as *Corner Tyles*, and have the same Dimensions on the convex Side. Their Wings are each 4 Inches broad and 8 long.

Pan, *Crooked*, or *Flemish Tyles*, are used in covering of Sheds, Lean to's, and all kinds of flat-roofed Buildings.—They are in Form of an oblong Parallelogram, as plain *Tyles*; but are bent breadth-wise forwards and backwards, in Form of an S, only one of the Arches is at least three Times as big as the other; which biggest Arch is always laid uppermost; and the lesser Arch of another *Tyle* lyes over the Edge of the great Arch of the former.—They have no Holes for Pins, but hang on the Laths by a Knot of their own Earth: They are usually 14½ Inches long, and 10½ broad.

Dormar, or *Dorman Tyles*, consist of a plain *Tyle*, and a triangular Piece of a plain *Tyle*, standing up at right Angles to one Side of a plain *Tyle*, and swept with an Arch of a Circle from the other End, which End terminates in a Point.—Of these *Tyles* there are two Kinds; the triangular Piece, in some, standing on the right, in others on the left Side of the plain *Tyle*.—And of each of these, again, there are two Kinds; some having a whole plain *Tyle*, others but half a plain *Tyle*. But in them all, the plain *Tyle* has two Holes for the Pins, at that End where the broad End of the triangular Piece stands.

Their Use is to be laid in the Gutters, betwixt the Roof and the Cheeks or Sides of the Dormars, the plain Part lying on the Roof, and the triangular Part standing perpendicularly by the Cheek of the Dormar.—They are excellent to keep out the Wet in those Places, and yet are not perhaps known any where but in *Sussex*.—The Dimensions of the plain *Tyle* Part are the same as those of a plain *Tyle*, and the triangular Part is of the same Length, and its Breadth at one end 7 Inches, and at the other nothing.

Scallop or *Astragal Tyles*, are in all respects like plain *Tyles*, only their lower Ends are in Form of an Astragal, *viz*. a Semicircle, with a Square on each Side.—They are us'd in some Places for Weather *Tyling*.

Travers Tyles, are a kind of irregular plain *Tyles*, having the Pin-holes broke out, or one of the lower Corners broke off.—These are laid with the broken End upwards, upon Rafters, where pinn'd *Tyles* cannot hang.

Flemish or *Dutch Tyles*, are of two Kinds, ancient and modern.—The first were used for Chimney-foot Foot-paces: They were painted with antick Figures, and frequently with Postures of Soldiers, some with Compartments, and sometimes with Moresque Devices; but came much short, both as to the Design and Colours of the modern ones.

The modern *Flemish Tyles* are commonly used, plaister'd up in the

Jaumbs of Chimneys, instead of Chimney-Cornerstones.—These are better glaz'd; and such as are painted (for some are only white) much better perform'd than the ancient ones.

But both Kinds seem to be made of the same whitish Clay as our white glazed Earthen Ware.—The ancient ones are five Inches three Quarters square, and about three Quarters of an Inch thick: The modern ones six Inches and half square, and three Quarters of an Inch thick.

TYLERY. See TUILERIE.

TYMPAN, or TYMPANUM, in Architecture, the Ground or Area of a Pediment; being that which is in a Level with the Naked of the Freeze. Or it is the Space included between the three Cornices of a triangular Pediment, or the two Cornices of a circular one. See PEDIMENT.

Sometimes the *Tympan* is cut out, and the Part fill'd with an Iron Lattice to give Light; and sometimes it is enrich'd with Sculpture, in Bas Relief, as in the West Front of St. *Paul's*, in the Temple of *Castor* and *Pollux* at *Naples*, *etc.*

The Word TYMPAN is also used for that Part of a Pedestal call'd the *Trunk* or *Dye*. See PEDESTAL. See also DYE.

TYMPAN, among Joyners, is also attributed to the Pannels of Doors.

TYMPAN, among Printers, is a Frame belonging to the Press, cover'd with Parchment, on which every Sheet is placed, in order to be printed off.

TYMPAN *of an Arch*, in Architecture, is a triangular Space or Table in the Corners or Sides of the Arch; usually hollow'd and enrich'd, sometimes with Branches of Laurel, Olive-tree, or Oak; or with Trophies, *etc.* sometimes with flying Figures, as Fame, *etc.* or sitting Figures, as the Cardinal Virtues. See ARCH.

U

UNION, among Painters, expresses a Symmetry and Agreement between the several Parts of a Painting; when, *e.g.* there is a deal of Relation and Connexion between them, both as to the Figures, and the Colouring: so that they apparently conspire to form one thing. See SYMMETRY.

When this *Union* is finely manag'd, some call it *Suavity*.

UPRIGHT, in Architecture, a Representation or Draught of the Front of a Building; call'd also, an *Elevation*. See ELEVATION.

V

VAGINA, in Architecture, is used for the lower Part of a *Terminus*; because resembling a Sheath, out of which the Statue seems to issue.

The *Vagina* is that long Part between the Base and the Capital; and is found in divers Manners, and with divers Ornaments.

VAN-Fosse, a Ditch dug without the Counterscarp, and running all along the Glacis; usually full of Water.

VASES, in Architecture, are certain Ornaments, plac'd Corniches, Socles, or Pedestals; representing the Vessel the Antients; particularly those used in Sacrifice, as *Præfericulum*, *Simpulum*, Incense-Pots, and Flower-Pots all which are occasionally enrich'd with Basso Relievo's.

They are commonly placed there to crown or finish facades, or Frontispieces. See CROWNING.

They are frequently call'd *Acroteria*; and are usually insulate. See ACROTERIA.

Vitruvius mentions a kind of *Theatrical Vases*, made in Brass, or Earthen Ware, call'd *Echeia*; which they dispose in private Places, under the Steps and Seats of the Theatre to aid and increase the Reflection and Resonance of the Actors Voices, *etc*. See THEATRE.

'Tis said, there are *Vases* of this Kind in the Cathedral Church of *Milan*.

VASE is particularly used in Architecture, to signify the Body of the *Corinthian* and *Composite* Capital; call'd the *Tambour*, or *Drum*; and sometimes the *Campana Bell*.

VAULT, *Fornix*, in Architecture, an arch'd Roof, so contriv'd, as that the several *Voussoirs*, or *Vault*-Stones whereof it consists, do by their Disposition sustain each other. See ARCH.

Vaults are to be prefer'd, on many Occasions, to Soffits, or flat Cielings, as they give a greater Rise and Elevation; and besides, are more firm and durable. See CIELING, ROOF, *etc*.

The Antients, *Salmasius* observes, had only three Kinds of *Vaults*: The first, the *Fornix*, made Cradle-wise; the second, *Testudo*, Tortoise-wise, call'd by the French *Cul de Four*; or Oven-wise: The third *Concha*, or Trumpet-wise, as growing narrower and narrower.

But the Moderns subdivide these three Sorts into a great many more, to which they give different Names, according to their Figures and Uses: Some are *Circular*, others *Elliptical*, etc.

The Sweeps of some, again, are larger, others less Portions of a Sphere: All above Hemispheres are call'd *high*, or *surmounted Vaults*; all that are less than Hemispheres, are *low*, or *surbased Vaults*, or *Testudines*, etc.

In some the Height is greater than the Diameter; in others 'tis less: There are others, again, quite flat, only made with Haunses; others Oven-like, or in form of a *Cul de Four*, etc. others growing wider as they lengthen, like a Trumpet.

There are likewise *Gothic Vaults, with Ogives, etc.*

Of *Vaults*, again, some are *single*, others *double*, *cross*, *diagonal*, *horizontal*, *ascending*, *descending*, *angular*, *oblique*, *pendent*, etc. See OGEE, and PENDENTIVE, *etc.*

Master Vaults, are those which cover the principal Parts of Buildings; in contradistinction to the *lesser*, or subordinate *Vaults*, which only cover some little Part; as a Passage, a Gate, *etc.*

Double Vault, is such a one as being built over another, to make the exterior Decoration range with the interior, leaves a Space between the Convexity of the one, and the Concavity of the other; as in the Dome of St. *Peter's* at *Rome*.

Vaults with Compartments, are such whose Sweep, or inner Face, is enrich'd with Pannels of Sculpture, separated by Plat-bands: These Compartments, which are of different Figures, according to the *Vaults*, and usually gilt, on a white Ground, are made with Stuc, on Brick Walls; as in the Church of St. *Peter's* at *Rome*; and with Plaister, on Timber *Vaults*.

Theory of VAULTS.

A Semicircular Arch, or *Vault*, standing on two *Piedroits*, or Imposts, and all the Stones that compose them, being cut and placed in such manner, as that their Joints, or Beds, being prolong'd, do all meet in the Centre of the *Vault*; 'tis evident, all the Stones must be in form of Wedges, *i.e.* must be wider and bigger a-top than at bottom: by virtue of which, they sustain each other, and mutually oppose the Effort of their Weight, which determines them to fall.

The Stone in the middle of the *Vault*, which stands perpendicular to the Horizon, and is call'd the *Key of the Vault*, is sustain'd on each side by two contiguous Stones, just as by two inclined Planes: and of consequence, the Effort it makes to fall, is not equal to its Weight.

But still, that Effort is greater, as the inclined Planes are less inclin'd; so that, if they were infinitely little inclined, *i.e.* if they were perpendicular to the Horizon, as well as the Key, it wou'd tend to fall with its whole Weight; and wou'd actually fall; but for the Mortar.

The second Stone, which is on the right or left of the Key-Stone, is sustain'd by a third; which, by virtue of the Figure of the *Vault*, is necessarily more inclin'd to the second, than the second is to the first; and of consequence, the second, in the Effort it makes to fall, employs a less Part of its Weight than the first.

For the same Reason, all the Stones, reckoning from the Key-Stone, employ still a less and less Part of their Weight, to the last; which resting on a horizontal Plane, employs no part of its Weight; or, which is the same thing, makes no effort to fall; as being entirely supported by the Impost.

Now, in *Vaults*, a great Point to be aim'd at, is, that all the *Voussoirs*, or Stones, make an equal effort towards falling: To effect this, it is visible, that as such (reckoning from the Key to the Impost) employ a still less and less Part of its whole Weight, the first, for Instance, only employing one half; the second, one third; the third, one fourth, *etc.* there is no other way to make those different Parts equal, but by a proportionable Augmentation of the whole, *i.e.* the second Stone must be heavier than the first; the third than the second, *etc.* to the last; which should be infinitely heavier.

M. *de la Hire* demonstrates, what that Proportion is, in which the Weights of the Stones of a Semicircular Arch, must be increased, to be *in Equilibrio*, or to tend with equal Forces to fall; which is the firmest Disposition a *Vault* can have.

Before him, the Architects had no certain Rule to conduct themselves by; but did all at random. Reckoning the Degrees of the Quadrant of the Circle from the Key-Stone to the Impost; the Extremity of each Stone will take up so much the greater Arch, as it is farther from the Key.

M. *de la Hire's* Rule is, to augment the Weight of each Stone, above that of the Key-Stone, as much as the Tangent of the Arch of the Stone, exceeds the Tangent of the Arch of half the Key. Now, the Tangent of the last Stone, of necessity becomes infinite, and of consequence its Weight should be so too: but as Infinity has no place in Practice, the Rule amounts to this, that the last Stones be loaded as much as possible, that they may the better resist the Effort which the *Vault* makes to separate 'em; which is call'd the *Shoot* or *Drift of the Vault*.

M. *Parent* has since determin'd the Curve, or Figure which the *Extrados*, or outside of a *Vault*, whose *Intrados*, or inside, is spherical, must have, that all the Stones may be *in Equilibrio*.

Key of a VAULT, is a Stone, or Brick, in the middle of the *Vault*, in form of a truncated Cone; serving to bind or fasten all the rest. See KEY.

The Reins of a VAULT, or the filling up, are the Sides which sustain it.

The Pendentive of a VAULT, is the Part suspended, between the Arches or Ogives. See PENDENTIVE.

The Impost of a VAULT, is the Stone whereon the first *Voussoir*, or Stone of the *Vault*, is laid. See IMPOSTS, *etc.*

VAULT, or VOLT, in the Manage, an Action of a Horse, wherein he turns round, or makes a Circle side-ways, going round a Centre.

There are some *Vaults* wherein the Horse makes two parallel Circles, the one with his fore-Feet, and the other with his hind; and others, wherein the Horse making Curvets and Capriols, his Haunches follow his Shoulders, and only make one Circle, or Oval, about a Pillar or Centre.

An *inverted*, or *backward Vault*, is that wherein the Horse, turning likewise sideways, has his Head towards the Centre, and his Tail towards the Circumference: So that the great Circle is form'd by the fore-Feet, and the little one by the hind-Feet.

A *Demi-Vault*, or *Half-Vault*, is when the Horse only goes half the Circle, and comes back again the same way.

Vault is also used for the Manages practis'd on the wooden Horse, to learn to mount and unmount with Ease and Expedition.

VENEERING, or VANEERING, a kind of Marquetry, or Inlaying; whereby, several thin Slices, or Leaves of fine Woods of different kinds, are applied and fasten'd on a Ground of some common Wood.

There are two Kinds of Inlaying; the one, which is the more ordinary, goes no further than the making of Compartments of different Woods; the other requires a deal more Art, and represents Flowers, Birds, and the like Figures.

The first Kind is what we properly call *Veneering*; the latter we have already describ'd, under the Article MARQUETRY.

The Wood intended for *Veneering*, is first saw'd out into Slices, or Leaves about a Line thick: In order to saw 'em, the Blocks or Planks are placed upright, in a kind of Sawing-Press.

These Slices are afterwards cut into narrow Slips, and fashion'd divers ways, according to the Design propos'd: Then the Joints being carefully adjusted, and the Pieces brought down to their proper thickness, with several Planes for the Purpose; they are glu'd down on a Ground or Block of dry Wood, with good strong *English* Glue.

The Pieces thus jointed and glu'd, the Work, if small, is put in a Press; if large, 'tis laid on the Bench, cover'd with a Board, and press'd down with Poles or Pieces of Wood, one end whereof reaches to the Ceiling of the Room, and the other bears on the Board.

When the Glue is quite dry, they take it out of the Press, and finish it; first with little Planes, then with divers Scrapers; some whereof resemble Rasps, which take off the Dents, *etc.* left by the Planes.

When sufficiently scrap'd, the Work is polish'd with the Skin of a Sea-Dog, Wax, and a Brush and Polisher of Shave-grass: which is the last Preparation.

VENTIDUCTS, in Building, call'd by the Italians Ventidotti, and by the French *Prisons des Vents*, or *Palais d'Eole*; are Spiracles, or subterraneous Places, where fresh, cool Winds being kept, are made to communicate, by means of Ducts, Funnels, or Vaults, with the Chambers or other Apartments of a House; to cool them in sultry Weather.

These are much in use in *Italy*, etc. See BUILDING, HOUSE, *etc.*

VERNACULAR, is applied to anything that is peculiar to some one Country.

VESTIBLE, VESTIBULUM, in Architecture, a kind of Entry into large Buildings; being an open Place before the Hall, or at bottom of the Staircase.

Vestibles only intended for Magnificence, are usually between the Court and the Garden: These are sometimes *simple*, that is, have their opposite Sides equally enrich'd with Arches; and sometimes their Plan is not contain'd under four equal Lines, or a circular one, but forms several Van-corps and Rear-corps, furnish'd with Pilasters, *etc.*

The *Romans* had Places call'd *Vestibles* at the Entrance of their Houses, to shelter People oblig'd to stand at the Door, from the Weather: We have still *Vestibles* of the like kind, in many old Churches, Houses, *etc.* call'd *Porches.*

Martinius derives the Word from *Vestæ stabulum*; by reason the Fore-part of the House was dedicated to *Vesta. Daviler* derives it from *Vestis*, and *ambulo*; by reason People there begin to let their Trains fall.

VESTRY, VESTIARIA, a Room adjoining to a Church, where the Priests Vestments and sacred Utensils are kept.

VILLA *Regis*, a Title antiently given to those Villages where the Kings of *England* had a Royal Seat, and held the Manor in their own Demesn.

VILLAGE, VILLA, or VILL, an Assemblage of Houses, inhabited chiefly by Peasants and Farmers, having usually a Church, but no Market.

The want of a Market distinguishes it from a *Town*, as the Church does from a Green, Street, *etc.*

The Word is *French*, form'd of *vil*, or *vilis*, low, mean, contemptible.

Among our *Saxon* Ancestors, *Vill*, or *Village* was used in the Sense of the Roman *Villa*; *viz.* for a Country Farm or Seat, furnish'd with convenient Out-houses, *etc.* for repositing the Fruits thereof.

Afterwards it came to be taken for a *Manor*, then for part of a Parish, or the Parish it self.

Hence, in several antient Law-Books, *Vill* and *Parish* are the same thing: Accordingly, *Fortescue*, *de Laudibus Leg. Ang.* writes, That the Boundaries of *Villages*, are not by Houses, Streets, or Walls, but by a large Circuit of Ground, within which may be divers Hamlets, Waters, Woods, *etc.*

Fleta makes this difference between a *Mansion*, a *Village*, and a *Manor*; that a Mansion may consist of one, or more Houses; tho it is only to be one Dwelling-place, without any other very near it: for if other Houses be contiguous, 'tis then a *Village*. A Manor may consist of one or more *Villages*.

For the better Government of *Villages*, the Lord of the Soil has usually a Power to hold a Court-Baron, every three Weeks.

VINEA, in Fortification. See MANTELETS.

VISUAL *Point*, in Perspective, is a Point in the horizontal Line, wherein the ocular Rays unite.

Thus, a Person standing in a strait long Gallery, and looking forwards, the Sides, Floor, and Ceiling seem to meet, and touch one another in a Point, or common Centre.

VIVO, in Architecture, the Shaft, or Fust of a Column.

See SHAFT, and COLUMN.

VOLUTE, VOLUTA, in Architecture, a kind of Scrowl, or spiral Contortion, used in the *Ionic* and *Composite* Capitals; whereof it is the principal Character and Ornament. See CAPITAL.

Some call it the *Ram's-Horn*, from its Figure, which bears a near resemblance thereto.

Most Architects suppose, that the Antients intended it to represent the Bark or Rind of a Tree, laid under the Abacus, and twisted thus at each Extreme where it is at Liberty: Others will have it a sort of Pillow, or Bolster laid between the Abacus and Echinus, to prevent the latter being broke by the Weight of the former, and the Entablature over it; and accordingly call it *Pulvinus*.

Others, after *Vitruvius*, will have it to represent the Curls or Tresses of a Woman's Hair. See IONIC *Order*.

There are also eight angular *Volutes* in the *Corinthian* Capital, accompany'd with eight other smaller ones, called *Helices*.

There are several Diversities in the *Volute*.—In some, the List or Edge, throughout all the Circumvolutions, is in the same Line, or Plane: such are the antique *Ionic Volutes*, and those of *Vignola*.—In others, the Spires or Circumvolutions fall back; in others project, or stand out. —Again, in some the Circumvolutions are oval; in others, the Canal of one Circumvolution is detach'd from the List of another, by a Vacuity or Aperture.—In others, the Round is parallel to the Abacus, and springs out from behind the Flower thereof.—In others, it seems to spring out of the Vase, from behind the Ovum, and rises to the Abacus, as in most of the fine *Composite* Capitals.

Consoles, Modillions, and other sorts of Ornaments, have likewise their *Volutes*. See CONSOLE, *etc*.

The *Volute* is a Part of great Importance to the Beauty of the Column.—Hence Architects have invented divers ways of delineating it.—The principal are that of *Vitruvius*, which was long lost, and at last restor'd by *Gouldman*; and that of *Palladio*.—*Daviler* prefers the former as the easier.—The manner thereof is as folows.

Divide the Altitude A B, (Tab. *Architect*. Fig. 12 *Volute*) into eight equal Parts, and assuming the Fifth Q P, for a Diameter, from the Centre G, with half the Diameter G P, describe a Circle for the Eye of the *Volute*.—Bissect the Radii G P and G Q in 1 and 4, and subdivide the half Parts G 1 and G 4 each into three equal Parts: Then, upon the right Line

1 4, construct a Square 1, 2, 3, 4; one of whose Sides 2, 3, continue on to D; another 3, 4, to E; and the third 1, 2, to C. From G draw right Lines, G 2 and G 3; which divide into three equal Parts—Then, thro' 6 and 10, draw 6 L and 10 N, parallel to 2 D. And thro' 11 and 7 draw 11 O and 7 K, parallel to 3 E; and thro' 5 and 9 draw 5 H, and 9 M, parallel to 1 C.—Lastly, from 12, 11, 10, 9, 8, *etc.* strike the Quadrants P O, O N, N M, M L, L K, *etc.* which will form the *Volute.*

VOUSSOIR, *Vault-Stone*, in Architecture, a stone proper to form the sweep of an Arch, being cut somewhat in manner of a truncated Cone, whose Sides, were they prolong'd, would terminate in a Centre, to which all the Stones of the Vault are directed. See VAULT.

W

WAINSCOT, in Building, the Timber-Work serving to line the Walls of a Room; being usually in Pannels, and painted, to serve in lieu of Hangings. See WALL, *etc.*

Even in Halls, 'tis common to have *Wainscot* breast high; by reason of the natural Humidity of Walls. See HALL.

Some Joiners put Charcoal behind the Pannels of the *Wainscot*, to prevent the Sweating of Stone and Brick-Walls from ungluing the Joints of the Pannels.—Others use Wool for the same purpose.—But neither the one nor the other is sufficient in some Houses: The only sure way, is by priming over the back-sides of the Joints with White Lead, Spanish Brown, and Linseed Oil.

Wainscotting with Norway *Oak*, the Workman finding Stuff, is valu'd at six or seven Shillings *per* Yard square.—*Plain square Wainscotting*, the Workman finding Deal, is valu'd at three Shillings, or three Shillings six Pence *per Yard.—Large bisection Wainscotting*, with *Dantzick* Stuff, is valued at six or seven Shillings *per* Yard; and *ordinary Bisection* Work at three Shillings six Pence *per* Yard.

Note, In taking Dimensions, they use a String, which they press into all the Mouldings; it being a Rule that they are to be paid for all where the Plane goes.

WALKS, in Gardening. See ALLEY.

To keep the Weeds from growing upon *Walks*, Mr. *Switzer* directs, that the Bottoms thereof be fill'd with Lime-Rubbish, or coarse Gravel, Flint-stones, or other rocky Stuff, eight or ten Inches deep; over which may be laid a like depth of Gravel, not too fine.

To keep 'em the drier, they are to be made round, or convex.—The usual Proportion is, that a *Walk* 20 Foot wide, be four Inches higher in the Middle, than at the Sides; a *Walk* 25 Foot wide, 5 Inches, *etc.*

After laying a *Walk*, it is to be roll'd, both lengthwise, and cross-wise; and to lay it the firmer, it must have three or four Water-Rollings; that is, when it rains so very fast, that the *Walk* swims with Water: which, when dry Weather comes, will bind as hard as Terrace.

To make the Gravel bind the better, some mix a little Lime with it; which being apt to stick to the Heels of the Shoes in wet Weather, others grind or pound Sea-shells, and lay a thin Coat thereof on the Gravel; which being roll'd, incorparates with the Gravel, and hinders its hanging to the Shoes. Others beat Smiths Cinders to Dust, and others Bricks; Strewing the Dust on the *Walks*, which dries up the Moisture, and gives 'em a Colour.

For Terrass-WALKS. See TERRASS.

WALL, in Architecture, *etc.* a Work of Stone, Brick, Wood, or the like; making the principal Part of a Building; as serving both to inclose it, and to support the Roof, Floors, *etc.* See BUILDING, HOUSE, *etc.*

Walls, tho built very thick and strong, and their Foundations laid deep, yet, if carried on strait in a Line, are inclined to lean, or fall; and such as are built crooked, tho thin and weak, are much more lasting. A *Wall* rais'd over a River, on Arches of Pillars, stands as firm as others, whose Foundation is entire.

Hence it appears, that a *Wall* built much thinner than usual, by only having at every 20 Foot's distance an Angle set out about two Foot, or more, in proportion to the Height of the *Wall*; or by having at the like distance, a Column or Pilaster erected along with it, six or eight Inches on each side over and above the thickness of the rest of the *Wall*: Such *Wall* will be much stronger than if five times the quantity of Materials were used in a strait Wall.

Walls are distinguish'd into divers Kinds, from the Matter whereof they consist; as *Plaister'd*, or *Mud-Walls*, *Brick-Walls*, *Stone-Walls*, *Flint* or *Boulder-Walls*, and *Boarded Walls*.—In all which, these general Rules are to be regarded.

1st, That they be built exactly perpendicular to the Ground-work.

2d, That the massiest and heaviest Materials be the lowest; as fitter to bear than be borne.

3d, That the *Walls*, as they rise, diminish proportionally in Thickness, for ease both of Weight and Expences.

4th, That certain Burses, or Lodges, of more strength than the rest, be interlaid, like Bows, to strengthen the whole Fabrick.

Mud and *Plaister'd Walls* are chiefly in ordinary Timber Buildings. —These *Walls*, being quarter'd and lath'd between the Timber, or sometimes lath'd over all, are plaister'd with Lome, (see LOME;) which being almost dry, is plaister'd over again with white Mortar. See MORTAR.

Brick Walls, are the most important and usual among us.—In these, particular Care is to be taken about the laying of the Bricks, *viz.* That in

Summer they be laid as wet, and in Winter as dry as possible; to make them bind the better with the Mortar: That in Summer, as fast as they are laid, they be cover'd up, to prevent the Mortar, *etc.* from drying too fast: that in Winter they be cover'd well to protect them from Rain, Snow, and Frost, which are all Enemies to the Mortar: That they be laid Point and Joint in the middle of *Walls*, as seldom as may be; but, good Bond made there, as well as on the outsides.—Care is likewise to be taken, that the Angles be firmly bound: In order to which, in working up the *Walls* of a Building, 'tis not advisable to raise any *Wall* above three Foot high, ere the next adjoining *Wall* be wrought up to it; that good Bond may be made in the progress of the Work.

Lastly, in Building a House in the City of *London*, the *Walls* are to be of such thicknesses as are enjoined by the Act of Parliament for rebuilding that City. See HOUSE.

Flint, or *Boulder-Walls*, are frequently used in divers Parts for Fence-*Walls* round Courts, Gardens, *etc.* and even for Walls of Out-houses. —Sir *Henry Wotton* observes, that the building of *Walls* of Flint, is a thing utterly unknown to the Antients; who observing in that Material a kind of metallic Nature, at least a Fusibilty, reserv'd it for nobler Uses.

These *Walls* are usually rais'd by a right and left-handed Man, who have a Hod of Mortar pour'd down on the Work, which they part betwixt them; each spreading it towards himself, and so they lay in the Flints. —The Mortar for this Work is to be very stiff.

WALLS, in Gardening, *etc.*—The Position, Matter, and Form of *Walls* for Fruit-Trees, are found to have a great Influence on the Fruit: Tho Authors differ as to the Preference.

The Rev[d]. Mr. *Lawrence* directs, that the *Walls* of a Garden be not built directly to face the four Cardinal Points, but rather between 'em, *viz.* South-East, South-West, North-East, and North-West: In which, the two former will be good enough for the best Fruit, and the two latter for Plums, Cherries, and baking Pears.

Mr. *Langford*, and some others, propose Garden-*Walls* to consist chiefly of Semicircles; each about six or eight Yards in Front, and including two Trees; and between every two Semicircles, a Space of two Foot of *plain Wall.*—By such a Provision, every Part of a *Wall* will enjoy a share of the Sun, one time with another; besides that the Warmth will be increas'd, by the collecting and reflecting of the Rays in the Semicircles; and the Trees within be skreen'd from injurious Winds.

As to the Materials of *Walls* for Fruit-Trees, Brick, according to Mr. *Switzer*, is the best; as being the warmest, and kindest for the ripening of Fruit, and affording the best conveniency for nailing.

Mr. *Lawrence*, however, affirms, on his own Experience, that *Mud-Walls*, made of Earth and Straw temper'd together, are better for the ripening of Fruit than either Brick or Stone *Walls*: He adds, that the Coping of Straw laid on such *Walls*, is of great advantage to the Fruit, in sheltering 'em from perpendicular Rains, *etc.*

M. *Fatio*, in a particular Treatise on the Subject, instead of the common perpendicular *Walls*, proposes to have the *Walls* built sloping, or reclining from the Sun; that what is planted against 'em may lie more expos'd to his perpendicular Rays; which must contribute greatly to the ripening of Fruit in our cold Climate.

The Angle of Reclination is to be that of the Latitude of the Place; that when the Sun is in the Meridian at the Equinoxes, his Rays may strike just perpendicularly.

Yet some others prefer perpendicular *Walls*, and even inclining ones, or such as hang fowards to the Sun; as such receive the Sun's Rays perpendicularly when he is low; as in Spring and Autumn, or in the Evening and Morning: which they imagine of more service than the greatest Heats of the Sun at Midsummer, upon reclining *Walls*.

Add, that in Autumn the Sun is most wanted to ripen Winter Pears; in order to which, they should be kept dry; which against sloping *Walls* cannot be; the Dews, *etc*. lying much longer thereon, than on those that are perpendicular.

One great advantage, however, of M. *Fatio's* sloping *Walls*, is, that Fruit-Trees, as Vines, *etc*. being planted against 'em, Melon-Glasses may be set on the Fruit; which will much forward its ripening.

WARDROBE, a Closet, or little Room, adjoining to a Bed-Chamber; serving to dispose and keep a Person's Apparel in; or for a Servant to lodge in, to be at hand to wait, *etc*.

WARDROBE, in a Prince's Court, is an Apartment wherein his Robes, wearing Apparel, and other Necessaries, are preserv'd; under the Care and Direction of proper Officers.

His Majesty has a *Great Wardrobe*; a *Removing Wardrobe*; and divers *Standing Wardrobes*, belonging to his Bed-chambers in each of his Palaces, *viz*. at *Whitehall*, *Kensington*, *Windsor*, *Hampton-Court*, and the *Tower*; each under its respective Keeper.

The *Removing Wardrobe* always attends on the King's Person; as also on Embassadors, at Chirstenings, Masques, Plays, *etc*.—It is under the Command of the Lord Chamberlain: The under-Officers are, a Yeoman, two Grooms, and three Pages.

The *Great Wardrobe*, is of great Antiquity—Antiently, it was kept near *Puddle-Wharf*, in a House purchas'd for that purpose by King *Edward* III. but since the Fire of *London*, it has been kept in *York-Buildings*.

The Master or Keeper thereof, is an Officer of great Dignity: high Privilages were confer'd on him by *Henry* VI. King *James* I. enlarg'd the same, and erected the Office into a Corporation.

WASHING, in Painting, is when a Design drawn with a Pen, or Crayon, has some one Colour laid over it with a Pencil, as Indian Ink, Bistre, or the like; to make it appear the more natural, by adding the

Shadows of Prominences, Apertures, *etc.* and by imitating the particular Matters whereof the Thing is supposed to consist.

Thus, they *wash* with a Pale red to imitate Brick and Tyle; with a pale Indian blue to imitate Water and Slate; with green for Trees and Meadows; with Saffron or *French* Berries, for Gold and Brass; and with several Colours for Marbles.

These *Washes* are usually given in equal Teints, or Degrees, throughout; which are afterwards brought down and soften'd over the Lights with fair Water, and strengthen'd with deeper Colours for the Shadows.

WATER-Scape, of the Saxon Waterschap, denotes an Aqueduct. See AQUEDUCT.

WATER-Table, in Architecture, is a sort of Ledge, left in Stone or Brick Walls, about 18 or 20 Inches from the Ground: from which Place, the Thickness of the Wall begins to abate. See Wall.

WELL-Hole, in Building, is the Hole left in a Floor for the Stairs to come up thro'. See STAIRS.

WINDOW, *q.d. Wind-Door*, an Aperture or open Place in the side of an House, to let in the Air and Light. See BUILDING, APERTIONS, LIGHT.

We have various Kinds and Forms of *Windows*; as Glass *Windows*, Wire *Windows*, Horn *Windows*, *etc.*—Arch'd *Windows*, Circular *Windows*, Elliptical *Windows*, Square and Flat *Windows*; Round *Windows*, Oval *Windows*, Gothic *Windows*, Regular *Windows*, Rustic *Windows*, *etc*. Sky-Lights.

Dormer WINDOWS, or *Lutherns*. See DORMER and LUTHERN.

Transom WINDOWS. See TRANSOM.

The chief Rules with regard to *Windows* are—1. That they be as few in Number, and as moderate in Dimensions, as may consist with other Respects; inasmuch as all Openings are Weakenings—2. That they be placed at a convenient Distance from the Angles, or Corners of the Building; because that part ought not to be open and infeebled, whose Office is to support and fasten all the rest of the Building—3. That Care be taken the *Windows* be all equal one with another in their Rank and Order; so that those on the right hand may answer to those on the left, and those above be right over those below: for this Situation of *Windows* will not only be handsome and uniform; but also the Void being upon the Void, and the Full upon the Full, it will be a strengthening to the whole Fabrick.

As to their Dimensions, Care is to be used neither to give them more, nor less Light than is needful; therefore regard is to be had to the Bigness of the Rooms which are to receive the Light. 'Tis evident that a great

Room needs more Light, and consequently a greater *Window* than a little Room; and *è contra.*

The Apertures of *Windows* in middle-sized Houses, may be four and a half, or five Feet between the Jambs; and in the greater Buildings, six and a half, or seven Feet; and their Height may be double the Length at the least—But in high Rooms, or larger Buildings, their Height may be a third, a fourth, or half their Breadth, more than double their Length.

Such are the Proportions for *Windows* of the first Story; and according to these must those in the upper Stories be for Breadth; but as to Height, they must diminish: the second Story may be one third part lower than the first, and the third Story one fourth part lower than the second.

WOOD, *Lignum*, a solid Substance, whereof the Trunks and Branches of Trees are form'd; and deriving its Growth from certain Juices in the Earth.

The *Wood* is all that part, in a Tree, included between the Bark and the Pith.

Dr. *Grew*, in his Anatomy of Plants, has discover'd, by means of the Microscope, that what we call *Wood* in a Vegetable, notwithstanding all its Solidity, is only an assemblage of infinite minute Canals, or hollow Fibres, some of which rise from the Root upwards, and are disposed in form of a Circle; and the others, which he calls *Insertions*, tend horizontally from the Surface to the Centre; so that they cross each other, and are interwove like the Threads of a Weaver's Web.

Woods are distinguish'd into divers Kinds; with regard to their Natures, Properties, Virtues, and Uses.

Of WOOD, consider'd according to its Qualities, whether Useful, Curious, Medicinal, *etc.* the Principal is that call'd *Timber*, used in building Houses, laying Floors, Roofs, Machines, *etc.*

WOODS valued on account of their Curiosity, are *Cedar*, *Ebony*, *Box*, *Calembouc*, etc. which, by reason of their extraordinary Hardness, agreeable Smell, or beautiful Polish, are made into Tables, Combs, Beads, *etc.*

WORK-*House*, a Place where indigent, vagrant, and idle People are set to work, and maintain'd with Clothing, Diet, *etc.*

Such is *Bridewell*, and several other Places about the City and Suburbs; particularly that in *Bishopsgate-street* for employing the poor Children of the City and Liberties, who have no Settlement; and that for the Parish of St. *Margaret's Westminster*, call'd the *Grey-Coat Hospital.*

At *Amsterdam* they have a famous *Work-House*, or House of Correction, call'd the *Rasphuyse*, which, by a Privilege granted in 1602, has alone the Right of Shaving and Cutting the Dyers Woods, as Brasil, Santal, Campeche, Sassafras, *etc.*

X

XYSTUS, in the antient Architecture—A *Xystus*, among the *Greeks*, was a long Portico, either open or cover'd over; wherein the *Athletæ* practis'd Wrestling, and Running.

The Gladiators who practised therein, were call'd *Xystici*.

Among the *Romans*, the *Xystus* was only an Alley, or double Row of Trees, meeting Arbor-wise a-top, and forming a Shade to walk under.

The Word is *Greek*, form'd of *to polish, shave, rub*.

Y

YARD, *Virga*, a long Measure used in *England*, and *Spain*; chiefly to measure Cloth, Stuffs, *etc*.

The *English Yard* contains three Foot.

It was first settled by *Henry* I. from the length of his own Arm.

The *English Yard* is just seven Ninths of the *Paris* Ell; so that nine *Yards* make seven Ells— To reduce Ells, therefore into *Yards*, say, If seven Ells give nine *Yards*, now many *Yards* will the given Number of Ells give?

Yards are converted into Ells *Flemish*, by adding a third part; into Ells *English*, by substracting a fifth part; or multiplying by 8, and casting off the right-hand Figure—Ells *English* are converted into *Yards*, by adding a fourth. To turn Ells *Flemish* into *Yards*, substract one Quarter.

The *Spanish Yard* chiefly used at *Sevil*, is, in some Places, call'd *Barra*—It contains $^{17}/_{24}$ of the *Paris* Ell; so that 17 Ells make 24 *Spanish Yards*.

Z

ZETA, or ZETECULA, a little withdrawing Chamber, with Pipe running along the Walls to receive from below either the cool Air or the Heat of warm Water.

ZOCCO, ZOCCOLO, ZOCLE, or SOCLE, in Architecture, a small kind of Stand or Pedestal; being a low, square Piece or Member, serving to support a Busto, Statue, or the like thing that needs to be rais'd.

The Word is *Italian*, form'd from *Soccus*, a Sandal, or high Shoe.

ZOCCO, or *Zocle*, is also used for a low, square Member serving to support a Column, or other part of a Building, instead of a Pedestal, Base, or Plinth. See PEDESTAL, PLINTH, *etc.*

A *continued* ZOCLE is a kind of continued Pedestal whereon a Structure is raised; but having no Base or Cornice. See PEDESTAL.

ZOOPHORUS, or ZOPHORUS, in the antient Architecture, the same thing with the *Frieze* in the Modern. See FRIEZE.

It was thus call'd in *Greek*, because antiently adorned with the Figures of Animals; from *Animal*, and *I bear*.

The *Greeks* sometimes also call the Zodiac, *Zoophorus*, because of the Signs or Constellations therein.

ZOOPHORIC *Column*, is a Statuary Column; or a Column that bears or supports the Figure of an Animal. See COLUMN.

References

The following references are the primary sources consulted for biographical and historical information and also for the quotations and citations to be found in the various texts. Given the nature of the interrelationships between the several volumes in our series *The Encyclopaedic Dictionary in the Eighteenth Century: Architecture, Arts and Crafts*, we have grouped together, for the readers interest and convenience, all the references consulted in the preparation of the individual volumes listed under the short title of each volume.

John Harris: *Lexicon Technicum*

Chapter 1

1. These ideas are outlined in the preface to Bacon's *The New Organon* for which see:
   R. L. Ellis and J. S. Spedding, Edited by John M. Robertson, *The Philosophical Works of Francis Bacon, Baron of Verulam, Viscount of St. Albans, and Lord High Chancellor of England*, George Routledge and Sons Ltd., New York: E. P. Dutton & Co., 1905.
2. The essay *Of Building* is adapted from the text given in note 1.
3. The essay *Of Gardens* is adapted from the text given in note 1.
4. The reader interested in Sir Francis Bacon's water gardens will find a description of these in:
   Paula Henderson, *Sir Francis Bacon's Water Gardens At Gorhambury*, The Journal of the Garden History Society, Vol. 20, Number 2, 1992, pp. 116-131.

Chapter 2

1. The biographical account of the life of Sir Henry Wotton is adapted from:
   Thomas Zouch, *The Lives of Dr. John Donne; Sir Henry Wotton; Mr. Richard Hooker; Mr. George Herbert; and Dr. Robert Sanderson, by Isaac Walton.* Volume I, York: Payne and Foss, and Wilson and Sons, 1817.

Chapter 4

1. Thompson Cooper, *John Harris*, The Dictionary of National Biography, Vol. IX, London: Oxford University Press, 1959-60 (Reprinted).
2. John Harris, *Lexicon Technicum: or, an Universal English Dictionary of ARTS and SCIENCES: Explaining not only The TERMS of ART, But The ARTS Themselves.* Volume I, London: MDCCIV.
3. For a discussion of eighteenth-century encyclopaedias, and in particular that of John Harris, see:
Lael Ely Bradshaw, *John Harris's Lexicon Technicum*, Voltaire Studies, no. 194, 1981: pp. 107-121.
4. Sir Henry Wotton, *The Elements of Architecture*, John Bill, London: 1642.
5. As we have noted in the text, the second volume of *Lexicon Technicum* was published as a 'Supplement' to the first as:
John Harris, *Lexicon Technicum: or an UNIVERSAL English Dictionary of ARTS and SCIENCES: Explaining not only The TERMS of ART, But The ARTS Themselves.* Volume II, London: MDCCX.

Ephraim Chambers: *Cyclopaedia*: *Volume I*

Chapter 1

1. Francis Espinasse, *Ephraim Chambers*, The Dictionary of National Biography, Vol. IV, London: Oxford University Press, 1908.
2. Gerald Le Grys Norgate, *John Senex*, The Dictionary of National Biography, Vol. XVIII, London: Oxford University Press, 1963-64.
3. Ephraim Chambers, *Prospectus*; *Proposals for Printing by Subscription*, *Cyclopaedia*, London, January 1710.
4. For a discussion of eighteenth-century encyclopaedias, and in particular that of Ephraim Chambers, see:
Lael Ely Bradshaw, *Ephraim Chambers' Cyclopaedia*, Voltaire Studies, no. 194, 1981: pp. 123-140
5. For a discussion of the work of Ephraim Chambers in its wider context see:
Robert Collison, *Encyclopaedias: their history throughout the ages*, New York: 1964.
6. The Rev. Alexander Gordon, *Abraham Rees*, The Dictionary of National Biography, Vol. XVI, London: Oxford University Press, 1909.
7. Ephraim Chambers, *CYCLOPAEDIA: or, an UNIVERSAL DICTIONARY of ARTS and SCIENCES.* Two volumes, London: MDCCXXVIII.

8. For a discussion of the transition of Chambers' *Cyclopaedia* into Diderot's *Encyclopédie* see:
Terence M. Russell, *Architecture in the Encyclopédie of Diderot and d'Alembert: The Letterpress Articles and Selected Engravings*, Aldershot: Scolar Press, Ashgate Publishing Ltd., 1993.
9. My understanding of the writings of Francis Bacon has been derived through consultation of the original texts (see Chapter 1, note 1 of John Harris) and:
F. H. Anderson, *The Philosophy of Francis Bacon*, Chicago, Illinois: The University of Chicago Press, 1948.
10. For some discussion of the endeavours to translate Chambers' *Cyclopaedia* into French see:
J. G. Alger, *John Mills*, The Dictionary of National Biography, Vol. XIII, London: Oxford University Press, 1909.

The Builder's Dictionary

Chapter 1

1. For a discussion of *The Builder's Dictionary* in the context of eighteenth-century architectural books and such precursors as Joseph Moxon's *Mechanick Exercises* see:
Eileen Harris assisted by Nicholas Savage, *British Architectural Books and Writers 1556-1785*, Cambridge: Cambridge University Press, 1990.
2. As we remark in the text, our specific reference here is to the work of Francis Bacon concerning his *Preparative towards a natural and Experimental History*. For the complete reference, see Chapter 1, note 1 of John Harris.
3. For Robert Ellis's introduction to the works of Sir Francis Bacon see Chapter 1, note 1 of John Harris.
4. The formal library-catalogue reference for the first edition is:
The Builder's Dictionary: *Or, Gentleman and Architect's COMPANION. Explaining not only the TERMS of ART In all the several PARTS of ARCHITECTURE, But also containing the THEORY and PRACTICE Of the Various Branches thereof, requisite to be known by Masons, Carpenters, Joiners, Bricklayers, Plaisterers, Painters, Glaziers, Smiths, Turners, Carvers, Statuaries, Plumbers, etc.* London: Printed for A. Bettesworth and C. Hitch, at the Red-Lion in Pater-noster-Row; and S. Austen, at the Angel and Bible in St. Paul's Church-Yard. MDCCXXXIV.
For the reader unable to locate a copy of the first edition, a facsimile is available in the following form:

The Builder's Dictionary: Or, Gentleman and Architect's Companion. Being A Complete UNABRIDGED REPRINT of the earlier work published by A. Bettesworth and C. Hitch. Washington: Association For Preservation Technology, 2 Vols., MCMLXXXI.

Samuel Johnson: *A Dictionary of the English Language*

Chapter 1

1. Quotations throughout Chapters 1 and 2, unless otherwise indicated, are taken from the opening chapters of Boswell's Life of Johnson in the following edition:
   James Boswell, *Life of Johnson*, London: H. Ingram & Co., Vol. I., 1851.
2. My understanding of Johnson has been greatly assisted by my reading of Mr. Wain's scholarly and moving account:
   John Wain, *Samuel Johnson*, London: MacMillan Ltd., 1974, reprinted 1980.
3. For insights into the everyday objects in Johnson's life and for portrayals of his contemporaries, I am indebted to Ms. Lane's richly illustrated study:
   Margaret Lane, *Samuel Johnson and his World*, London: Hamish Hamilton Ltd., 1975.

Chapter 2

1. Information relating to Johnson and his London circle has, in part, been derived from:
   Leslie Stephen, *Samuel Johnson*, The Dictionary of National Biography, Vol. X, London: Oxford University Press, 1963-64 (Reprinted).
2. For an overview of Johnson at the time of his work on the *Dictionary*, I have benefited from consulting:
   Sir Sydney Castle Roberts, *Samuel Johnson*, Chicago and London: Encyclopaedia Britannica, 1972.
3. Samuel Johnson, *The Plan of a Dictionary of the English Language*, London: 1747.
4. My understanding of many matters bearing upon the creation of Johnson's *Dictionary* has been aided by the views of:
   J. D. Fleeman, *The Genesis of Johnson's Dictionary*, London: Longman, 1990 (?).
5. A description of Johnson's working method based upon the most recent research is given by:
   Allen Reddick, *The Making of Johnson's Dictionary 1746-1773.* Cambridge: Cambridge University Press, 1990.

6. Robert Harrison, *Richard Bathurst*, The Dictionary of National Biography, Vol. I, London: Oxford University Press, 1963-64 (Reprinted).
7. Cecil J. L. Price, *Philip Dormer Stanhope, Fourth Earl of Chesterfield*, Chicago and London: *Encyclopaedia Britannica*, 1972.

Chapter 3

1. The tables have been compiled from, and the letterpress articles have been extracted from, the first edition of the *Dictionary*: Samuel Johnson, *A DICTIONARY OF THE ENGLISH LANGUAGE: In Which The WORDS are deduced from their ORIGINALS, and ILLUSTRATED in their DIFFERENT SIGNIFICATIONS by EXAMPLES from the best WRITERS.* London: Printed by W. Strahan, For J. and P. Knapton [etc.], 2 Vols., MDCCLV.
2. For an assessment of Johnson's accomplishment placed in a wider context see: Brian O'kill, *The Lexicographic Achievement of Johnson*, London: Longman, 1990 (?).

Encyclopaedia Britannica

Chapter 1

1. For an outline of the lives of Bell, Macfarquhar and Smellie and their contribution to the creation and publication of the first edition of Encyclopaedia Britannica I have benefited from the following: John Armitage, *Andrew Bell, Colin Macfarquhar and William Smellie*, Chicago and London: Encyclopaedia Britannica, 1972.
2. Bernard Barham Woodward, *William Smellie*, The Dictionary of National Biography, Vol. XVIII, London: Oxford University Press, 1959-60 (Reprinted).
3. *Encyclopaedia Britannica, Or, A Dictionary Of Arts and Sciences, Compiled Upon A New Plan In Which The different Sciences and Arts are digested as they occur in the order of the Alphabet*, Edinburgh: Printed for A. Bell and C. Macfarquhar . . . at his Printing Office, Nicolson-Street. MDCCLXXI.

Additional Works by the same Author

As stated in the Preface, the present work *The Encyclopaedic Dictionary in the Eighteenth Century: Architecture, Arts and Crafts* has its origins in previous bibliographical researches undertaken by the author. For the reader interested in these matters, the following publication details are provided:

Architecture in the Encyclopedié of Diderot and d'Alembert: The Letterpress Articles and selected Engravings. Scolar Press, Aldershot 1993. pp. xii, 224. 100 Plates and Frontispiece. This book traces the origins and publication history of the *Encyclopédie* and provides a detailed inventory, of 5,000 architectural writings abstracted from the original 17 folio volumes of text and 11 folio volumes of plates.

The Built Environment: A Subject Index 1800-1960. 4 Volumes. Gregg Publishing, London 1988. The books in this index were researched over a ten-year period from the extensive British Library Catalogue, with the approval of the copyright authorities and received substantial government financial support. The index gives detailed citations to more than 70,000 printed books across the whole realm of the built environment.

Volume 1:	Town Planning and Urbanism Architecture Gardens and Landscape Design	1120 pp
Volume 2:	Environmental Technology Constructional Engineering Building and Materials	960 pp
Volume 3:	Decorative Art and Industrial Design International Exhibitions and Collections Recreational and Performing Arts	1152 pp
Volume 4:	Public Health Municipal Services Community Welfare	1232 pp

Bibliographical Index: A Subject Guide to Printed Books in the Fields of Architecture, Building, Constructional Engineering, Environment, Technology, Planning, Landscape Architecture, Design and allied Arts and Crafts. University Microfilms International (UMI) Ann Arbor, Michigan USA 1982. This work was compiled over a ten-year period from the British Library Catalogue with the consent of the copyright authorities. Each volume has an introduction in French, German, Italian, and Spanish. The 32 volumes which constitute the index are each substantial monographs which are published (in the USA) in both microfiche and conventional hard-back book form. The titles in the series are listed below together with the number of pages for each volume. The number of bibliographical entries contained within each volume is shown in brackets: each citation is listed under a subject heading with details of the author, book title and all relevant publication details.

	Pages	Citations
ARCHITECTURE		
Vol. 1 Architectural history and theory	178	(2579)
Vol. 2 Domestic architecture	115	(1322)
Vol. 3 Ecclesiastical and sacred architecture	170	(2162)
Vol. 4 Egyptian, Greek and Roman antiquities	128	(1660)
Vol. 5 National architecture	177	(2397)
STUDIES OF BUILDING TYPES		
Vol. 1 Cinema art and craft	117	(1561)
Vol. 2 Factories and workshops	100	(1214)
Vol. 3 Hospitals and hospital work	107	(1209)
Vol. 4 Libraries	200	(2643)
Vol. 5 Prisons	94	(1106)
Vol. 6 Theatres and the theatre	223	(3165)
Vol. 7 Miscellaneous buildings	217	(2949)
BUILDING AND CONSTRUCTIONAL ENGINEERING		
Vol. 1 Architectural engineering	152	(1911)
Vol. 2 Building	280	(3555)
Vol. 3 Structure and materials	288	(3571)
PLANNING, HOUSING AND PUBLIC HEALTH		
Vol. 1 Hygiene and welfare	261	(3644)
Vol. 2 Public health	262	(3615)
Vol. 3 The home and household management	92	(1206)
Vol. 4 Town planning and urbanism	179	(2168)
Vol. 5 Working class housing	223	(2168)
ENVIRONMENT AND TECHNOLOGY		
Vol. 1 Acoustics, lighting and colour	170	(2017)
Vol. 2 Electricity and electrical engineering	218	(2849)
Vol. 3 Heating, ventilation and gas engineering	214	(2754)
Vol. 4 Irrigation, drainage and sanitation	161	(2019)
Vol. 5 Water and water supply	171	(2092)
DESIGN, FURNITURE AND DECORATION		
Vol. 1 Aesthetics	85	(1171)
Vol. 2 Design, decoration and industrial art	218	(2977)
Vol. 3 Furniture and interior design	174	(2038)
Vol. 4 International and industrial exhibitions	113	(981)
Vol. 5 Museums, art galleries and collections	228	(3112)
GARDENS AND LANDSCAPE ARCHITECTURE		
Vol. 1 Gardens and gardening	283	(3934)
Vol. 2 Landscape architecture, botanic parks and gardens	141	(1711)

Index

TAB. ARCHITECTURE
Mouldings or Members
TUSCAN
DORIC
IONIC
CORINTHIAN
COMPOSITE
Forms of Masonry
Masonry
Compound Masonry
Modillion of Vignola
Cornich
Frieze
Architrave
Capital
Shaft
Base
Dye
ORDER Intire
Pediment
Attic Order
ARCH
Persian Order
Caryatic Order
Archivault
Diminution of a Column
Volute
Eye of ye Volute
Intercolumniations
Window
Rustic Door
Console
Cartouche
Festoon
Rose
Frette
No III

Plate: *Architecture*.

The five orders of architecture, namely: Tuscan, Doric, Ionic, Corinthian, and Composite – see text.

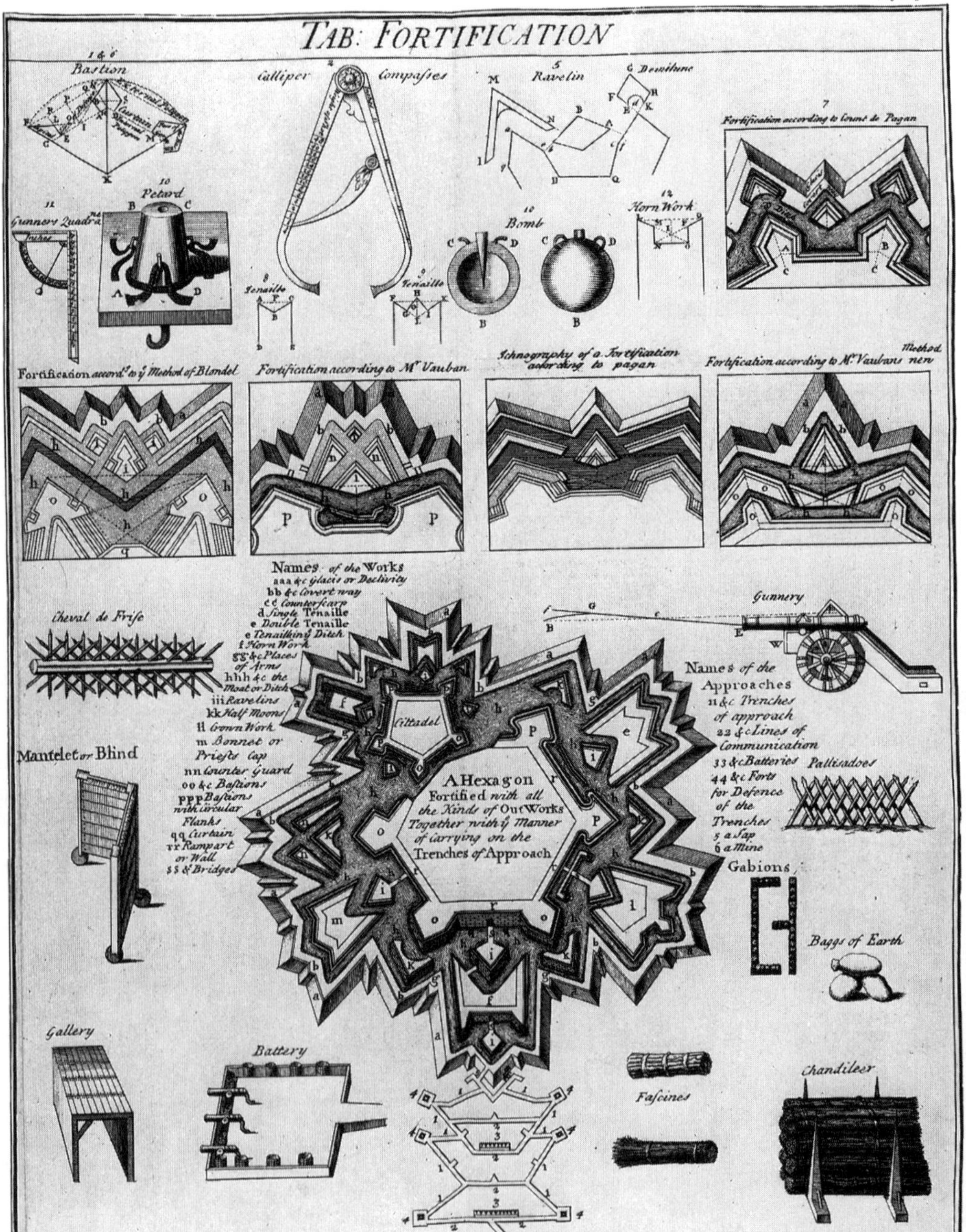
TAB: FORTIFICATION
Bastion
Calliper
Compasses
Ravelin
Demilune
Fortification according to Count de Pagan
Petard
Gunners Quadrant
Bomb
Horn Work
Tenaille
Tenaille
Fortification accord.^g to y^e Method of Blondel
Fortification according to M^r. Vauban
Ichnography of a Fortification according to pagan
Fortification according to M^r. Vaubans new Method
Names of the Works
aaa &c Glacis or Declivity
bb &c Covert way
cc Counterscarp
d Single Tenaille
e Double Tenaille
e Tenailling ye Ditch
f Horn Work
gg &c Places of Arms
hhh &c the Moat or Ditch
iii Ravelins
kk Half Moons
ll Crown Work
m Bonnet or Priests Cap
nn Counter Guard
oo &c Bastions
ppp Bastions with Circular Flanks
qq Curtain
rr Rampart or Wall
ss &c Bridges
Cheval de Frise
Gunnery
Citadel
A Hexagon Fortified with all the Kinds of Out Works Together with ye Manner of Carrying on the Trenches of Approach
Names of the Approaches
11 &c Trenches of approach
22 &c Lines of Communication
33 &c Batteries
44 &c Forts for Defence of the Trenches
5 a Sap
6 a Mine
Mantelet or Blind
Pallisadoes
Gabions
Baggs of Earth
Gallery
Battery
Fascines
Chandileer

Plate: *Fortification*.

A hexagonal system of fortification is illustrated showing the *outworks* and scheme of defences – see text.

Tab: Perspective

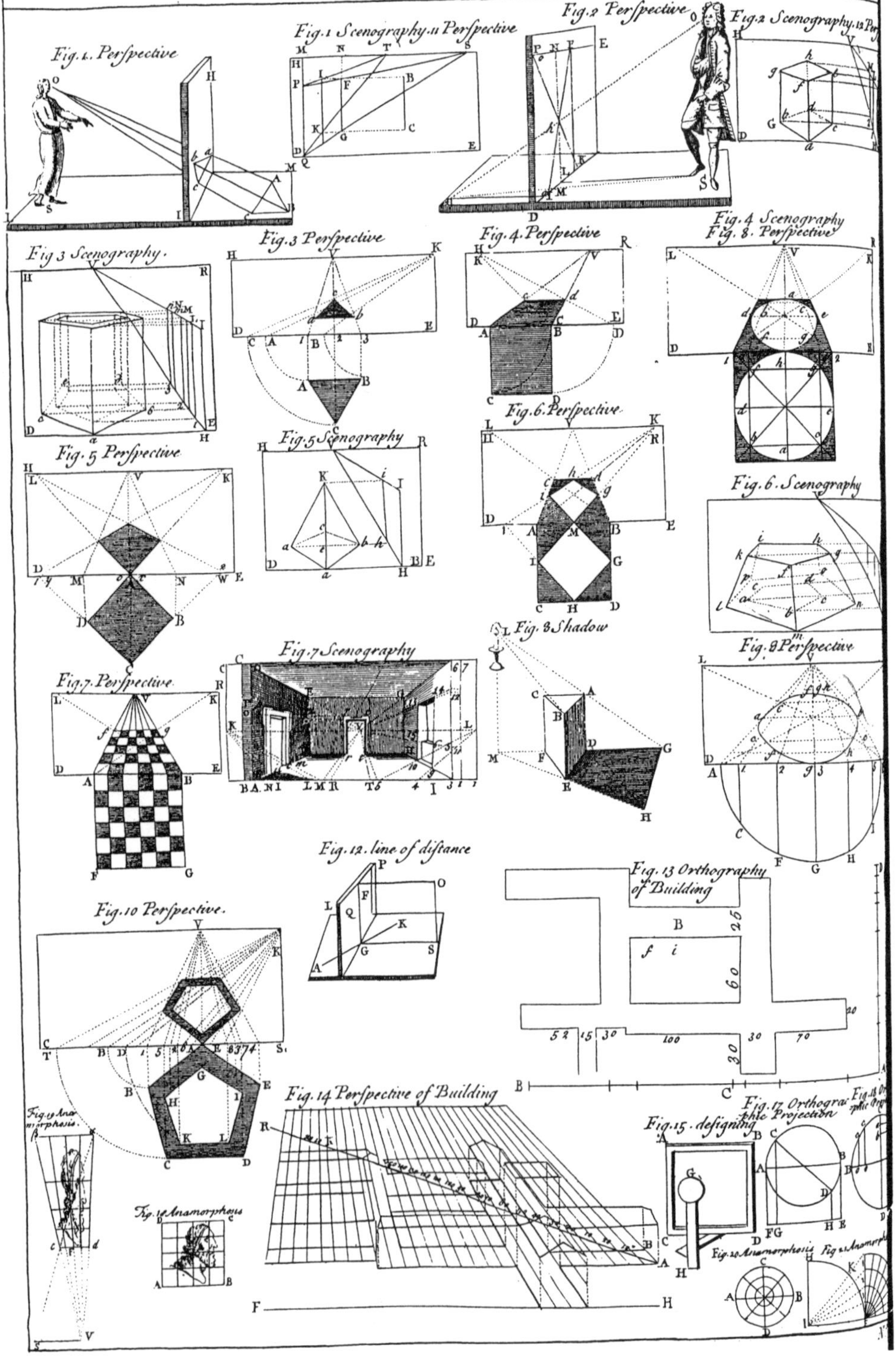

Plate: *Perspective*.

The principles of perspective drawing – see text.

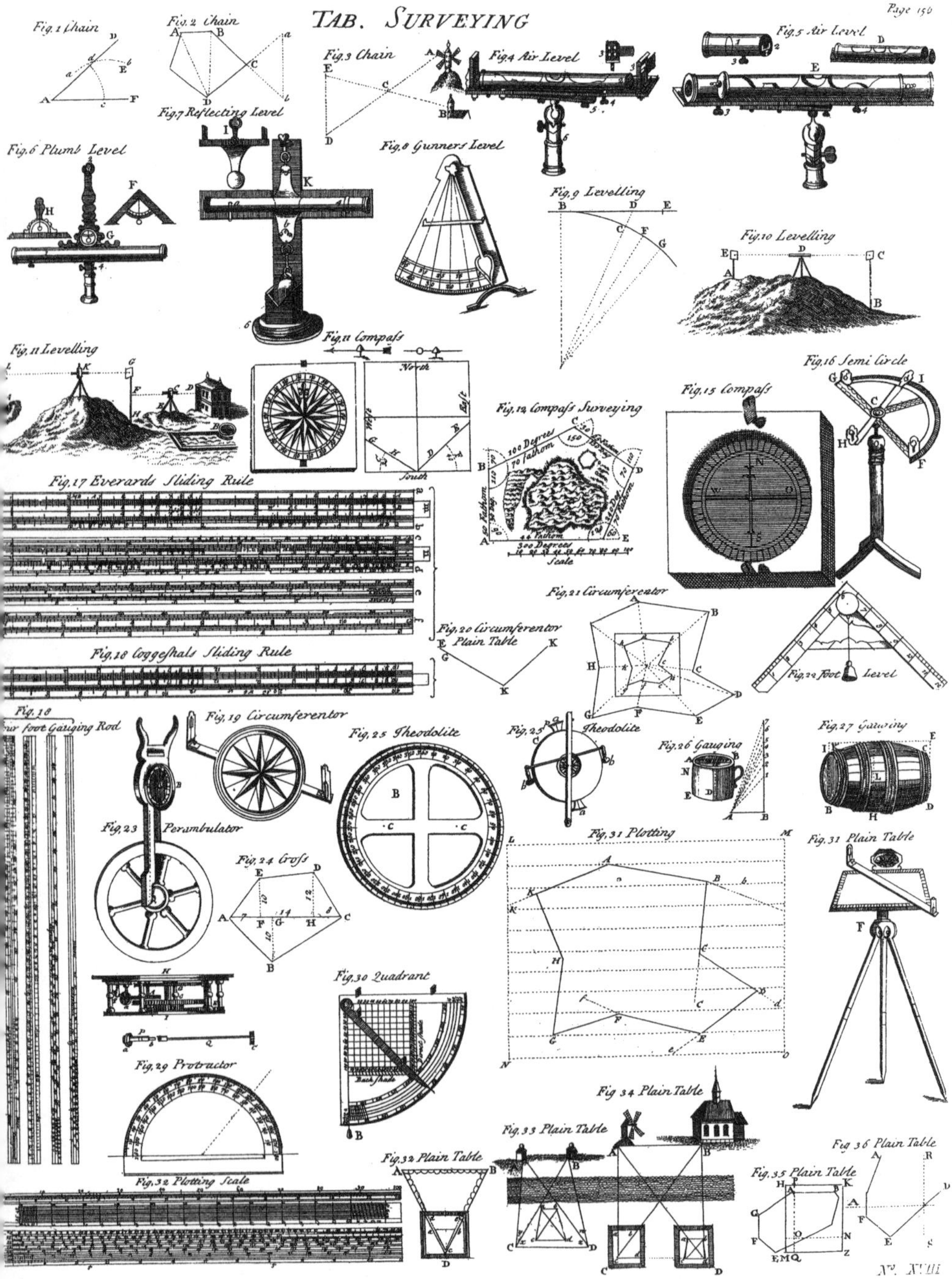
TAB. SURVEYING
Fig. 1 Chain
Fig. 2 Chain
Fig. 3 Chain
Fig. 4 Air Level
Fig. 5 Air Level
Fig. 6 Plumb Level
Fig. 7 Reflecting Level
Fig. 8 Gunners Level
Fig. 9 Levelling
Fig. 10 Levelling
Fig. 11 Levelling
Fig. 11 Compass
North
South
East
West
Fig. 14 Compass Surveying
100 Degrees
70 Fathom
Scale
Fig. 15 Compass
Fig. 16 Semi Circle
Fig. 17 Everards Sliding Rule
Fig. 18 Coggeshals Sliding Rule
Fig. 20 Circumferentor Plain Table
Fig. 21 Circumferentor
Fig. 22 Foot Level
Fig. 18
Four foot Gauging Rod
Fig. 19 Circumferentor
Fig. 23 Perambulator
Fig. 24 Gross
Fig. 25 Theodolite
Fig. 25 Theodolite
Fig. 26 Gauging
Fig. 27 Gauging
Fig. 31 Plotting
Fig. 31 Plain Table
Fig. 30 Quadrant
Fig. 29 Protractor
Fig. 32 Plotting Scale
Fig. 32 Plain Table
Fig. 33 Plain Table
Fig. 34 Plain Table
Fig. 35 Plain Table
Fig. 36 Plain Table

Plate: *Surveying*.

The instruments of the surveyor together with portrayals of their application – see text.